Masterclass in Neuroendocrinology

Volume 13

Series Editors

Mike Ludwig, Centre for Discovery Brain Sciences, The University of Edinburgh, Edinburgh, UK

Rebecca Campbell, School of Biomedical Sciences, University of Otago, Dunedin, New Zealand

Masterclass in Neuroendocrinology is published in collaboration with the International Neuroendocrine Federation (INF) https://www.inf-neuroendocrinology.org/. This series aims to illustrate and promote the use of the latest and most impactful technologies in basic and clinical research in the exciting field of neuroendocrinology. Neuroendocrinology is defined as the study of the control of endocrine function by the brain and the actions of hormones on the brain.

The series is made up of volumes that highlight different topics in neuroendocrinology that are of current interest and encompass the study of both normal and abnormal function, and the developmental origins of disease. They include the study of neural networks in the brain that regulate and form neuroendocrine systems, and also include the study of behaviours and mental states that are influenced or regulated by hormones. They also necessarily include the study of peripheral physiological systems that are regulated by neuroendocrine mechanisms. Some recent examples include: Developmental Neuroendocrinology, Neuroendocrine Clocks and Calendars, and Neuroanatomy of Neuroendocrine Systems.

Each book

- is edited by leading experts in the field
- is written by a team of internationally respected researchers
- is focused on an area of neuroendocrinology research
- demonstrates the present state of knowledge and understanding in the field of neuroendocrinology and highlights the translation of such knowledge to improve human health and well-being
- includes assessments of different experimental approaches, both *in vivo* and *in vitro*, and of how the resulting data are interpreted.

The Series is aimed at a wide audience, as neuroendocrinology integrates neuroscience and endocrinology. It is intended for established scientists, clinicians and early-career researchers.

Founding Series Co-Editors: William E. Armstrong and John A. Russell

Jan Pieter Konsman • Teresa M. Reyes
Editors

Neuroendocrine-Immune System Interactions

Springer

Editors
Jan Pieter Konsman
CNRS UMR5164 ImmunoConcept
University of Bordeaux
Bordeaux, France

Teresa M. Reyes
College of Medicine
University of Cincinnati
Cincinnati, OH, USA

ISSN 2662-2068 ISSN 2662-2076 (electronic)
Masterclass in Neuroendocrinology
ISBN 978-3-031-21357-1 ISBN 978-3-031-21358-8 (eBook)
https://doi.org/10.1007/978-3-031-21358-8

This Springer imprint is published by the registered company Springer Nature Switzerland AG
The registered company address is: Gewerbestrasse 11, 6330 Cham, Switzerland

Series Preface

This series began as a joint venture between the International Neuroendocrine Federation and Wiley-Blackwell and now is continuing with Springer Nature as publisher for the federation. The broad aim of the series is to provide established researchers, trainees, and students with authoritative up-to-date accounts of the present state of knowledge and prospects for the future across a range of topics in the burgeoning field of neuroendocrinology. The series is aimed at a wide audience as neuroendocrinology integrates neuroscience and endocrinology. We define neuroendocrinology as the study of the control of endocrine function by the brain and the actions of hormones on the brain. It encompasses the study of normal and abnormal function and the developmental origins of disease. It includes the study of the neural networks in the brain that regulate and form neuroendocrine systems and also includes the study of behaviors and mental states that are influenced or regulated by hormones. In addition, it includes the understanding and study of peripheral physiological systems that are regulated by neuroendocrine mechanisms. While neuroendocrinology embraces many issues of concern to human health and well-being, research in reductionist animal models is required to fully understand these issues.

Contemporary research in neuroendocrinology involves the use of a wide range of techniques and technologies, from the subcellular and systems level to the whole-organism level. A particular aim of the series is to provide expert advice and discussion about experimental or technical protocols in neuroendocrinology research and to further advance the field by giving information and advice about novel techniques, technologies, and interdisciplinary approaches.

To achieve our aims, each book focuses on a particular theme in neuroendocrinology. For each book, we recruit editors, who are leaders in their field, to engage an international team of experts to contribute chapters in their individual areas of expertise. The mission of each contributor is to provide an update of current knowledge and recent discoveries, and to discuss new approaches, "gold standard" protocols, translational possibilities, and future prospects. Authors are asked to write for a wide audience, to use references selectively, and to consider use of video clips and explanatory text boxes; each chapter is peer reviewed and has a Glossary. In all of these efforts, we are guided by an Advisory Editorial Board.

The Masterclass Series is open-ended; books in the series published to date are:

- *Neurophysiology of Neuroendocrine Neurons* (2014, ed. WE Armstrong & JG Tasker)
- *Neuroendocrinology of Stress* (2015, ed. JA Russell & MJ Shipston)
- *Molecular Neuroendocrinology: From Genome to Physiology* (2016, ed. D Murphy & H Gainer)
- *Computational Neuroendocrinology* (2016, ed. DJ MacGregor & G Leng)
- *Neuroendocrinology of Appetite* (2016; ed. SL Dickson & JG Mercer)
- *The GnRH Neuron and its Control* (2018; ed. AE Herbison & TM Plant)
- *Model Animals in Neuroendocrinology* (2019, ed. M Ludwig & G Levkowitz).

The first books of the series published by Springer Nature are:

- *Neurosecretion: Secretory Mechanisms* (2020, ed. JR Lemos & G Dayanithi)
- *Developmental Neuroendocrinology* (2020, ed. S Wray & S Blackshaw)
- *Neuroendocrine Clocks and Calendars* (2020, ed. FJP Ebling & HD Piggins)
- *Glial-Neuronal Signaling in Neuroendocrine Systems* (2021, ed. JG Tasker, JS Bains, & JA Chowen)
- *Neuroanatomy of Neuroendocrine Systems* (2021, ed. V Grinevich & A Dobolyi).

In development are *Neuroendocrinology of Pregnancy and Lactation* (ed. P Brunton & D Grattan) and *Cardiovascular Neuroendocriology* (ed. T Cunningham & G Yosten).

Feedback and suggestions are welcome.

Series Editors,

Edinburgh, UK — Mike Ludwig
Dunedin, New Zealand — Rebecca Campbell

International Neuroendocrine Federation – https://www.inf-neuroendocrinology.org/

Volume Preface

This book is the 13th volume in the *Masterclass in Neuroendocrinology* book series. To introduce this book, we first address the question: why a volume on neuroendocrine and immune system interactions? Our answer is based on the recent history of the study of these systems and their components. First, parts of the hypothalamus and pituitary and hormones like glucocorticosteroids, now considered to be part of neuroendocrine axes, have long been known to have effects on immune responses (MacGregor et al. 1969; Pierpaoli and Sorkin 1969; Pierpaoli et al. 1969; Kalden et al. 1970; Gisler and Schenkel-Hulliger 1971; Tyrey and Nalbandov 1972; Isakovic and Jankovic 1973; Jankovic and Isakovic 1973). Second, some immune responses, and in particular, but not only, innate immune or inflammatory responses to the administration of microbial fragments or other antigens, are accompanied by altered neuroendocrine responses (Besedovsky et al. 1977; Spinedi et al. 1992; Stenzel-Poore et al. 1993; Frederic et al. 1993; Shurin et al. 1997). These findings convinced many scientists and physicians that it was important to understand how neuroendocrine and immune systems and their components interact in health and disease in a more interdisciplinary manner (Besedovsky and Sorkin 1977; Dunn 1988; Bateman et al. 1989; Dantzer and Kelley 1989).

Additionally, some of the intercellular messenger molecules used by neuroendocrine and immune systems are the same. For example, some peptide hormones/neuropeptides, such as adrenocorticotropin hormone (ACTH) and corticotropin-releasing hormone, were found to have effects on or even to be synthesized by immune cells (Blalock et al. 1985; Karalis et al. 1997). Conversely, cytokines, typically associated with immune responses, such as interleukin-1, can also be present in and act on neuroendocrine systems, for example the paraventricular nucleus of the hypothalamus (Lechan et al. 1990; Licinio et al. 1991; Huitinga et al. 2000; Watt and Hobbs 2000; Blandino et al. 2013). Thus, some of the entities studied by a specific discipline (e.g., neuroendocrinology or immunology) turned out to be non-specific for that domain.

Although distinct scientific fields are important for the progress of science, "their emergence can restrict access by outsiders" (Casadevall and Fang 2015). One way to foster progress through and improve access to emerging fields would be "to promote interdisciplinary interaction between fields" (Casadevall and Fang 2015). For interdisciplinarity to occur, it is not sufficient for an experimental method, technique, or tool developed in one scientific discipline to be adopted by another. Rather, we believe that interdisciplinary approaches can "advance fundamental understanding ... *beyond the scope* of a single discipline or area of research practice" (emphasis added (Institute of Medicine 2005, p. 2). Indeed, the complexity of the research question is probably the main driving force for interdisciplinary research with the primary aim of furthering understanding (Smith 2004; Alvargonzalez 2011; Mazzocchi 2019). Therefore, the interdisciplinary research discussed in this volume largely used techniques and tools established in other disciplinary research fields and is "the *science of connections* or *interactions* [between neuroendocrine and immune systems]" (Straub 2015, p. xiii). Accordingly, methods and techniques will be discussed in the context of the questions and hypotheses addressed in the different chapters composing this volume.

This volume is organized into four parts with several chapters, each of which can be read rather independently as many questions are treated by several chapters. The first part deals with concepts relevant to a better understanding of interactions between neuroendocrine systems and immune systems, such as the history (Chap. 1) and philosophy (Chap. 2) of science as well as some evolutionary context in which such interactions occur, for example during sickness after infection (Chap. 3). The second part covers the main regulatory pathways of interaction from classic pathways like peripheral cytokines activating the hypothalamo–pituitary–adrenal axis (Chap. 4) and corticosteroid and catecholamines regulating immune responses (Chaps. 5 and 6) to the brain choroid plexus and cerebrospinal fluid constituting an original terrain of neuroendocrine–immune interactions (Chap. 7) and autoimmune antibodies possibly regulating neuroendocrine responses (Chap. 8). The third part presents factors modulating neuroendocrine–immune system interactions, including the importance of the early postnatal time period (Chap. 9), as well as the impact of sex differences (Chap. 10), biological clocks (Chap. 11), and the gut microbiota (Chap. 12) on neuroendocrine–neuroimmune interactions. The last part examines these interactions in disease conditions (chronic inflammatory conditions and cancer in Chap. 13 and eating disorders in Chap. 14). Collectively, the chapters in this series will provide context for understanding neuroimmune–neuroendocrine system interactions and illustrate research in this domain, without being exhaustive.

To conclude, the volume editors would like to dedicate this book to Dr. Paul Sawchenko, Salk Institute for Biological Studies, for his seminal contributions to the study of neuroimmune–neuroendocrine interactions and for inspiring the careers of the editors (and countless others) with his generous, supportive, and collaborative approach to rigorous experimental science.

Bordeaux, France Jan Pieter Konsman
Cincinnati, OH, USA Teresa M. Reyes

References

Alvargonzalez D (2011) Multidisciplinarity, interdisciplinarity, transdisciplinarity, and the sciences. Int Stud Philos Sci 25:387–403

Bateman A, Singh A, Kral T, Solomon S (1989) The immune-hypothalamic-pituitary-adrenal axis. Endocr Rev 10:92–112

Besedovsky H, Sorkin E (1977) Network of immune-neuroendocrine interactions. Clin Exp Immunol 27:1–12

Besedovsky H, Sorkin E, Felix D, Haas H (1977) Hypothalamic changes during the immune response. Eur J Immunol 7:323–325

Blalock JE, Smith EM, Meyer WJ 3rd (1985) The pituitary-adrenocortical axis and the immune system. Clin Endocrinol Metab 14:1021–38

Blandino P Jr, Hueston CM, Barnum CJ, Bishop C, Deak T (2013) The impact of ventral noradrenergic bundle lesions on increased IL-1 in the PVN and hormonal responses to stress in male sprague dawley rats. Endocrinology 154:2489–500

Casadevall A, Fang FC (2015) Field Science--the Nature and Utility of Scientific Fields. mBio 6:e01259-15

Dantzer R, Kelley KW (1989) Stress and immunity: an integrated view of relationships between the brain and the immune system. Life Sci 44:1995–2008

Dunn AJ (1988) Nervous system-immune system interactions: an overview. J Receptor Res 8:589–607

Frederic F, Oliver C, Wollman E, Delhaye-Bouchaud N, Mariani J (1993) IL-1 and LPS induce a sexually dimorphic response of the hypothalamo-pituitary-adrenal axis in several mouse strains. Eur Cytokine Netw 4:321–9

Gisler RH, Schenkel-Hulliger L (1971) Hormonal regulation of the immune response. II. Influence of pituitary and adrenal activity on immune responsiveness in vitro. Cell Immunol 2:646–657

Huitinga I, Van Der Cammen M, Salm L, Erkut Z, Van Dam A, Tilders F, Swaab D (2000) IL-1beta immunoreactive neurons in the human hypothalamus: reduced numbers in multiple sclerosis. J Neuroimmunol 107:8–20

Institute of Medicine (2005) Facilitating interdisciplinary research. The National Academies Press, Washington, DC

Isakovic K, Jankovic, BD (1973) Neuro-endocrine correlates of immune response. II. Changes in the lymphatic organs of brain-lesioned rats. Int Arch Allergy Appl Immunol 45:373–84

Jankovic BD, Isakovic K (1973) Neuro-endocrine correlates of immune response. I. Effects of brain lesions on antibody production, Arthus reactivity and delayed hypersensitivity in the rat. Int Arch Allergy Appl Immunol 45:360–72

Kalden JR, Evans MM, Irvine WJ (1970) The effect of hypophysectomy on the immune response. Immunology 18:671–679

Karalis K, Muglia LJ, Bae D, Hilderbrand H, Majzoub JA (1997) CRH and the immune system. J Neuroimmunol 72:131–6

Lechan RM, Toni R, Clark BD, Cannon JG, Shaw AR, Dinarello CA, Reichlin S (1990) Immunoreactive interleukin-1 beta localization in the rat forebrain. Brain Res 514:135–40

Licinio J, Wong ML, Gold PW (1991) Localization of interleukin-1 receptor antagonist mRNA in rat brain. Endocrinology 129:562–4

Macgregor RR, Sheagren JN, Lipsett MB, Wolff SM (1969) Alternate-day prednisone therapy. Evaluation of delayed hypersensitivity responses, control of disease and steroid side effects. N Engl J Med 280:1427–31

Mazzocchi F (2019) Scientific research across and beyond disciplines: Challenges and opportunities of interdisciplinarity. EMBO Rep 20(6):e47682

Pierpaoli W, Baroni C, Fabris N, Sorkin E (1969) Hormones and immunological capacity. II. Reconstitution of antibody production in hormonally deficient mice by somatotropic hormone, thyrotropic hormone and thyroxin. Immunology 16: 217–30

Pierpaoli W, Sorkin E (1969) Effect of growth hormone and anti-growth hormone serum on the lymphatic tissue and the immune response. Antibiotica Chemotherapia 15:122–34

Shurin G, Shanks N, Nelson L, Hoffman G, Huang L, Kusnecov AW (1997) Hypothalamic-pituitary-adrenal activation by the bacterial superantigen staphylococcal enterotoxin B: role of macrophages and T cells. Neuroendocrinology 65: 18–28

Smith JT (2004) Interdisciplinarity and complexity: an evolving relationship. E:CO 6:2–10

Spinedi E, Suescun MO, Hadid R, Daneva T, Gaillard RC (1992) Effects of gonadectomy and sex hormone therapy on the endotoxin-stimulated hypothalamo-pituitary-adrenal axis: evidence for a neuroendocrine-immunological sexual dimorphism. Endocrinology 131:2430–6

Stenzel-Poore M, Vale WW, Rivier C (1993) Relationship between antigen-induced immune stimulation and activation of the hypothalamic-pituitary-adrenal axis in the rat. Endocrinology 132:1313–8

Straub RH (2015) The origin of chronic inflammatory systemic disease and their sequelae, Elsevier, Amsterdam

Tyrey L, Nalbandov AV (1972) Influence of anterior hypothalamic lesions on circulating antibody titers in the rat. Am J Physiol 222:179–85

Watt, J. A. & Hobbs, N. K. 2000. Interleukin-1beta immunoreactivity in identified neurons of the rat magnocellular neurosecretory system: evidence for activity-dependent release. J Neurosci Res 60:478–89

Contents

About the Editors

Jan Pieter Konsman received a master's degree in neurobiology from the University of Groningen, the Netherlands. He obtained his PhD from the Bordeaux University in France for his work on immune-to-brain communication. He then did his postdoctoral training at the University of LinKöping in Sweden working on the distribution of the interleukin-1 receptor in the rodent brain. Dr. Konsman next took up a permanent position at the Centre National de Recherche Scientifique in France working on neuroimmune substrates of disease-associated anorexia and encephalopathy. He also holds a master's degree in history and philosophy of science and is interested in interdisciplinary approaches of the mind–body problem.

Teresa M. Reyes received her PhD in psychology from the University of Wisconsin-Madison in which she examined how circulating cytokines affect memory and behavior in a non-human primate model. For her postdoctoral training, she joined the Laboratory of Neuronal Structure and Function, at the Salk Institute for Biological Studies in La Jolla, CA. Here she extended her training to neuroanatomical studies of neuroimmune interactions using rodent models. Dr. Reyes is currently on the faculty of the University of Cincinnati, College of Medicine, Department of Pharmacology and Systems Physiology. Dr. Reyes's research program is focused on identifying the mechanisms that underlie the effect of adversity in early life on brain development, with a specific interest in neuroimmune interactions and cognition.

Part I

Concepts

A History of Immune and Neuroendocrine System Interactions

1

Jan Pieter Konsman

Abstract

The purpose of this chapter is to provide a twentieth century history of neuroendocrine and immune systems and the interactions of their components. The ideas of immune and neuroendocrine structure–function relationships emerged in the life sciences once the cell had been recognized as the fundamental unit of life, in large part due to the use of improved microscopic and tissue-staining techniques. In addition, throughout the twentieth century, the study of immunity and neuroendocrinology has been guided by the idea of receptor molecules showing specificity for certain biological components. Interestingly, the very notions of neuroendocrine and immune systems, reminiscent of those still used today, were only explicitly formulated in the 1970s. While initial thinking about neuroendocrine–immune interactions in the 1970s–1980s was mostly framed in terms of systems, subsequent physiological and evolutionary research indicated that these interactions can also occur at the organ, tissue, cellular, and molecular levels and that the very labels "immune" and "neuroendocrine" need to be used with caution in present-day and future research.

Keywords

Biological systems · History of science · Intercellular communication

J. P. Konsman (✉)
IMMUNOlogy from CONcepts and ExPeriments to Translation, CNRS UMR 5164, University of Bordeaux, Bordeaux, France
e-mail: jan-pieter.konsman@u-bordeaux.fr

J. P. Konsman, T. M. Reyes (eds.), *Neuroendocrine-Immune System Interactions*, Masterclass in Neuroendocrinology 13,
https://doi.org/10.1007/978-3-031-21358-8_1

1.1 Introduction: Why and What History?

The question of why the history of science is important can receive many answers, ranging from mastering the literature and knowing the origin of a technique to describing the concepts at work in a field. One reason for considering the history of interactions between the immune and neuroendocrine systems from the second half of the nineteenth century onward and in particular in the twentieth century is that the notions of immune and neuroendocrine have been linked to the idea of the cell being the fundamental unit of life during that period, with the help of different kinds of techniques and instruments. An attempt is thus made to give space both to the history of ideas and to that of techniques.

1.1.1 History of Techniques Prior to WWII

Improvements in microscopic and coloration techniques at the end of the nineteenth century turned out to be revealing for immune and neuroendocrine cells, as several of the histological stains made particular cells stand out. As is well known, Ramon y Cajal modified the silver stain developed by Camillo Golgi and applied it to many parts of the central and peripheral nervous systems. Thus, in the last decade of the nineteenth century, he showed that visceral ganglia of rodents, using Meissner plexus as an example, contain cells with short and long cellular ramifications that seemed to terminate between smooth muscle cells and close to gland cells in the intestine (Ramon and Cajal 1893). In these innervated gastrointestinal glands, the granule-rich cell type that stained with eosin, with chromium salts (hence entero-chromaffin cells), and with silver nitrate were proposed to be endocrine cells, early in the twentieth century (Ciaccio 1906; Masson 1914, 1928). Another important line of research that has its origins in observations of stained tissue was the characterization of hypothalamic neurons projecting to the posterior part of the pituitary. Even though these connections had been hypothesized before, Ramon y Cajal showed their existence using Golgi staining in 1894 (Ramon and Cajal 1911, pp. 488–490).

A more physiological research tradition at the end of the nineteenth century can be considered to have started with the publication by Oliver and Schafer of a short report of the effects of pituitary extracts on blood pressure. They found that pituitary extract, just like adrenal extract, rapidly increases blood pressure (Oliver and Schafer 1895). This finding could be reproduced using posterior pituitary extracts and was considered to not be mediated by the sympathetic nervous system or adrenaline (Dale 1906, 1909). In parallel, the effects of total removal of the pituitary were being studied in dogs (Paulescu 1907; Crowe et al. 1909) and partial hypophysectomy was even proposed as an experimental treatment for patients suffering from acromegaly (Cushing 1909), which, in adults, is characterized by increased bone size in the hands, feet, and face. In laboratory rats, hypophysectomy-induced adrenal, ovarian, testicular, and thyroid atrophy as well as dwarfism could be reversed by administration of anterior pituitary extract (Smith 1930). These findings stimulated other groups to study the effects of different fractions of anterior pituitary extracts and led to the isolation of adrenotropic, gonadotropic, growth and thyrotropic hormone-containing fractions (Anderson and Collip 1933, 1934; Collip 1933; Collip et al. 1933a, b).

While the influences of the pituitary on other glands in the body became progressively clear using lesion, isolation, and replacement techniques, the effects did not seem to be fully specific to the pituitary as some pituitary-sparing ventral hypothalamic lesions also induced genital atrophy (Camus and Roussy 1920; Bailey and Bremer 1921; Smith 1926; Harris 1937). Geoffrey Harris therefore compared the effects of electrical stimulation of the *tuber cinereum* region of the hypothalamus to those of pituitary stimulation and found that both stimulated ovulation in rabbits (Harris 1937, p. 392). In discussing these findings, he raised the possibility that "the hypothalamus controls the secretion of hormones . . . from the anterior lobe [of the pituitary]" (Harris 1937, p. 392).

Using an improved microscope to observe tissues stained with, among other things, aniline dyes, Elie Metchnikoff was able to develop his "comparative pathology of inflammation" and to distinguish different leukocytes at the end of the nineteenth century (Metchnikoff 1893). Some years earlier, Paul Ehrlich had developed many of these stains and had also used them to characterize different types of leukocytes (Ehrlich 1880). In the final decades of the nineteenth century, it was shown that body fluids of animals inoculated with bacteria could lyze bacterial cells, neutralize bacterial toxins, or precipitate bacterial products (Nuttall 1888; Von Behring and Kitasako 1890; Pfeiffer 1894). Importantly, the bactericidal activity could also be studied in culture dishes. Such in vitro approaches established that blood serum contains more bactericidal activity than peritoneal exudate (Bordet 1895, 1909). In addition, two active components could be distinguished, a "bactericidal substance," which could be inactivated by heating to 55 °C, and a "preventive substance," the activity of which resisted heating in different laboratories, as shown by Jules Bordet and Paul Ehrlich (Bordet 1909, pp. 75–80).

At the very end of the nineteenth century, Paul Ehrlich advanced one of the first formulations of his side-chain theory (see below) to provide a hypothetical chemical explanation of the specificity of antibody–antigen reactions. In 1900, Karl Landsteiner decided to systematically study the thus far anecdotal reports of human sera lyzing human red blood cells and identified three types, A, B, and C (that later became our blood groups with Landsteiner's type C being our group AB) with types A and B lyzing red blood cells of other type and type C lyzing no other cell types (Landsteiner 1901, 1961). Later in his career, Landsteiner undertook a series of studies in which he varied the chemical properties of the antigen and this led him to conclude in 1928 that "the steric configuration of antigenic groups is one of the factors determining serological specificity" (Landsteiner and Van Der Scheer 1928, p. 320). When he was awarded the Nobel Prize for Medicine or Physiology in 1930 for his work on blood groups, Landsteiner indicated links between his two lines of research as well as the implications of the existence of blood groups for transfusions (Landsteiner 1930). So while the chemical nature of immune reactions would remain elusive for some decades more, progress was made by the systematic study of some variables, independently of Ehrlich's influential side-chain theory.

1.1.2 History of Ideas Prior to WWII

The germ theory formulated by Pasteur and like-minded scientists and physicians at the end of the nineteenth century was subsequently developed to link specific tissue lesions to specific microbes. In this respect, the emergence of the cell theory in the second half of the nineteenth century can be proposed as a starting point for considering the history of the neuroendocrine and immune systems. In his "remarks on microorganisms" and "their relation to disease," Joseph Lister referred to the work of both Robert Koch and Louis Pasteur and echoed Pasteur's question on what parts of bacteria were required to confer "immunity from further attacks of ... disease" (Lister 1880, p. 364). At the turn of the nineteenth century, Lister discussed how application of his "antiseptic system" allowed the eradication of "hospital gangrene" and gave credit to "Pasteur [who] saw the analogy between the immunity to fowl-cholera produced by its attenuated virus and the protection afforded against small-pox by vaccination" (Lister 1896, pp. 418, 421). Finally, he addressed "a subject which, though not bacteriological, has intimate relations with bacteria" and cited observations by Elie Metchnikoff that indicated "that the microbes of infective diseases are subject to [a] process of devouring and digestion, carried on both by the white corpuscles and by cells that line the blood vessels" as a means of defense (Lister 1896, pp. 427–428).

In the early twentieth century, Ernest Starling introduced the term "hormone" to refer to chemical messengers that "have to be carried from the organ where they are produced to the organ which they affect by means of the blood stream" to meet "the continually recurring physiological needs of the organism" (Starling 1905, p. 340). He continued to promote the concept after the first World War and proposed, in a lecture entitled *The wisdom of the body,* that in addition to the growth-promoting secretion of the anterior pituitary, "[t]he posterior lobe ... forms one or more substances which, circulating in the blood, have the most diverse influences on various parts of the body" (Starling 1923, p. 689). Thus, he raised the idea that the posterior pituitary could communicate not only through cellular connections with the hypothalamus, but also by humoral means, by releasing mediators into the bloodstream. A decade later, Harvey Cushing considered that there "appear to be two routes, both probably under nervous control, whereby the products of pituitary secretion are transported to the tuberal and possibly to other nuclei of the anterior hypothalamus: (i) from *pars anterior* by way of the 'hypophysio-portal' vessels; and (2) from *pars intermedia* by way of the tissue" (Cushing 1933, pp. 540–541).

These considerations can be taken to be part of a more general debate in the first half of the twentieth century regarding the modalities of intercellular communication, in particular between nervous and endocrine signal transmission. For example, histochemical stains seemed to indicate the presence of gland-like nerve cells in several invertebrate ganglia and in parts of the vertebrate central nervous system (Scharrer 1928; Scharrer and Scharrer 1937). Interestingly, similarities between the effects of administration of adrenal extracts and electrical stimulation of the sympathetic nervous system had repeatedly been pointed out, for example, by John Langley and Walter Cannon (Langley 1901; Cannon 1914, 1927, pp. 36–38). In addition, it was reported that systemic administration of brain and intestinal tissue extracts in rabbits lowered their blood pressure in ways that could not be explained by the action of acetylcholine or adenosine (Von Euler and Gaddum 1931). These findings not only indicated the discovery of a new substance, designated Substance P, but also the possibility that the same intercellular messengers exist in the brain and the intestine. In this respect, another important, but often overlooked, possibility that gastrointestinal tract epithelia constitute a diffuse endocrine organ in close interaction with nerves, was put forward by Friedrich Feyrter (1939). Feyrter compared ductal cells of the pancreas and enterochromaffin cells of the gastrointestinal tract and observed that both responded similarly to different kinds of stains (among which was Masson's silver stain). Based on these observations, he proposed that endocrine-like signals could not only be produced by specialized glands, such as the pancreas and pituitary, but also by surface epithelia. In addition, Feyrter suggested, based on Cajal's and Masson's findings mentioned above, that the cells of the diffuse endocrine organ of the gastrointestinal tract were connected to local neural fibers.

1.2 Post-World War II Success Stories of Immunology and Neuroendocrinology

1.2.1 Connecting the Hypothalamus and Pituitary

Immediately after World War II (WWII) it seemed established that the release of the so-called antidiuretic hormone from the posterior pituitary in response to emotional stress was linked to "stimulation ... of the supraoptic, and possibly of the paraventricular nuclei [of the hypothalamus], whose axons pass down the stalk to the pars nervosa [of the pituitary]" (Verney 1947, p. 99). This hypothesis of hypothalamic, neuron-derived hormonal messengers being released in extracerebral organs was subsequently corroborated by findings confirming the anatomical continuity between cell bodies of hypothalamic nuclei and the posterior pituitary, based on a new histological stain and the effects of interruptions of the pituitary stalk connecting the two (Bargmann and Scharrer 1951; Hild 1951).

In the 1930s, Geoffrey Harris had postulated, based on stimulation experiments, that "the hypothalamus controls the secretion of hormones ... from the anterior lobe [of the pituitary]" (Harris 1937, p. 392). A decade later, in a review article, he first argued that regarding direct nerve supply and control of endocrine glands, the posterior pituitary and adrenal medulla are exceptions in that developmentally they may be considered as extensions of the central and peripheral nervous systems (Harris 1948). Harris next summarized evidence from lesion and stimulation studies that indicated neural control of the anterior pituitary or adenohypophysis, but pointed out that "[t]he mechanism whereby this neural control is exerted is uncertain" (Harris 1948, p. 157). Furthermore, Harris cited reports, including his own work, indicating "a true portal system of blood vessels in the pituitary stalk" with blood flowing from the median eminence of the hypothalamus to the anterior pituitary (Green and Harris 1949, p. 360), raising the possibility that "nervous stimuli might cause the liberation of some substance into the capillary sinusoids of the median eminence" and that "this substance [is] transported via the hypophysial portal vessels to excite or inhibit" pituitary secretions (Harris 1948, pp. 168, 169). Finally, he proposed that the neuronal and hormonal links of the hypothalamus with, respectively, the posterior or neurohypophysis and the anterior pituitary or adenohypophysis mediate different functions (Harris 1951a, b). Thus, Harris's work laid the foundation for envisioning some new structure–function relationships in biology.

1.2.2 Study of Immunity Between Chemistry and Biology

Immunology had taken a chemical turn before WWII with debates about the nature of the chemical bond or affinity that occurred between bacterial toxins and antibodies. The idea that chemistry was essential to the advancement of immunology was widespread after WWII. For example, one could read in a medical journal that: "The phenomena of immunity are essentially chemical, and we might reasonably expect that some of its problems will soon be solved if a more sustained chemical and biochemical attack is made on them" (Wormall 1948, p. 333). Chemical approaches were not only expected to further the understanding of antigen–antibody binding, but also that of phagocytosis, blood groups, and complement reactions (Wormall 1948). Indeed, one important way in which chemistry allowed immunology to make progress was through the development of new quantitative methods for the study of antigen–antibody, blood group, and complement reactions (Mayer 1951). Another physicochemical technique that proved useful to further characterize various protein immune components was John Cohn's chemical fractionation of blood plasma (Cohn et al. 1944). However, chemistry in the immediate post-WWII years did not really provide many answers to immunological problems encountered in the clinic, such as autoimmunity or host rejection of blood transfusion and organ and tissue transplantations.

During the 1940s, the Australian virologist Frank Macfarlane Burnet started to make important conceptual contributions to the field of immunology. In his *Biological aspects of infectious disease*, Burnet stated that "a conflict between man and his parasites... in constant environment, would tend to result in a virtual equilibrium," but that because "[m]an lives in an environment constantly being changed by his own activities ... few of his diseases have attained such an equilibrium" (Burnet 1940, p. 23). Another interesting aspect of Burnet's thinking emerged after observing amoebic digestion (like Metchnikoff more than half a century earlier). Thus, Burnet remarked that: "The fact that the one is digested, the other not, demands that in some way or other the living substance of the amoeba can distinguish between the chemical structure characteristic of "self" and any sufficiently different chemical structure which is recognized as "not-self" (Burnet 1940, p. 29). He developed this idea further in *The production of antibodies* and pointed out that: "It is an obvious physiological necessity and a fact fully established by experiment that the body's own cells should not provoke antibody formation"

(Burnet and Fenner 1949, p. 85). Finally, he noted that "[i]mmunology has through most of its history been remote from the general stream of biological discovery and generalization" and urged that "immunological phenomena and interpretations must be given due weight in any future formulations of the nature of living process" (Burnet and Fenner 1949, pp. 132–133). Thus, Burnet, in the course of the 1940s, elaborated a vision of immunology that was complementary to the chemical discourse of the time.

1.2.3 Characterization of Some Hypothalamic Releasing Factors and Pituitary Hormones

An influential vision of neuroendocrinology was articulated by Geoffrey Harris in his 1955 *Neural control of the pituitary gland* monograph in which he suggested that neuroendocrine neurons were motor or effector neurons. He also proposed functional criteria for determining whether an endogenous substance constitutes a releasing factor at the level of the anterior pituitary (Harris 1955). Interestingly, the same year, Roger Guillemin and Barry Rosenberg proposed "[t]he possibility of investigating this problem by simple *in vitro* techniques" involving culturing fragments of the pituitary gland along with those of the ventral hypothalamus (Guillemin and Rosenberg 1955, p. 599). In parallel, it was established that adrenocorticotropic hormone (ACTH), the active principle secreted by the anterior pituitary to control adrenal cortical structure, weight, and secretion, corresponds to a 20 kDalton protein (Morris 1951). Around the same time, it was also shown that some steroids secreted by the adrenal cortex, such as 17-hydroxy-1 1-dehydrocorticosterone, soon to become better known as cortisone or cortisol, have the same beneficial effects as ACTH on arthritis (Hench et al. 1949). Moreover, Vincent du Vigneaud and colleagues found in 1954 that a cyclic nine-amino acid peptide they synthesized had the same physicochemical and biological properties as the "oxytocic substance of the neurohypophysis" (see above) and therefore concluded "the identity of the synthetic product with natural oxytocin" (Du Vigneaud et al. 1954, p. 3115). The same group established that oxytocin differs by only two amino acids from vasopressin and that the posterior pituitary hormone that increases blood pressure also promotes water retention (see above; Katsoyannis and Du Vigneaud 1958). Interestingly, Murray Saffran and Andrew Schally also identified a peptide in hypothalamic and posterior pituitary extracts, dubbed Corticotropin-Releasing Factor (CRF), which was different from vasopressin but induced ACTH secretion from rat anterior pituitary tissue in vitro (Saffran and Schally 1955a, b; Schally et al. 1958). Furthermore, arterial perfusion of isolated adrenal glands also allowed to establish that vasopressin, but not oxytocin, has the same effects as ACTH on hydrocortisone secretion (Hilton et al. 1960). Thus, while the 1950s witnessed a consolidation of findings obtained by classical experimental approaches that allowed formulation of the first conceptual neuroendocrinological framework, it also saw the development of new in vitro and biochemical methods relevant for the future development of neuroendocrinology.

1.2.4 Tolerance and Antibody Production

Besides his theoretical contributions, Burnet was also well-versed in biochemical, genetic, and microscopic approaches and often combined these to address scientific questions. For example, chick embryos allowed him to study clumping or agglutination of blood cells when mixed with fluid of influenza-infected individuals and to use this phenomenon to generate hypotheses "on the process of interaction between the virus and the cell" (Burnet 1952, p. 229). Chick embryo preparations had long been used and were known to allow grafting and growth of mammalian cells and tumors (Murphy 1913), a phenomenon that Burnet coined tolerance (Burnet 1941, p. 45). Interestingly, Burnet subsequently showed that chick embryos are also unable to produce antibodies after inoculation of influenza virus (Burnet et al. 1950).

Burnet's concept of immature immune tolerance was progressively supplemented with Peter Medawar's ideas on this matter. Medawar was one of the physicians who had explored the possibility of skin transplantation to treat burns during WWII (Gibson and Medawar 1943). In a 1948 article discussing the contribution of tissue culture methods to elucidating "the nature of immunity against transplanted skin," he clearly stated the problem by remarking that: "When skin is grafted from one human being or one rabbit to another, a 'defense' mechanism is called into action that leads, in due course, to the complete destruction of the foreign grafted tissue" (Medawar 1948, p. 239). Medawar next indicated that this reaction "varies with the antigenic relationship between donor and recipient" and that "[skin transplantation immunology] conformed ... with the pattern of immunity created by bacterial and other crudely foreign antigens" (Medawar 1948, p. 239). A couple of years later, with colleagues, he claimed to have found "a 'laboratory' solution of the problem of how to make tissue homografts immunologically acceptable to hosts, which would normally react against them" (Billingham et al. 1953, p. 603). Medawar and colleagues interpreted their findings, based on experiments with mouse feti and chick embryos, as showing "that mammals and birds never develop, or develop to only a limited degree, the power to react immunologically against foreign homologous tissue cells to which they have been exposed sufficiently early in foetal life" (Billingham et al. 1953, p. 603). The work by Burnet and Medawar was awarded the 1960 Nobel prize for medicine or physiology for "for [the] discovery of acquired immunological tolerance" (Committee 1960).

While the multiple binding sites of antibodies (Marrack et al. 1951) and "the structure of antigen-antibody aggregates and complement fixation" (Marrack 1955, p. 369) were being progressively understood in the 1950s, this was less the case for the mechanism of antibody production (Stallybrass 1950). In 1941, Burnet had postulated (1) that "production of antibody is quite certainly not a multiplication of antibody molecules in the blood plasma, but a cellular phenomenon," (2) that "each antigen molecule makes contact with a cell 'sensitized' by previous contact with antigen [and] sets going a change in the cell so that after a suitable latent period the cell liberates a series of antibody molecules," and (3) that the logarithmic character of this response "is because the entities concerned are either themselves multiplying or are being produced by multiplying agent" (Burnet 1941, p. 23). Niels

Jerne next proposed his natural selection theory of antibody formation, according to which antigen is "a selective carrier of spontaneously circulating antibody to a system of cells which can reproduce this antibody" so that "[a]ntigen, secondarily introduced into the circulation, now meets a larger concentration of specific molecules and carries a larger quantity of these, selected for the better-fitting ones, to the antibody-producing apparatus" (Jerne 1955, pp. 849–850). Accordingly, he proposed that: "In the absence of antigen no directional pressure is imposed upon globulin synthesis, and it seems reasonable to assume that a great variety of configurations, due, perhaps, to various amino acid sequences at the specific sites of the globulin molecules, may develop at random" (Jerne 1955, p. 851). Burnet postulated, in turn, that "when antigen-natural antibody contact takes place on the surface of a lymphocyte the cell is activated to settle in an appropriate tissue, spleen, lymph node or local inflammatory accumulation, and there undergo proliferation to produce a variety of descendants" and referred to this idea as the "clonal selection hypothesis" (Burnet 1957, pp. 68–69).

In the early 1960s, important structural elements relevant to antibody function were unraveled by Gerald Edelman and colleagues, combining ultracentrifugal fractionation under different reducing conditions, chromatography and electrophoretic separation, and ion exchange chromatography-based amino acid analysis (Edelman and Poulik 1961). These authors concluded (1) that the different subunits of antibodies were held together by disulfide bonds, (2) that "antibodies of different specificity consist of different types of polypeptide chains," (Edelman et al. 1961, p. 1757) and (3) that "[t]he antigenic cross-reactivity among the classes would be accounted for by the general structural similarity of their [light] chains" of around 20,000 (Edelman and Benacerraf 1962, p. 1039). While broadly subscribing to Burnet's views on antibody formation, some authors speculated that "[t]he genic diversity of the precursors of antibody-forming cells arises from a high rate of spontaneous mutation during their lifelong proliferation" (Lederberg 1959, p. 1650). More elaborate hypotheses accounting for the degree of antibody diversity by invoking the combined effects of gene duplication, point mutations, and somatic recombination were thus formulated at the end of the 1960s (Edelman and Gally 1967; Smithies 1967).

1.2.5 Neuroendocrinology as a Discipline

The development of electron microscopy and ultracentrifugation allowed scientists with a background in anatomy as well as those with a training in biochemistry to advance their studies in the 1960s. Using electron microscopy, sometimes combined with ultracentrifugation, different groups obtained evidence indicating that nerve endings are indeed present in the posterior pituitary (Labella and Sanwal 1965; Rodriguez and Dellmann 1970). Observations of preparations under the electron microscope also enabled various independent confirmations of the existence of neurosecretory vessels in the median eminence in proximity to the portal vessels running to the anterior pituitary (Rinne 1960; Mazzuca 1964; Wittkowski 1967).

Interestingly, using an array of biochemical approaches, including centrifugation, a protein isolated from the hypophyseal portal vessel system was also found to stimulate adrenocorticotropic hormone release and could therefore be considered a CRF (Porter and Rumsfeld 1959). A physiological "tour de force" was realized by Averill and colleagues, who showed that hypophyseal portal vessel blood of rats, in response to electrical stimulation of the hypothalamus, contained a Thyrotropin Releasing Factor (TRF) that promoted Thyroid Stimulating Hormone production after administration into the rabbit pituitary circulation (Averill et al. 1966). In parallel, several groups were engaged in a "biochemical race" to extract, purify, and identify this TRF, with Andrew Schally's group proposing, based on extracts of 100,000 porcine hypothalami, a substance that was 30% composed of amino acids (Schally et al. 1966), while Roger Guillemin's group claimed to have obtained, from 270,000 sheep hypothalami, a compound that was 80% composed of three amino acids (Guillemin et al. 1965). But beyond, or perhaps in part because of, this scientific dispute, neuroendocrinology started to enjoy some recognition, with review articles in major scientific and medical journals and books being published under that banner (Reichlin 1963a, b, c; Scharrer and Scharrer 1963). Furthermore, the journal *Neuroendocrinology* was launched in the mid-1960s as an outlet for articles "with the unifying concepts and the common denominators of neuroendocrinology" after "[n]euroendocrinology has gained recognition as a field of research in its own right" (Scharrer 1965).

As noted above, the starting points of many of the anatomical and biochemical approaches were the effects of lesion of glands or injection of gland extracts on water and mineral physiology, growth, sexual maturity and lactation. Progressively, after WWII, these effects were thought to be mediated by functional axes formed by the hypothalamus, the pituitary (anterior and posterior) and endocrine glands, for example, of the adrenal and thyroid. Indeed, one encounters the first mention of a hypothalamo–pituitary–adrenal (HPA) axis on PubMed in the late 1960s, that of the hypothalamo–pituitary–gonadal (HPG) axis in the early 1970s, and that of a hypothalamo–pituitary–thyroid axis in the mid-1970s. In addition, a new physiological and behavioral concept progressively linked to the HPA axis was put forward by Hans Selye from the 1950s onward, in the form of the stress response (Selye 1950, 1976). In parallel with these hypothalamus-controlled neuroendocrine axes, it was shown, using anatomical and cytochemical approaches, that enteroendocrine cells share characteristics with pituitary corticotroph cells and pancreatic islet cells, known to secrete polypeptide hormones (Pearse 1968, 1969; Pearse and Polak 1971). These observations indicate that enteroendocrine and pituitary cells may share a secretory mechanism for polypeptide hormones and thus raises the question of the physiological functions of such mechanisms in the gastrointestinal tract.

1.2.6 Antibodies as Tools

In the early 1950s, the improved method developed by Albert Coons and Melvin Kaplan for conjugation of the fluorescent marker fluorescein to antibodies made it possible for this group to localize antigens and antibodies in tissues (Coons and Kaplan 1950; Coons et al. 1950, 1951, 1955). Over the following decade, this technique became very popular (Coons 1961). Antibody conjugation approaches were subsequently expanded to enzymes, which enabled more permanently stained tissues to be obtained and circumvented the problems of fluorophore fading and the autofluorescence of tissues (Nakane and Pierce 1967). In the 1970s, the use of different fluorescent labels conjugated to different antibodies also allowed the sorting and concentration of cell populations (Bonner et al. 1972; Julius et al. 1972). All of these approaches were further improved after monoclonal antibodies with predefined specificity became available (Koehler and Milstein 1975). The fluorescence-activated cell sorter (FACS) was to become an essential tool in immunology and even gave rise to the cluster of differentiation (CD) international classification of white blood cells.

Not surprisingly, given the peptidergic nature of many hormones, many attempts were undertaken to develop antibody-based detection techniques of these mediators in bodily fluids and tissue extracts (Yalow and Berson 1960; Utiger et al. 1962; Felber 1963; Spitzer 1968). Thus, the so-called radioimmunoassays were employed to study the effects of potential hypothalamic releasing factors on pituitary contents of growth hormone or ACTH (Rodger et al. 1969; Brazeau et al. 1973; Rivier et al. 1973). Alternatively, these assays were also used to detect luteinizing hormone-releasing hormone and TRH in the hypothalamus and the cerebrospinal fluid (Ishikawa 1973; Brownstein et al. 1974; Palkovits et al. 1974). Furthermore, immunohistochemical techniques employing antisera raised against polypeptide hormones demonstrated the presence of such immunoreactive material not only in enteroendocrine cells, but also in the brain, beyond the hypothalamus (Barry et al. 1973; Leonardelli et al. 1973; Polak et al. 1974a, b; Gross 1976; Gross and Baker 1977; Fuxe et al. 1977). Thus, these new antibody-based techniques made it possible to refine both biochemical and anatomical approaches in neuroendocrinology.

1.2.7 The Recognition of Neuroendocrine Systems

The 1970s were a decade during which neuroendocrinology was the object of important conceptual developments and also received important marks of scientific respectability. Indeed, the book series Progress in Brain Research dedicated two volumes to neuroendocrinology between 1970 and 1973, the first on the *Pituitary, adrenal and the brain* and the second on *Drug effects on neuroendocrine regulation*. In the latter volume, John Porter looked at the history of neuroendocrinology and asked "what precisely is a neuroendocrine system?" (Porter 1973, p. 1). He proposed that "[a] neuroendocrine system consists of a neural cell or cells which secrete into the extracellular fluid a substance which upon reaching other cells modify their behavior" (Porter 1973, pp. 2–3). Porter also specified that his "definition does not exclude the possibility that non-neuronal cells may also be neurosecretory cells" and that "it does not require that the secretory product be transported through blood"

(Porter 1973, p. 3), raising the possibility that cerebrospinal fluid could constitute a medium for neuroendocrine signals.

Another sign of scientific recognition was the tribune in the journal *Nature* offered to Anthony Pearse, who had previously pointed out anatomical and cytochemical similarities between enteroendocrine cells, pancreatic islet cells, and pituitary corticotroph cells. In his review, Pearse first reminded the reader of the discovery of the same substance, denoted Substance P, in the brain and intestine by Von Euler and colleagues in the 1930s, before relating his own recent work as well as that of others showing that the same is true for other peptides, such as somatostatin and vasoactive-intestinal peptide, and reiterating the question of a possible common embryologic origin (Pearse 1976). In a subsequent chapter title, he proposed the term "diffuse neuroendocrine system" and speculated about the different modes of action that the same peptides could have in different biological contexts (Pearse 1978, p. 49).

The ultimate recognition of the scientific community for neuroendocrinology came in 1977 with the Nobel Prize for physiology or medicine, shared between Roger Guillemin and Andrew Schally "for their discoveries concerning the peptide hormone production of the brain" and Rosalyn Yalow "for the development of radioimmunoassays of peptide hormones" (Commitee 1978). In his Nobel lecture entitled Peptides in the brain: the new endocrinology of the neuron, Guillemin also mentioned the presence of somatostatin and other peptides in the brain and intestine and joined Pearse in wondering about their possible paracrine and endocrine modes of action (Guillemin 1978). Finally, at the turn of the decade, Pearse's colleagues Julia Pollak and Steve Bloom elaborated on the idea of "the neuroendocrine design of the gut," proposed a couple of years earlier (Makhlouf 1974; Polak and Bloom 1979b), and referred to the "diffuse neuroendocrine system" as a "powerful controlling system" (Polak and Bloom 1979a, p. 1400). So, important theoretical progress was made by the proposal of a definition of neuroendocrine systems that included both the hypothalamo–pituitary–end organ axes as well as gastrointestinal enteroendocrine cells.

1.2.8 A Formulation of the Immune System

In 1968, an important distinction of lymphocyte populations was proposed between bone marrow-derived cells (or B-cells), which become antibody producing cells, and thymus-derived cells (later shortened to T-cells), which help in antibody production (Miller and Mitchell 1968; Mitchell and Miller 1968a, b). Interestingly, in the early 1970s, Niels Jerne updated his natural selection theory of antibody formation by postulating that the selection process "of mutant cells expressing . . . genes that have been modified by spontaneous random somatic mutation" occurs in the thymus (Jerne 1971, p. 1). Subsequently, he published a series of theoretical articles in which he proposed a conceptual view of the immune system. While the very term "immune system" had been used before, including in publications on the ontogeny of the thymus, it was often not elaborated beyond a mention (Metcalf and Brumby

1966; Tyan 1968; Tyan and Herzenberg 1968). Jerne, instead, argued that "[t]he immune system is comparable in the complexity of its functions to the nervous system" in that both "respond adequately to an enormous variety of signals," and "[b]oth systems . . . learn from experience and build up a memory" (Jerne 1973). In a follow-up paper, Jerne concluded that "the immune system, even in the absence of antigens . . . achieves a dynamic steady state as its elements interact between themselves" (Jerne 1974, p. 383). These "theories concerning the specificity in development and control of the immune system" were the main motivation for awarding Jerne the Nobel Prize in 1984 (Committee 1984). So, like for the neuroendocrine system, important theoretical progress led to the explicit proposal of what the immune system is and does.

1.3 A History of Neuroendocrine–Immune Interactions

Historically speaking, it is interesting to note that most of the work on interactions between the neuroendocrine and immune systems occurred once these adjectives were applied to systems, even though (neuro)endocrine and immune cells had long been known. This may seem even more surprising given that some of the leading immunologists, like Niels Jerne, had remarked concerning the immune and nervous systems that "[t]he two systems penetrate most other tissues of our body, but they seem to be kept separate from each other by the so-called blood-brain barrier" (Jerne 1974, p. 387), and that "a population of lymphocytes . . . in appropriate tissue culture fluid" stimulated with antigen "will produce specific antibody molecules, in the absence of any nerve cells (Mishell and Dutton 1967)" (Jerne 1985, p. 852). However, it can also be argued that it was necessary to have some clearer understanding of the systems' structure and function before their interactions could be envisioned.

1.3.1 Conditioning Immune Responses

Conditioning of a behavioral or physiological response to neutral stimuli (which are "conditions [that] do not trigger a response [initially]") is considered to be related to the presence of a central nervous system (Ginsburg and Jablonka 2021, p. 5). Interestingly, classical conditioning of leukocyte and antibody responses had already been shown in the early twentieth century (Metalnikov and Chrorine 1926, 1928). However, probably because many of the follow-up studies were done in the Soviet Union and Eastern Europe, the findings of such experiments were only noticed in the Western world after the 1970s when Robert Ader and Nicholas Cohen showed that the immunosuppressive effects of certain drugs could be conditioned (Ader and Cohen 1975, 1982; Pacheco-López et al. 2007). Not surprisingly, a review article published in 1985 stated that "[t]he traditional view that the nervous and immune systems are functionally independent . . . is being challenged" by the "possib[ility] to change the activity of the immune system by means of Pavlovian conditioning, just as it is possible to condition other physiological events influenced by the autonomic nervous system or neuroendocrine substances" (Brittain and Wiener 1985, p. 181).

1.3.2 Stressing Corticotropic Influences on Immune Responses

While the anti-inflammatory effects of glucocorticoids had been known since the 1950s (Gordon and Katsh 1949; Glyn 1998), there were also some indications that could be interpreted as suggesting that ACTH might influence immune cell counts and antibody production (Harris 1951b; Mayer 1951; Morris 1951). Interestingly, in the early 1970s, it was shown that prior hypophysectomy leads to a subsequent depression of spleen immune responses to antigen in vitro (Gisler and Schenkel-Hulliger 1971). However, these effects have not been easy to reproduce in vivo (Kalden et al. 1970). Similarly, the reported consequences of lesions of the hypothalamus on antibody responses and their dependence on the pituitary and corticosteroids have been highly variable (Stein et al. 1976; Cross et al. 1980, 1982). Nevertheless, a view started to emerge according to which the neuroendocrine and immune systems interact, after it was shown that the administration of two classic antigens in rodents, namely sheep red blood cells and hemocyanin, resulted in increased electrophysiological activity in the hypothalamus (Besedovsky et al. 1977). It has been argued that such a vision "bring[s] the self-regulated immune system into conformity with other body systems" and "is based on the existence of afferent-efferent pathways between immune and neuroendocrine structures" (Besedovsky and Sorkin 1977, p. 1). Thus, in the same year that Jerne announced that "a population of lymphocytes ... in appropriate tissue culture fluid" stimulated with antigen "will produce specific antibody molecules, in the absence of any nerve cells" (Jerne 1985, p. 852), others affirmed that "hormones, neurotransmitters, and neuropeptides [released] in the microenvironment of immunologic cells" can provide "external immunoregulatory signals imposed upon autoregulatory mechanisms" (Besedovsky et al. 1985, p. 750s).

This emerging framework of neuroendocrine–immune interactions also provided a potential biological substrate for the reported effects of stress on disease. Furthermore, it had become clear, between the 1950s and 1970s, that acute and chronic stress were associated with increased corticosteroid production in animals, including humans (Hale et al. 1957; Mason et al. 1961; Treiman et al. 1970; Weiss 1970; Arguelles et al. 1972; Bassett et al. 1973; Tache et al. 1976). Not surprisingly, in the second half of the 1970s, several authors therefore concluded that "psychosocial processes influence the susceptibility to some infections, to some neoplastic processes, and to some aspects of humoral and cell-mediated immune responses" (Stein et al. 1976, p. 439) and that "[t]hese psycho-social effects may be related to hypothalamic activity, the autonomic nervous system, and neuro-endocrine activity" (Miller 1977, p. 413). However, given that "[t]he term "stress" has been used in so many different ways," it is important to specify, for example, that the "focus [is] on psychological stress rather than physical stresses such as starvation or exposure to extreme cold" (Rogers et al. 1979, p. 147, 153). Thus, it could be concluded at the end of the decade that "[t]here seems little doubt that different psychological states . . . can influence the immune system" and that "[t]he questions now are really what the mechanisms are, and how clinically significant they might be" (Rogers et al. 1979, p. 158). Regarding the latter, "[t]he predominant hypothesis has been that CNS change leads to immunologic change through the mechanism of hypothalamic-pituitary hormonal stimulation" (Rogers et al. 1979, p. 158). Interestingly, the idea that psychological stress alters the immune system through activation of the neuroendocrine system has been a working hypothesis for many years (Stein et al. 1985; Tecoma and Huey 1985).

1.3.3 Shared Markers and Multilevel Neuroendocrine–Immune Interactions

In addition to immune–neuroendocrine interactions formulated at the systems level, the findings obtained by different experimental approaches also seemed to indicate other kinds of relationships. For example, in spite of the fact that the thymus since the early 1970s had been considered an immune organ, it was found a couple of years later that congenitally athymic (nude) mice and neonatally thymectomized mice show endocrine changes indicating that "the thymus may well have a basic role in the organization of the adult hypothalamus-pituitary axis for thyroid and sexual functions" (Pierpaoli and Besedovsky 1975, p. 323). Moreover chromogranin, a marker that had been used to propose the notion of the diffuse neuroendocrine system, was also detected in tissues of the spleen, lymph nodes and thymus, thought to be part of the immune system (Angeletti and Hickey 1985). Other indications that labels such as immune, neuroendocrine, or neuronal may not neatly characterize our epistemic categories once and for all can also be found in the literature. Thus, it turned out in the early 1980s that immune cells were not only sensitive to mediators like ACTH, which by then was considered a(n) (neuro)endocrine messenger molecule, but could also synthesize them during infection (Johnson et al. 1982, 1984; Blalock and Smith 1985). This illustrates that the initial context in which a molecule was discovered (neuroendocrine or immune) is often not the only biological condition in which it plays a role. Conversely, it was shown that interleukin-1, which was, as its name indicates, thought of as a messenger molecule between leukocytes, can stimulate both pituitary mRNA expression and secretion of ACTH as well as hypothalamic production of corticotropin-releasing hormone (Berkenbosch et al. 1987; Bernton et al. 1987; Brown et al. 1987; Sapolsky et al. 1987). This illustrates not only the occurrence of neuroendocrine–immune interactions between molecular and cellular elements of these systems, but also that these neuroendocrine–immune interactions may occur in parallel with endocrine–immune or lower-level interactions, and thus raises the question of the circumstances under which neuroendocrine–immune interactions occur.

Another example of neuroendocrine–immune interactions concerned the central nervous system action of some interleukins or lymphokines, as they were called in the 1970s and early 1980s, before being grouped under the wider-ranging name of cytokines. The first step, in hindsight, was the recognition that brain cells, and in particular glial cells, are capable of producing interleukin-1 in vitro and in vivo in response to administration of bacterial lipopolysaccharide or local injury (Fontana et al. 1982; Coceani et al. 1988; Hetier et al. 1988). These findings led not only to a revision of the immune-privileged status of the brain, but also to the possibility that interleukin-1 action in or on the brain could play roles other than those in response to local injury or infection. Indeed, peripheral administration of purified interleukin-1, obtained after exposure of a macrophage cell line to bacterial lipopolysaccharide, not only stimulates thymocyte proliferation, but also induces fever (Duff and Durum 1983). The possibility that central interleukin-1 action could play a role in fever induction or neuroendocrine activation has been repeatedly addressed (Fontana et al. 1984; Hooghe-Peters et al. 1991). But one of the earliest CNS-regulated host responses to be induced by bacterial fragments, which was shown to both be mimicked and mediated by central interleukin-1, was increased sleep (Krueger et al. 1984; Shoham et al. 1987; Imeri et al. 1993; Takahashi et al. 1996).

While some of the somnogenic effects of central interleukin-1 are mediated and modulated by hypothalamic releasing factors (Krueger 1990; Krueger and Obál 1993), and can therefore be qualified as neuroendocrine–immune interactions, it turned out that interleukin-1's presumed mode of action may also be considered as neuroendocrine. First, brain interleukin-1 expression and its cerebrospinal fluid concentrations increase during sleep, in comparison to the awakened state of animals, even in the absence of exposure to microbial fragments (Lue et al. 1988; Taishi et al. 1998). Moreover, central inhibition of interleukin-1 action has been shown to reduce sleep and its rebound after sleep deprivation (Opp and Krueger 1994; Takahashi et al. 1996, 1997). Finally, infusion of interleukin-1 into different parts of the ventricular and subarachnoid cerebrospinal fluid systems allowed the demonstration that its maximal sleep-stimulating effects occur at the site where prostaglandin D2, an already known somnogenic substance of which the production can be induced interleukin-1, promotes sleep (Terao et al. 1998). Altogether, it thus seems that interleukin-1, a mediator classically associated with the immune system, plays a role in the physiological regulation of sleep in the brain, through an endocrine-like mode of action that does not involve the systemic blood circulation, but rather the cerebral ventricular cerebrospinal fluid.

Finally, evolution-inspired considerations can be interpreted to urge some reframing of the way neuroendocrine–immune interactions are understood. Based on immunoreactivity and immunoneutralization studies indicating that "hormonal peptides and neuropeptides ... are native to unicellular organisms," "a common phylogenetic origin for the endocrine system and nervous system of vertebrates" was proposed in the early 1980s (Le Roith et al. 1982). The subsequent detection of mediators, which were thought to be hormones and neuropeptides, such as ACTH and beta-endorphin, in immune(−like) cells of mammals, amphibians, and gastropods exposed to bacterial fragments gave rise to several hypotheses (Blalock et al. 1985; Ottaviani et al. 1991, 1992). These varied from the suggestion that "the immune and neuroendocrine systems represent a totally integrated circuit by virtue of sharing a common set of hormones" (Blalock et al. 1985, p. 858s) to the proposal that the "immune and neuroendocrine systems share a common evolutionary origin" (Ottaviani et al. 1991, p. 215). The idea of a shared common origin between the neuroendocrine and immune systems could also explain why certain markers thought to indicate neuroendocrine function can be found in immune cells (Day and Salzet 2002). Thus, it has been proposed that it is necessary "to redefine previous 'neuroendocrine' concepts to include the notion that activation of specific genetic switches can lead to the expression of a partial or full neuroendocrine phenotype in a variety of cell types, including immune cells" (Day and Salzet 2002, p. 447). So while initial thinking about neuroendocrine–immune interactions seemed to be formulated in terms of systems in the 1970s–1980s, subsequent physiologic and evolutionary research on some of the mediators involved led to the conclusion that these interactions could occur at the organ, tissue, cellular, and molecular levels as well and that the very labels "immune" and "neuroendocrine" might need to be revised. A recent illustration of this is the finding that in *Hydra*, an organism that lacks immune cells, neuroendocrine-like cells secrete peptides that control bacterial growth on its epithelial surface (Augustin et al. 2017).

1.4 Perspectives for Future Research

One of the reasons for looking back at the history of science is the belief that one might learn from it. Regarding the history of science itself, the hope is that this chapter has shown that the evolution of scientific fields is driven neither exclusively by technological progress nor solely by conceptual developments. Instead, the histories of immunology and neuroendocrinology and that of the study of neuroendocrine–immune interactions presented here illustrate that both technological progress and conceptual development have played important, but variable, roles throughout the twentieth century. For example, on the one hand, the techniques and technologies for generating antibodies have been instrumental in gaining a better understanding of the tissue distribution and changes in body fluid concentrations of molecules of interest in all of these fields. But, on the other hand, the very emergence of the notions of immune and neuroendocrine systems could only occur after certain theoretical considerations regarding systems, regulation, and function had taken place. Furthermore, much of the initial excitement and interest regarding neuroendocrine–immune interactions seemed to be fed by the challenges it posed to the current ideas of structure–function relationships of and regulation by immune and neuroendocrine systems.

One of the important general lessons that science can learn from its history is that the categories, concepts, and labels that a field of science proposes to make sense of a part of the world at one point should not be taken to be definitive. One example of this is the expansion of the label "neuroendocrine" from functional axes involving the hypothalamus, the pituitary, and some peripheral glands to more localized organizations in the gastrointestinal system. The gut itself is also a good example of how the functions associated with it have evolved over time. Indeed, throughout most of the twentieth century, the gut's function was perceived as being mainly digestive, with its secretions allowing the metabolism of food and the endocrine signaling of hunger and satiety to the brain. However, with the progressive realization that many, if not most, of the body's immune cells are associated with the gastrointestinal tract, the gut has also been ascribed a role in the organism's defense against infection. But in the light of the recently recognized importance of gut microbiota for the physiology of their multicellular hosts, it is to be expected that this will not only lead to a reconsideration of the gut's functions and that of the local endocrine, immune, and nervous systems, but also to an increased interest in local neuroendocrine–immune interactions in the gut (see also Chap. 2).

Another, somewhat related, lesson is that many regulatory processes in biology are functionally redundant to some extent. Thus, it is important to keep in mind that a regulatory response that seems to be top-down, in the sense that it involves entities at a perceived higher level of organization influencing lower-level entities, does not exclude other forms of regulation. For example, in terms of neuroendocrine–immune interactions, the pro-inflammatory cytokine interleukin-1 had been shown in the 1980s to activate the HPA axis both at the level of hypothalamic corticotropin-releasing hormone-expressing neurons and at the level of pituitary ACTH-producing endocrine cells. Both sites of action of interleukin-1 can, in turn, give rise glucocorticoid release from the adrenal. However, in the first decade of the twenty-first century, it was shown that interleukin-1 can act directly on the adrenal to promote corticosteroid production (Engstrom et al. 2008). Therefore, an increase in circulating glucocorticoid concentrations under inflammatory conditions alone cannot be taken to reflect HPA-axis activation or interaction between neuroendocrine and immune systems as such. In addition, these findings raise questions regarding the "fine-tunedness" of these different regulatory processes and the contexts in which they are activated.

These considerations are also important to bear in mind when employing and implementing new technological approaches. For example, many "omics" approaches, using high-throughput technologies allowing for the measurement of thousands of transcripts, proteins, or metabolites, rely on the gene ontology [GO] database for interpretation of findings. However, and although "[i]deally, GO would contain a complete description of all gene functions" (Soldatos et al. 2015, Table 3), it has been shown regarding functional annotations "that advanced methods … significantly outperformed a straightforward application of function transfer by local [gene] sequence similarity" (Jiang et al. 2016, p. 2). Moreover, it has been proposed that "interactions between GO terms [should be] based on further experimental data that cover a wide range of biological functions," for example "using a well-recognized cell biology textbook" to gain more insight into the functional role of subcellular processes (Hansen et al. 2017, p. 2, 10). Such cell biology textbook-based assessment of interactions can be complemented by taking into account neuroendocrine–immune interactions, of which a few were described in this chapter.

Among the newly emerging scientific approaches that can inform and can be informed by neuroendocrine–immune interactions are three-dimensional in vitro systems, such as organoids and organs on a chip. Although many of these approaches have been focused on studying interactions between cells of the same system (Morsink et al. 2020; Tambalo and Lodato 2020), other groups have proposed three-dimensional in vitro systems that contain gut epithelial, immune cells and microbiota (Ambrosini et al. 2020; Li et al. 2021) or neurons, glial and endothelial cells (Chukwurah et al. 2019; Makrygianni and Chrousos 2021). While these systems can never replace in vivo studies, they can be expected to prove useful in determining how neuroendocrine–immune interactions regulating bodily functions may take place at lower levels of perceived organization.

In conclusion, the now long histories of the neuroendocrine and immune systems and their interactions indicate that progress in our understanding occurs when conceptual and technological innovations take place and are both considered. It is hoped that this "lesson" will carry over to the immediate future and promote continued collective exploration of the richness of interactions between these systems and their elements in an ecologically-relevant context.

1.5 Key References

Besedovsky et al., *Journal of Immunology,* 1985. This review provides a discussion of the ways in which the immune and neuroendocrine systems can interact.

Burnet, *The Australian Journal of Science,* 1957. This historical article underlines the important conceptual work that was necessary to start and explain specific antibody production.

Guillemin, *Science,* 1978. This article contains the text of Guillemin's Nobel lecture in which he discusses the role of peptides in the brain in the context of the "new endocrinology of the neuron."

Harris, *Physiological Reviews*, 1948. This historical review lays the conceptual groundwork for how the hypothalamus influences pituitary gland hormonal secretion.

Jerne, *The EMBO Journal/Science,* 1985. This article contains the text of Jerne's Nobel lecture in which he summarizes several decades of research on specific immunity using language as a metaphor.

Pearse, *Nature,* 1976. This historical review article proposes that peptides both in the brain and in the intestine can function as hormones.

References

Ader R, Cohen N (1975) Behaviorally conditioned immunosuppression. Psychosom Med 37:333–340

Ader R, Cohen N (1982) Behaviorally conditioned immunosuppression and murine systemic lupus erythematosus. Science 215:1534–1536

Ambrosini YM, Shin W, Min S, Kim HJ (2020) Microphysiological engineering of immune responses in intestinal inflammation. Immune Network 20:e13

Anderson EM, Collip JB (1933) Thyreotropic hormone of anterior pituitary. Proc Soc Exp Biol Med 30:680–683

Anderson EM, Collip JB (1934) Studies on the physiology of the thyreotropic hormone of the anterior pituitary. J Physiol 82:11–25
Angeletti RH, Hickey WF (1985) A neuroendocrine marker in tissues of the immune system. Science 230:89–90
Arguelles AE, Martinez MA, Hoffman C, Ortiz GA, Chekherdemian M (1972) Corticoadrenal and adrenergic overactivity and hyperlipidemia in prolonged emotional stress. Hormones 3:167–174
Augustin R, Schroder K, Murillo Rincon AP, Fraune S, Anton-Erxleben F, Herbst EM, Wittlieb J, Schwentner M, Grotzinger J, Wassenaar TM, Bosch TCG (2017) A secreted antibacterial neuropeptide shapes the microbiome of hydra. Nat Commun 8:698
Averill RL, Salaman DF, Worthington WC (1966) Thyrotrophin releasing factor in hypophyseal portal blood. Nature 211:144–145
Bailey P, Bremer F (1921) Experimental diabetes insipidus. Arch Intern Med 28:773–803
Bargmann W, Scharrer E (1951) The site of origin of the hormones of the posterior pituitary. Am Sci 39:255–259
Barry J, Dubois MP, Poulain P (1973) LRF producing cells of the mammalian hypothalamus. A fluorescent antibody study. Z Zellforsch Mikrosk Anat 146:351–366
Bassett JR, Cairncross KD, King MG (1973) Parameters of novelty, shock predictability and response contigency in corticosterone release in the rat. Physiol Behav 10:901–907
Berkenbosch F, Van Oers J, Del Rey A, Tilders F, Besedovsky H (1987) Corticotropin-releasing factor-producing neurons in the rat activated by interleukin-1. Science 238:524–526
Bernton EW, Beach JE, Holaday JW, Smallridge RC, Fein HG (1987) Release of multiple hormones by a direct action of interleukin-1 on pituitary cells. Science 238:519–521
Besedovsky H, Sorkin E (1977) Network of immune-neuroendocrine interactions. Clin Exp Immunol 27:1–12
Besedovsky H, Sorkin E, Felix D, Haas H (1977) Hypothalamic changes during the immune response. Eur J Immunol 7:323–325
Besedovsky HO, Del Rey AE, Sorkin E (1985) Immune-neuroendocrine interactions. J Immunol 135:750s–754s
Billingham RE, Brent L, Medawar PB (1953) Actively acquired tolerance of foreign cells. Nature 172:603–606
Blalock JE, Smith EM (1985) A complete regulatory loop between the immune and neuroendocrine systems. Fed Proc 44:108–111
Blalock JE, Harbour-Mcmenamin D, Smith EM (1985) Peptide hormones shared by the neuroendocrine and immunologic systems. J Immunol 135:858s–861s
Bonner WA, Hulett HR, Sweet RG, Herzenberg LA (1972) Fluorescence activated cell sorting. Rev Sci Instrum 43:404–409
Bordet J (1895) Cintribution à l'étude du sérum chez les animaux vaccinés. Lamertin, H, Bruxelles
Bordet J (1909) Studies in immunity. Wiley, New York
Brazeau P, Vale W, Burgus R, Ling N, Butcher M, Rivier J, Guillemin R (1973) Hypothalamic polypeptide that inhibits the secretion of immunoreactive pituitary growth hormone. Science 179:77–79
Brittain RW, Wiener NI (1985) Neural and Pavlovian influences on immunity. Pavlov J Biol Sci 20: 181–194
Brown SL, Smith LR, Blalock JE (1987) Interleukin 1 and interleukin 2 enhance proopiomelanocortin gene expression in pituitary cells. J Immunol 139:3181–3183
Brownstein MJ, Palkovits M, Saavedra JM, Bassiri RM, Utiger RD (1974) Thyrotropin-releasing hormone in specific nuclei of rat brain. Science 185:267–269
Burnet FM (1940) Biological aspects of infectious disease. Macmillan, New York
Burnet FM (1941) The production of antibodies. Macmillan, Melbourne
Burnet FM (1952) Haemagglutination in relation to host cell-virus interaction. Annu Rev Microbiol 6:229–246
Burnet FM (1957) A modification of Jerne's theory of antibody production using the concept of clonal selection. Aust J Sci 20:67–69

Burnet FM, Fenner F (1949) The production of antibodies. Macmillan, Melbourne
Burnet FM, Stone JD, Edney M (1950) The failure of antibody production in the chick embryo. Aust J Exp Biol Med Sci 28:291–297
Camus J, Roussy G (1920) Experimental researches on the pituitary body: diabetes insipidus, glycosuria and those dystrophies considered as hypophyseal in origin. Endocrinology 4:507–522
Cannon WB (1914) The emergency function of the adrenal medulla in pain and the major emotions. Am J Phys 33:356–372
Cannon WB (1927) Bodily changes in pain, fear and rage. D. Appleton & Company, New York
Chukwurah E, Osmundsen A, Davis SW, Lizarraga SB (2019) All together now: modeling the interaction of neural with non-neural systems using organoid models. Front Neurosci 13:582
Ciaccio C (1906) Sur une nouvelle espéce cellulaire dans les glandes de Lieberkühn. C R Soc Biol 60:76–77
Coceani F, Lees J, Dinarello CA (1988) Occurrence of interleukin-1 in cerebrospinal fluid of the conscious cat. Brain Res 446:245–250
Cohn EJ, Oncley JL, Strong LE, Hughes WL, Armstrong SH (1944) Chemical, clinical, and immunological studies on the products of human plasma fractionation. I. the characterization of the protein fractions of human plasma. J Clin Invest 23:417–432
Collip JB (1933) The anterior pituitary lobe: fractionation of active principles. Lancet 224:1208–1209
Collip JB, Anderson EM, Thomson DL (1933a) The adrenotropic hormone of the anterior pituitary lobe. Lancet 222:347
Collip JB, Selye H, Thomson DL (1933b) Gonad-stimulating hormones in hypophysectomised animals. Nature 131:56
Commitee NP (1978) The Nobel Prize in Physiology or Medicine 1977 [Online]. https://www.nobelprize.org/prizes/medicine/1977/summary/. Accessed 1 Nov 2021
Committee NP (1960) The Nobel Prize in Physiology or Medicine 1960 [Online]. https://www.nobelprize.org/prizes/medicine/1960/summary/. Accessed 1 Nov 2021
Committee NP (1984) The Nobel Prize in Physiology or Medicine 1984 [Online]. https://www.nobelprize.org/prizes/medicine/1984/jerne/facts/. Accessed 1 Nov 2021
Coons AH (1961) The beginnings of immunofluorescence. J Immunol 87:499–503
Coons AH, Kaplan MH (1950) Localization of antigen in tissue cells; improvements in a method for the detection of antigen by means of fluorescent antibody. J Exp Med 91:1–13
Coons AH, Snyder JC, Sheever FS, Murray E (1950) Localization of antigen in tissue cells; antigens of Rickettsiae and mumps virus. J Exp Med 91:31–38
Coons AH, Leduc EH, Kaplan MH (1951) Localization of antigen in tissue cells. VI. The fate of injected foreign proteins in the mouse. J Exp Med 93:173–188
Coons AH, Leduc EH, Connolly JM (1955) Studies on antibody production. I. A method for the histochemical demonstration of specific antibody and its application to a study of the hyperimmune rabbit. J Exp Med 102:49–60
Cross RJ, Markesbery WR, Brooks WH, Roszman TL (1980) Hypothalamic-immune interactions. I. The acute effect of anterior hypothalamic lesions on the immune response. Brain Res 196:79–87
Cross RJ, Brooks WH, Roszman TL, Markesbery WR (1982) Hypothalamic-immune interactions. Effect of hypophysectomy on neuroimmunomodulation. J Neurol Sci 53:557–566
Crowe SJ, Cushing H, Homans J (1909) Effects of hypophyseal transplantation following total hypophysectomy in the canine. Q J Exp Physiol 2:389–400
Cushing H (1909) III. Partial hypophysectomy for acromegaly: with remarks on the function of the hypophysis. Ann Surg 50:1002–1017
Cushing H (1933) Posterior pituitary activity from an anatomical standpoint. Am J Pathol 9(539–548):19
Dale HH (1906) On some physiological actions of ergot. J Physiol 34:163–206
Dale HH (1909) The action of extracts of the pituitary body. Biochem J 4:427–447

Day R, Salzet M (2002) The neuroendocrine phenotype, cellular plasticity, and the search for genetic switches: redefining the diffuse neuroendocrine system. Neuro Endocrinol Lett 23:447–451
Du Vigneaud V, Ressler C, Swan JC, Roberts CW, Katsoyannis PG (1954) The synthesis of oxytocin. J Am Chem Soc. 76:12
Duff GW, Durum SK (1983) The pyrogenic and mitogenic actions of interleukin-1 are related. Nature 304:449–451
Edelman GM, Benacerraf B (1962) On structural and functional relations between antibodies and proteins of the gamma-system. Proc Natl Acad Sci U S A 48:1035–1042
Edelman GM, Gally JA (1967) Somatic recombination of duplicated genes: an hypothesis on the origin of antibody diversity. Proc Natl Acad Sci U S A 57:353–358
Edelman GM, Poulik MD (1961) Studies on structural units of the gamma-globulins. J Exp Med 113:861–884
Edelman GM, Benacerraf B, Ovary Z, Poulik MD (1961) Structural differences among antibodies of different specificities. Proc Natl Acad Sci U S A 47:1751–1758
Ehrlich P (1880) Methodologische Beiträge zur Physiologie und Pathologie der verschiedenen Formen der Leukocyten. Z Klin Med 1:553–560
Engstrom L, Rosen K, Angel A, Fyrberg A, Mackerlova L, Konsman JP, Engblom D, Blomqvist A (2008) Systemic immune challenge activates an intrinsically regulated local inflammatory circuit in the adrenal gland. Endocrinology 149:1436–1450
Felber JP (1963) ACTH antibodies and their use for a radio-immunoassay for ACTH. Experientia 19:227–229
Feyrter F (1939) Uëber diffuse endokrine epitheliale Organe. J. A. Barth, Leipzig
Fontana A, Kristensen F, Dubs R, Gemsa D, Weber E (1982) Production of prostaglandin E and an interleukin-1 like factor by cultured astrocytes and C6 glioma cells. J Immunol 129:2413–2419
Fontana A, Weber E, Dayer JM (1984) Synthesis of interleukin 1/endogenous pyrogen in the brain of endotoxin-treated mice: a step in fever induction? J Immunol 133:1696–1698
Fuxe K, Hokfelt T, Said SI, Mutt V (1977) Vasoactive intestinal polypeptide and the nervous system: immunohistochemical evidence for localization in central and peripheral neurons, particularly intracortical neurons of the cerebral cortex. Neurosci Lett 5:241–246
Gibson T, Medawar PB (1943) The fate of skin homografts in man. J Anat 77:299–314
Ginsburg S, Jablonka E (2021) Evolutionary transitions in learning and cognition. Philos Trans R Soc Lond Biol Sci 376:20190766
Gisler RH, Schenkel-Hulliger L (1971) Hormonal regulation of the immune response. II. Influence of pituitary and adrenal activity on immune responsiveness in vitro. Cell Immunol 2:646–657
Glyn J (1998) The discovery and early use of cortisone. J R Soc Med 91:513–717
Gordon AS, Katsh GF (1949) The relation of the adrenal cortex to the structure and phagocytic activity of the macrophagic system. Ann N Y Acad Sci 52:1–30
Green JD, Harris GW (1949) Observation of the hypophysio-portal vessels of the living rat. J Physiol 108:359–361
Gross DS (1976) Distribution of gonadotropin-releasing hormone in the mouse brain as revealed by immunohistochemistry. Endocrinology 98:1408–1417
Gross DS, Baker BL (1977) Immunohistochemical localization of gonadotropin-releasing hormone (GnRH) in the fetal and early postnatal mouse brain. Am J Anat 148:195–215
Guillemin R (1978) Peptides in the brain: the new endocrinology of the neuron. Science 202:390–402
Guillemin R, Rosenberg B (1955) Humoral hypothalamic control of anterior pituitary: a study with combined tissue cultures. Endocrinology 57:599–607
Guillemin R, Sakiz E, Ward DN (1965) Further purification of Tsh-releasing factor (Trf) from sheep hypothalamic tissues, with observations on the amino acid composition. Proc Soc Exp Biol Med 118:1132–1137

Hale HB, Sayers G, Sydnor KL, Sweat ML, Van Fossan DD (1957) Blood adrenocorticotrophic hormone and plasma corticosteroids in men exposed to adverse environmental conditions. J Clin Investig 36:1642–1646

Hansen J, Meretzky D, Woldesenbet S, Stolovitzky G, Iyengar R (2017) A flexible ontology for inference of emergent whole cell function from relationships between subcellular processes. Sci Rep 7:17689

Harris GW (1937) The induction of ovulation in the rabbit, by electrical stimulation of the hypothalamo-hypophysial mechanism. Proc R Soc Lond B 122:374–394

Harris GW (1948) Neural control of the pituitary gland. Physiol Rev 28:139–179

Harris GW (1951a) Neural control of the pituitary gland. I. The neurohypophysis. Br Med J 2:559–564

Harris GW (1951b) Neural control of the pituitary gland. II. The adenohypophysis, with special reference to the secretion of A.C.T.H. Br Med J 2:627–634

Harris GW (1955) Neural control of the pituitary gland. Edward Arnold, London

Hench PS, Kendall EC, Slocumb CH, Polley HF (1949) The effect of a hormone of the adrenal cortex (17-hydroxy-11-dehydrocorticosterone: compound E) and of pituitary adrenocortical hormone in arthritis: preliminary report. Ann Rheum Dis 8:97–104

Hetier E, Ayala J, Denefle P, Bousseau A, Rouget P, Mallat M, Prochiantz A (1988) Brain macrophages synthesize interleukin-1 and interleukin-1 mRNAs in vitro. J Neurosci Res 21: 391–397

Hild W (1951) Experimental studies on the morphology of the neurosecretory pathway following severance of the pituitary stalk, super- and dehydration of the tissues, and overload of osmoregulation. Virchows Archiv fuer pathologisch Anatomie und Physiologie und fuer klinische Medizin 319:526–546

Hilton JG, Scian LF, Westermann CD, Nakano J, Kruesi OR (1960) Vasopressin stimulation of the isolated adrenal glands: nature and mechanism of hydrocortisone secretion. Endocrinology 67: 298–310

Hooghe-Peters E, Velkeniers B, Vanhaelst L, Hooghe R (1991) Interleukin-1, interleukin-6: messengers in the neuroendocrine immune system? Pathol Res Pract 187:622–625

Imeri L, Opp MR, Krueger JM (1993) An IL-1 receptor and an IL-1 receptor antagonist attenuate muramyl dipeptide- and IL-1-induced sleep and fever. Am J Physiol 265:R907–R913

Ishikawa H (1973) Study on the existence of TRH in the cerebrospinal fluid in humans. Biochem Biophys Res Commun 54:1203–1209

Jerne NK (1955) The natural-selection theory of antibody formation. Proc Natl Acad Sci USA 41: 849–857

Jerne NK (1971) The somatic generation of immune recognition. Eur J Immunol 1:1–9

Jerne NK (1973) The immune system. Sci Am 229:52–60

Jerne NK (1974) Towards a network theory of the immune system. Ann Immunol (Paris) 125C:373–389

Jerne NK (1985) The generative grammar of the immune system. EMBO J 4:847–852

Jiang Y, Oron TR, Clark WT, Bankapur AR, D'andrea D, Lepore R, Funk CS, Kahanda I, Verspoor KM, Ben-Hur A, Koo Da CE, Penfold-Brown D, Shasha D, Youngs N, Bonneau R, Lin A, Sahraeian SM, Martelli PL, Profiti G, Casadio R, Cao R, Zhong Z, Cheng J, Altenhoff A, Skunca N, Dessimoz C, Dogan T, Hakala K, Kaewphan S, Mehryary F, Salakoski T, Ginter F, Fang H, Smithers B, Oates M, Gough J, Toronen P, Koskinen P, Holm L, Chen CT, Hsu WL, Bryson K, Cozzetto D, Minneci F, Jones DT, Chapman S, Bkc D, Khan IK, Kihara D, Ofer D, Rappoport N, Stern A, Cibrian-Uhalte E, Denny P, Foulger RE, Hieta R, Legge D, Lovering RC, Magrane M, Melidoni AN, Mutowo-Meullenet P, Pichler K, Shypitsyna A, Li B, Zakeri P, Elshal S, Tranchevent LC, Das S, Dawson NL, Lee D, Lees JG, Sillitoe I, Bhat P, Nepusz T, Romero AE, Sasidharan R, Yang H, Paccanaro A, Gillis J, Sedeno-Cortes AE, Pavlidis P, Feng S, Cejuela JM, Goldberg T, Hamp T, Richter L, Salamov A, Gabaldon T, Marcet-Houben M, Supek F, Gong Q, Ning W, Zhou Y, Tian W, Falda M, Fontana P, Lavezzo E,

Toppo S, Ferrari C, Giollo M et al (2016) An expanded evaluation of protein function prediction methods shows an improvement in accuracy. Genome Biol 17:184
Johnson HM, Smith EM, Torres BA, Blalock JE (1982) Regulation of the in vitro antibody response by neuroendocrine hormones. Proc Natl Acad Sci U S A 79:4171–4174
Johnson HM, Torres BA, Smith EM, Dion LD, Blalock JE (1984) Regulation of lymphokine (gamma-interferon) production by corticotropin. J Immunol 132:246–250
Julius MH, Masuda T, Herzenberg LA (1972) Demonstration that antigen-binding cells are precursors of antibody-producing cells after purification with a fluorescence-activated cell sorter. Proc Natl Acad Sci U S A 69:1934–1938
Kalden JR, Evans MM, Irvine WJ (1970) The effect of hypophysectomy on the immune response. Immunology 18:671–679
Katsoyannis PG, Du Vigneaud V (1958) The synthesis of the histidine analog of the Vasopressins. Arch Biochem Biophys 78:555–562
Koehler G, Milstein C (1975) Continuous cultures of fused cells secreting antibody of predefined specificity. Nature 256:495–497
Krueger JM (1990) Somnogenic activity of immune response modifiers. Trends Pharmacol Sci 11: 122–126
Krueger JM, Walter J, Dinarello CA, Wolff SM, Chedid L (1984) Sleep-promoting effects of endogenous pyrogen (interleukin-1). Am J Physiol 246:R994–R999
Krueger JM, Obál F (1993) Growth hormone-releasing hormone and interleukin-1 in sleep regulation. FASEB J 7:645–652
Labella FS, Sanwal M (1965) Isolation of nerve endings from the posterior pituitary gland. Electron microscopy of fractions obtained by centrifugation. J Cell Biol 25(Suppl):179–193
Landsteiner K (1901) Ueber Agglutinationserscheinungen normalen menschlichen Blute. Wien Klin Wochenschr 14:1132–1134
Landsteiner K (1930) On individual differences in human blood [Online]. NobelPrize.org. https://www.nobelprize.org/uploads/2018/06/landsteiner-lecture.pdf. Accessed 1 Nov 2021
Landsteiner K (1961) On agglutination of normal human blood. Transfusion 1:5–8
Landsteiner K, Van Der Scheer J (1928) Serological differentiation of steric isomers. J Exp Med 48: 315–320
Langley JN (1901) Observations on the physiological action of extracts of the supra-renal bodies. J Physiol 27:237–256
Le Roith D, Shiloach J, Roth J (1982) Is there an earlier phylogenetic precursor that is common to both the nervous and endocrine systems? Peptides 3:211–215
Lederberg J (1959) Genes and antibodies. Science 129:1649–1653
Leonardelli J, Barry J, Dubois MP (1973) Demonstration by fluorescent antibody technic of a substance immunologically related to LH-RF in hypothalamus and median eminence in mammals. C R Hebd Séances Acad Sci 276:2043–2046
Li XG, Chen MX, Zhao SQ, Wang XQ (2021) Intestinal models for personalized medicine: from conventional models to microfluidic primary intestine-on-a-chip. Stem Cell Rev Rep. 18(6): 2137–2151
Lister J (1880) Remarks on micro-organisms: their relation to disease. Br Med J 2:363–365
Lister J (1896) Address by the president. Science 4:409–429
Lue FA, Bail M, Jephthah-Ochola J, Carayanniotis K, Gorczynski R, Moldofsky H (1988) Sleep and cerebrospinal fluid interleukin-1-like activity in the cat. Int J Neurosci 42:179–183
Makhlouf GM (1974) The neuroendocrine design of the gut. The play of chemicals in a chemical playground. Gastroenterology 67:159–184
Makrygianni EA, Chrousos GP (2021) From brain organoids to networking assembloids: implications for neuroendocrinology and stress medicine. Front Physiol 12:621970
Marrack J (1955) The structure of antigen-antibody aggregates and complement fixation. Annu Rev Microbiol 9:369–386
Marrack JR, Hoch H, Johns RG (1951) The valency of antibodies. Br J Exp Pathol 32:212–230

Mason JW, Brady JV, Polish E, Bauer JA, Robinson JA, Rose RM, Taylor ED (1961) Patterns of corticosteroid and pepsinogen change related to emotional stress in the monkey. Science 133: 1596–1598
Masson P (1914) La glande endocrine de l'intestin chez l'homme. Comp Rend Acad Sci 158:59–61
Masson P (1928) Carcinoids (Argentaffin-cell tumors) and nerve hyperplasia of the appendicular mucosa. Am J Pathol 4(181–212):19
Mayer MM (1951) Immunochemistry. Annu Rev Biochem 20:415–440
Mazzuca M (1964) Fine structure of the pericapillary nerve endings of the hypothalamo-hypophysial tract at the level of the infundibulum in the Guinea pig. C R Seances Soc Biol Fil 158:2090–2092
Medawar PB (1948) Tests by tissue culture methods on the nature of immunity to transplanted skin. Q J Microsc Sci 89:239–252
Metalnikov S, Chrorine V (1926) Rôle des reflexes conditionnels dans l'immunité. Ann Inst Pasteur 40:893–900
Metalnikov S, Chrorine V (1928) Rôle des reflexes conditionnels dans la formation des anticorps. C R Soc Biol 99
Metcalf D, Brumby M (1966) The role of the thymus in the ontogeny of the immune system. J Cell Physiol 67(1):149–168
Metchnikoff E (1893) Lectures on the comparative pathology of inflammation. Kegan Paul, Trench, Trübner & Co., London
Miller TR (1977) Psychophysiologic aspects of cancer: the James Ewing lecture. Cancer 39:413–418
Miller JF, Mitchell GF (1968) Cell to cell interaction in the immune response. I. Hemolysin-forming cells in neonatally thymectomized mice reconstituted with thymus or thoracic duct lymphocytes. J Exp Med 128:801–820
Mishell RI, Dutton RW (1967) Immunization of dissociated spleen cell cultures from normal mice. J Exp Med 126:423–442
Mitchell GF, Miller JF (1968a) Cell to cell interaction in the immune response. II. The source of hemolysin-forming cells in irradiated mice given bone marrow and thymus or thoracic duct lymphocytes. J Exp Med 128:821–837
Mitchell GF, Miller JF (1968b) Immunological activity of thymus and thoracic-duct lymphocytes. Proc Natl Acad Sci U S A 59:296–303
Morris CJ (1951) The adrenocorticotropic hormone (A.C.T.H.) of the pituitary gland. Lancet 1: 161–164
Morsink MAJ, Willemen NGA, Leijten J, Bansal R, Shin SR (2020) Immune organs and immune cells on a Chip: an overview of biomedical applications. Micromachines (Basel) 11
Murphy JB (1913) Transplantability of tissues to the embryo of foreign species : its bearing on questions of tissue specificity and tumor immunity. J Exp Med 17:482–493
Nakane PK, Pierce GB Jr (1967) Enzyme-labeled antibodies for the light and electron microscopic localization of tissue antigens. J Cell Biol 33:307–318
Nuttall GHF (1888) Experimente ueber die bacterienfeindlichen Einfluesse des thierischen Korpers. Z Hyg Infektionskr 4:353–394
Oliver G, Schafer EA (1895) The physiological effects of extracts of the suprarenal capsules. J Physiol 18:230–276
Opp MR, Krueger JM (1994) Anti-interleukin-1 beta reduces sleep and sleep rebound after sleep deprivation in rats. Am J Physiol 266:R688–R695
Ottaviani E, Cossarizza A, Ortolani C, Monti D, Franceschi C (1991) ACTH-like molecules in gastropod molluscs: a possible role in ancestral immune response and stress. Proc Biol Sci 245: 215–218
Ottaviani E, Trevisan P, Pederzoli A (1992) Immunocytochemical evidence for ACTH- and beta-endorphin-like molecules in phagocytic blood cells of urodelan amphibians. Peptides 13:227–231

Pacheco-López G, Niemi MB, Engler H, Schedlowski M (2007) Neuro-immune associative learning. In: Bermúdez-Rattoni F (ed) Neural plasticity and memory: from genes to brain imaging. CRC Press, Boca Raton
Palkovits M, Arimura A, Brownstein M, Schally AV, Saavedra JM (1974) Luteinizing hormone-releasing hormone (LH-RH) content of the hypothalamic nuclei in rat. Endocrinology 95:554–558
Paulescu CN (1907) Recherches sur la physiologie de l'hypophyse du cerveau: L'hypophysectomie et ses effets. J Physiol Pathol Gen 9:441–456
Pearse AG (1968) Common cytochemical and ultrastructural characteristics of cells producing polypeptide hormones (the APUD series) and their relevance to thyroid and ultimobranchial C cells and calcitonin. Proc R Soc Lond Biol Sci 170:71–80
Pearse AG (1969) The cytochemistry and ultrastructure of polypeptide hormone-producing cells of the APUD series and the embryologic, physiologic and pathologic implications of the concept. J Histochem Cytochem 17:303–313
Pearse AG (1976) Peptides in brain and intestine. Nature 262:92–94
Pearse AG (1978) Diffuse neuroendocrine system: peptides common to brain and intestine and their relationship to the APUD concept. In: Hughes, J. (ed) Centrally acting peptides. Palgrave Macmillan, London
Pearse AG, Polak JM (1971) Neural crest origin of the endocrine polypeptide (APUD) cells of the gastrointestinal tract and pancreas. Gut 12:783–788
Pfeiffer R (1894) Weitere Untersuchungen uber das Wesen der Choleraimmunitat und uber specifischbaktericide Process. Z Hyg Infektionskr 18:1–16
Pierpaoli W, Besedovsky HO (1975) Role of the thymus in programming of neuroendocrine functions. Clin Exp Immunol 20:323–338
Polak JM, Bloom SR (1979a) The diffuse neuroendocrine system. Studies of this newly discovered controlling system in health and disease. J Histochem Cytochem 27:1398–1400
Polak JM, Bloom SR (1979b) The neuroendocrine design of the gut. Clin Endocrinol Metab 8:313–330
Polak JM, Pearse AG, Adams C, Garaud JC (1974a) Immunohistochemical and ultrastructural studies on the endocrine polypeptide (APUD) cells of the avian gastrointestinal tract. Experientia 30:564–567
Polak JM, Pearse AG, Garaud JC, Bloom SR (1974b) Cellular localization of a vasoactive intestinal peptide in the mammalian and avian gastrointestinal tract. Gut 15:720–724
Porter JC (1973) Neuroendocrine systems: the need for precise identification and rigorous description of their operations. Prog Brain Res 39:1–6
Porter JC, Rumsfeld HW Jr (1959) Further study of an ACTH-releasing protein from hypophyseal portal vessel plasma. Endocrinology 64:948–954
Ramon Y, Cajal S (1893) Sur les ganglions et les plexus nerveux de l'intestin. C R Soc Biol 39:217–223
Ramon Y, Cajal S (1911) Histologie du système nerveux de l'homme et des vertébrés. Maloine, Paris
Reichlin S (1963a) Neuroendocrinology. N Engl J Med 269:1182–1191
Reichlin S (1963b) Neuroendocrinology. N Engl J Med 269:1296–1303
Reichlin S (1963c) Neuroendocrinology. N Engl J Med 269:1246–1250
Rinne UK (1960) Neurosecretory material around the hypophysial portal vessels in the median eminence of the rat. Studies on its histological and histochemical properties and functional significance. Acta Endocrinol 35:1–108
Rivier C, Vale W, Guillemin R (1973) An in vivo corticotropin-releasing factor (CRF) assay based on plasma levels of radioimmunoassayable ACTH. Proc Soc Exp Biol Med 142:842–845
Rodger NW, Beck JC, Burgus R, Guillemin R (1969) Variability of response in the bioassay for a hypothalamic somatotrophin releasing factor based on rat pituitary growth hormone content. Endocrinology 84:1373–1383

Rodriguez EM, Dellmann HD (1970) Ultrastructure and hormonal content of the proximal stump of the transected hypothalamo-hypophysial tract of the frog (Rana pipiens). Z Zellforsch Mikrosk Anat 104:449–470
Rogers MP, Dubey D, Reich P (1979) The influence of the psyche and the brain on immunity and disease susceptibility: a critical review. Psychosom Med 41:147–164
Saffran M, Schally AV (1955a) In vitro bioassay of corticotropin: modification and statistical treatment. Endocrinology 56:523–532
Saffran M, Schally AV (1955b) The release of corticotrophin by anterior pituitary tissue in vitro. Can J Biochem Physiol 33:408–415
Sapolsky R, Rivier C, Yamamoto G, Plotsky P, Vale W (1987) Interleukin-1 stimulates the secretion of hypothalamic corticotropin-releasing factor. Science 238:522–524
Schally AV, Saffran M, Zimmermann B (1958) A corticotrophin-releasing factor: partial purification and amino acid composition. Biochem J 70:97–103
Schally AV, Bowers CY, Redding TW, Barrett JF (1966) Isolation of thyrotropin releasing factor (TRF) from porcine hypothalamus. Biochem Biophys Res Commun 25:165–169
Scharrer E (1928) Die Lichtempflindlickeit Blinder Elritzen. Z Vgl Physiol 7:1–38
Scharrer E (1965) Editorial. Neuroendocrinology 1:1–3
Scharrer E, Scharrer B (1937) Ueber Druesen-Nervenzellen und neurosekretorische Organe bei Wirbellosen und Wirbeltieren. Biol Rev 12:185–216
Scharrer E, Scharrer B (1963) Neuroendocrinology. Columbia University Press, New York
Selye H (1950) Stress and the general adaptation syndrome. Br Med J 1:1383–1392
Selye H (1976) Forty years of stress research: principal remaining problems and misconceptions. Can Med Assoc J 115:53–56
Shoham S, Davenne D, Cady AB, Dinarello CA, Krueger JM (1987) Recombinant tumor necrosis factor and interleukin 1 enhance slow-wave sleep. Am J Physiol 253:R142–R149
Smith PE (1926) Ablation and transplantation of the hypophysis of the rat. Anat Rec 32:221
Smith PE (1930) Hyposphysectomy and a replacement therapy in the rat. Am J Anat 45:205–273
Smithies O (1967) Antibody variability. Somatic recombination between the elements of "antibody gene pairs" may explain antibody variability. Science 157:267–273
Soldatos TG, Perdigao N, Brown NP, Sabir KS, O'donoghue SI (2015) How to learn about gene function: text-mining or ontologies? Methods 74:3–15
Spitzer R (1968) The clinical importance of polypeptide hormone assays. Clin Biochem 1:216–223
Stallybrass CO (1950) The mechanism of antibody production. Proc R Soc Med 43:137–142
Starling EH (1905) On the chemical correlations of the functions of the body. Lancet 166:339–341
Starling EH (1923) The wisdom of the body: the Harveian oration, delivered before the Royal College of Physicians of London on St. Luke's Day, 1923. Br Med J 2:685–690
Stein M, Schiavi RC, Camerino M (1976) Influence of brain and behavior on the immune system. Science 191:435–440
Stein M, Keller SE, Schleifer SJ (1985) Stress and immunomodulation: the role of depression and neuroendocrine function. J Immunol 135:827s–833s
Tache Y, Du Ruisseau P, Tache J, Selye H, Collu R (1976) Shift in adenohypophyseal activity during chronic intermittent immobilization of rats. Neuroendocrinology 22:325–336
Taishi P, Chen Z, Obál FJR, Hansen MK, Zhang J, Fang J, Krueger JM (1998) Sleep-associated changes in interleukin-1beta mRNA in the brain. J Interferon Cytokine Res 18:793–798
Takahashi S, Kapás L, Fang J, Seyer JM, Wang Y, Krueger JM (1996) An interleukin-1 receptor fragment inhibits spontaneous sleep and muramyl dipeptide-induced sleep in rabbits. Am J Physiol 271:R101–R108
Takahashi S, Fang J, Kapás L, Wang Y, Krueger JM (1997) Inhibition of brain interleukin-1 attenuates sleep rebound after sleep deprivation in rabbits. Am J Physiol 273:R677–R682
Tambalo M, Lodato S (2020) Brain organoids: human 3D models to investigate neuronal circuits assembly, function and dysfunction. Brain Res 1746:147028
Tecoma ES, Huey LY (1985) Psychic distress and the immune response. Life Sci 36:1799–1812

Terao A, Matsumura H, Saito M (1998) Interleukin-1 induces slow-wave sleep at the prostaglandin D2-sensitive sleep-promoting zone in the rat brain. J Neurosci 18:6599–6607
Treiman DM, Fulker DW, Levine S (1970) Interaction of genotype and environment as determinants of corticosteroid response to stress. Dev Psychobiol 3:131–140
Tyan ML (1968) Studies on the ontogeny of the mouse immune system. I. Cell-bond immunity. J Immunol 100:535–542
Tyan ML, Herzenberg LA (1968) Studies on the ontogeny of the mouse immune system. II. Immunoglobulin-producing cells. J Immunol 101:446–450
Utiger RD, Parker ML, Daughaday WH (1962) Studies on human growth hormone. I. A radio-immunoassay for human growth hormone. J Clin Invest 41:254–261
Verney EB (1947) The antidiuretic hormone and the factors which determine its release. Proc R Soc Lond Biol Sci 135:25–106
Von Behring E, Kitasako S (1890) Ueber das Zustandekommen der Diphtherie-Immunitat and der Tetanus-Immunitat bei Thieren. Dtsch Med Wochenschr 16:1113–1114
Von Euler US, Gaddum JH (1931) An unidentified depressor substance in certain tissue extracts. J Physiol 72:74–87
Weiss JM (1970) Somatic effects of predictable and unpredictable shock. Psychosom Med 32:397–408
Wittkowski W (1967) Synaptic structures and elementary granules in the neurohypophysis of the Guinea pig. Z Zellforsch Mikrosk Anat 82:434–458
Wormall A (1948) Contributions of chemistry to the problems of immunology. Br Med Bull 5:333–338
Yalow RS, Berson SA (1960) Immunoassay of endogenous plasma insulin in man. J Clin Invest 39: 1157–1175

Philosophical Perspectives on Neuroendocrine–Immune Interactions: The Building Block Model and Complementary Neuro-Endocrine-Immune-Microbiota Systems Approaches

2

Gregor P. Greslehner, Federico Boem, Lynn Chiu, and Jan Pieter Konsman

Abstract

The study of the interactions between the neuroendocrine and immune systems is a highly interdisciplinary research endeavor, in which the boundaries between the systems being studied become blurred. We address a common scientific perspective in dealing with intertwined complex systems, namely the conceptual approach in science that treats each system (e.g., nervous, immune, endocrine systems) as separate units or "building blocks" with unique functions that correspond to specific structures. While there are merits to this way of decomposing complex systems, there are several reasons why such an approach is limited when trying to recompose a physiological system that is engaged in intricate co-functioning and that is the result of co-development, and co-evolution, not just between these systems, but with the gut microbiota as well. Our suggestion is to take an alternative ecological evolutionary developmental approach to the

Author's major contributions: GPG first draft, sections 2.2 & 2.5; FB section 2.3, edited draft; LC section 2.1 & 2.4, edited draft; JPK section 2.4, edited draft

G. P. Greslehner
Department of Philosophy, University of Vienna, Vienna, Austria

F. Boem (✉)
Section Philosophy, University of Twente, Enschede, The Netherlands

L. Chiu
Department of Evolutionary Biology, University of Vienna, Vienna, Austria

J. P. Konsman
IMMUNOlogy from CONcepts and ExPeriments to Translation, CNRS UMR 5164, University of Bordeaux, Bordeaux, France

J. P. Konsman, T. M. Reyes (eds.), *Neuroendocrine-Immune System Interactions*, Masterclass in Neuroendocrinology 13,
https://doi.org/10.1007/978-3-031-21358-8_2

neuro-endocrine-immune-microbiota system (NEIMS) as a whole, which can serve as complementary to the predominant building block perspective.

Keywords

Biological systems gut · Microbiota · Neuroimmune interactions · Philosophy of science

In trying to analyze the natural world, scientists are seldom aware of the degree to which their ideas are influenced both by their way of perceiving the everyday world and by the constraints that our cognitive development puts on our formulations. At every moment of perception of the world around us, we isolate objects as discrete entities with clear boundaries while we relegate the rest to a background in which the objects exist.

That tendency [. . .] is one of the most powerful influences on our scientific understanding. As we change our intent, also we identify anew what is object and what is background.—Richard C. Lewontin (1929–2021)[1]

Until recently, many scientists viewed immune cells and the central nervous system (CNS) as a deadly mix. [...] Decades of research on this autoimmune disorder [multiple sclerosis] opened a window into how the immune system and the CNS interact, but more recent research efforts have revealed the exceptionally broad scope of communication between the two. We now know that the immune system is very likely a key player in many neurological diseases and, surprisingly, that immune-CNS interactions may not all be bad.—(Mueller et al. 2016, p. 760)

2.1 Introduction

Neuroimmunology is a relatively new and interdisciplinary field. The term "neuroimmunology" has been around for several decades, but the canon of topics that can be categorized as such continues to evolve. Other terms—like "psychoneuroimmunology" and "psycho-neuro-endocrino-immunology" have been used as well, indicating both the interconnectedness of these various systems and the disciplines studying them.[2] For reasons of simplicity, "neuroimmunology" will be employed here as the most inclusive term that refers to the study of interactions between the neuroendocrine and immune systems.

[1] "It's Even Less in Your Genes," *The New York Review*, May 26, 2011 issue, https://www.nybooks.com/articles/2011/05/26/its-even-less-your-genes/ [accessed: 28-01-2022].

[2] Historically, various terms have been used reflecting slightly different approaches, from psychoneuroimmunology and neuroimmunomodulation, which, at least initially, were rather top-down in nature, to immunopsychiatry, being more bottom-up (Konsman 2019; Pariante 2015). However, in all cases, some interactions between neuroendocrine and immune systems are invoked or assumed. Therefore, we propose to use the term "neuroimmune" as shorthand for these different approaches while being well aware that neuroimmunology refers to a scientific field.

The biological systems in question, however, are anything but simple. From a methodological point of view, in order to deal with complex systems and study them in ways that are appropriate to arrive at empirically meaningful statements, researchers tend to dissect or decompose them into smaller units (Bechtel and Richardson 2010). The overall organization of the organism has thus been traditionally broken down into the nervous, endocrine, and immune systems, studied in isolation as structurally or functionally distinct biological systems. Even though host microbiota are not historically considered a bodily "system," they are nevertheless increasingly thought of as constituting an ecological system within the host body. [3] In this case, the designation of "system" is also driven by structural–functional considerations, as well as the distinct genealogical lineages of macro- versus microorganismal cells.

Another tendency is to assume a natural hierarchy between these decomposed systems. The brain, for instance, as a complex system is supposed to deal with regulating complexities internal and external to the body. Two metaphoric images of the brain have been distinguished, one "as governor and [the other] as transducer" with "the former treat[ing] the brain as the executive control center of the body, [and] the latter as an interface between the organism and reality at large" (Fuller 2014). In several domains, the "governor vision" or "master vision" of the brain is dominant, for example, regarding cognition and emotion when they are considered to involve both body and brain, but to be driven by the latter (Colombetti and Zavala 2019; Facchin et al. 2021). Furthermore, historically the "brain as governor" vision has proven fruitful regarding the idea of neuroendocrine systems by postulating and establishing an important role for the hypothalamus in controlling pituitary–endocrine gland functional axes (see Chap. 1). It is therefore not surprising that initially the fields studying neuroimmune interactions, such as psychoneuroimmunology and neuroimmunomodulation, were largely motivated by the "top-down" idea that the brain controlled immune responses via neuroendocrine mechanisms.

However, these fields have been progressively enriched by a network or systems approach, in which the immune system can influence brain function as well. The latter even seems to have become the main working hypothesis of the more recent field of immunopsychiatry (Konsman 2019). Many authors have noticed that some important functional features seem to be shared between the immune and nervous systems, for example memory. This can further contribute to questions, such as, whether the nervous system controls the immune system, or the other way round, what kind of interactions exist between these systems, etc. (for a general and

[3] Microorganisms were initially studied (mainly in medicine) in their role as pathogens (consider, for instance, the studies of Koch and Pasteur). This means that they were studied in isolation without consideration of the structures they usually assume in nature, i.e. biofilms. Indeed, biofilms were first proposed by microbial ecologists and then adapted and adopted in clinical settings. Microorganisms are now seen as systems in virtue of their ecological nature. This is another example of how the way to look at things can "change" the nature of the objects of scientific investigations, including the methods to study those entities.

informative discussion of conceptual questions about such relations, see Pradeu (2020, pp. 54 ff.)).

It is now widely accepted that the immune system and the neuroendocrine system contribute to an integrated physiology that regulates the homeostasis of the whole organism (Ader and Cohen 1975; Besedovsky et al. 1985; Besedovsky and Rey 1996; Blalock 1994), see also Ader (2000), Ader and Kelley (2007) and many more reviewed in Ashley and Demas (2017). We seem to have moved (at least partially) beyond the idea that these systems are isolated and separated and only interacting under pathological situations (e.g., the neuroinflammation). When approaching functional aspects of how these systems operate in an organismic context, we are starting to recognize that one cannot neglect the ample interactions and crosstalk between these systems, with important contributions coming from entities and activities that usually are not considered to be part of the same system.

Yet how should we think about the re-integration, or recomposition of the neuroendocrine and immune systems? A bottom-up approach might investigate how neuroimmune circuits and networks cluster and overlap. Such a "connectome," similar to genetic regulatory networks and circuits, can be formed on the basis of a "common language" of molecules and receptors (e.g., cytokines or chemokines, neurotransmitters, hormones) or interconnected feedback loops. Some evidence seems to point in that direction. The three types of molecules are usually thought of as being unique for each system—cytokines (immune system), neurotransmitters (nervous system), and hormones (endocrine system), but also seem to share evolutionary and developmental origins (Petrovsky 2001). Evolution might have 'picked out' pathways that criss-cross all three systems instead of evolving each system separately, up to a certain point, after which they evolved together (Ashley and Demas 2017; Verburg-van Kemenade et al. 2017).

From this bottom-up perspective, systems biology becomes an attractive methodological approach. The aim of systems biology is to "understand how functional properties and behavior of living organisms are brought about by the interactions of their constituents" (Boogerd et al. 2007, p. 3) with the constituents being mainly molecules such as mRNA, proteins, and metabolites, including for neuroendocrinology and particularly immunology (Boonen et al. 2009; Eiden et al. 2020; Gardy et al. 2009; Germain et al. 2011; Gottschalk et al. 2013). Systems biology is often based on high-throughput technologies allowing the measurement of thousands of transcripts, proteins, or metabolites and bioinformatics approaches to generate hypotheses regarding functional behaviors of interest. Therefore, one might think that, focusing on understanding and dissecting the structural patterns detected by these techniques might be the key to comprehending these systems in a unified solution.

However, it would be naïve to think that these approaches can, by themselves, recompose the various objects of scientific investigation in the systemic nature of biological systems. Merely elucidating the structure of networks is not enough to understand their mechanisms and function. Similar problems have frustrated big data-omics approaches in other areas of biology. While trying to understand the genetic regulatory networks that govern developmental processes, molecules and pathways form an intractable "hairball" of nodes and edges that does not lend itself to scientific understanding, explanation, or prediction of the mechanisms involved (Jaeger 2017). Rather, the connectome approach needs to be supplemented with a framework that elucidates the structure, the function, and the mechanisms and processes. For instance, some computational methods and bioinformatic tools can provide a complementary look at the **reductionism** of certain compartmentalized and mechanistic investigations (see Boem 2016; Ratti 2016; Leonelli 2019)—however, they do not constitute, *per se*, a privileged point of view that is somehow superior to others. [4]

Another widely followed approach lies somewhere in between simplistic reductionism and intangible **holism**. This middle ground keeps the relative autonomy of each system (nervous, endocrine, immune) as largely independent modules but acknowledges the rich and overlapping interactions that form a comprehensive whole. This *building block model* does take structure, function, and process into account, assigning each block a unique structure, function, and types of processes. They are then integrated like Lego blocks, with each block retaining their unique features as they interact to maintain the homeostasis and functioning of the whole organism. However, there are multiple problems with this approach, which we will address below.

We suggest addressing these issues from a somewhat different angle, one that does not presuppose different systems as given units or building blocks. When we take a re-integrated neuro-endocrine-immune-microbiota systems (**NEIMS**) approach, we include additional elements of co-development, co-functioning, co-evolution, and ecological context.

The "normal" physiological role of neuroimmune interactions beyond disease might not be so "surprising"—as (Mueller et al. 2016) proclaimed in the epigraph—when we approach these systems with a more integrative perspective that takes their interactions, not their distinctiveness, as starting points. By doing so, we extend the roles of each system into the other, which in turn could trigger a rethink of what they are and what they do. There are important methodological consequences to this

[4] It is essential here to recall the recent developments in computational biology and systems biology. The tools used in computational and system biology by these research sectors allow one to consider and analyze vast amounts of data and to examine and manipulate some properties by abstracting vast amounts of them from the objects of scientific investigation. This practice has led some to believe that this was the main way to deal with the complexity of biological phenomena and to avoid simplistic reductionism. The discussion of these aspects is too extensive to be fully reported here. However, it is enough for us to point out that, while it is certainly true that, while these approaches have made it possible to build new "privileged observation points" on biological phenomena (not otherwise investigable), it is also important to remember that complexity is not just a question of quantity or computational capacity, but something that (unlike "complication") is inherently irreducible. By providing a complementary look at the reductionism of certain compartmentalized and mechanistic investigations, computational methods will enable researchers to have a more authentic picture of their field of investigation, because even if they employ reductionist methods (the use of which is often a harbinger of discoveries), they will be able to give a broader and more legitimate meaning to their results within the general framework.

reorientation: when each system is conceived as distinct, one might think that they could be studied separately—as they have been for quite a while. For example, it was widely believed that the brain was "immuno-privileged," in the sense of being outside the immune system's reach, because of the existence of the so-called blood-brain barrier that, in a way, seemed to neatly separate the immune and central nervous systems. Many are now reconsidering the barrier metaphor misleading, with questioning starting after the discovery of glial cells as immunocompetent cells in the brain.

This chapter proceeds as follows. In Sect. 2.2, we critically examine the "building block model." Sect. 2.3 criticizes the view that each block can be investigated independently of each other by examining the immune system; this section also offers an extended view of the immune system and invites us to rethink its structure and function. Section 2.4 is concerned with how the building block model conceptualizes the integration of multiple systems, illustrated by the mammalian gut. From these examples, we outline a NEIMS perspective driven by the developmental, ecological, and evolutionary relationships between the neuro-endocrine-immune-microbiota systems. In Sect. 2.5, we conclude with general lessons for dealing with neuro-endocrine-immune-microbiota systems.

2.2 Biological Systems and Their Functions: It's a Difficult Relationship, Not Just a Sum of Building Blocks

What we call the "building block model" is a way of practicing science and a conceptual model of how the biological world is organized. It acknowledges that systems interact with each other but treats each of them as separable subsystems that can be re-integrated in an additive fashion as Lego blocks snapped together. Researchers each study their own domains but come together only when interactions between systems concern issues that affect diverse systems at the same time.

The underlying assumption of the building block model is a 'one system–one function' mapping, where each system corresponds to a unique and main function. This way of thinking is pervasive in the life sciences, like, e.g. the famous one gene–one enzyme hypothesis (Beadle and Tatum 1941). Similarly, it is all too common to think along the lines of other one-to-one mappings, such as one gene–one disease, one cell–one function, one tissue–one function, one pathogen–one disease, etc.

In traditional neuroimmunology and neuroendocrinology, the building block model is apparent. Historically, the nervous system, the endocrine system, and the immune system were studied largely in isolation from each other, and each attributed a main function. The function of the nervous system would be information processing or cognition, that of the endocrine system, control of body metabolism, growth, and reproduction, that of the immune system, defense, and that of the gut (which we discuss below), digestion. While these might not always be clear-cut cases, the general notion seems to be widespread and inherent in the division of the fields that study these systems. When it comes to integrating the immune system with the nervous system (and many other systems), the expectation was often that it

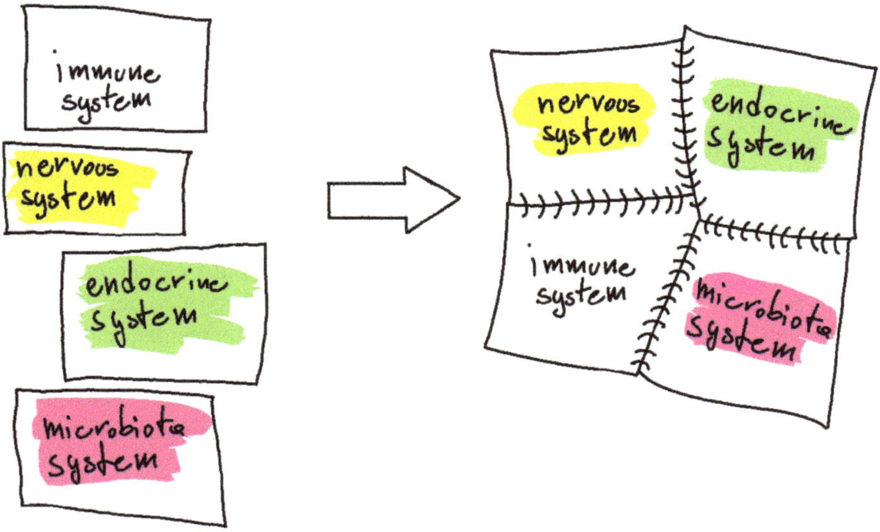

Fig. 2.1 Illustration of the building block model that systems (and their functions, in color) could be taken as independent units and be stacked or stitched together

just consisted of adding one system to another, thus ending up with a combination of the two respective main functions of these individual systems. For instance, neuroimmunology would be assumed to be the study of the immune defense of the nervous system (which is indeed how the field started) (Yoo and Mazmanian 2017; Rankin and Artis 2018; Fung et al. 2017). Furthermore, many introductory overview articles or textbooks on neuroimmunology emphasize the interactions between systems only in the context of pathologies and disease, for instance, when components of the nervous system are involved in fighting pathogens or, conversely, when immune components are involved in diseases of the nervous system.

According to the building block model, together with assumptions about independent developmental and evolutionary origins, (causally) linear and additive systems, often as (structurally and functionally) modular subsystems could be put together like pieces—building blocks (Fig. 2.1). Each building block is prescribed a unique, main function.

Yet it is not sufficient merely to acknowledge the presence of multi-directional interactions between systems. Doing so still attributes specific functions to each system. By focusing primarily (or even exclusively) on the role of the immune system as fighting disease and the nervous system mostly as a cognitive or "governing" organ, we may miss out on many of the other important functional roles these systems have as a result of their physiology, development, and evolution.

The building block model does not just make problematic assumptions about the nature and boundaries of systems, but also assumes that we know what systems there are in the first place. How do we demarcate something like a system, and according to which aspects? Decomposing and recomposing according to various functional and evolutionary aspects raise the question of where to localize these functions in the first place. All these questions will play a crucial role in addressing the questions surrounding neuro-endocrine-immune-microbiota systems (NEIMS) and how to integrate them.

To go beyond thinking of various systems being brought together like building blocks, we need new guidelines to take the integrative nature of these systems seriously without lumping them into a holistic, intangible network. One can better understand these systems by putting aside their "traditional functions" to see how their functionalities instead emerge from their interactions, and by examining how their respective structures and functions originate from a process of co-development and ecological interactions. NEIMS, as an ecological, developmental, and evolutionary approach, should serve as a guide for how these systems should be studied and conceptualized, to facilitate a better understanding of these systems, elaborating new hypotheses and experiments to test them, including the potential for novel therapeutic approaches.

The NEIMS approach also has therapeutic consequences. For instance, it has become clear over the years that the properties of the blood-brain interfaces cannot be reduced to the tight junction molecules sealing brain endothelial cells, but are determined by the interplay of astrocytes, pericytes, and microglia/macrophages, that is, it includes parts of what is traditionally considered the immune system. As a result, the blood-brain barrier is now being conceived of as one aspect of a blood-brain interface that functions more as a border than a barrier, recognizing among others the importance of the meninges constituting another aspect of this interface (Rustenhoven and Kipnis 2019). Furthermore, it has recently been proposed that the bona fide macrophages of the brain meninges, choroid plexus, and perivascular spaces, which share many cell markers (Faraco et al. 2017), should be referred to as "border-associated macrophages" (Pedragosa et al. 2018; Van Hove et al. 2019). By acknowledging the structural and dynamic properties of the blood-brain interface, it may indeed be more appropriate to use the "border" metaphor rather than that of a "barrier" (Badaut et al., Blood-brain borders: a new proposed concept to address limitations of historical blood-brain barrier terminology, *submitted*).

Decomposition, a central assumption of the building block model, is often a legitimate research strategy when dealing with complex systems (Bechtel and Richardson 2010). Many researchers, including philosophers, have helped come up with different criteria for complex systems and how to deal with them (Ladyman and Wiesner 2020). But there is always an inherent risk involved in dividing a system into components. For instance, there is the possibility of erroneously attributing fixed properties to a dynamic, complex system.[5]

[5]This does not necessarily imply that a particular way of dividing subsystems was the result of inattention, the lack of a critical attitude or methodological negligence, but instead reflects the need to be mindful of the nature and necessity of boundaries. Even if past divisions have successfully worked for certain experimental results, the adopted partition still calls for continuous, additional justification.

Each partitioning—with respect to such features—comes with certain trade-offs and biases. We cannot treat them as objective demarcations that would allow us to 'carve nature at its joints,' as this results in classifications that can work operationally but are nevertheless dependent on the research concern and experimental design. For instance, we tend to classify and conceive the nervous system (and its cells) as mainly associated with "perception" and "cognition" because those interests are dominant and central in these research fields. Thus, those interests played a crucial role in shaping the way scientific investigations have been pursued.

The NEIMS approach is a much more integrative approach that takes the interwoven components, their interactions, co-development, and co-evolution explicitly into account. It starts with the recognition that there might not be a "main function" associated with each subsystem. Furthermore, the specific function a subsystem plays depends on its relationship with other subsystems and the larger organismal environment.

Combining the nervous and immune systems is not just about putting together the functions of cognition and defense. In the next sections, we argue against the building block model by looking at how the function of the immune system needs to be understood in light of new findings. In addition, we look at what happens when we add other systems, including the endocrine system and microbiota in their ecological context.

2.3 Rethinking Individual Building Blocks: The Case of the Immune System

Between the eighteenth and nineteenth centuries, the notion of system became progressively employed in anatomy to indicate a combination of bodily tissues or organs having the same characteristics in physiology, like the vascular and nervous systems (Keel 1982; Moulin 1991). In the twentieth century, the philosopher Joseph Woodger affirmed that "the machine theory of the organism tacitly acknowledges an organization above the chemical level" in the sense that "the organism is analysable into organs systems, organs, tissues, cells and cell-parts" (Woodger 1929, pp. 292–293). Thus, "according to the 'standard' hierarchical model of physiology, each living organism comprises organ systems" and "[e]ach organ system, in turn, is composed of individual organs" (Ashrafian 2018). Although the relationships between functions and organ systems have already been challenged by findings outlined above, indicating that functions attributed to the immune systems involve mediators classically associated with the neuroendocrine system and vice versa, the very notion of systems has also changed in biology. Indeed, different forms of systems biology exist between twentieth century systems biology "considering whole living systems, which include their organization [and] the dynamics within systems and the interplay between different levels" and "today's systems biology, which is often a bottom-up approach from molecular dynamics to cellular behavior" (Drack and Wolkenhauer 2011).

Traditionally, the immune system is viewed as being formed by a set of certain cell types, sharing distinctive and similar tasks that are extremely specialized. It is characterized as a system of defense, as a means to protect the organism, the 'self,' from external threats, that is, the 'non-self' (including tumor cells, which would become 'strangers' to the organism). The lexicon adopted of protection and threat already guides us to conceive of immunity as a sort of 'defense system.' This conceptual, and terminological, stance is explicit in the historical development of immunology as a research field (Burnet 1961).

2.3.1 Argument 1: The Immune System Beyond Defense

The first critical problem of the building block model regarding the immune system is its 'static' characterization, mainly as a defense system. However, it has progressively been shown that the activities of the immune system go far beyond the defensive functions and are involved in various other biological processes, from those related to development, metabolism, and tissue repair, to the regulation of the nervous system (e.g., affecting signaling pathways, producing neuroendocrine mediators, facilitating synapsis activity) and the preservation of homeostasis, both locally and organismically (Dantzer 2018; Rankin and Artis 2018; Pradeu 2020). Once the immune system is recognized as having functions that go beyond those of defense, it is easier to understand why it is problematic to use simple self/non-self and internal/external distinctions to define the immune system.

For instance, from a genetic point of view, there is a plethora of exogenous elements that are not perceived as such but rather allowed or tolerated within the boundaries of the extended immune system. First of all, during mammalian reproduction, the fetus constitutes an entity external to the mother, yet it is tolerated by its immune system. Other cases showing the pervasiveness and flexibility of immunological tolerance include, for example, the cases of chimeras, organisms that are composed of cells of different individuals (both of the same and of different species), and those of genetic mosaicism, i.e., the presence, in a multicellular organism, of two different genetic lines derived from a single zygote.[6]

2.3.2 Argument 2: The Immune System as an Ecological and Whole-Body System

Differently from other systems of the organism, the immune system is not spatially circumscribed, tied to a specific organ, or pertaining only to certain tissues. On the

[6]In chimerism, each cell population retains its own characteristics, given that they are genetically distinct because they originate from different zygotes. Thus, the resulting organism is a mixture of differently matched regions. On the other hand, in genetic mosaicism, different cells, with a diverse genetic heritage, arise from the same zygote.

contrary, its structure is ‘sparse,’ ‘diffuse,’ highly dynamic, and usually studied according to its functionality. [7] Dissecting and understanding the functionality and the organization of the immune system is a crucial task for both scientific and clinical investigations. Yet immune cells can differ drastically in functionality despite sharing a common developmental route. It has been shown that functional fluctuations in immune cells’ activities and specialization can be due to individual genetic variability and epigenetic modifications. Moreover, the activity of the immune system presents an impressive range of variability, not only between distinct single individuals, but also within the same person during the different phases of its life (Poon and Farber 2020). Furthermore, immune functions cannot be fully disjointed or separately considered from their crosstalk with specific tissues and cell types (Farber 2021). These characteristics showcase the systemic and plastic nature of immune cells, as the context and the types of interactions determine their capacities and activities.

More fundamentally, the context-dependency of immune cell functionality corroborates the idea that the immune system presents an ecological character. Subgroups of immune cells that share their origin, mature and develop functionally in different ways, in strict relation to the microenvironment in which they operate (Poon and Farber 2020). The immune system, due to its widespread and capillary nature, is thought to interact with all the other bodily systems and constitutes a complex functional network, thus resembling in some respects an ecosystem that comprises the organism itself, in its totality (Poon and Farber 2020).

Herein lies a second clue to the limitations of the building block model. From both a developmental and structural point of view, the immune system is composed of a specific macro-type of cells: leukocytes. However, from a functional perspective, the activities of the immune system, from the defensive ones to the more general and regulatory ones, do not end with the operational possibilities of the leukocytes alone but must instead be understood as the product of the interaction of these cells with other cells of the organism, such as epithelial cells, and with symbiotic residents (we will discuss this second aspect in the next section). The interaction of the immune system with the organism involves at least two levels. On the one hand, there is a local immune activity, determined by the interaction of single compartments of leukocytes linked to specific tissues or areas. On the other hand, there is a global crosstalk that involves the whole organism both through the blood flow and also the interaction of the immune system with other systems, such as the endocrine and nervous system (Poon and Farber 2020). Therefore, we can argue that functionally, it is the organism as a whole that is the system of the immune system.[8]

---

[7]Even though this structural comprehension of the immune system is possible, and although immune cells share developmental origins (and evolutionary as well), their definition as such relies more upon their capabilities and mutual interactions.

[8]This does not entail a naïve (and not very useful) form of reductionism for which every organismic activity is attributable to the immune system. On the contrary, if the immune system extends to the whole body, this means that one cannot ignore the immune system to provide a complete

Interestingly, as some recent research suggests, the functionally global interactive nature of the immune system can also be seen as that of a "social" network, layered on different levels (Bergthaler and Menche 2017). A recent study (based on proteomic analysis of immune cells' activities under different stimuli) has shown how the relational and multi-level architecture of the immune system exhibits a particularly high level of intercellular specialization, with divisions and subdivisions resulting in distinct cells performing tailored and context-dependent tasks. Just like any complex system, the overall functionality of the system extends the simple sum of the activities taken individually, exhibiting distinctly emergent properties (Rieckmann et al. 2017).

It is also interesting to note how some studies show that even different cell types, traditionally not associated with the immune system, can become a functional part of it. For example, some cell types, usually classified as structural (such as fibroblasts or epithelial cells), in certain contexts exhibit the ability to communicate with immune cells in order to modulate and coordinate their activity. Indeed, these structural cells can even secrete molecules both attracting and activating immune cells (Gomes and Teichmann 2020). For instance, studies have shown how epithelial cells in the lungs can detect pathogens' activities and produce molecules interacting with dendritic cells pushing them toward T-cell activation. Moreover, they can promote fine-grained modulation by recruiting specific subtypes of immune cells (through the production of chemokines), providing a more tailored immune response (Schleimer et al. 2007). Other recent work on airway epithelium (an area obviously exposed to the external environment) has shown how epithelial cells interface and collaborate with specific subpopulations of immune cells, both in coordinating defense activities and in maintaining tissue functionality, including their repair (Hewitt and Lloyd 2021).

Particularly interesting, still from a systemic perspective, is that in the lungs the 'perceptive possibility' of the organism in recognizing the lack of uniformity (such as pathogenic activities) rests on different capacities and 'sensory' pathways. This is the case of the neuroimmune response which acts in concert with the pulmonary epithelium. In particular, the lungs are also densely innervated by sensory neurons that are able to activate, differently depending on the stimulus, specific populations of leukocytes (e.g., the T-helper 2 cells), which, in turn, communicate with neuro-endocrine cells responsible for the production of mucus. Conversely, cholinergic neurons are able to stimulate a different immune response that entails the production of cytokines by other immune cells (Hewitt and Lloyd 2021).

This phenomenon is not confined to the lungs alone. The cells of the epithelium of different tissues (from the intestinal tract to the skin) have shown the ability to contribute to the function of the immune system, not just activating the "usual" immune compartment but also orchestrating the entire "immune" activity (Larsen et al. 2020). Moreover, recent studies have shown how tissue-related epithelial stem

---

explanation of many activities of the organism itself. It will be the main task, for future science, to clarify the aspects and mechanisms by which this happens.

cells preserve an epigenetic memory of their immune activities, potentially altering further inflammation responses and tissue repair (Naik et al. 2018).

2.3.3 Argument 3: How to Distinguish the Immune System from Other Systems

Finally, in the building block model, the functions of regulating and coordinating internal bodily functions and defense against external threats have classically been attributed to the (neuro)endocrine and immune systems, respectively. The two systems seem to be distinguished mainly by their different cell types, but also by different modes of intercellular signaling. One important way by which cells signal to each other is through direct contact. Another important manner of cell signaling is through the release of molecules into the extracellular space. Depending on the amount released and the expression of receptors for signaling molecules, this may result in signaling to neighboring or distant cells, which have been labeled paracrine and endocrine signaling, respectively (Hartenstein 2006). The endocrine system has been characterized by hormone-mediated signaling, which is "defined by their ability to send signals to and from different tissues at long distances" (Kodis et al. 2012). Neurons as a cell type, on the other hand, are often associated with synaptic signaling, which involves a hybrid form of close contact and paracrine signaling.

However, over the past decades, it has become clear that neuroendocrine responses, for example, activation of the hypothalamus–pituitary–adrenal axis, occur in response to detection of microbes or their constituents (Besedovsky and del Rey 1989; Rivier et al. 1989) and that molecules typically associated with the immune system, such as pro-inflammatory cytokines, can play an important role in regulating body metabolism (Peluso and Palmery 2016; Wallenius et al. 2002). Moreover, removal of neurons and endocrine cells from Hydra epithelium modifies local bacterial populations (Fraune et al. 2009). Furthermore, mammalian neuropeptides released in an endocrine-like manner have antimicrobial properties and can regulate innate immunity against microbes (Aresti Sanz and El Aidy 2019; Brogden et al. 2005). Finally, the immune system has been proposed to recognize antigenic discontinuity, that is, by the "sudden modification of molecular motifs" (Pradeu et al. 2013), and microbial function in addition to pathogen-associated molecular patterns (Greslehner 2020), which opens the broader perspective of the immune system being able to detect tissue function.

It is important to note that many neurons also release intercellular signaling molecules, e.g., neuropeptides, outside synaptic clefts (a phenomenon that has been termed volume transmission) into the extracellular space, including into the cerebrospinal fluid or blood and can therefore be labeled neuroendocrine (Hartenstein 2006). Phylogenetic and ontogenetic findings indicate that the label neuroendocrine can be applied not only to chains of hormone release regulated by neurons in the hypothalamus, but also to gastrointestinal neuroendocrine cells and nervous fibers targeting such cells (Falkmer 1993; Hartenstein 2006; Modlin et al. 2006). Hence, "the gut neuroendocrine system is viewed as a syncytium of neural and endocrine cells" (Modlin et al. 2006). Thus, it seems that neuroendocrine systems, as a whole, can be investigated in different ways based on their mode of intercellular signaling.

Classic examples of ligand–receptor contact-mediated intercellular signaling occur between immune cells, but paracrine and endocrine intercellular signaling is also widespread among immune system components. Consequently, the immune system does not seem to correspond to one particular mode of intercellular signaling either. Although "[t]he classical model of immunity posits that the immune system reacts to pathogens and injury and restores homeostasis," recent models of immunity have proposed that "effector immune responses [. . .] are induced by an antigenic discontinuity; that is, by the sudden modification of molecular motifs" (Pradeu et al. 2013) or that "the healthy immune system is always active and in a state of dynamic equilibrium between antagonistic types of response" (Eberl 2016). Therefore, it seems that the notion of the immune system refers more to certain stimulus–response modes than a particular intercellular communication mode.

To conclude, an account based on a classical "causal role function," like heart contractions contributing to blood circulation (Brigandt 2017) no longer allows a clear characterization of the immune system as a defense system itself nor does it offer a clear distinction between the neuroendocrine and immune. As we have seen, molecular structure does not necessarily determine whether a molecule should be considered part of the neuroendocrine or immune system. Furthermore, it is possible to extend the notion of the immune system in its functionality, not only to the compartment of cells traditionally ascribed to it but also to other cells. It is good to remember that there is no 'immune system' as something that is out there in nature, waiting to be discovered. In the real world, there is no clear-cut phenomenon but rather complexity, which comes inherently altogether (Hacking et al. 1983). On the contrary, science investigates some phenomena that it manages to isolate analytically, and to which it attributes properties in order to better understand them. However, it is the progress of research itself that shows that these categories are not fixed and that they can be modified precisely because of new discoveries or new ways of seeing those phenomena.

2.4 Rethinking the Recomposition of Multiple Building Blocks: The Enteric Nervous, Immune, Endocrine, and Microbiota Systems

Discussions around neuroendocrine and immune integration have largely centered on the central nervous system (see, for instance, Mašek et al. 2003). The enteric nervous system has received less attention in neuroimmunology, yet its integration with gut immune and endocrine systems presents an important case study in understanding neuroimmune–endocrine integration. By taking an ecological evolutionary developmental approach to the gut system, we illustrate the inadequacy of the building block model of mapping individual functions to individual subsystems and challenge the idea that neuroimmune interactions are relevant only in the context of disease and disorders. With this example, we can illustrate how an integrated NEIMS approach emerges from co-developing, co-functioning, and co-evolving nervous, immune, endocrine, and gut microbiota systems. But first, why focus on the gut?

2.4.1 Proposal 1: The Gut Complex as a Unit of Organization

The gut, especially the largely autonomous mid/lower gastrointestinal (GI) system, is a remarkable organ from a neuroimmune perspective. In mammals, for instance, in part due to the enormous surface area of the intestines, the gut is the largest immune and lymphoid organ (News & Highlights 2008; Vijay-Kumar et al. 2014), the largest endocrine organ (Ahlman and Nilsson 2001; Rehfeld 2004), and contains one of the largest set of intrinsic and interneuron connections amongst the autonomous nervous systems in the body (Kulkarni et al. 2018; Furness and Stebbing 2018). The enteric mucosal immune system, the enteric endocrine system, and the enteric nervous system are complex systems operating intimately with each other. Altogether, the 'gut complex' as a border is exposed to food, to toxins, and to the trillions of microorganisms residing in our intestines, it manages immune tolerance and produces localized inflammatory responses to infections.

The operations of these gut subsystems are each relatively autonomous from their respective systemic counterparts. Starting with the gut immune system, the gut mucosal immune system consists of anatomically distinct microcompartments that house gut mucosa-associated lymphoid tissues (GALT) with a unique repertoire of lymphocytes (with IgA as the dominant antibody type) (Janeway et al. 2001; Wershil and Furuta 2008). GALT compartments and structures, such as Peyer's patches and isolated lymphoid follicles, are sites of local adaptive immune responses that require microbial signals to fully develop (Silva-Sanchez and Randall 2020).

The enteric nervous system and its extensive local connections make it also relatively autonomous compared to other nervous systems. In organisms with a central nervous system (CNS), it is well-known that the enteric nervous system (ENS) is a unique neural network, distinct from but interacting with the CNS. Unlike other peripheral parts of the nervous system, it has interneurons that form a richly interconnected and largely autonomous system with sparse connections to the CNS (Furness and Stebbing 2018). Evolutionary comparative work of the enteric nervous system across the animal kingdom shows that it is not derived developmentally or evolutionarily from the CNS. Instead, it is evolutionarily conserved across animals with or without the CNS and has been proposed to constitute one of the first nervous systems in evolution, rather than a second or derived brain (Furness and Stebbing 2018). Even though there are bidirectional pathways linking the gut to the CNS, the complex internal processes of the gut illustrate that gut function is not ultimately driven by the CNS but in large part mediated and controlled by the local regulatory networks and reflexes in the gut. The gut complex is thus an important causal center beyond the inputs from the CNS.

Finally, the gut endocrine system can be considered to correspond to enteroendocrine cells (EEC), which are scattered throughout that gut lining (Gribble and Reimann 2019) and differentiating into more than 15 subtypes (Posovszky 2017). EECs form an important bridge between the epithelium and the enteric nervous system through "synapse"-like structures (Kulkarni et al. 2018). As outlined in Chap. 1, it has been shown, using anatomical and cytochemical approaches, that enteroendocrine cells share characteristics with pituitary corticotroph cells and pancreatic islet cells, all known to secrete polypeptide hormones. Moreover, at the end of the 1970s, the idea of the gut containing a "diffuse neuroendocrine system" started to emerge (Polak and Bloom 1979, p. 1400).

Increasingly, the microbiota has started to be treated as a unit. This has a twofold implication. On the one hand, it means that associated bacteria may possess discrete collective boundaries as if they were organs. This can explain why microbial activities are increasingly associated with specific organismic functions (e.g., digestion, immune modulation, etc.). On the other hand, the microbiota can be now considered an object of scientific investigation as a unit (Raman et al. 2019; Cani and Van Hul 2020).

The functioning of the gut requires all four systems—the gut nervous, immune, endocrine, and microbiota systems—to work together to serve as an interface with the outside world, processing food, expelling toxins, and managing trillions of resident bacteria, viruses, protists, fungi, and sometimes, helminths. In addition, the gut complex has also been proposed to be involved in emotions, sensory processing, cognition, (social) behavior, general motivational state and attitudes, etc. (Mayer 2011).

2.4.2 Proposal 2: The Building Block Model and Ecological Models for the Gut Complex

A building block model of the gut would treat each of these systems as modules with distinct cellular and tissue structure and functioning. Gut motility, for instance, is then thought of as the main function of the enteric nervous system. Gut secretion is the main function of the gut endocrine system while gut defense is the main function of the gut mucosal immune system. Resident commensal microbiota are assigned the roles of probiotics that supplement the body's metabolism and modulate the immune system. Treating the gut as consisting of building blocks implies that we can infer the specific contributions of each system through knock-out and gain-of-function techniques to block or enable parts of the system. A building block model also implies that we can safely study each system in isolation. The control and regulation of the gastrointestinal system would be seen as the 'stitched together' combined efforts of these distinct physiological systems. A common 'language' might then allow communication between these systems, for instance, via neurotransmitters, hormones, or cytokines such as interleukin-6, substance P, or leptin.

The building block model, however, is inadequate. Ecological developmental biology (**eco-evo-devo**) is a subfield of evolutionary developmental biology (evo-devo) that examines how phenotypes arise from the processing of environmental signals and cues throughout development, and how such processes in turn affect their evolution (Gilbert and Epel 2015). Taking an eco-evo-devo angle to the gut complex strongly supports the idea that the gut is not the coming together of multiple independently originating systems, but a *multi-system complex* that has co-developed and co-evolved from the very beginning. These findings compel us to recompose the gut complex as a system of its own with an intermingled NEIMS.

First of all, the subsystems of the gut complex require input from each other to fully develop as a *co-developed unit*. Each physiological system does not differentiate and develop on its own but co-develops as a 'gut complex.' Signaling molecules from the enteric glial cells of the ENS are implicated in the development of gut immune compartments and immune cells (Yoo and Mazmanian 2017). Conversely, the local immune system and the microbiota are needed for the full development and maturation of the enteric nervous system (De Vadder et al. 2018; Kabouridis and Pachnis 2015; Obata and Pachnis 2016; Vuong et al. 2020) as well as its innervation of gut tissue (Kang et al. 2021). Furthermore, the development and differentiation of enteroendocrine cellular subtypes are highly plastic and dependent on the nutritional and microbiota content of the lumen (Posovszky 2017).

Second, in organisms with nervous, immune, and endocrine systems, the integration between these three with the microbiota is likely to be ancient and evolutionarily conserved as a co-evolved unit. It is important in this respect to remember that some sort of neuroendocrine system is already present in organisms without immune system elements and that this system regulates bacterial growth (Augustin et al. 2017). Not surprisingly, ever since the appearance of so-called immune cells and mediators during evolution, interactions between elements of endocrine, immune, and nervous systems have been documented to occur (Panerai and Ottaviani 1995; del Rey and Besedovsky 2017).

Finally, abundant research has focused on how the proper functioning of each system relies on the others. As the functioning of each subsystem is context-dependent on the state of the other systems as well as the situation of the body as a whole, they constitute a *co-functioning unit*. An example of this are the EECs. They are considered hormone-secreting endocrine cells. We now know that they influence other physiological systems (Posovszky 2017). On the luminal-facing end, some types of enteroendocrine cells are in direct contact with the gut content, detecting and integrating signals from food, toxins, the immune system, and the gut microbiota, etc., with their microvilli. They can secrete into the lumen peptide mediators that regulate gut motility and metabolism but also immune responses and defense (Wikoff et al. 2009; O'Mahony et al. 2015; Kuwahara et al. 2020; Cani and Knauf 2016; Gribble and Reimann 2016; Psichas et al. 2015). In addition, they can regulate gut function and communicate with the CNS by secreting hormones into the bloodstream (the classic hormones include gastrin from the stomach and cholecystokinin and secretin from the small intestine), by acting on local sensory nerves or by synaptical communication with enteric glia via neuropod structures (Kaelberer et al. 2020). Neuropods are associated with local nerve terminals, suggesting that these cells are part of a local neuroendocrine system (Sharkey et al. 2018). The gut endocrine system is thus a part of the gut nervous and immune circuitries, bridging them with luminal microbiota. The neural system of the gut extends beyond just the enteric neurons to also include enteric glial cells as well as enteroendocrine cells (Bohórquez and Liddle 2015).

Co-functioning is especially prominent when we consider the gut microbiota. Consider, for instance, that microorganisms are involved in the regulation of peristalsis of food. The short-chain fatty acids secreted by gut microbes stimulate the release of serotonin (De Vadder et al. 2018). Gut microbiota is implicated in the development and functioning of the intestine (Hooper 2004) and its immune (Belkaid and Hand 2014; Rhee et al. 2004), endocrine (Wikoff et al. 2009; O'Mahony et al. 2015; Watnick and Jugder 2020), and enteric nervous systems (De Vadder et al. 2018; Hyland and Cryan 2016).

The co-development, co-evolution, and co-functioning of the gut nervous, immune, endocrine, and microbiota systems support the idea that together, they form a 'gut complex.' From the NEIMS perspective, functions emerge from intermingling of physiological systems, which do not possess their typical functions prior to these interactions. Furthermore, the microbiota is now gradually recognized as an intrinsic, yet also 'ecological,' factor modulating the physiological responses of the body organism, with most of the literature focused on the immune system (Belkaid and Hand 2014; Boem et al. 2020; Chiu et al. 2017). This intrinsicity, therefore, suggests considering the activity of the microbiota as close to an endogenous regulator: i.e. the microbiota can be seen as a functional part of the immune system itself (Amedei and Boem 2018; Belkaid and Hand 2014; Fung et al. 2017; Zheng et al. 2020). The eco-evo-devo perspective of NEIMS provides a new way to think about the four systems and their relationships.

So far, we have shown how new insights from recent literature challenge the neuroimmunology building block model. New findings push us to rethink the 'one structure-one function' assumption. In the previous section, we argued that the immune system is not just a defense system. Here, with the gut complex as an illustrative example, we argued that the nervous, immune, and endocrine systems do not come together like Lego pieces, with each system exhibiting its main functions prior to their interactions. Instead, from an eco-evo-devo perspective, the respective functionality of enteric nervous, immune, endocrine, and microbiota systems emerges from their interactions.

2.4.3 Proposal 3: A NEIMS Approach

A NEIMS approach invites us to rethink the functionality of the entire neuro-endocrine-immune-microbiota system complex. Once we look at NEIMS as a whole, there are many potential ways in which one can recompose the gut complex and to assign function to NEIMS as a whole. One way is to look at other structural systems (e.g., nervous system) and their preferred associated function (e.g., cognition) and propose that the gut complex as a whole is also capable of it. There are indeed previously underappreciated—sometimes not even considered—potential functions of the gut complex that suggest that the gut is not the mere sum of distinct building blocks 'dedicated' to digestion. One possibility is to think of the gut complex as a proto-cognitive organ. It is an example of how using different criteria for systems can lead to the proposal that the gut complex as a unified entity that involves all four categories of host physiological and microbial systems.

The gut has a wide diversity of sensory functions. Instead of analyzing them separately under their respective historical systems (e.g., the gut endocrine system, the gut nervous system, the gut immune and tissue defense system), Furness and colleagues have long argued that we should treat gut detection as a function of an integrated gut sensory organ (Furness et al. 1999, 2013; Furness and Cottrell 2017). These detection features can be interpreted as an example of integration between the enteric nervous system, the gut mucosal immune system, the enteroendocrine system, and the gut microbiota cells.

The gut has an enormous capacity to sense and distinguish between nutrients, irritants, and microorganisms in its internalized milieu and to respond differentially to those (Mayer 2011; Breer et al. 2012; Collins et al. 2012; Latorre et al. 2016). It is important to point out first that the gut can exhibit at least six stereotypical patterns of motor and secretion behaviors that can be observed in different physiological and pathological conditions and that can be modulated by numerous signals (Wood 2004; Schemann et al. 2020). While most of the studies establishing these patterns and their modulation have been done in vertebrates that possess central nervous systems capable of influencing the gut through the peripheral nervous system, thus making it sometimes hard to determine the part played by the gut, it is important to consider *Cnidaria*, like hydras, box jellies, jellyfish, corals, and sea anemones, which can display several gut motor patterns despite possessing only nerve nets with the most important being situated around the central body cavity.

Signal integration is "[t]he capacity to combine information from multiple sources," and valence, "[t]he capacity of a [biological system] to assign a value to the summary of information about its surroundings at a given moment, relative to its own current state" (Lyon 2015, p. 4). It should be noted that sensing of gut luminal contents seem to be categorized as beneficial or threatening based on gut signals and responses and can be integrated with systemic hormonal signals at the level of gut endocrine cells and nerve fibers (Dockray 2003; Wood 2004; Holzer and Holzer-Petsche 2009; Mayer 2011; Brookes et al. 2013; Sclafani 2013; Neunlist and Schemann 2014; Maniscalco and Rinaman 2018; Sharkey et al. 2018; Han et al. 2018; Lu et al. 2021). Behavior can be considered as "[t]he capacity of a [biological system] to adapt via changing its spatial, structural or functional relation to its external or internal milieu" (Lyon 2015, p. 4). Accordingly, it can be argued that the gut's modification of motility and secretion in response to nutrients, irritants, and infectious microorganisms constitutes local behavioral responses, but also that gut responses can give rise to behavioral modifications at the level of the whole organism (Wood 1999; Stephen 2001; Khan and Collins 2006; Chen et al. 2009; Mikkelsen 2010; Akiho et al. 2011; Brookes et al. 2013; Skibicka and Dickson 2013; Latorre et al. 2016; Furness 2016; Yang and Chiu 2017; Serna-Duque and Esteban 2020). Thus, the gut is clearly capable of discriminating, integrating, evaluating, and responding to different kinds of stimuli.

Enteroendocrine cells can respond to gut luminal contents through multiple receptors, including through pattern recognition receptors that recognize particular microbial molecules (Mayer 2011; Sharkey et al. 2018). It is also important to keep in mind that signaling molecules in the gut are not necessarily specific for one single cell type as enteroendocrine cells, in addition to mast cells, can also release serotonin (Sharkey et al. 2018). In terms of anticipation, it is clear that changes in feeding patterns are met with responses both at the level of the enteric nervous system and gut neuroendocrine–immune interactions (Schemann et al. 2020). While it is clear that local mast cells play an important role in Pavlovian conditioning of vertebrate gut responses (Wood 2004), it is still an open question as to whether gut components like the enteric nervous system alone are solely capable of this kind of associative learning or whether it always involves the CNS (Schemann et al. 2020). In this respect, it is interesting to consider that classical conditioning occurs in animals without a brain, such as sea anemones. Moreover, these *Cnidaria* have also been shown to display habituation and sensitization (Cheng 2021). Finally, it is beyond doubt though that gastrointestinal infections and inflammation can give rise to long-lasting enteric neuroplasticity in vertebrates (Schemann et al. 2020). Taken together, these findings indicate that the gut can be considered as capable of minimal cognition or having proto-cognitive capacities, which is in line with proposals of proto-cognition developed by Lyon (2015).

Instead of focusing on the nervous and immune systems as historical cognitive systems, the possibility that interactions between neuroendocrine and immune components confer minimal cognitive capacities to the gut should be considered in animals that have both systems. This gives rise to the questions of the precise contributions of the gut immune and neuroendocrine cells and how the gut engages and interacts with cognitive systems in animals with a central nervous system. Indeed, the enteric nervous system and the gut immune and neuroendocrine system have co-developed, co-constructed, co-evolved as part of an integrated biological unit that may exhibit scaffolding and niche construction that also need to be considered regarding gut microbiota. [9]

2.5 Conclusion and Future Outlook

After this conceptual tour through a philosophical perspective on the interactions between the neuroendocrine and immune system and how to deal with NEIMS—instead of a building block model—, let us conclude with a couple of take-home messages which we would like the reader to integrate into their thinking (see also Box 2.1). We hope they will be helpful not just as another building block piece of

[9]Finally, although not addressed here, the organization of the gut between the endocrine, immune, and nervous components may also play a role in the emergence of proto-awareness in animals—a topic to be discussed in the future.

knowledge, but as conceptual tools for thinking about these systems and approaching them experimentally.

First, the microbiota is a necessary ecological component of our physiology. The microbiota is involved in the constitution of the whole organism during development and functioning.[10] It is known that the microbiota can affect (either positively or negatively) the predisposition to the onset of pathologies (such as autoimmune diseases), and that it is fundamental for the maturation of secondary lymphoid structures, in the strengthening of the intestinal epithelium, and in the homeostasis of tissues of interest. Other known functions concern the modulation of the immune system, such as the regulation of the inflammatory response and the activation of immune cells related to specific tissues. Next, it carries out an action (both directly and indirectly) against other microorganisms with pathogenic potential. For instance, not only is it now widely recognized that microbiota plays a role in the modulation and the development of the immune system (Belkaid and Hand 2014) but it can directly interact with other bacterial species determining their capability of interacting with the host. Thus, by preventing some species from inhabiting the host, the so-called colonization resistance explains, in an ecological way, how commensal microorganisms constitute a functional extension of the immune system itself (Amedei and Boem 2018; Ronai et al. 2020).

Taking the microbiota into account may show that "systems biology could ultimately turn out to be more like an ecological problem than one of molecular biology" (Nicholson and Wilson 2003, p. 669). The gut is an outward-facing organ that serves as an interface between the organism and the environment. The involvement of the gut microbiota in the gut neuro-immuno-endocrine system thus introduces an ecological perspective to internal physiology. The diverse ecosystems of microbiota are engaged with the gut in ecological relationships (Costello et al. 2012; Coyte et al. 2015; Dethlefsen et al. 2007), sometimes in ways that integrate with host physiological systems (see Chiu et al. 2017) for a co-immunity between host immune systems and the microbiota). The gut complex constitutes an extended developmental system that includes environmental cues and stimuli (Griffiths and Gray 1994, 2004).

The symbiotic bond from a functional but also an evolutionary point of view implies that the microbiota of *Homo sapiens* has co-evolved with our species making the individual referred to as the single human a functional and evolutionary unit,

---

[10]From a systemic point of view, however, it is important to remember that the microbiota is not only involved in the modulation of the local or tissue-specific response. Indeed, treatment with antibiotics has, in fact, shown how the decrease of the microbiota (both in terms of its number and composition/diversity) has a global impact on the activities of the immune system at the level of the whole organism. This occurs both in situations of inflammatory and defensive response and concerning the general and regulatory functions of the immune system in its broadest sense. There are now countless studies concerning the potential contribution of the microbiota (with either protective or promoting effects) with conditions such as chronic, metabolic, autoimmune, neurodegenerative diseases, and cancer.

composed of the networks of interactions formed by the various actors of the symbiotic association.

This means that the concept of symbiosis itself has been updated. Individual beings such as animals or plants, in their purely genetic dimension, can no longer be considered complete and autonomous. Accordingly, singular organisms should now be seen as functional wholes, resulting from this interactive network. Indeed, the developmental biologist Scott Gilbert described this fact by saying that "we have never been individuals" (Gilbert et al. 2012). Because of that, one now cannot now ignore the fact that what was thought to be single and well defined is actually the product of a multitude of different agents, in an ecological and dynamic relationship: i.e. the holobiont (Bordenstein and Theis 2015). Beyond the purely theoretical debates on the nature of the holobiont, it is important to keep in mind that it is not only the object of experimental but also of therapeutic investigations.

The result is that the "holobiont" (the macro-organism and its microorganisms) can be considered as a new unit of biological organization which is privileged, from the point of view of functions and criteria regarding individuality. This means, given the eco-systemic nature of the holobiont, that the relations between these physiological systems and its extensions can only also be analyzed from an ecological point of view (Martin et al. 2011; Chiu et al. 2017; Ronai et al. 2020; Schneider 2021).

Ideally, readers will appreciate this eco-evo-devo perspective and change their perspective of how to look at experimental data—especially data inconsistent with the building block model, which would be hard to interpret otherwise. By adopting a different perspective of that of the building block model, the study of neuroendocrine–immune system interactions could become a 'block-buster'.

Box 2.1 Philosophical Key Points

- Conventional dismantling of systems in terms of structure and function is often biased by conceptual and methodological limitations.
- Division of systems and division of labor are often necessary and helpful, but we also need re-integration (for which the building blocks approach often will not do).
- Do not shy away from considering functions classically associated with one system for another and propose new functions by observing systems as a whole.
- Do not take the systems as a given, having a clear 'main task.' Systems cannot be combined like building blocks in an additive way, resulting in the mere sum of these structures and functions. Nor are functions determined once and for all.
- Do not ask: who is controlling whom? Which system regulates the other? Many of these questions are ill-posed when framed through the lens of the building block model.

(continued)

Box 2.1 (continued)

- Do not just focus on disease. Lip-service is often being paid through the phrase "in health and disease." However, it is far from clear what health is and putting them as polar opposites might not be the most helpful way for understanding how these systems inter- and co-act.
- There is not just bidirectional communication or regulation between different systems. Proper re-integration of systems takes more than just putting them together.
- Thus, multidisciplinary and interdisciplinary approaches are needed. One cannot just treat the systems nor the disciplines studying them as building blocks.
- Re-integration might be difficult and uncomfortable at times, forcing one to question what seems to be established common knowledge within fields.
- Still, such an approach is more promising than a reductionistic enterprise, in which one tries to find the 'most fundamental units.'
- This should be kept in mind for training and educating future scientists, including relevant inputs from and collaborations with philosophers of science.

2.6 Key References

Bechtel, W., & Richardson, R. C. (2010). *Discovering Complexity: Decomposition and Localization as Strategies in Scientific Research*. MIT Press.

This book is important as it reminds scientists that decomposition and localization of phenomena to yield "building blocks" are heuristic strategies in the proposal of mechanisms that can only be applied under certain conditions, which are probably not met for many living systems.

Furness, J. B., & Stebbing, M. J. (2018). The first brain: Species comparisons and evolutionary implications for the enteric and central nervous systems. *Neurogastroenterology & Motility*.

This review compares the central and enteric nervous systems, describes their relationships, and makes the important point that from the point of view of evolution the latter is, metaphorically speaking, the "first brain."

Gilbert, S. F., Sapp, J., & Tauber, A. I. (2012). A Symbiotic View of Life: We Have Never Been Individuals. *The Quarterly Review of Biology*.

This position paper questions the historical notion of biological individual against the light of the numerous findings of symbiotic relationships between multicellular organisms and microorganisms.

Pradeu, T. (2020). *Philosophy of Immunology*. Cambridge University Press.

This book is important not only because it deals with the question of biological individuality, but also because it proposes a view of the functions of the immune systems that goes beyond defense, providing an overview over several philosophical issues in and from immunology.

References

Ader R (2000) On the development of psychoneuroimmunology. Eur J Pharmacol 405(1–3): 167–176. https://doi.org/10.1016/S0014-2999(00)00550-1

Ader R, Cohen N (1975) Behaviorally conditioned immunosuppression. Psychosom Med 37(4): 333–340. https://doi.org/10.1097/00006842-197507000-00007

Ader R, Kelley KW (2007) A global view of twenty years of brain, behavior, and immunity. Brain Behav Immunity 21(1):20–22. https://doi.org/10.1016/j.bbi.2006.07.003

Ahlman H, Nilsson O (2001) The gut as the largest endocrine organ in the body. Ann Oncol 12: S63–S68. https://doi.org/10.1093/annonc/12.suppl_2.S63

Akiho H, Ihara E, Motomura Y, Nakamura K (2011) Cytokine-induced alterations of gastrointestinal motility in gastrointestinal disorders. World J Gastrointest Pathophysiol 2(5):72–81

Amedei A, Boem F (2018) I've gut a feeling: microbiota impacting the conceptual and experimental perspectives of personalized medicine. Int J Mol Sci 19(12):3756. https://doi.org/10.3390/ijms19123756

Aresti Sanz J, El Aidy S (2019) Microbiota and gut neuropeptides: a dual action of antimicrobial activity and neuroimmune response. Psychopharmacology 236(5):1597–1609. https://doi.org/10.1007/s00213-019-05224-0

Ashley NT, Demas GE (2017) Neuroendocrine-immune circuits, phenotypes, and interactions. Horm Behav 87:25–34. https://doi.org/10.1016/j.yhbeh.2016.10.004

Ashrafian H (2018) The case for a distinctive philosophy of physiology and pathophysiology. J Basic Clin Physiol Pharmacol 29(3):237–240. https://doi.org/10.1515/jbcpp-2017-0143

Augustin R, Schröder K, Murillo Rincón AP, Fraune S, Anton-Erxleben F, Herbst E-M, Wittlieb J, Schwentner M, Grötzinger J, Wassenaar TM, Bosch TCG (2017) A secreted antibacterial neuropeptide shapes the microbiome of Hydra. Nat Commun 8(1):698. https://doi.org/10.1038/s41467-017-00625-1

Beadle GW, Tatum EL (1941) Genetic control of biochemical reactions in neurospora. Stud Hist Phil Biol Biomed Sci 27(11):499–506

Bechtel W, Richardson RC (2010) Discovering complexity: decomposition and localization as strategies in scientific research. MIT Press, Cambridge, MA

Belkaid Y, Hand TW (2014) Role of the microbiota in immunity and inflammation. Cell 157(1): 121–141. https://doi.org/10.1016/j.cell.2014.03.011

Bergthaler A, Menche J (2017) The immune system as a social network. Nat Immunol 18(5): 481–482. https://doi.org/10.1038/ni.3727

Besedovsky HO, del Rey A (1989) Mechanism of virus-induced stimulation of the hypothalamus-pituitary-adrenal axis. J Steroid Biochem 34(1–6):235–239. https://doi.org/10.1016/0022-4731(89)90087-3

Besedovsky HO, Rey AD (1996) Immune-neuro-endocrine interactions: facts and hypotheses. Endocr Rev 17(1):64–102. https://doi.org/10.1210/edrv-17-1-64

Besedovsky HO, del Rey AE, Sorkin E (1985) Immune-neuroendocrine interactions. *J Immunol 135*(2 Suppl):750s–754s

Blalock JE (1994) The syntax of immune-neuroendocrine communication. Immunol Today 15(11): 504–511. https://doi.org/10.1016/0167-5699(94)90205-4

Boem F (2016) Orienteering tools: biomedical research with ontologies. HUMANA.MENTE J Philos Stud 9(30):37–65

Boem F, Nannini G, Amedei A (2020) Not just "immunity": how the microbiota can reshape our approach to cancer immunotherapy. Immunotherapy 12(6):407–416. https://doi.org/10.2217/imt-2019-0192

Bohórquez DV, Liddle RA (2015) The gut connectome: making sense of what you eat. J Clin Investig 125(3):888–890. https://doi.org/10.1172/JCI81121

Boogerd FC, Bruggeman FJ, Hofmeyr J-HS, Westerhoff HV (2007) Towards philosophical foundations of systems biology: introduction. In: Boogerd FC, Bruggeman FJ, Hofmeyr J-HS, Westerhoff HV (eds) Systems biology: philosophical foundations. Elsevier, New York, pp 3–19

Boonen K, Creemers JW, Schoofs L (2009) Bioactive peptides, networks and systems biology. Bioessays 31(3):300–314. https://doi.org/10.1002/bies.200800055

Bordenstein SR, Theis KR (2015) Host biology in light of the microbiome: ten principles of holobionts and hologenomes. PLoS Biol 13(8):e1002226. https://doi.org/10.1371/journal.pbio.1002226

Breer H, Eberle J, Frick C, Haid D, Widmayer P (2012) Gastrointestinal chemosensation: chemosensory cells in the alimentary tract. Histochem Cell Biol 138(1):13–24. https://doi.org/10.1007/s00418-012-0954-z

Brigandt I (2017) Bodily parts in the structure-function dialectic. In: Biological individuality. University of Chicago Press, Chicago, pp 249–274. https://doi.org/10.7208/9780226446592-011

Brogden KA, Guthmiller JM, Salzet M, Zasloff M (2005) The nervous system and innate immunity: the neuropeptide connection. Nat Immunol 6(6):558–564. https://doi.org/10.1038/ni1209

Brookes SJH, Spencer NJ, Costa M, Zagorodnyuk VP (2013) Extrinsic primary afferent signalling in the gut. Nat Rev Gastroenterol Hepatol 10(5):286–296. https://doi.org/10.1038/nrgastro.2013.29

Burnet FM (1961) Immunological recognition of self: such recognition suggests a relationship with processes through which functional integrity is maintained. Science 133(3449):307–311. https://doi.org/10.1126/science.133.3449.307

Cani PD, Knauf C (2016) How gut microbes talk to organs: the role of endocrine and nervous routes. Mol Metab 5(9):743–752. https://doi.org/10.1016/j.molmet.2016.05.011

Cani PD, Van Hul M (2020) Microbial signatures in metabolic tissues: a novel paradigm for obesity and diabetes? Nat Metab 2(3):211–212. https://doi.org/10.1038/s42255-020-0182-0

Chen C-Y, Asakawa A, Fujimiya M, Lee S-D, Inui A (2009) Ghrelin gene products and the regulation of food intake and gut motility. Pharmacol Rev 61(4):430–481. https://doi.org/10.1124/pr.109.001958

Cheng K (2021) Learning in Cnidaria: a systematic review. Learn Behav 49(2):175–189. https://doi.org/10.3758/s13420-020-00452-3

Chiu L, Bazin T, Truchetet M-E, Schaeverbeke T, Delhaes L, Pradeu T (2017) Protective microbiota: from localized to long-reaching co-immunity. Front Immunol 8. https://doi.org/10.3389/fimmu.2017.01678

Collins SM, Surette M, Bercik P (2012) The interplay between the intestinal microbiota and the brain. Nat Rev Microbiol 10(11):735–742. https://doi.org/10.1038/nrmicro2876

Colombetti G, Zavala E (2019) Are emotional states based in the brain? A critique of affective brainocentrism from a physiological perspective. Biol Philos 34(5):45. https://doi.org/10.1007/s10539-019-9699-6

Costello EK, Stagaman K, Dethlefsen L, Bohannan BJM, Relman DA (2012) The application of ecological theory toward an understanding of the human microbiome. Science 336(6086):1255–1262. https://doi.org/10.1126/science.1224203

Coyte KZ, Schluter J, Foster KR (2015) The ecology of the microbiome: networks, competition, and stability. Science 350(6261):663–666. https://doi.org/10.1126/science.aad2602

Dantzer R (2018) Neuroimmune interactions: from the brain to the immune system and vice versa. Physiol Rev 98(1):477–504. https://doi.org/10.1152/physrev.00039.2016

De Vadder F, Grasset E, Mannerås Holm L, Karsenty G, Macpherson AJ, Olofsson LE, Bäckhed F (2018) Gut microbiota regulates maturation of the adult enteric nervous system via enteric serotonin networks. Proc Natl Acad Sci 115(25):6458–6463. https://doi.org/10.1073/pnas.1720017115

del Rey A, Besedovsky HO (2017) Immune-neuro-endocrine reflexes, circuits, and networks: physiologic and evolutionary implications. In: Savino W, Guaraldi F (eds) *Frontiers of hormone research*, vol 48. Karger AG, Basel, pp 1–18. https://doi.org/10.1159/000452902
Dethlefsen L, McFall-Ngai M, Relman DA (2007) An ecological and evolutionary perspective on human–microbe mutualism and disease. Nature 449(7164):811–818. https://doi.org/10.1038/nature06245
Dockray GJ (2003) Luminal sensing in the gut: an overview. J Physiol Pharmacol 54(Suppl 4):9–17
Drack M, Wolkenhauer O (2011) System approaches of Weiss and Bertalanffy and their relevance for systems biology today. Semin Cancer Biol 21(3):150–155. https://doi.org/10.1016/j.semcancer.2011.05.001
Eberl G (2016) Immunity by equilibrium. Nat Rev Immunol 16(8):524–532. https://doi.org/10.1038/nri.2016.75
Eiden LE, Gundlach AL, Grinevich V, Lee MR, Mecawi AS, Chen D, Buijs RM, Hernandez VS, Fajardo-Dolci G, Zhang L (2020) Regulatory peptides and systems biology: a new era of translational and reverse-translational neuroendocrinology. J Neuroendocrinol 32(5):e12844. https://doi.org/10.1111/jne.12844
Facchin M, Viola M, Zanin E (2021) Retiring the "Cinderella view": the spinal cord as an intrabodily cognitive extension. Biol Philos 36(5):45. https://doi.org/10.1007/s10539-021-09822-1
Falkmer S (1993) Phylogeny and ontogeny of the neuroendocrine cells of the gastrointestinal tract. Endocrinol Metab Clin N Am 22(4):731–752
Faraco G, Park L, Anrather J, Iadecola C (2017) Brain perivascular macrophages: characterization and functional roles in health and disease. J Mol Med 95(11):1143–1152. https://doi.org/10.1007/s00109-017-1573-x
Farber DL (2021) Tissues, not blood, are where immune cells function. Nature 593(7860):506–509. https://doi.org/10.1038/d41586-021-01396-y
Fraune S, Abe Y, Bosch TCG (2009) Disturbing epithelial homeostasis in the metazoan Hydra leads to drastic changes in associated microbiota. Environ Microbiol 11(9):2361–2369. https://doi.org/10.1111/j.1462-2920.2009.01963.x
Fuller S (2014) Neuroscience, neurohistory, and the history of science: a tale of two brain images. Isis 105(1):100–109. https://doi.org/10.1086/675552
Fung TC, Olson CA, Hsiao EY (2017) Interactions between the microbiota, immune and nervous systems in health and disease. Nat Neurosci 20(2):145–155. https://doi.org/10.1038/nn.4476
Furness JB (2016) Integrated neural and endocrine control of gastrointestinal function. In: Brierley S, Costa M (eds) The enteric nervous system: 30 years later. Springer, New York, pp 159–173. https://doi.org/10.1007/978-3-319-27592-5_16
Furness JB, Cottrell JJ (2017) Signalling from the gut lumen. Anim Prod Sci 57(11):2175–2187. https://doi.org/10.1071/AN17276
Furness JB, Stebbing MJ (2018) The first brain: species comparisons and evolutionary implications for the enteric and central nervous systems. Neurogastroenterol Motil 30(2):e13234. https://doi.org/10.1111/nmo.13234
Furness JB, Kunze WAA, Clerc N (1999) II. The intestine as a sensory organ: Neural, endocrine, and immune responses. Am J Physiol 277(5):G922–G928. https://doi.org/10.1152/ajpgi.1999.277.5.G922
Furness JB, Rivera LR, Cho H-J, Bravo DM, Callaghan B (2013) The gut as a sensory organ. Nat Rev Gastroenterol Hepatol 10(12):729–740. https://doi.org/10.1038/nrgastro.2013.180
Gardy JL, Lynn DJ, Brinkman FSL, Hancock REW (2009) Enabling a systems biology approach to immunology: focus on innate immunity. Trends Immunol 30(6):249–262. https://doi.org/10.1016/j.it.2009.03.009
Germain RN, Meier-Schellersheim M, Nita-Lazar A, Fraser IDC (2011) Systems biology in immunology: a computational modeling perspective. Annu Rev Immunol 29:527–585. https://doi.org/10.1146/annurev-immunol-030409-101317

Gilbert SF, Epel D (2015) Ecological developmental biology: the environmental regulation of development, health, and evolution, 2nd edn. Sinauer Associates, Sunderland

Gilbert SF, Sapp J, Tauber AI (2012) A symbiotic view of life: we have never been individuals. Q Rev Biol 87(4):325–341. https://doi.org/10.1086/668166

Gomes T, Teichmann SA (2020) An antiviral response beyond immune cells. Nature 583(7815): 206–207. https://doi.org/10.1038/d41586-020-01916-2

Gottschalk RA, Martins AJ, Sjoelund V, Angermann BR, Lin B, Germain RN (2013) Recent progress using systems biology approaches to better understand molecular mechanisms of immunity. Semin Immunol 25(3):201–208. https://doi.org/10.1016/j.smim.2012.11.002

Greslehner GP (2020) Not by structures alone: can the immune system recognize microbial functions? Stud Hist Philos Biol Biomed Sci 84:101336. https://doi.org/10.1016/j.shpsc.2020.101336

Gribble FM, Reimann F (2016) Enteroendocrine cells: chemosensors in the intestinal epithelium. Annu Rev Physiol 78(1):277–299. https://doi.org/10.1146/annurev-physiol-021115-105439

Gribble FM, Reimann F (2019) Function and mechanisms of enteroendocrine cells and gut hormones in metabolism. Nat Rev Endocrinol 15(4):226–237. https://doi.org/10.1038/s41574-019-0168-8

Griffiths PE, Gray RD (1994) Developmental systems and evolutionary explanation. J Philos 91(6): 277–304. https://doi.org/10.2307/2940982

Griffiths PE, Gray RD (2004) The developmental systems perspective. In: Pigliucci M, Preston K (eds) Phenotypic integration: studying the ecology and evolution of complex phenotypes. Oxford University Press, Oxford, pp 409–430

Hacking I, Hacking EUPI, Hacking J (1983) Representing and intervening: introductory topics in the philosophy of natural science. Cambridge University Press, Cambridge

Han W, Tellez LA, Perkins MH, Perez IO, Qu T, Ferreira J, Ferreira TL, Quinn D, Liu Z-W, Gao X-B, Kaelberer MM, Bohórquez DV, Shammah-Lagnado SJ, de Lartigue G, de Araujo IE (2018) A neural circuit for gut-induced reward. Cell 175(3):665–678.e23. https://doi.org/10.1016/j.cell.2018.08.049

Hartenstein V (2006) The neuroendocrine system of invertebrates: a developmental and evolutionary perspective. J Endocrinol 190(3):555–570. https://doi.org/10.1677/joe.1.06964

Hewitt RJ, Lloyd CM (2021) Regulation of immune responses by the airway epithelial cell landscape. Nat Rev Immunol 21(6):347–362. https://doi.org/10.1038/s41577-020-00477-9

Holzer P, Holzer-Petsche U (2009) Pharmacology of inflammatory pain: local alteration in receptors and mediators. Diges Dis 27(Suppl 1):24–30. https://doi.org/10.1159/000268118

Hooper L (2004) Bacterial contributions to mammalian gut development. Trends Microbiol 12(3): 129–134. https://doi.org/10.1016/j.tim.2004.01.001

Hyland NP, Cryan JF (2016) Microbe-host interactions: Influence of the gut microbiota on the enteric nervous system. Dev Biol 417(2):182–187. https://doi.org/10.1016/j.ydbio.2016.06.027

Jaeger J (2017) The importance of being dynamic: systems biology beyond the hairball. In: Green S (ed) Philosophy of systems biology, vol 20. Springer, New York, pp 135–146. https://doi.org/10.1007/978-3-319-47000-9_13

Janeway CA Jr, Travers P, Walport M, Shlomchik MJ (2001) The mucosal immune system. Immunobiology: the immune system in health and disease, 5th edn. Garland Science, New York

Kabouridis PS, Pachnis V (2015) Emerging roles of gut microbiota and the immune system in the development of the enteric nervous system. J Clin Investig 125(3):956–964. https://doi.org/10.1172/JCI76308

Kaelberer MM, Rupprecht LE, Liu WW, Weng P, Bohórquez DV (2020) Neuropod cells: the emerging biology of gut-brain sensory transduction. Annu Rev Neurosci 43(1):337–353. https://doi.org/10.1146/annurev-neuro-091619-022657

Kang Y-N, Fung C, Vanden Berghe P (2021) Gut innervation and enteric nervous system development: a spatial, temporal and molecular tour de force. Development 148(3):dev182543. https://doi.org/10.1242/dev.182543

Keel O (1982) Les conditions de la décomposition « analytique » de l'organisme: Haller, Hunter, Bichat. Etudes Philos 1:37–62
Khan WI, Collins SM (2006) Gut motor function: immunological control in enteric infection and inflammation. Clin Exp Immunol 143(3):389–397. https://doi.org/10.1111/j.1365-2249.2005.02979.x
Kodis EJ, Smindak RJ, Kefauver JM, Heffner DL, Aschenbach KL, Brennan ER, Chan K, Gamage KK, Lambeth PS, Lawler JR, Sikora AK, Vercruysse NR, Deppmann CD (2012) First messengers. In: ELS, 1st edn. Wiley, Boca Raton, FL. https://doi.org/10.1002/9780470015902.a0024167
Konsman J (2019) Inflammation and depression: a nervous plea for psychiatry to not become immune to interpretation. Pharmaceuticals 12(1):29. https://doi.org/10.3390/ph12010029
Kulkarni S, Ganz J, Bayrer J, Becker L, Bogunovic M, Rao M (2018) Advances in enteric neurobiology: the "brain" in the gut in health and disease. J Neurosci 38(44):9346–9354. https://doi.org/10.1523/JNEUROSCI.1663-18.2018
Kuwahara A, Matsuda K, Kuwahara Y, Asano S, Inui T, Marunaka Y (2020) Microbiota-gut-brain axis: enteroendocrine cells and the enteric nervous system form an interface between the microbiota and the central nervous system. Biomed Res 41(5):199–216. https://doi.org/10.2220/biomedres.41.199
Ladyman J, Wiesner K (2020) What is a complex system? Yale University Press, New Haven
Larsen SB, Cowley CJ, Fuchs E (2020) Epithelial cells: liaisons of immunity. Curr Opin Immunol 62:45–53. https://doi.org/10.1016/j.coi.2019.11.004
Latorre R, Sternini C, De Giorgio R, Greenwood-Van Meerveld B (2016) Enteroendocrine cells: a review of their role in brain–gut communication. Neurogastroenterol Motil 28(5):620–630. https://doi.org/10.1111/nmo.12754
Leonelli S (2019) The challenges of big data biology. elife 8:e47381. https://doi.org/10.7554/eLife.47381
Lu VB, Gribble FM, Reimann F (2021) Nutrient-induced cellular mechanisms of gut hormone secretion. Nutrients 13(3):883. https://doi.org/10.3390/nu13030883
Lyon P (2015) The cognitive cell: Bacterial behavior reconsidered. Front Microbiol 6:264. https://www.frontiersin.org/article/10.3389/fmicb.2015.00264
Maniscalco JW, Rinaman L (2018) Vagal interoceptive modulation of motivated behavior. Physiology 33(2):151–167. https://doi.org/10.1152/physiol.00036.2017
Martin LB, Hawley DM, Ardia DR (2011) An introduction to ecological immunology. Funct Ecol 25(1):1–4. https://doi.org/10.1111/j.1365-2435.2010.01820.x
Mašek K, Slánský J, Petrovický P, Hadden JW (2003) Neuroendocrine immune interactions in health and disease. Int Immunopharmacol 3(8):1235–1246. https://doi.org/10.1016/S1567-5769(03)00015-8
Mayer EA (2011) Gut feelings: the emerging biology of gut–brain communication. Nat Rev Neurosci 12(8):453–466. https://doi.org/10.1038/nrn3071
Mikkelsen HB (2010) Interstitial cells of Cajal, macrophages and mast cells in the gut musculature: morphology, distribution, spatial and possible functional interactions. J Cell Mol Med 14(4):818–832. https://doi.org/10.1111/j.1582-4934.2010.01025.x
Modlin IM, Champaneria MC, Bornschein J, Kidd M (2006) Evolution of the diffuse neuroendocrine system—clear cells and cloudy origins. Neuroendocrinology 84(2):69–82. https://doi.org/10.1159/000096997
Moulin AM (1991) Le Dernier Langage de La Médecine: Histoire de l'immunologie de Pasteur Au Sida. 1re éd. Pratiques Théoriques. Presses universitaires de France, Paris
Mueller KL, Hines PJ, Travis J (2016) Neuroimmunology. Science 353(6301):760–761. https://doi.org/10.1126/science.353.6301.760
Naik S, Larsen SB, Cowley CJ, Fuchs E (2018) Two to Tango: dialog between immunity and stem cells in health and disease. Cell 175(4):908–920. https://doi.org/10.1016/j.cell.2018.08.071

Neunlist M, Schemann M (2014) Nutrient-induced changes in the phenotype and function of the enteric nervous system. J Physiol 592(14):2959–2965. https://doi.org/10.1113/jphysiol.2014.272948

News & Highlights (2008) Anonymous. Mucosal Immunol 1(4):246–247. https://doi.org/10.1038/mi.2008.17

Nicholson JK, Wilson ID (2003) Understanding "global" systems biology: metabonomics and the continuum of metabolism. Nat Rev Drug Discov 2(8):668–676. https://doi.org/10.1038/nrd1157

O'Mahony SM, Clarke G, Borre YE, Dinan TG, Cryan JF (2015) Serotonin, tryptophan metabolism and the brain-gut-microbiome axis. Behav Brain Res 277:32–48. https://doi.org/10.1016/j.bbr.2014.07.027

Obata Y, Pachnis V (2016) The effect of microbiota and the immune system on the development and organization of the enteric nervous system. Gastroenterology 151(5):836–844. https://doi.org/10.1053/j.gastro.2016.07.044

Panerai AE, Ottaviani E (1995) Immunoendocrine reshaping with age. Int Rev Immunol 12(1):75–84. https://doi.org/10.3109/08830189509056703

Pariante CM (2015) Psychoneuroimmunology or immunopsychiatry? Lancet Psychiatry 2(3):197–199. https://doi.org/10.1016/S2215-0366(15)00042-5

Pedragosa J, Salas-Perdomo A, Gallizioli M, Cugota R, Miró-Mur F, Briansó F, Justicia C, Pérez-Asensio F, Marquez-Kisinousky L, Urra X, Gieryng A, Kaminska B, Chamorro A, Planas AM (2018) CNS-border associated macrophages respond to acute ischemic stroke attracting granulocytes and promoting vascular leakage. Acta Neuropathol Commun 6(1):76. https://doi.org/10.1186/s40478-018-0581-6

Peluso I, Palmery M (2016) The relationship between body weight and inflammation: lesson from anti-TNF-α antibody therapy. Hum Immunol 77(1):47–53. https://doi.org/10.1016/j.humimm.2015.10.008

Petrovsky N (2001) Towards a unified model of neuroendocrine–immune interaction. Immunol Cell Biol 79(4):350–357. https://doi.org/10.1046/j.1440-1711.2001.01029.x

Polak JM, Bloom SR (1979) The diffuse neuroendocrine system. Studies of this newly discovered controlling system in health and disease. J Histochem Cytochem 27(10):1398–1400. https://doi.org/10.1177/27.10.512327

Poon MML, Farber DL (2020) The whole body as the system in systems immunology. IScience 23(9):101509. https://doi.org/10.1016/j.isci.2020.101509

Posovszky C (2017) Development and anatomy of the enteroendocrine system in humans. In: Wabitsch M, Posovszky C (eds) *Endocrine development*, vol 32. Karger AG, Basel, pp 20–37. https://doi.org/10.1159/000475729

Pradeu T (2020) Philosophy of immunology. Cambridge University Press, Cambridge

Pradeu T, Jaeger S, Vivier E (2013) The speed of change: towards a discontinuity theory of immunity? Nat Rev Immunol 13:764–769

Psichas A, Reimann F, Gribble FM (2015) Gut chemosensing mechanisms. J Clin Invest 125(3):908–917. https://doi.org/10.1172/JCI76309

Raman AS, Gehrig JL, Venkatesh S, Chang H-W, Hibberd MC, Subramanian S, Kang G, Bessong PO, Lima AAM, Kosek MN, Petri WA, Rodionov DA, Arzamasov AA, Leyn SA, Osterman AL, Huq S, Mostafa I, Islam M, Mahfuz M et al (2019) A sparse covarying unit that describes healthy and impaired human gut microbiota development. Science 365(6449):eaau4735. https://doi.org/10.1126/science.aau4735

Rankin LC, Artis D (2018) Beyond host defense: emerging functions of the immune system in regulating complex tissue physiology. Cell 173(3):554–567. https://doi.org/10.1016/j.cell.2018.03.013

Ratti E (2016) The end of 'small biology'? Some thoughts about biomedicine and big science. Big Data Soc 3(2):205395171667843. https://doi.org/10.1177/2053951716678430

Rehfeld JF (2004) A centenary of gastrointestinal endocrinology. Horm Metab Res 36(11/12):735–741. https://doi.org/10.1055/s-2004-826154

Rhee K-J, Sethupathi P, Driks A, Lanning DK, Knight KL (2004) Role of commensal bacteria in development of gut-associated lymphoid tissues and preimmune antibody repertoire. J Immunol 172(2):1118–1124. https://doi.org/10.4049/jimmunol.172.2.1118

Rieckmann JC, Geiger R, Hornburg D, Wolf T, Kveler K, Jarrossay D, Sallusto F, Shen-Orr SS, Lanzavecchia A, Mann M, Meissner F (2017) Social network architecture of human immune cells unveiled by quantitative proteomics. Nat Immunol 18(5):583–593. https://doi.org/10.1038/ni.3693

Rivier C, Chizzonite R, Vale W (1989) In the mouse, the activation of the hypothalamic-pituitary-adrenal axis by a lipopolysaccharide (endotoxin) is mediated through interleukin-1. Endocrinology 125(6):2800–2805. https://doi.org/10.1210/endo-125-6-2800

Ronai I, Greslehner GP, Boem F, Carlisle J, Stencel A, Suárez J, Bayir S, Bretting W, Formosinho J, Guerrero AC, Morgan WH, Prigot-Maurice C, Rodeck S, Vasse M, Wallis JM, Zacks O (2020) "Microbiota, symbiosis and individuality summer school" meeting report. Microbiome 8(1): 117. https://doi.org/10.1186/s40168-020-00898-7

Rustenhoven J, Kipnis J (2019) Bypassing the blood-brain barrier. Science 366(6472):1448–1449. https://doi.org/10.1126/science.aay0479

Schemann M, Frieling T, Enck P (2020) To learn, to remember, to forget—how smart is the gut? Acta Physiol 228(1):e13296. https://doi.org/10.1111/apha.13296

Schleimer RP, Kato A, Kern R, Kuperman D, Avila PC (2007) Epithelium: at the interface of innate and adaptive immune responses. J Allergy Clin Immunol 120(6):1279–1284. https://doi.org/10.1016/j.jaci.2007.08.046

Schneider T (2021) The holobiont self: understanding immunity in context. Hist Philos Life Sci 43(3):99. https://doi.org/10.1007/s40656-021-00454-y

Sclafani A (2013) Gut–brain nutrient signaling. Appetition vs. satiation. *Appetite* 71:454–458. https://doi.org/10.1016/j.appet.2012.05.024

Serna-Duque JA, Esteban MÁ (2020) Effects of inflammation and/or infection on the neuroendocrine control of fish intestinal motility: a review. Fish Shellfish Immunol 103:342–356. https://doi.org/10.1016/j.fsi.2020.05.018

Sharkey KA, Beck PL, McKay DM (2018) Neuroimmunophysiology of the gut: advances and emerging concepts focusing on the epithelium. Nat Rev Gastroenterol Hepatol 15(12):765–784. https://doi.org/10.1038/s41575-018-0051-4

Silva-Sanchez A, Randall TD (2020) Anatomical uniqueness of the mucosal immune system (GALT, NALT, iBALT) for the induction and regulation of mucosal immunity and tolerance. In: Mucosal vaccines. Elsevier, New York, pp 21–54. https://doi.org/10.1016/B978-0-12-811924-2.00002-X

Skibicka KP, Dickson SL (2013) Enteroendocrine hormones—central effects on behavior. Curr Opin Pharmacol 13(6):977–982. https://doi.org/10.1016/j.coph.2013.09.004

Stephen J (2001) Pathogenesis of infectious diarrhea. Can J Gastroenterol 15(10):669–683. https://doi.org/10.1155/2001/264096

Van Hove H, Martens L, Scheyltjens I, De Vlaminck K, Pombo Antunes AR, De Prijck S, Vandamme N, De Schepper S, Van Isterdael G, Scott CL, Aerts J, Berx G, Boeckxstaens GE, Vandenbroucke RE, Vereecke L, Moechars D, Guilliams M, Van Ginderachter JA, Saeys Y, Movahedi K (2019) A single-cell atlas of mouse brain macrophages reveals unique transcriptional identities shaped by ontogeny and tissue environment. Nat Neurosci 22(6):1021–1035. https://doi.org/10.1038/s41593-019-0393-4

Verburg-van Kemenade BML, Cohen N, Chadzinska M (2017) Neuroendocrine-immune interaction: evolutionarily conserved mechanisms that maintain allostasis in an ever-changing environment. Dev Comp Immunol 66:2–23. https://doi.org/10.1016/j.dci.2016.05.015

Vijay-Kumar M, Chassaing B, Kumar M, Baker M, Singh V (2014) Mammalian gut immunity. Biomed J 37(5):246. https://doi.org/10.4103/2319-4170.130922

Vuong HE, Pronovost GN, Williams DW, Coley EJL, Siegler EL, Qiu A, Kazantsev M, Wilson CJ, Rendon T, Hsiao EY (2020) The maternal microbiome modulates fetal neurodevelopment in mice. Nature 586(7828):281–286. https://doi.org/10.1038/s41586-020-2745-3

Wallenius V, Wallenius K, Ahrén B, Rudling M, Carlsten H, Dickson SL, Ohlsson C, Jansson J-O (2002) Interleukin-6-deficient mice develop mature-onset obesity. Nat Med 8(1):75–79. https://doi.org/10.1038/nm0102-75
Watnick PI, Jugder B-E (2020) Microbial control of intestinal homeostasis via enteroendocrine cell innate immune signaling. Trends Microbiol 28(2):141–149. https://doi.org/10.1016/j.tim.2019.09.005
Wershil B, Furuta G (2008) 4. Gastrointestinal mucosal immunity. J Allergy Clin Immunol 121(2): S380–S383. https://doi.org/10.1016/j.jaci.2007.10.023
Wikoff WR, Anfora AT, Liu J, Schultz PG, Lesley SA, Peters EC, Siuzdak G (2009) Metabolomics analysis reveals large effects of gut microflora on mammalian blood metabolites. Proc Natl Acad Sci 106(10):3698–3703. https://doi.org/10.1073/pnas.0812874106
Wood JD (1999) Enteric nervous control of motility in the upper gastrointestinal tract in defensive states. Dig Dis Sci 44(8 Suppl):44S–52S
Wood JD (2004) Enteric neuroimmunophysiology and pathophysiology. *Gastroenterology 127*(2): 635–657. https://doi.org/10.1053/j.gastro.2004.02.017
Woodger JH (1929) Biological principles: a critical study. Harcourt
Yang NJ, Chiu IM (2017) Bacterial signaling to the nervous system through toxins and metabolites. J Mol Biol 429(5):587–605. https://doi.org/10.1016/j.jmb.2016.12.023
Yoo BB, Mazmanian SK (2017) The enteric network: interactions between the immune and nervous systems of the gut. Immunity 46(6):910–926. https://doi.org/10.1016/j.immuni.2017.05.011
Zheng D, Liwinski T, Elinav E (2020) Interaction between microbiota and immunity in health and disease. Cell Res 30(6):492–506. https://doi.org/10.1038/s41422-020-0332-7

The Behavioural Immune System of Lower Vertebrates

3

Krzysztof Rakus, Magdalena Widziolek,
B. M. Lidy Verburg-van Kemenade, and Magdalena Chadzinska

Abstract

The immune system evolves under the ever-changing selective pressure of pathogens and its effective functioning involves not only immune-related cells and mediators. It also requires neuroendocrine-immune interaction, including regulation of specific organismal behaviour. Mechanisms of the behavioural immune system result in a change of psychological behaviour, comprising sickness behaviour: fever, decreased appetite, as well as depressed locomotor, social and exploratory activities. Behavioural changes are induced by inflammatory mediators released upon infection, which predominantly activate glial cells such as astrocytes and microglia. This phenomenon is evolutionarily conserved and can be observed in all vertebrates. In the oldest and most numerous vertebrate group, teleost fish, inflammatory cytokines (e.g. IL-1β, TNF-α and type I IFNs) induce behavioural fever and several other behavioural changes. Pathogens developed several mechanisms to reach the brain, including passive diffusion through injured endothelia, infection of endothelial cells or a Trojan horse tactic in phagocytes. Therefore local infection in fish is often connected with neuroinflammation and microglia activation.

Among vertebrates, the zebrafish model provides a useful tool for in vitro and in vivo studies of microglia at a single-cell level. The advantages of this model are

K. Rakus · M. Widziolek · M. Chadzinska (✉)
Department of Evolutionary Immunology, Institute of Zoology and Biomedical Research, Faculty of Biology, Jagiellonian University, Kraków, Poland
e-mail: krzysztof.rakus@uj.edu.pl; magdalena.widziolek@uj.edu.pl; magdalena.chadzinska@uj.edu.pl

B. M. L. Verburg-van Kemenade
Cell Biology and Immunology Group, Wageningen University, Wageningen, The Netherlands

J. P. Konsman, T. M. Reyes (eds.), *Neuroendocrine-Immune System Interactions*, Masterclass in Neuroendocrinology 13,
https://doi.org/10.1007/978-3-031-21358-8_3

its small size, rapid external development and embryo transparency, which create ample possibilities for genetic manipulation.

Keywords

Inflammation · Sickness behaviour · Microglia activation · Fish

3.1 Introduction

Today, over 1.3 million animal species exist around the world, including >66,000 vertebrates. The group of vertebrate species includes primitive jawless lampreys and hagfishes, and jawed vertebrates such as the cartilaginous fish (e.g. sharks, rays, and ratfish) and the abundant and diverse group of bony fishes (Teleosts), tetrapods, which include amphibians (e.g. salamanders, and anurans: frogs and toads, collectively 7300 species), reptiles (turtles, tuataras, lizards and snakes and crocodiles, collectively over 10,000 species), birds (10,425 species) and mammals (5513 species). Teleost fish represent more than one-half of all extant vertebrate species (almost 33,000 species). They are the earliest vertebrates to have developed neural and endocrine pathways that resemble the classical mammalian pathway. Teleosts are also the oldest to have developed a complex innate and adaptive immune response with specific T- and B-cells and antibodies. Moreover, fish display bi-directional neuroendocrine-immune interaction, and we hypothesize that this has greatly contributed to their enormous evolutionary success and their adaptive radiation.

The immune system evolves under the ever-changing selective pressure of pathogens. Therefore, constant adaptation is required to acquire an efficient protection from, and response to, microbial threats to ultimately obtain homeostasis. The defence mechanisms first of all comprise the physical barriers of the epithelium, including chemical barriers, e.g. mucosal defence molecules that may obliterate an array of pathogens. Secondly, the innate and adaptive immune system will enable detection of intruding pathogens so as to subsequently mobilize an adequate physiological reaction to encapsulate or kill the microbes or to kill infected cells in order to prevent intracellular viral replication. However complex, adaptive and efficient this defence may be, pandemic outbreaks remain a continuous threat to animal populations.

Moreover, we believe that an effective immune reaction cannot be based solely on an efficiently functioning immune system, but must involve the entire body, especially the nervous and endocrine systems, also collectively referred to as the neuroendocrine system. Neuroendocrine-immune interaction in a bi-directional fashion has indeed been accepted over the last decades as an important mechanism to ensure a concerted physiological response to pathogen challenge. The physiological importance of this phenomenon is supported by its evolutionary conservation (Verburg-van Kemenade et al. 2017).

One of the best-described examples of such interactions is sickness behaviour, considered as a part of behavioural immune system (BIS) (Shakhar 2019). The behavioural immune system is not merely reactive but also includes proactive reactions. Mechanisms of the behavioural immune system will initiate physiological responses to perceptual signals that indicate the presence of pathogens in the near environment, including their presence in conspecific animals. This may result in a change of psychological behaviour, for instance inducing disease avoidance behaviour or influence social or mating preferences. Sickness behaviour comprises lethargy, hypersomnia, reduced eating and drinking behaviour as well as social withdrawal and decreased libido. This sickness behaviour takes place during the most infectious period of the illness and is triggered mostly by signalling from the innate immune system (Dantzer 2006). It also includes behavioural fever that is induced upon infection and/or inflammation. Though having clear disadvantages to the animal in terms of loss of strength, higher predation risk and lower reproduction rate, it may also save energy and may largely be beneficial as it reduces the transmission of dangerous pathogens (Verburg-van Kemenade et al. 2017). Saving energy is important at this stage as the immune defence mechanisms are extremely costly to the animals. For example a 1 °C rise in body temperature during fever in endothermic (warm-blooded) animals requires a 10–12.5% increase in metabolic rate (Evans et al. 2015).

Thus, behavioural changes that occur after infection imply a connection between the central nervous and immune systems. Macrophages and granulocytes are the first immune cells to recognize infections and their activation induces an array of defence reactions. Pathogens and the inflammatory mediators released by macrophages and granulocytes not only stimulate the adaptive and specific immune response, but they also signal to the brain through receptors present on microglia, astrocytes, neuroendocrine cells or neurons. Vice versa, immune cells may be sensitive to neuroendocrine signals through glucocorticoid, opioid and adrenergic receptors as well as other endocrine receptors that were identified on different leukocytes (Verburg-van Kemenade et al. 2017).

This review will focus on lower vertebrates, with emphasis on sickness behaviour in teleost fish. We focus on the evolutionarily conserved infection/inflammation-induced behavioural changes in lower/ectothermic animals. We describe: (i) the main leukocyte populations and immune mediators involved in the response to pathogens and in the neuroendocrine-immune crosstalk in ectothermic vertebrates, (ii) examples of infection/inflammation-induced changes in animal behaviour and the mechanisms involved in this phenomenon, as well as their adaptive function, (iii) the potential of the zebrafish model for present and future studies of infection/inflammation-induced behavioural changes.

3.2 Cells and Mediators Involved in the Immune Response of Lower Vertebrates

Both viral and bacterial infections induce an inflammatory response. This is a multistep process that starts when resident macrophages recognize the infection and induce the chemotactic recruitment and activation of immune cells to eliminate the infectious microbes and the infected cells.

3.2.1 Pathogen Recognition and Intracellular Response to Infection

In all vertebrates, the initiation of the host innate immune response is based on recognition of the non-self-signature of invading pathogens, the so-called pathogen-associated molecular patterns (PAMPs). These PAMPs are recognized by pattern recognition receptors (PRRs) localized on different types of cells, including leukocytes and epithelial cells. PRRs can also recognize damage/danger-associated molecular patterns (DAMPs), which are present in abnormal locations or atypical molecular complexes induced by tissue destruction and/or cellular stress (Mojzesz et al. 2020).

Currently, five major groups of PRRs have been discovered: (i) Toll-like receptors (TLRs), (ii) nucleotide-binding oligomerization domain (NOD)-leucine-rich repeats (LRR)-containing receptors (NLRs), (iii) retinoic acid-inducible gene 1 (RIG-1)-like receptors (RLRs) which belong to the large family of DExD/H-box RNA helicases, (iv) C-type lectin receptors (CLRs) and (v) cytosolic DNA sensors. They detect PAMPs based on their structure (glycoproteins, lipopolysaccharides, proteoglycans, and nucleic acid motifs) and sub-cellular localization. Therefore, they are either associated with cellular and endosomal membranes (e.g. TLRs) or are localized in the cytosol (e.g. NLRs, RLR). Secreted forms of PRRs can also be present extracellularly in the bloodstream and in interstitial fluids (Mojzesz et al. 2020).

The main effect of ligand-related PRR activation is intracellular activation of the Map kinase, NFkB and IRF pathways with consequent production of pro-inflammatory cytokines, type I interferons (IFNs) and anti-viral proteins such as Mx proteins, viperin and ISG-15 (Mojzesz et al. 2020). In mammals activation of NLRs and AIM2 receptors from the cytosolic DNA sensors leads to the formation of multiprotein signalling complexes known as inflammasomes. The inflammasomes activate caspase-1-mediated processing and activate pro-interleukins IL-1β and IL-18. Moreover, through gasdermin D cleavage, they induce cell death called pyroptosis (Morimoto et al. 2021). Recently, numerous inflammasome-related genes have been identified in lower vertebrates, mainly fish. Moreover, their involvement in pyroptosis induction has been confirmed (Morimoto et al. 2021). By these mechanisms, PAMP recognition activates leukocytes and induces the expression of pro-inflammatory mediators (cytokines and enzymes).

3.2.2 The Course of Inflammation and the Key Players Involved

The resident macrophages recognize the microbial invasion and then initiate the response by the production and release of chemokines and pro-inflammatory cytokines. Neutrophilic granulocytes are the first effector cells to be recruited. Neutrophils arrive in great numbers and play a vital role in the clearance of pathogens during the first hours of infection. Monocytes quickly follow to infiltrate, and they differentiate into macrophages. Macrophages play an essential role during all phases of the inflammatory response. The recruited monocytes subsequently differentiate and polarize into classically activated M1 macrophages upon PAMP and cytokine stimulation, predominantly by interferon γ (INF-γ) with or without tumour necrosis factor α (TNF-α). M1 macrophages produce high levels of interleukin1 (IL-1) and interleukin 12 (IL-12), reactive oxygen intermediates (ROI) and nitric oxide (NO) and thereby promote the inflammation and clearance of pathogens. Alternative polarization occurs later during the inflammatory process and different phenotypes of M2 macrophages (M2a, M2b, M2c or M2d) are identified. These phenotypes all contribute with activities to avoid additional host tissue damage: suppression of the inflammation, enhanced phagocytosis of apoptotic cells, induction of tissue repair and parasite elimination. Where classically activated macrophages are Th1 driven, alternatively activated macrophages are typically Th2 driven and predominantly produce interleukin 4, 10 and 13 (IL-4, IL-10, IL-13) and transforming growth factor β (TGF-β). Classically or alternatively activated macrophages, moreover, have a distinctive difference in their L-arginine metabolism. In classically activated macrophages L-arginine is converted by inducible nitric oxide synthase (iNOS) into NO and L-citrulline. Alternatively activated macrophages show high arginase activity that converts L-arginine into urea and L-ornithine, a precursor for proline and polyamines, which promote cell proliferation and collagen production, thereby helping to orchestrate the resolution of inflammation and tissue repair to avoid collateral damage (Maciuszek et al. 2020).

The inflammatory process in lower vertebrates, especially in teleost fish, has been described for multiple species. These in vitro and in vivo studies revealed that the inflammatory process is evolutionarily conserved and the principal cells and the functional homologues for the important cytokines in teleost fish have been identified and characterized.

Although cytokine molecules may have low sequence identity to their mammalian counterparts, though they are not always paralogs and although few antibodies to teleost cytokines are available, expression studies and studies with recombinant proteins reveal a great level of functional conservation.

The most important cytokines involved in the initiation of inflammation and for induction of disease behaviour are IL-1β, IL-6, IL-12 and TNF-α, and different molecules from the chemokine family.

The IL-1 or β-trefoil cytokine family in mammals has 11 members, of which only two (IL-1β and IL-18) are found in bony fish (e.g. Huising et al. 2004). However, an additional novel family member forms a teleost-specific group (nIL-1Fm). Multiple copies of teleost IL-1β genes are found that have a different exon-intron organization and are classified into Type I and Type II. Based on their phylogenetic trees, they are both closely related to but are separate from the mammalian IL-1β clade. Type I is conserved among jawed vertebrates and is considered the ancestral type. Cyprinids have only type I IL-1β. After PAMP or DAMP stimulation pro-IL-1β is synthesized and released. Both intra- and extracellularly this precursor molecule can be cleaved, predominantly by caspase 1, also called IL-1β converting enzyme (ICE). Although the typical ICE cutting site is missing in teleost fish, caspases are indicated to cut the precursor into smaller peptides. The mechanisms of IL-1β processing seem complicated and must be strictly regulated during infection. The size of the cleavage products is different from that in mammals and moreover differs in different species of teleosts. For instance, we found that in carp pro-IL-1β is cleaved into intermediate and mature products of 24 and 15 kD, while in zebrafish pro-IL-1β is cleaved into 22 and 18 kD products (Verburg-van Kemenade et al. 1995). IL-1β binds to two different receptors: IL-1R1 and IL-1R2, of which the second is a decoy receptor and might in this way have an inhibitory function, just like nIL-1Fm, which might act as an inhibitor through competition at the IL-1R1. Recombinant IL-1β molecules are pro-inflammatory for leukocytes and macrophages. In vitro studies revealed induction of a number of pro-inflammatory genes encoding TNF-α, IL-1, IL-6, IL-8 and COX-2 by recombinant IL-1β via the NFkB and MAPK pathways. In vivo administration of recIL-1β induces local inflammatory responses in fish, which indicates that its diverse functions are conserved (Zou and Secombes 2016).

TNF-α is an important member of the B-jellyrol cytokines or tumour necrosis factor superfamily of ligands and receptors. They are type II membrane proteins with an N-terminal intracellular domain, a transmembrane domain and a C-terminal extracellular TNF homology domain with the family signature motif. TNF-α is released as homotrimers to react with target cells. During the release process, the TNF-converting enzyme (TACE) cleaves the pro-TNF from the membrane and this process is conserved within vertebrates. The TNF homologues/paralogues in fish can be categorized into three phylogenetic groups: the type I and II TNF-α group and the TNF-N group. The genes encoding type I TNF-α and TNF-N show a conserved synteny with those in humans, while type II TNF-α genes are teleost-specific. TNF-α functions are also evolutionarily conserved among vertebrates. Like IL-1β, TNF-α is rapidly synthesized and released after infection, and induces IL-1β, CXCL8, TNF-α and COX-2 expression via the NFkB intracellular pathway. It is also important for bactericidal activity through production of ROS and increases phagocytosis and macrophage survival. Moreover, it is involved in the regulation of leukocyte homing, proliferation and migration. In contrast to TNF-α, TNF-N is encoded by a single-copy gene and its functions have yet to be characterized (Zou and Secombes 2016).

The group of type I helical cytokines includes the IL-2, the IL-6 and the IL-12 subfamily, the β-chain cytokines and the colony-stimulating factors (CSFs) (Huising et al. 2006). Homologs and paralogs of all have been identified and studied in fish, but their pleiotropic functional characteristics still need extensive study.

The IL-12 subfamily in fish consists of heterodimers of an α chain and a β chain. IL-12 is formed by a p40 chain and a p35 chain. Different paralogs of the p40 chain exist that are differentially regulated. IL-12 is expressed in macrophages during inflammation, and induces the expression of other cytokines, including the Th1 signature cytokine IFN-γ (Zou and Secombes 2016).

Another cytokine from the pro-inflammatory cascade is interleukin-6 (IL-6), but it also has anti-inflammatory effects, such as stimulation of IL-10 and IL-1ra expression. In mammals, IL-6 is involved in multiple cellular processes, including regulation of cell proliferation and differentiation and inhibition of apoptosis. IL-6 also regulates the acute-phase reaction, haematopoiesis and antibody production. It binds to the membrane and the soluble receptor (IL-6R) as well as to gp130, which transduces the IL-6 signal. The *il-6* gene has been identified in teleost species and, as in other vertebrates and in fish, IL-6 has four long α-helices (A, B, C, and D). In fish, IL-6 induces cell proliferation and antimicrobial activity (Zou and Secombes 2016).

The group of type II helical cytokines in mammals includes cytokines of the IL-10 subfamily and the interferons: type I IFNs (e.g., IFN-α, β), type II IFNs (IFN-γ) and type III IFNs (IFN-λ/IL28s). The first IFN-like molecules have been described in teleost fish, but a recent study by Redmond et al. (2019) discovered ancestral IFN orthologs in a cartilaginous fish. They discovered that the four major lineages of genes encoding the superfamily of class II α-helical cytokines diverged by multiple gene duplications in the ancestor of jawed vertebrates and that the anti-viral interferons IFN1 and IFN3 are likely to be sister groups, while IFN2 is a sister to the IL-10 family. Interestingly, in mammals type I IFN genes are intronless, while type III IFN genes have introns. In fish however, all IFN genes have introns. Studies in amphibians, reptiles (including birds) and in mammals revealed that intronless IFN genes originated from retroposition event(s), which likely occurred during the early evolution of tetrapods (Boudinot et al. 2016; Redmond et al. 2019).

The variation of type I IFN gene numbers in fish species is impressive, the most complex system described being in salmonids (currently 11 type I IFN genes identified in Atlantic salmon). Based on cysteine motifs, fish type I IFNs fall into two subgroups, which use two different receptors. Group I comprises IFN sequences with two conserved cysteines with one S-S bridge, while group II sequences have four conserved cysteines and two conserved S-S bridges (Boudinot et al. 2016). Group I in zebrafish consists of IFNφ1 and IFNφ4 while group II consists of IFNφ2 and IFNφ3. Moreover, zebrafish possess about 100 IFN-stimulated genes (ISGs) that are orthologous to human ISGs (Boudinot et al. 2016).

In fish, up-regulation of type I IFN genes was, for example observed in zebrafish upon infection with the Chikungunya Virus (CHIKV) (Palha et al. 2013) and Tilapia-like virus (TiLV) infection (Rakus et al. 2020).

Type II IFN (INF-γ) functions to induce the classical activation of phagocytes, thus skewing towards a Th1-like profile of immune activation. This function appears evolutionarily well conserved in teleost fish. In mammals only one IFN-γ exists, but in teleost fish two IFN-γ gene sequences, in two phylogenetic clusters, can be distinguished, which may be the result of a tandem gene duplication. IFN-γ2 shares more structural features with the mammalian IFN-γ molecules including a comparable signal peptide, the IFN-γ signature motif and a 6-helix secondary structure. In addition to these two IFN-γ molecules, another member of the IFN family, with moderate homology, is found in fish: IFN-γrel. Expression analysis as well as functional analysis with carp IFN-γ molecules revealed that the classical IFN-γ function is associated with the IFN-γ2 cluster. Carp IFN-γ2, in contrast to carp IFN-γ1, induced a strong classical pro-inflammatory reaction in phagocytes. A synergistic response with LPS (Gram-negative bacterial endotoxin) was observed for the induction of *inos* expression and of NO release, and for the expression of genes encoding CXCL9–11-like chemokines, the pro-inflammatory cytokines IL-1β, TNF-α and the IL-12 subunits p35 and p40. IFN-γ2 also induces formation of neutrophil extracellular traps (NETs), structures that are formed from DNA, histones and granular components and are secreted by activated neutrophils. IFN-γrel in fish has different functions in different species, mostly regulating anti-bacterial and anti-viral immunity (Zou and Secombes 2016).

IL-10 is an anti-inflammatory cytokine that is, among others, produced by M2 macrophages. Like many cytokines it has pleiotropic functions. IL-10 inhibits PAMP- and DAMP-mediated induction of the pro-inflammatory cytokines TNF-α, IL-1β, IL-12 and IFN-γ. It enhances B cell survival, proliferation and antibody production. IL-10 is a homodimer molecule and signals through a receptor complex of two IL-10 receptor-1 (IL-10R1) and two IL-10 receptor-2 (IL-10R2) proteins. IL-10 was identified in several teleost species, IL-10R1 in Cyprinids and IL-10R2 in carp. Consistent with the mammalian orthologs, in teleost fish IL-10 has an anti-inflammatory function, inhibiting ROS and NO production, as well as down regulating expression of pro-inflammatory cytokines TNF-α, IL-1β and CXCL8. In carp, moreover, promotion of B cell differentiation and antibody secretion by IL-10 was shown in an antigen-dependent manner (Piazzon et al. 2016).

During the immune response, chemokines belonging to different families guide cell migration by chemotaxis to the focus of inflammation. The different leukocyte types are recruited during distinct phases of the infection and react to different chemokines. Polymorphonuclear leukocytes like neutrophils, basophils and eosinophils are mainly recruited by CXC chemokines typically characterized by an ELR signature, e.g. CXCL8. Monocytes and lymphocytes are attracted by CXC chemokines that lack an ELR signature (CXCL9-11) and by CC chemokines (Bird and Tafalla 2015).

Recruitment of the different leukocytes during different phases of inflammation has also been studied in teleost fish. Several CXC chemokines, e.g. CXCL 12 and 14, which are functional during development, have orthologous molecules in teleost fish. In fish, CXCL8_L1 (former name CXCa) does not form a true ortholog but is most similar to CXCL8 and moreover has the typical functional features of CXCL8. Moreover, in Cyprinids a second gene encoding CXCL8 (CXCL8_L2) was discovered. Both CXCL8 molecules, like their mammalian counterparts, are important in attracting neutrophils and monocytes and are predominantly expressed and active during the early phases of inflammation. However, unlike mammalian CXCL8, fish CXCL8 molecules do not contain the ELR motif. During the later stages of inflammation, CXCb molecules are predominantly expressed. CXCb molecules also do not have true orthologues to mammalian CXC, but are clearly mostly related to mammalian CXCL9, 10 and 11, in structure, in function and in their expression profiles.

The CC chemokine family is the largest chemokine family, and it underwent extensive species-specific intrachromosomal duplications. The number of CC molecules therefore varies in different species, with up to 81 CC chemokines in zebrafish (Bird and Tafalla 2015).

The inflammatory response is also connected with activation of several enzymes such as cyclooxygenase-2 (COX-2), matrix metalloproteinase 9 (MMP-9) or neutrophil elastase (NE). Inducible cyclooxygenase-2 (COX-2) catalyses the first step in the synthesis of prostanoids - prostaglandins (PGD2, PGE2, PGF2α, PGI2), prostacyclin and thromboxanes (i.e. TxA2) - and plays a key role in inflammation. It has been identified in several fish species, including rainbow trout and zebrafish, which express two differentially regulated COX-2 forms (COX-2a and -2b). For example changes in *cox-2* expression were observed in fish both in vivo upon bacterial and parasite infections and in vitro (e.g. in LPS-stimulated head kidney leukocytes). Moreover, the infection-induced up-regulation of *cox-2* expression could be regulated by activator protein-1 (AP-1) and nuclear transcription factor kappa-B (NF-κB) pathways (Wang et al. 2016).

Upon activation with pro-inflammatory cytokines and chemokines, circulating leukocytes emigrate from vessels and migrate through the extracellular matrix (ECM). A crucial step in this process is degradation of the vessel basement membranes and ECM macromolecules. In mammals both neutrophil elastase (NE) and gelatinase B (matrix metalloproteinase 9, MMP-9) participate in this process and efficiently hydrolyse components of basement membranes and ECM, e.g. denatured collagen (gelatine), fibronectin and elastin (Vandooren et al. 2013). MMP-9 belongs to the zinc-dependent endopeptidases synthesized in a latent zymogen form and converted to an active MMP-9 protein by cleavage after release. MMP-9 synthesis and activity are strictly regulated by its natural inhibitors, such as tissue inhibitor of metalloproteinase 1 (TIMP-1) or α2-macroglobulin. The main sources of MMP-9 are activated phagocytes. Next to leukocyte diapedesis and chemotaxis, MMP-9 is also involved in the regulation of cytokine and chemokine activity, for example it activates pro-IL-1β and increases the potency of CXCL8/IL-8 by truncation. Moreover, it regulates haematopoietic stem cell mobilization (Vandooren et al. 2013). Knowledge of the biological function of piscine MMP-9 is still limited. Gelatine-degrading enzymes similar to mammalian MMP-9 have been described in several fish species. We and others found that expression of MMPs changes in fish upon sterile inflammation, bacterial and viral infection as well as during liver hyperplasia and at the earliest stages of granuloma formation during tuberculosis (Pedersen et al. 2015). In turn, NE is a proteolytic enzyme belonging to the chymotrypsin-like family of the serine-proteinases. In mammals it is located in azurophilic granules within neutrophilic granulocytes while upon neutrophil activation, degranulation is triggered and NE is liberated into the extracellular space. Soluble NE participates in neutrophil diapedesis and bacterial killing upon neutrophil degranulation or as a component of neutrophil extracellular traps. The activated enzyme can be completely neutralized by proteinase inhibitors such as A1AT (Gramegna et al. 2017). Teleost fish do express homologues of NE; however, further examination must be performed to confirm the expression of these enzymes at protein level and to describe their functionality (Havixbeck and Barreda 2015).

Cytokines and chemokines can also be found in brain tissue, where they may influence neuronal growth and function or regulate the local immunity. For instance, CXCL12 is crucially required for cerebellar development besides mediating B cell lymphopoiesis and bone marrow myelopoiesis. As the CNS is evolutionarily older than the specialized adaptive immune system, we earlier concluded that the ancestral CXC chemokine system, mediating chemotaxis in the brain, was later recruited by the immune system. Also, genes encoding type I IFNs and ISGs are readily induced in the brain and have multiple effects on brain function and pathology. Cytokines may either be produced in peripheral cells and reach the brain by passage through the blood-brain barrier, or they may be locally produced, e.g. by glial cells that originate from macrophage stem cells. Being the resident macrophage cells, they form the main active immune defence in the central nervous system (CNS). They are constantly scavenging the CNS for plaques, damaged neurons and synapses and infectious agents. Microglia are very sensitive to small pathological invasions in the CNS and they can produce pro-inflammatory cytokines, like IL-1β, TNF-α, IL-2, IL-6, IL-12 and IFN-γ. Above mentioned features of behaviour immunity like fever, reduced appetite, locomotion or sexual activity may all be influenced by cytokine activity.

3.3 Systemic Inflammation vs. Neuroinflammation in Ectothermic Vertebrates

As mentioned above, many types of infection can initiate both systemic and local inflammation, including neuroinflammation. In the latter case, immune-related molecules in the brain are up-regulated and can affect permeability of the blood-brain barrier (BBB), which is critical to protection of the CNS from peripheral infections. In mammals, BBB is formed by a monolayer of endothelial cells with three specific components: (i) tight junctions ensure high electrical resistance and elimination of intercellular gaps; (ii) a very limited number of pinocytotic vesicles; and (iii) a limited number of fenestrae (Banks and Erickson 2010).

The BBB in teleost fish exhibits features similar to those of the BBB of higher vertebrates. In a zebrafish model, it was shown that from 3 days post-fertilization onward, brain endothelial cells of cerebral microvessels already show immunoreactivity to Claudin-5 and Zonula Occludens-1 (ZO-1), implying the presence of tight junctions in these cells (Jeong et al. 2008).

In the brain, up-regulation of immune molecules can be induced directly by brain infection, e.g. by neurotrophic RNA viruses such as Influenza virus or Zika virus, but it can also be induced by DAMPs and/or circulating pro-inflammatory molecules that can reach the brain tissue by different routes.

In humans, infectious aetiology has been proposed as a causative agent for the neuroinflammation associated with the progression of neurodegenerative diseases such as Alzheimer's disease (Dominy et al. 2019). This includes infections with viruses such as herpes simplex virus (HSV-1) and bacteria like the oral pathogen *Poprhyromonas gingivalis,* responsible for the development of chronic periodontal disease (Dominy et al. 2019). In rodents, standard models for neuroinflammation include LPS treatment, either by chronic local brain injection or by systemic treatment. In this model, neuroinflammation is induced by activation of microglia, which in turn produce many pro-inflammatory cytokines (Bardou et al. 2014).

Infections can lead to CNS colonization by pathogens that developed different mechanisms to reach the brain by (i) passive diffusion of viral particles through injured endothelia (ii) infection of endothelial cells, (iii) virus transcytosis by endosomal vesicles, and (iv) a Trojan horse tactic in phagocytes (Ayala-Nunez and Gaudin 2020). For example in humans, HIV virus uses monocytes/macrophages to infiltrate the brain parenchyma and infect other CNS cells, therefore monocytes are used as a biomarker in HIV-associated neurocognitive dementia (HAND). In addition, some of the HIV virus proteins induce MMP-9 production by astrocytes which can degrade junctional proteins of the BBB endothelium causing vascular leakage (Borrajo et al. 2021).

In response to harmful stimuli, brain resident cells can produce pro-inflammatory molecules which may increase blood-brain barrier permeability and affect further progress of the inflammation. In addition, age-related BBB breakdown might increase penetration to the brain of pathogens and inflammatory molecules, leading to cognitive decline. Importantly, mammals have a very limited capacity to regenerate the CNS. This is in contrast to teleost fish, where zebrafish neuroinflammation plays a crucial role in brain regeneration (Kyritsis et al. 2014).

Much less is known about the implications of fish infections for BBB dysfunction and neuroinflammation. Nevertheless, certain pathogens, such as *Pseudomonas aeruginosa,* disrupt the BBB in silver catfish, and *Streptococcus agalactiae* does so in zebrafish larvae (Baldissera et al. 2018). In addition, foci of tilapia lake virus (TiLV) infection were found along the blood vessels and the brain ventricle in the Nile Tilapia (*Oreochromis niloticus*), which suggests that the virus enters the brain via the circulatory system (Dinh-Hung et al. 2021).

With the zebrafish infection model, it was shown that some viruses such as Chikungunya virus (CHIKV) can effectively cross the BBB by infecting the brain microvasculature endothelial cells, while other viruses such as Sindbis virus (SINV) can enter the CNS through axonal transport in the peripheral nerves (Passoni et al. 2017).

Putative ways of pro-inflammatory molecules crossing the BBB have also been described. Cytokines can pass the BBB through a saturable influx transport (SIT) or a retrograde axonal transport system. They can also diffuse through capillaries with fenestrated endothelial cells in the circumventricular organs (Yarlagadda et al. 2009). Moreover, pro-inflammatory molecules may increase BBB permeability through activation and destruction of tight junctions of microvascular endothelial cells that form the BBB (Pan et al. 2006). Interestingly endothelial cells of BBB can secrete several cytokines, either spontaneously or after stimulation that can act at both peripheral tissues and within the CNS (Vadeboncoeur et al. 2003).

As soon as pro-inflammatory mediators get access to the brain, they predominantly activate glial cells such as astrocytes and microglia. In this context, the microglia cells are of special interest. These are the resident brain macrophages that are derived from primitive peripheral macrophages that migrated to the brain during early development. These cells are the main source of the pro-inflammatory IL-1β in the CNS and therefore they are involved in the neuroinflammatory processes that may lead to brain pathologies such as Alzheimer's or Parkinson diseases in humans (Cuoghi and Mola 2007).

3.3.1 Microglia Activation

Microglia cells were first described and named in 1919 by Pio del Rio-Hortega, whose landmark publication of 1932 accurately described important features of this cell type (Del Rio-Hortega 1932). Microglia cells are rapidly activated in response to inflammation, to infection and to injury of the CNS, where they provide innate immune protection. Even upon a minor infection, microglia change their gene expression profile and change their morphology from a ramified phenotype, resting but dynamically scanning the brain, to amoeboid motile cells, and they subsequently migrate to the affected site. In the healthy brain, these cells are responsible for many processes, ranging from phagocytosis of apoptotic neurons to control of behaviour through synaptic pruning. In this context, it is important to note that microglia may function as a double-edged sword: they play a crucial role both in neuroprotective and in neurodegenerative processes. This serves to maintain brain homeostasis, which is accomplished with different stimuli and through interactions with other glial cells or neurons. Production of pro-inflammatory cytokine-like molecules, similar in function to mammalian IL-1β, IL-6 and TNF-α, has been also found in invertebrates, demonstrating that this is an evolutionarily well-conserved mechanism (Schlegelmilch et al. 2011).

Although microglia have been intensively studied in vitro, a prerequisite to understanding the dynamic processes and interactions they undergo is a study in the wider context of the CNS of the living organism. Microglial functions have been studied mainly in mammals, but few non-mammalian models exist that enable us to approach fundamental questions about the biology of these cells. The leech *Hirudo medicinalis* was the first in vivo model in which microglial cells were studied after staining with silver carbonate. This simple technique revealed that they have phagocytic properties and a ramified phenotype similar to their vertebrate counterparts. The leech model was then exploited for the first time to visualize the microglial response to neuronal injury in vivo and this led to the discovery of the role of NO in guiding microglia towards the site of the injury (Sieger and Peri 2013).

Among vertebrates, teleost fish models provide a good tool to study in vitro and in vivo microglia at a single-cell level. A few fish species from teleostean taxa were used to study the role of microglia during optic nerve repair, including puffer fish (*Takifugu rubripes*), goldfish (*Carassius auratus*), different species of Oreochromis and *Tetraodon fluviatilis* (Cuoghi and Mola 2007). However, zebrafish (*Danio rerio*) undoubtedly became the most intensively used teleost model for neurodevelopmental and neurodegenerative studies. This small fish was introduced as a model organism by George Streisinger, the founding father of zebrafish research. The advantages of this model are its small size, rapid external development and embryo transparency (Streisinger et al. 1981), creating ample possibilities for genetic manipulations.

As in mammals and birds, in zebrafish primitive macrophages migrate from the yolk sac region to the brain during the early stages of development (Xu et al. 2016). In zebrafish this migration occurs as early as 2 days post-fertilization, possibly in response to increased levels of neuronal apoptosis that takes place in the developing embryonic brain between 2 and 3 dpf (Xu et al. 2016). These cells then rapidly mature into early microglia over the next 24 h, and in the healthy brain complete phagocytosis of dead neurons is possible by 6 dpf (Xu et al. 2016). During the process of maturation, larval microglia significantly up-regulate the expression of microglial core genes such as *apoeb, p2ry12, hexb* and *csf1ra* at 3dpf (Mazzolini et al. 2020). In addition, microglial maturation in zebrafish larvae was characterized by down-regulation of L-plastin and up-regulation of *apoE* and this coincided with typical morphological changes (Mazzolini et al. 2020) and the expression of typical microglial marker genes, such as *p2ry12* (Sieger and Peri 2013). In adult fish, embryonic microglial cells are replaced by definitive microglia that derive from aorta-gonad-mesonephros (AGM) haematopoietic stem cells. In adult zebrafish, two types of microglial cells exist that show different transcriptome profiles and a different response to bacterial infection. The predominant population of microglia (expressing the *ccl34b.1* marker) are ameboid cells, which are phagocytic and widely distributed in the brain, where they play a major role in brain clearance of invading pathogens, dying cells and cellular debris created during brain injury. Upon bacterial challenge, these cells up-regulate expression of pro-inflammatory cytokines, antimicrobial effectors and neutrophil-recruiting chemokines to resolve the bacterial infection. The second population (lacking *ccl34b.1* marker) is mostly located in the white matter and has a regulatory function. They have ramified protrusions but have limited mobility and phagocytosis capability (Wu et al. 2020). Upon bacterial challenge, these cells up-regulate genes that are important for the recruitment of T cells and peripheral macrophages, and they are crucial for the resolution of inflammation and tissue regeneration.

Despite intensive studies of peripheral populations of macrophages and their responses upon various viral, bacterial and fungal infections in zebrafish, the behaviour of microglial cells is poorly examined. During systemic infection of larvae with *Mycobacterium marinum* a reverse traffic of microglial cells from the brain was observed. When the early larvae were infected by injection in the hindbrain, the peripheral macrophages were mobilized to the infection site (Davis et al. 2002).

Recently, microglia were also isolated from Nile Tilapia and shown to form extracellular traps to combat bacterial infection of *Weissella cibaria*, which show tropism towards the CNS (Eto et al. 2021).

In mammals, activation of microglia after viral infection is well documented. In fish, however, little is known about the response of microglia to viral brain infection, although many viruses were found to show neurotropism, and thus potentially affect microglial function. Some viruses, such as spring viremia of carp virus (SVCV), preferentially infect macrophages and thereby significantly decrease their number at the initial stage of infection. This causes increased infiltration of neutrophils into the head region (Sullivan et al. 2021). Microglia also appeared crucial to fighting Nervous necrosis virus (NNV) infection in Asian Groupers (*Epinephelus* spp.). This virus is responsible for viral nervous necrosis disease (VNN) and attacks the CNS, thereby causing vacuolization of the brain and retina (Munday et al. 2002). Infection with this virus resulted in changes of fish behaviour such as lethargy, spiral swimming and loss of equilibrium. In addition, infected fish showed an anorectic morphology and pale skin (Munday et al. 2002). Chiang et al. disclosed the important role of microglia with a tissue culture of giant grouper (*Epinephelus lanceolatus*) brains that were infected with NVV. The authors hypothesized that NVV infection activates microglial cell proliferation and the secretion of pro-inflammatory cytokines that are subsequently responsible for neuronal cell death in infected brains (Chiang et al. 2017).

The zebrafish was also successfully used to study many fish and human viruses that are located in the central nervous system (CNS) and show neurotropism (Sullivan et al. 2021).

Recently, we found direct evidence of microglia activation upon systemic TiLV infection in zebrafish larvae and in adult fish (Fig. 3.1). TiLV shows prolonged neurotropism and persists in the brain of infected adult zebrafish for at least 90 days, even if the virus is not detectable in other peripheral organs such as spleen, kidney and liver (Mojzesz et al. 2021).

3.3.2 Methods/Markers to Study Microglia

Although the role of microglia in the inflammatory processes and during infections is well studied in mammals, including humans, little is known about the mechanisms of infection and defense of brain tissue by microglia in teleost fish. Here we describe the methods that can be successfully exploited to study the microglial response to viral infection and neuroinflammation in teleost fish, emphasizing the great advantages of the zebrafish model.

Microglia are often identified in brains of transgenic zebrafish by co-expression of the pan-leukocyte marker Lcp1/L-plastin and the pan-macrophage markers mpeg1 or mfap4. Though widely used for live imaging and immunostaining, these are not microglia-specific markers (Mazzolini et al. 2020; Wu et al. 2020). Apolipoprotein E (ApoE) more specifically labels microglia (Peri and Nüsslein-Volhard 2008) although in mammals, astrocytes, oligodendrocytes and neurons can also, to some extent, express apolipoprotein E. It is therefore recommended to use ApoE in combination with other macrophage or leukocyte markers. Despite this, transgenic zebrafish with fluorescent protein (e.g. GFP) expression under the control of the ApoE locus has been used to study microglia (Peri and Nüsslein-Volhard 2008). This model remains problematic, however, because not all brain macrophages

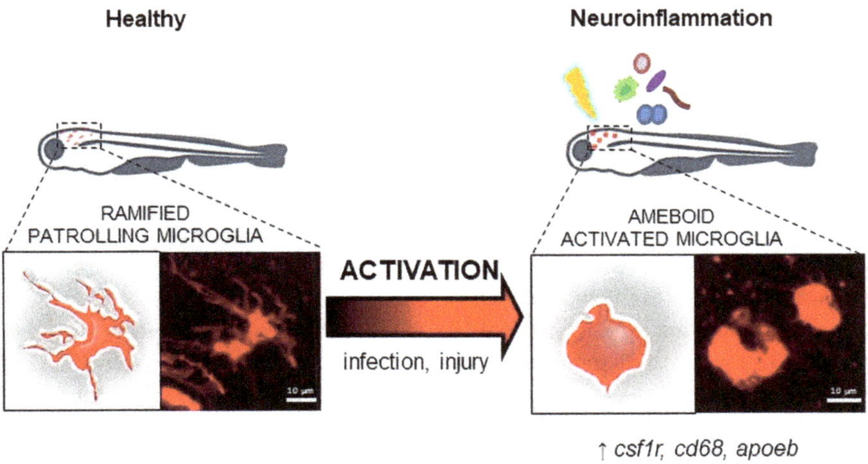

Fig. 3.1 Microglia activation during neuroinflammation. Stimulation of ramified, resting but patrolling microglia (left panel) in the brain by pathogens, injury or apoptotic cells and cellular debris leads to change in microglial morphology and activity. Activated microglia (right panel) retract branched filopodia/lamellipodia and become ameboid, highly motile and phagocytic. Active microglia also change their gene expression pattern. Prospective genes that are up-regulated in activated microglia in zebrafish larvae and/or adult fish during TiLV infection include *csf1r, cd68* and *apoeb*. Photographs were captured using confocal microscopy and presented as maximum intensity projections of Z-stack images of midbrain region of healthy (left panel) or viral-induced neuroinflammed zebrafish larvae (right panel) (Mojzesz et al. 2021). Zebrafish larvae scheme was obtained from scidraw.io (doi.org/10.5281/zenodo.3926121)

express ApoE, making it difficult to distinguish microglia and peripheral migrating macrophages.

The distinction between resident microglia and peripheral macrophages in zebrafish is an emerging area of study. Recently, the 4C4 antibody has been found to specifically recognize zebrafish microglia in the brain and CNS (Mazzolini et al. 2020).

In mice, both in vitro and in vivo studies showed that a purinergic receptor, P2Y12, is a typical microglial marker that is not expressed on peripheral macrophages (Sieger and Peri 2013). As its expression decreases during microglial activation, it is mainly useful for marking resting ramified microglia, but not the ameboid-activated ones. In zebrafish a transgenic line expressing p2y12:GFP has been generated and used to distinguish brain microglia from peripheral macrophages during brain injury (Herzog et al. 2019). This transgenic line allows the differentiation of brain microglia (P2Y12 positive) from peripheral macrophages (P2Y12 negative), but some populations of skin macrophages also express this marker.

Microglia can be isolated from larval or adult zebrafish brains using fluorescence-activated cell sorting (FACS). Brains of transgenic zebrafish reporter lines, e.g. mpeg1:mCherry with red fluorescent macrophages are homogenized and mpeg1+ fluorescent cells are sorted for further studies. Prior to sorting, homogenates can also be treated with microglia-specific antibodies such as 4C4 (Mazzolini et al.

2020). Transcriptomic analysis can then be performed on isolated microglial cells at different time-points. Activation of different pathways can thus be established (Mazzolini et al. 2020).

Ex vivo zebrafish microglia can be studied in larval stages as well as in adult fish using cryo-section and immunofluorescent antibody staining of the desired markers. Although the number of antibodies against zebrafish markers is rapidly growing, their commercial availability is still limited. Therefore, whole mount in situ hybridization (WISH) is also used with antisense RNA probes (Wu et al. 2020). Here the expression of host genes can be studied upon infection, but viral tropism can also be tracked (Sullivan et al. 2021).

For functional analysis of microglia, genes can be depleted in single (−/+) or double (−/−) zebrafish mutants using genetic engineering (e.g. TALEN or CRISPR/Cas9 methods). Patterns of gene expression can be studied by RNA sequencing (Mazzolini et al. 2020).

Another commonly used method to study gene function in microglia of zebrafish is targeted knockdown with morpholino antisense oligonucleotides, which transiently block gene expression. Morpholinos can also be used to deplete microglia/macrophages in zebrafish when directed against the pu.1 transcription factor, which is essential for myeloid cell fate (Peri and Nüsslein-Volhard 2008).

Microglial activation and proliferation are studied with the conditional neuronal ablation model in which metronidazole treatment causes ablation of transgenic neurons expressing NTR (Oosterhof et al. 2017). Viral neurotropism, interactions with or activation of microglia and viral replication can be studied in vivo in the live transparent larvae with viruses that express fluorescent proteins.

In adult zebrafish, microglia can be analysed directly using acute brain slice culture and time-lapse imaging. For this purpose, transgenic fish with microglia expressing fluorescent proteins are anesthetized on ice and the brains are dissected and microtome sectioned into 200 μm slices. Brain slices are then immediately collected and transferred into culture inserts and kept in culture medium in a thermal chamber at 28 °C. Tracking the movement, activation and phagocytosis of live microglia can be then performed in real-time using confocal microscopy (Wu et al. 2020).

3.4 Infection-Induced Changes in Behaviour of Ectothermic Vertebrates

During infection or injury, the host's defense mechanisms are activated, and this induces not only profound physiological changes (for example induction of fever), but also behavioural changes. These behavioural changes, so-called sickness behaviour, are triggered by pro-inflammatory cytokines, secreted in response to acute infections and/or tissue injury. They include lethargy, increased sleepiness, reduced food intake and mobility and reduced social, exploration and sexual behaviours. This pattern of behaviour evolved to conserve the energy necessary to fight infection and to facilitate the healing process, but also to minimize the risk of exposing the host to other pathogens or to predation (Dantzer 2006).

3.4.1 Fever

Fever is defined as a state of core temperature elevation above its normal range. It is an evolutionarily conserved defence mechanism that develops during many infectious and inflammatory diseases (Evans et al. 2015). In mammals, in the most classical way, fever is induced by pro-inflammatory cytokines such as IL-1β, TNF-α, and IL-6. These cytokines are also known as endogenous pyrogens, i.e. factors causing fever, and are secreted by leukocytes in response to a wide range of exogenous pyrogens such as different PAMPs (Evans et al. 2015; Netea et al. 2000). Among them, IL-6 appears to be the major mediator in sustaining fever. IL-6 knockout mice or mice injected with IL-6 neutralizing antibodies were unable to develop fever in response to stimulation with LPS, despite increased synthesis of IL-1β and TNF-α (Kozak et al. 1998). Interferons (IFN) are also included among endogenous pyrogens. While IFN-γ can induce fever by inducing the synthesis of IL-1β and TNF-α, which then activates the synthesis of IL-6, IFN-α has a direct influence on the development of fever (Netea et al. 2000). Intravenous or intraventricular administration of recombinant IFN-α causes fever in rabbits, rodents, cats and humans, independent of the activation of other endogenous pyrogens such as IL-1β (Dinarello 1999).

In humans, endogenous pyrogens, produced locally with increasing concentration, migrate via the bloodstream to the brain, where they penetrate the preoptic anterior hypothalamic area (POA) either by active transendothelial transport or by diffusion through the organum vasculosum laminae terminalis (OVLT). In the brain, these endogenous pyrogens induce the expression of cyclooxygenase 2 (COX-2), which is involved in the synthesis of prostaglandins E2 (PGE2), the final mediator of fever (Evans et al. 2015). Alternatively, vascular endothelial cells within the POA can also recognize PAMP structures in the blood via PRR receptors or react to circulating cytokines, leading to a direct activation of PGE2 synthesis (Evans et al. 2015). PGE2 may also be secreted in peripheral tissues, mainly by hepatic and pulmonary macrophages (Evans et al. 2015). PGE2 secreted outside the brain can bind to albumin, which protects it against enzymatic degradation and transport it through the bloodstream to the brain (Ivanov and Romanovsky 2004). A neural pathway is also suggested for the induction of fever. It requires PGE2 stimulation of peripheral sensory nerves, i.e. the vagal and trigeminal nerves. This neural stimulation results in release of norepinephrine (NE) in the POA, hence stimulating synthesis of PGE2. Regardless of its origin (peripheral or cerebral), PGE2 binds to PGE2 receptors (EP3) on heat-sensitive neurons located in the preoptic hypothalamus, precisely in the median preoptic nucleus (MnPO), raising the temperature set-point. This triggers the release of NE which elicits intrinsic thermogenesis in brown adipose tissue (non-shivering thermogenesis) and heat conservation by peripheral vasoconstriction. Moreover, the disinhibition of rMR neurons excites the somatomotor neurons which release acetylcholine (ACh) and activate shivering in skeletal muscles, resulting in an increase in body temperature (Evans et al. 2015; Rakus et al. 2017a). The role of PGE2 in inducing fever in mammals was described in 1971 by Milton and Wendlandt (1971), who showed that microinjection of PGE2 into the third ventricle of cats and rabbits causes an increase in body temperature. Subsequent studies using PGE2 synthesis inhibitors confirmed that PGE2 plays a key role in the development of fever in mammals (Ivanov and Romanovsky 2004).

In mammals, there are so-called endogenous antipyretics that reduce or inhibit fever but do not affect normal body temperature by themselves. These compounds include, for example anti-inflammatory cytokines, mainly IL-10. Studies carried out in a mouse model showed that mutant mice lacking the gene encoding IL-10 reacted with a higher fever than control mice. Furthermore, injection of recombinant IL-10 inhibited the development of fever in response to LPS (Evans et al. 2015).

Interestingly, fever develops not only in endothermic vertebrates such as mammals, but also in ectothermic vertebrates (fish, amphibians, reptiles) or even invertebrates (e.g. bees) (Rakus et al. 2017a). In most cases, ectothermic animals have a body temperature close to the environmental temperature. However, in response to infection or injection of exogenous pyrogens, ectotherms can increase their body temperature by behavioural regulation, which causes the animals to migrate to a warmer environment. This phenomenon is known as behavioural fever and is defined as an acute change in thermal preference driven by pathogen infection (Evans et al. 2015; Rakus et al. 2017a). While descriptions of fever in humans have been known since antiquity, the first report of behavioural fever in ectotherms was published in 1974 (Vaughn et al. 1974). Desert iguanas (*Dipsosaurus dorsalis*) injected with killed Gram-negative *Aeromonas hydrophila* bacteria tend to migrate to a warmer environment, which results in an increase in body temperature of approximately 2 °C (Vaughn et al. 1974). Subsequent studies in different species of ectothermic vertebrates showed that behavioural fever develops as a result of infection with various types of bacteria, viruses or fungi, as well as injection of inactivated bacteria or various PAMPs such as lipopolysaccharide (LPS) or poly (I:C) (a synthetic analogue of double-stranded RNA mimicking viral infections) (reviewed by Rakus et al. 2017a). Although most of the research on behavioural fever involves experiments conducted in laboratory conditions, this phenomenon has also been investigated under natural conditions. For example Richards-Zawacki (2010) showed that wild Panamanian golden frogs (*Atelopus zeteki*) can raise their body temperature by migrating to an environment with elevated temperature in response to natural infection with the pathogenic fungus *Batrachochytrium dendrobatidis*.

Heat-sensitive neurons have been described in the preoptic anterior hypothalamus of ectothermic vertebrates, and damage of this part of the brain completely inhibited the development of behavioural fever in toads. Additionally, an evolutionarily conserved role of PGE2 in the development of behavioural fever in ectothermic vertebrates was demonstrated both in studies aimed at induction of fever by injection of PGE2 into the brain and in studies aimed at suppression of fever (caused by an earlier bacterial infection or LPS injection) by administering COX inhibitors such as sodium salicylate or indomethacin (reviewed by Rakus et al. 2017a). Moreover, in zebrafish an increased level of plasma PGE2 was correlated with the induction of behavioural fever in poly(I:C)-stimulated animals (Boltana et al. 2013). However, little is known about the role of endogenous pyrogens in induction of behavioural fever in ectothermic vertebrates. Myhre et al. (1977) demonstrated the pyrogenic effect of blood plasma obtained from frogs *Rana esculenta* infected with the pathogenic bacteria *M. ranae*. Also, the lizard *Dipsosaurus dorsalis* developed behavioural fever after injection of supernatant collected from ex vivo incubated leukocytes, isolated from lizards pre-injected with dead *A. hydrophila*, while injection of denatured supernatants did not affect body temperature (Bernheim and Kluger 1977). A recent study showed a significant role of TNF-α in the development of behavioural fever during cyprinid herpesvirus-3 (CyHV-3) infection of common carp (*Cyprinus carpio* L.) (Rakus et al. 2017b). Administration of anti-TNF-α antibodies during CyHV-3 infection in carp caused a complete inhibition of behavioural fever, which led to the death of infected fish. These results suggest that behavioural fever in ectotherms and fever in endotherms are evolutionarily and functionally related through their common cytokine mediators that originated more than 400 million years ago (Rakus et al. 2017b). Interestingly, CyHV-3 contains a gene (ORF12) encoding a soluble TNF-α binding receptor, which has been shown to delay the development of behavioural fever. This allows the virus to replicate longer in the body of infected fish, since migration of infected carp into the water at higher temperatures (above 30 °C) completely blocks virus replication. These results demonstrate that CyHV-3 can alter the expression of behavioural fever in its host via the expression of a single gene, thus favouring its replication (Rakus et al. 2017b).

Elevation of body temperature reduces the growth rate and survival of the pathogens but also affects the activity of the host immune system. Boltana et al. (2013) demonstrated that zebrafish injected with poly(I:C) express behavioural fever which promotes the emergence of a highly specific anti-viral immune response by induction of the transcription of specific genes in the brain of zebrafish (Boltana et al. 2013). Moreover, it was suggested that behavioural fever plays a role in inhibition of the anti-inflammatory reflex and provides conditions for increased anti-viral responses (Boltana et al. 2013). In Atlantic salmon infected with infectious pancreatic necrosis virus (IPNV), behavioural fever resulted in higher proportions of CD4+ T cells in pronephros tissue and significantly higher expression of inflammation-related immune genes (Boltana et al. 2018).

3.4.2 Eating 'Disorders'

Another behavioural change exhibited by animals during sickness is disordered eating. The most common symptoms are reduced food and fluid intake, reduced frequency of meals and changes in food quality preferences. In mammals, this phenomenon is dependent on IL-1β, TNF-α, IFN-α and PGE2 (Andreasson et al. 2007). Injection of recombinant human IL-1β into the brains of rats reduced the amount of food consumed by the animals and the time spent eating, while the administration of an IL-1β receptor antagonist (IL-1Ra) blocked the IL-1β-induced suppression of food and water intake (Plata-Salamán 1994). Also, injection of recombinant murine TNF-α into the brains of rats resulted in reduction of the amount of food consumed, and this effect was dose-dependent. Moreover, IFNα-inhibited the activity of glucose-sensitive neurons in the nucleus of the lateral hypothalamic area (LHA). In rats, IFNα increased the activity of the ventromedial hypothalamus (VMH) (Dafny et al. 1996). In humans, IFNα therapy causes anorexia and this effect disappears about a week after the end of therapy. Loss of appetite induced by IFNα seems to be independent of the fever induced by this cytokine, although an increase in temperature after IFN administration has also been found to modulate the activity of glucose-sensitive VMH and LH neurons (Alam et al. 2013).

As in mammals, food intake in fish is considered to be under the control of a central feeding system (located in the brain) and several specific protein molecules including neuropeptides and gastrointestinal peptides, as well as hormones and blood metabolites, are involved in controlling this process (Volkoff 2016). Reduced food intake is a common observation during infection of the fish, and this phenomenon appears to be regulated by cytokines. In goldfish, both central and peripheral injections of LPS, which is a potent immunostimulant and induces the expression of IL-1β and TNF-α, caused a pronounced decrease in food intake. It is likely that pro-inflammatory cytokines mediated this effect (Volkoff and Peter 2004). We also observed a strong effect of TiLV infection on zebrafish appetite. TiLV-infected fish showed reduction in food intake starting from 5–6 dpi, which resulted in weight loss measured until 21 dpi. At the same time, a high up-regulation of type I IFNs and IL-1β was observed in the brains of infected fish (Mojzesz et al. 2021).

3.4.3 Locomotor, Social and Exploratory Behaviour

As mentioned before, mounting an immune response against pathogens is an energy-demanding process, and therefore some of the biological processes not related to immune function may be compromised during sickness. For example locomotor activities, social, exploratory and sexual behaviours, mating and reproductive success, growth and development are often reduced during pathogen and/or parasite challenge of the animals.

Very recently, the interaction between the immune system and social and exploratory behaviour has been described in zebrafish (*Danio rerio*). Adult zebrafish were discriminated according to their exploratory behaviour (high and low responders to novelty, HRN and LRN respectively) or according to their social behaviour (preference and no preference for conspecifics). Subsequently, gene expression of specific cytokines in the brain was studied to verify their immunological status. In the case of exploratory behaviour, the time spent in the presence of a novel object was significantly lower in fish classified as HRN. Interestingly, these fish showed higher expression of *il-1β* and reduced expression of *il-10* compared to fish from the LRN group. In the case of social behaviour, zebrafish exhibiting a "no preference" profile spent less time in the segment close to their conspecifics. In these fish, *ifn-γ* expression was significantly reduced as compared to fish from the "preference" group (Kristen et al. 2018a). Altogether these results indicate that the expression of specific cytokine genes in the brain of fishes correlates with their behaviour pattern, and it is most likely that the immune system modulates fish behaviour (Kristen et al. 2018a). This was later confirmed by studying the social and exploratory behaviour of fish inoculated with inactivated *Aeromonas hydrophila* bacteria, which induce an inflammatory response. In the *A. hydrophila*-injected fish, the locomotor activity, social preference and exploratory behaviour towards a new object was reduced compared to the control fish, while the expression of pro-inflammatory cytokines (*il-1β*, *il-6* and *tnf-α*) in the brain was up-regulated (Kristen et al. 2018b).

Very recently we studied changes in the behaviour of zebrafish during viral infection with TiLV. We observed that at day 11 post-infection, TiLV-infected adult zebrafish showed significantly lower velocity (cm/s) and distance travelled (cm) as compared to mock-infected fish. Moreover, TiLV-infected fish spent significantly more time (s) at the bottom zone of the tank, while mock-infected fish preferred to stay in the top zone of the tank (Mojzesz et al. 2021).

3.5 Perspectives

The last decades of genomic research have confirmed that fish express almost all neuropeptides, hormones and cytokines orthologous or homologous to their mammalian counterpart. Many functional studies demonstrated an evolutionarily conserved role for fish cytokines in the development of an immune response against various pathogens. Moreover, the neuroendocrine and immune systems interact in a bi-directional fashion. Cytokines induce changes in fish behaviour, including behavioural fever, they change food intake and decrease locomotor and social activity (Fig. 3.2). Zebrafish provide an elegant model for inflammation/infection-induced behavioural changes. They have behavioural characteristics similar to other vertebrates, and clear individual differences in behaviour. Several behavioural tests are standardized for scientific use in this species, and the use of transgenic zebrafish lines or mutants allows the study of the specific functions of the genes and molecules involved.

3.6 Key References

1. Cuoghi B, Mola L (2007) First comprehensive review of teleost microglia, which shows morphological and histochemical features of the three principal microglia types in the teleost central nervous system.
2. Dantzer R (2006) Description of sickness behaviour in mammals, the molecules involved in its regulation and the signaling pathways of peripherally released cytokines to the brain.

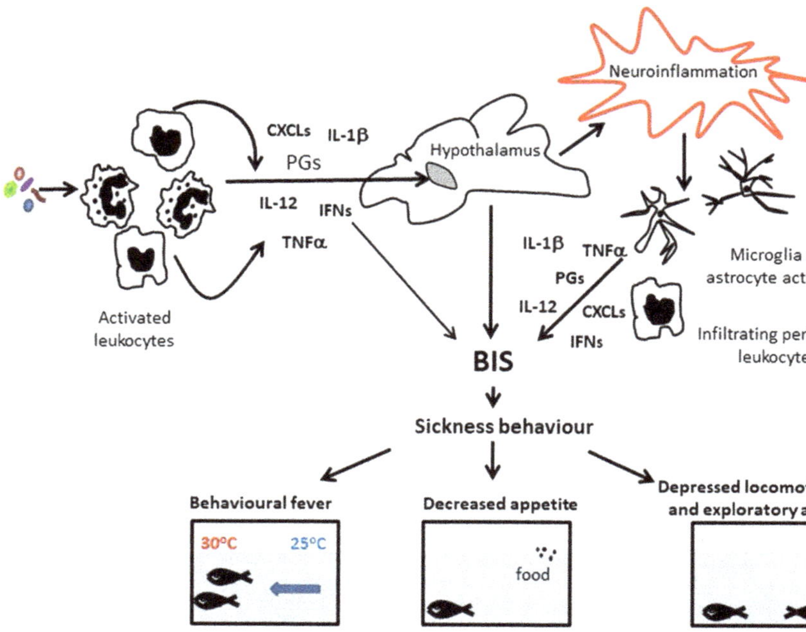

Fig. 3.2 Infection-induced sickness behaviour in fish. Pathogens induce leukocytes to migrate towards the focus of inflammation. They also induce leukocytes to produce pro-inflammatory mediators: interleukins (IL-1β, IL-6, IL-12), TNF-α, interferons (IFNs), chemokines (CXCLs) and prostaglandins (PGs). These mediators activate local inflammation and can signal to the brain and activate glia cells (microglia and astrocytes). Both pathogens and pro-inflammatory mediators can also induce neuroinflammation that implicates brain infiltration with peripheral leukocytes. Pro-inflammatory signals stimulate the behavioural immune system (BIS) and induce sickness behaviour: (i) behavioural fever, (ii) decrease of appetite, (iii) depression of locomotor activity and social and exploratory activities of fish

3. Evans SS, Repasky EA, Fisher DT (2015) Summary of the current knowledge of the mechanisms of induction of fever in mammals, the pyrogenic function of IL-6 and the involvement of the adrenergic-regulated thermogenesis in defining immune cell function.
4. Rakus K, Ronsmans M, Vanderplasschen A (2017a) Description of the evolutionarily conserved mechanisms common to fever in mammals and behavioural fever in ectothermic vertebrates.
5. Shakhar K (2019) Description of the function of sickness behaviour as the mechanism evolved to reduce the contact of an infected host and to improve animal fitness.
6. Zou J, Secombes CJ (2016) Comprehensive recent description of fish cytokines: their orthology, structure and function.

References

Alam I, Ullah N, Alam I, Ali I (2013) The effects and underlying mechanism of interferon therapy on body weight and body composition. Pak J Pharm Sci 26:1251–1257

Andreasson A, Arborelius L, Erlanson-Albertsson C, Lekander M (2007) A putative role for cytokines in the impaired appetite in depression. Brain Behav Immun 21:147–152

Ayala-Nunez NV, Gaudin R (2020) A viral journey to the brain: current considerations and future developments. PLoS Pathog 16:e1008434

Baldissera MD, Souza CF, Santos RCV, Baldisserotto B (2018) Blood-brain barrier breakdown and myeloperoxidase activity in silver catfish experimentally infected with *Pseudomonas aeruginosa*. J Fish Dis 41:209–213

Banks WA, Erickson MA (2010) The blood-brain barrier and immune function and dysfunction. Neurobiol Dis 37:26–32

Bardou I, Kaercher RM, Brothers HM, Hopp SC, Royer S, Wenk GL (2014) Age and duration of inflammatory environment differentially affect the neuroimmune response and catecholaminergic neurons in the midbrain and brainstem. Neurobiol Aging 5:1065–1073

Bernheim HA, Kluger MJ (1977) Endogenous pyrogen-like substance produced by reptiles. J Physiol 267:659e666

Bird S, Tafalla C (2015) Teleost chemokines and their receptors. Biology (Basel) 4:756–784

Boltana S, Rey S, Roher N, Vargas R, Huerta M, Huntingford FA, Goetz FW, Moore J, Garcia-Valtanen P, Estepa A, MacKenzie S (2013) Behavioural fever is a synergic signal amplifying the innate immune response. Proc R Soc B Biol Sci 280:20131381

Boltana S, Aguilar A, Sanhueza N, Donoso A, Mercado L, Imarai M, Mackenzie S (2018) Behavioral fever drives epigenetic modulation of the immune response in fish. Front Immunol 9:1241

Borrajo A, Spuch C, Penedo MA, Olivares JM, Agís-Balboa RC (2021) Important role of microglia in HIV-1 associated neurocognitive disorders and the molecular pathways implicated in its pathogenesis. Ann Med 53:43–69

Boudinot P, Langevin C, Secombes CJ, Levraud JP (2016) The peculiar characteristics of fish type I interferons. Viruses 8:298

Chiang YH, Wu YC, Chi SC (2017) Interleukin-1β secreted from betanodavirus-infected microglia caused the death of neurons in giant grouper brains. Dev Comp Immunol 70:19–26

Cuoghi B, Mola L (2007) Microglia of teleosts: facing a challenge in neurobiology. Eur J Histochem 51(4):231–240

Dafny N, Prieto-Gomez B, Dong WQ, Reyes-Vazquez C (1996) Interferon modulates neuronal activity recorded from the hypothalamus, thalamus, hippocampus, amygdala and the somatosensory cortex. Brain Res 734:269–274

Dantzer R (2006) Cytokine, sickness behavior, and depression. Neurol Clin 24:441–460

Davis JM, Clay H, Lewis JL, Ghori N, Herbomel P, Ramakrishnan L (2002) Real-time visualization of mycobacterium-macrophage interactions leading to initiation of granuloma formation in zebrafish embryos. Immunity 17(6):693–702

Del Rio-Hortega P (1932) Microglia. In: Penfield W (ed) Cytology and cellular pathology of the nervous system. Hoeber, New York, pp 482–534

Dinarello CA (1999) Cytokines as endogenous pyrogens. J Infect Dis 179:294–304

Dinh-Hung N, Sangpo P, Kruangkum T, Kayansamruaj P, Rung-Ruangkijkrai T, Senapin S, Rodkhum C, Dong HT (2021) Dissecting the localization of tilapia tilapinevirus in the brain of the experimentally infected Nile tilapia, *Oreochromis niloticus* (L.). J Fish Dis 44:1053. https://doi.org/10.1111/jfd.13367

Dominy SS, Lynch C, Ermini F, Benedyk M, Marczyk A, Konradi A, Nguyen M, Haditsch U, Raha D, Griffin C, Holsinger LJ, Arastu-Kapur S, Kaba S, Lee A, Ryder MI, Potempa B, Mydel P, Hellvard A, Adamowicz K, Hasturk H, Walker GD, Reynolds EC, Faull RLM, Curtis MA, Dragunow M, Potempa J (2019) Porphyromonas gingivalis in Alzheimer's disease brains:

evidence for disease causation and treatment with small-molecule inhibitors. Sci Adv 5: eaau3333

Eto SF, Fernandes DC, Funnicelli MIG, Alecrim JVC, Souza PG, Carvalho FCA, Belo MAA, Pizauro JM (2021) Microglia extracellular traps in *Oreochromis niloticus* infected with *Weissella cibaria*. Fish Shellfish Immunol 113:148–153

Evans SS, Repasky EA, Fisher DT (2015) Fever and the thermal regulation of immunity: the immune system feels the heat. Nat Rev Immunol 15:335–349

Gramegna A, Amati F, Terranova L, Sotgiu G, Tarsia P, Miglietta D, Calderazzo MA, Aliberti S, Blasi F (2017) Neutrophil elastase in bronchiectasis. Respir Res 18:211

Havixbeck JJ, Barreda DR (2015) Neutrophil development, migration, and function in teleost fish. Biology (Basel) 4:715–734

Herzog C, Pons Garcia L, Keatinge M, Greenald D, Moritz C, Peri F, Herrgen L (2019) Rapid clearance of cellular debris by microglia limits secondary neuronal cell death after brain injury *in vivo*. Development 146:dev174698

Huising MO, Stet RJ, Savelkoul HF, Verburg-van Kemenade BM (2004) The molecular evolution of the interleukin-1 family of cytokines; IL-18 in teleost fish. Dev Comp Immunol 28:395–413

Huising MO, Kruiswijk CP, Flik G (2006) Phylogeny and evolution of class-I helical cytokines. J Endocrinol 189:1–25

Ivanov AI, Romanovsky AA (2004) Prostaglandin E2 as a mediator of fever: synthesis and catabolism. Front Biosci 9:1977–1993

Jeong JY, Kwon HB, Ahn JC, Kang D, Kwon SH, Park JA, Kim KW (2008) Functional and developmental analysis of the blood-brain barrier in zebrafish. Brain Res Bull 75:619–628

Kozak W, Kluger MJ, Soszynski D, Conn CA, Rudolph K, Leon LR, Zheng H (1998) IL-6 and IL-1 beta in fever. Studies using cytokine-deficient (knockout) mice. Ann N Y Acad Sci 856:33–47

Kristen K, Fior D, Kreutz LC, Barcellos LJG (2018a) First description of behavior and immune system relationship in fish. Sci Rep 8:846

Kristen K, Soares SM, Koakoski G, Kreutz LC, Barcellos LJG (2018b) Characterization of sickness behavior in zebrafish. Brain Behav Immun 73:596–602

Kyritsis N, Kizil C, Brand M (2014) Neuroinflammation and central nervous system regeneration in vertebrates. Trends Cell Biol 24:128–135

Maciuszek M, Klak K, Rydz L, Verburg-van Kemenade BML, Chadzinska M (2020) Cortisol metabolism in carp macrophages: a role for macrophage-derived cortisol in M1/M2 polarization. Int J Mol Sci 21:8954

Mazzolini J, Le Clerc S, Morisse G, Coulonges C, Kuil LE, van Ham TJ, Zagury JF, Sieger D (2020) Gene expression profiling reveals a conserved microglia signature in larval zebrafish. Glia 68:298–315

Milton AS, Wendlandt S (1971) Effects on body temperature of prostaglandins of the A, E and F series on injection into the third ventricle of unanaesthetized cats and rabbits. J Physiol 218: 325–336

Mojzesz M, Rakus K, Chadzinska M, Nakagami K, Biswas G, Sakai M, Hikima JI (2020) Cytosolic sensors for pathogenic viral and bacterial nucleic acids in fish. Int J Mol Sci 21:7289

Mojzesz M, Widziolek M, Adamek M, Orzechowska U, Podlasz P, Prajsnar TK, Pooranachandran N, Pecio A, Michalik A, Surachetpong W, Chadzinska M, Rakus K (2021) Tilapia lake virus-induced neuroinflammation in zebrafish: microglia activation and sickness behavior. Front Immunol 12:760882

Morimoto N, Kono T, Sakai M, Hikima JI (2021) Inflammasomes in Teleosts: structures and mechanisms that induce pyroptosis during bacterial infection. Int J Mol Sci 22:4389

Munday BL, Kwang J, Moody N (2002) Betanodavirus infections of teleost fish: a review. J Fish Dis 25:127–142

Myhre K, Cabanac M, Myhre G (1977) Fever and behavioural temperature regulation in the frog Rana esculenta. Acta Physiol Scand 101:219–229

Netea MG, Kullberg BJ, van der Meer JWM (2000) Circulating cytokines as mediators of fever. Clin Infect Dis 31:178–184

Oosterhof N, Holtman IR, Kuil LE et al (2017) Identification of a conserved and acute neurodegeneration-specific microglial transcriptome in the zebrafish. Glia 65:138–149

Palha N, Guivel-Benhassine F, Briolat V, Lutfalla G, Sourisseau M, Ellett F, Wang CH, Lieschke GJ, Herbomel P, Schwartz O, Levraud JP (2013) Real-time whole-body visualization of Chikungunya virus infection and host interferon response in zebrafish. PLoS Pathog 9: e1003619

Pan W, Cain C, Yu Y, Kastin AJ (2006) Receptor-mediated transport of LIF across blood-spinal cord barrier is upregulated after spinal cord injury. J Neuroimmunol 174:119–125

Passoni G, Langevin C, Palha N, Mounce BC, Briolat V, Affaticati P, De Job E, Joly JS, Vignuzzi M, Saleh MC, Herbomel P, Boudinot P, Levraud JP (2017) Imaging of viral neuroinvasion in the zebrafish reveals that Sindbis and chikungunya viruses favour different entry routes. Dis Model Mech 10:847–857

Pedersen ME, Vuong TT, Rønning SB, Kolset SO (2015) Matrix metalloproteinases in fish biology and matrix turnover. Matrix Biol 44–46:86–93

Peri F, Nüsslein-Volhard C (2008) Live imaging of neuronal degradation by microglia reveals a role for v0-ATPase a1 in phagosomal fusion *in vivo*. Cell 133:916–927

Piazzon MC, Lutfalla G, Forlenza M (2016) IL10, a tale of an evolutionarily conserved cytokine across vertebrates. Crit Rev Immunol 36:99–129

Plata-Salamán CR (1994) Meal patterns in response to the intracerebroventricular administration of interleukin-1 beta in rats. Physiol Behav 55:727–733

Rakus K, Ronsmans M, Vanderplasschen A (2017a) Behavioral fever in ectothermic vertebrates. Dev Comp Immunol 66:84–91

Rakus K, Ronsmans M, Forlenza M, Boutier M, Piazzon MC, Jazowiecka-Rakus J, Gatherer D, Athanasiadis A, Farnir F, Davison AJ, Boudinot P, Michiels T, Wiegertjes GF, Vanderplasschen A (2017b) Conserved fever pathways across vertebrates: a herpesvirus expressed decoy TNF-α receptor delays behavioral fever in fish. Cell Host Microbe 21:244–253

Rakus K, Mojzesz M, Widziolek M, Pooranachandran N, Teitge F, Surachetpong W, Chadzinska M, Steinhagen D, Adamek M (2020) Antiviral response of adult zebrafish (Danio rerio) during tilapia lake virus (TiLV) infection. Fish Shellfish Immunol 101:1–8

Redmond AK, Zou J, Secombes CJ, Macqueen DJ, Dooley H (2019) Discovery of all three types in cartilaginous fishes enables phylogenetic resolution of the origins and evolution of interferons. Front Immunol 10:1558j

Richards-Zawacki CL (2010) Thermoregulatory behaviour affects prevalence of chytrid fungal infection in a wild population of Panamanian golden frogs. Proc R Soc Lond B Biol Sci 277: 519–528

Schlegelmilch T, Henke K, Peri F (2011) Microglia in the developing brain: from immunity to behaviour. Curr Opin Neurobiol 21:5–10

Shakhar K (2019) The inclusive behavioral immune system. Front Psychol 10:1004

Sieger D, Peri F (2013) Animal models for studying microglia: the first, the popular, and the new. Glia 61:3–9

Streisinger G, Walker C, Dower N, Knauber D, Singer F (1981) Production of clones of homozygous diploid zebra fish (*Brachydanio rerio*). Nature 291:293–296

Sullivan C, Soos BL, Millard PJ, Kim CH, King BL (2021) Modeling virus-induced inflammation in zebrafish: a balance between infection control and excessive inflammation. Front Immunol 12:636623

Vadeboncoeur N, Segura M, Al-Numani D, Vanier G, Gottschalk M (2003) Pro-inflammatory cytokine and chemokine release by human brain microvascular endothelial cells stimulated by Streptococcus suis serotype 2. FEMS Immunol Med Microbiol 35:49–58

Vandooren J, Van den Steen PE, Opdenakker G (2013) Biochemistry and molecular biology of gelatinase B or matrix metalloproteinase-9 (MMP-9): the next decade. Crit Rev Biochem Mol Biol 48:222–272

Vaughn LK, Bernheim HA, Kluger MJ (1974) Fever in the lizard *Dipsosaurus dorsalis*. Nature 252:473–474

Verburg-van Kemenade BM, Weyts FA, Debets R, Flik G (1995) Carp macrophages and neutrophilic granulocytes secrete an interleukin-1-like factor. Dev Comp Immunol 19:59–70
Verburg-van Kemenade BML, Cohen N, Chadzinska M (2017) Neuroendocrine-immune interaction: evolutionarily conserved mechanisms that maintain allostasis in an ever-changing environment. Dev Comp Immunol 66:2–23
Volkoff H (2016) The neuroendocrine regulation of food intake in fish: a review of current knowledge. Front Neurosci 10:540
Volkoff H, Peter RE (2004) Effects of lipopolysaccharide treatment on feeding of goldfish: role of appetite-regulating peptides. Brain Res 998:139–147
Wang T, Mai K, Ai Q (2016) A review of cyclooxygenase-2 role in fish. Austin J Nutr Metab 3: 1037
Wu S, Nguyen LTM, Pan H, Hassan S, Dai Y, Xu J, Wen Z (2020) Two phenotypically and functionally distinct microglial populations in adult zebrafish. Sci Adv 6:eabd1160
Xu J, Wang T, Wu Y, Jin W, Wen Z (2016) Microglia colonization of developing zebrafish midbrain is promoted by apoptotic neuron and lysophosphatidylcholine. Dev Cell 38:214–222
Yarlagadda A, Alfson E, Clayton AH (2009) The blood brain barrier and the role of cytokines in neuropsychiatry. Psychiatry (Edgmont) 6:18–22
Zou J, Secombes CJ (2016) The function of fish cytokines. Biology (Basel) 5:23

Part II

Regulatory Pathways

Cytokines and Hypothalamo-Pituitary-Adrenal Axis Activation: Now a Classic of Immune-Neuroendocrine System Interactions

4

Jan Pieter Konsman

Abstract

Precise descriptions of neuroendocrine and immune systems, along with a more integrative orientation of research in the life sciences, created the conditions in the last quarter of the twentieth century to envision interactions between these systems. In particular, it was shown that pro-inflammatory cytokines such as interleukin-1, produced in response to host detection of bacterial fragments, can activate the hypothalamo-pituitary-adrenal axis, resulting in the release of corticosteroids, which, in turn, have anti-inflammatory effects. Given the existence of the blood-brain barrier that prevents hydrophilic molecules such as cytokines from passively entering the brain parenchyma, research efforts have focused on elucidating so-called cytokine-to-brain signaling pathways. The findings obtained indicate that several mechanisms, including prostaglandin-induced neuronal activation, vagal afferents and ascending catecholaminergic brainstem projections, can be involved in hypothalamo-pituitary-adrenal axis activation after systemic interleukin-1 administration, depending on both experimental and physiological conditions. Most recently, both homeostatic-physiological stressors and emotional-psychological stressors have been proposed to give rise to inflammatory responses, suggesting that crosstalk between innate immune mediators and hypothalamo-pituitary-adrenal axis hormones will continue to be at the forefront of research.

Keywords

Cytokines · Hypothalamo-pituitary axis · Stress

J. P. Konsman (✉)
IMMUNOlogy from CONcepts and ExPeriments to Translation, CNRS UMR 5164, University of Bordeaux, Bordeaux, France
e-mail: jan-pieter.konsman@u-bordeaux.fr

J. P. Konsman, T. M. Reyes (eds.), *Neuroendocrine-Immune System Interactions*,
Masterclass in Neuroendocrinology 13,
https://doi.org/10.1007/978-3-031-21358-8_4

4.1 Introduction: Neuroendocrine and Immune Systems

4.1.1 Post-WWII Conceptual Developments Leading to Neuroendocrine and Immune Systems

Although the notions of neuroendocrine and immune systems seem natural to present day scientists and physicians, it took, in fact, considerable conceptual and experimental work to first describe these systems. As pointed out in Chap. 1, an influential vision of neuroendocrine systems was articulated in Geoffrey Harris' 1955 book *Neural control of the pituitary gland,* in which he suggested that neuroendocrine neurons could be considered motor or effector neurons (Harris 1955). He also specified functional criteria for determining whether an endogenous substance constitutes a releasing factor at the level of the anterior pituitary. These proposals, along with experimental findings, led to the idea of functional axes formed by the hypothalamus, the pituitary (anterior and posterior), and endocrine glands, for example, the adrenal and thyroid. Thus, one can encounter mention of the hypothalamo-pituitary-adrenal (HPA) axis, the hypothalamo-pituitary-gonadal (HPG) axis, and the hypothalamo-pituitary-thyroid (HPT) axis from the late 1960s onwards. In the early 1970s, John Porter proposed that "[a] neuroendocrine system consists of a neural cell or cells which secrete into the extracellular fluid a substance which upon reaching other cells modify their behavior" (Porter 1973, pp. 2–3).

A conceptual view of the immune system was also developed in the 1970s by Niels Jerne. He proposed that "the immune system, even in the absence of antigens ... achieves a dynamic steady state as its elements interact between themselves" (Jerne 1974, p. 383), suggesting possible regulation of immune responses. While much of twentieth-century research into immunity has focused on antibodies when considering extracellular molecules, another category started to emerge during its last quarter in the form of cytokines. Cytokines can be defined as "cell-surface associated or secreted proteins that interact with specific cell-surface receptors resulting in the mobilization and or modulation of target cells" (Oppenheim 2018, p. 1). From the point of view of immunology, cytokines correspond to a broadening of the category of interleukins that were viewed as messengers between leukocytes. The very idea of interleukins, and later, cytokines, as intercellular messengers progressively altered the antigen-antibody stimulus-responses vision of the immune system.

As a category of intercellular signaling molecules, they are distinguished from hormones, in that cytokines often act locally and at much lower concentrations. In addition, cytokines typically have many different biological properties (pleiotropism), and, in part, as a consequence, different cytokines may induce similar biological effects (redundancy) (Dinarello 2007; Oppenheim 2018). The history of interferons illustrates the changing framework of immunology well, in that interferon was initially considered an antiviral molecule, but over time, with other effects and related molecules being discovered, became a prototypical cytokine (Vilcek 2006; Billiau and Matthys 2009). Another illustration of categories in immunology changing over time is that of different names referring to functional effects, for example, "lymphocyte-activating factor" and "endogenous pyrogen," which most likely corresponded to one and the same molecule, coined interleukin-1 (IL-1) (Oppenheim and Gery 1993).

4.1.2 Post-WWII Tools to Study Neuroendocrine and Immune System Components

In the early 1950s, conjugation of the fluorescent marker fluorescein to antibody made it possible to localize antigens and antibodies in tissues (Coons and Kaplan 1950; Coons et al. 1950, 1951, 1955). Antibody conjugation approaches were subsequently expanded to enzymes, to obtain more permanently stained tissues and to circumvent the problems of fluorophore fading and tissue autofluorescence (Nakane and Pierce 1967). Not surprisingly, given the peptidergic nature of many hormones, many attempts were undertaken to develop antibody-based detection techniques for these mediators in bodily fluids and tissue extracts (Yalow and Berson 1960; Utiger et al. 1962; Felber 1963; Spitzer 1968). Thus, radioimmunoassays were employed to study the effects of potential hypothalamic-releasing factors on pituitary contents of growth hormone or adrenocorticotropin hormone (Rodger et al. 1969; Brazeau et al. 1973; Rivier et al. 1973). A variant of radioimmunoassays in which the radioactive isotope was replaced by an enzyme gave rise to enzyme-linked immunosorbent assays (ELISAs) (Aydin 2015). Thus, these antibody-based techniques made it possible to refine both biochemical and anatomical approaches. Moreover, In the 1970s, the use of different fluorescent labels also allowed sorting and concentrating of different cell populations (Bonner et al. 1972; Julius et al. 1972). All of the approaches were further improved after monoclonal antibodies with predefined antigen specificity became available (Koehler and Milstein 1975).

Other technical developments, of which the impact can hardly be overstated and which have given rise to different technologies, are those of the cloning of genes and the expression of recombinant proteins. Indeed, very few proteins are naturally produced in sufficient quantities and purity to allow their isolation, analysis, or calibration (Hartley 2006). To obtain such quantities, cloning of the gene encoding the protein of interest and its expression by cell types different from those naturally expressing the protein is often necessary (Hartley 2006). It turned out that cytokines, including interferons and IL-1, were amongst the first proteins to be thus produced (Oppenheim and Gery 1993; Dinarello 2007). This, in turn, made it easier to generate antibodies and calibration standards for immunoassays. In addition, the cloning of genes encoding for hormones or cytokines allowed for the detection of their expression in tissues by in situ hybridization. Naturally, the new techniques were not only employed for better observation and description of immune and neuroendocrine systems and their components, but also favored the emergence of new tools of intervention. Thus, the classic approaches of lesions of organs or glands or administration of their extracts were enriched by strategies using neutralizing antibodies, recombinant forms of naturally occurring functional receptor antagonists, such as alpha-helical Corticotropin-Releasing Hormone and IL-1 receptor antagonist, and animals that are deficient for certain genes (for example, knock-out mice).

4.2 Regulation of Immune and Neuroendocrine Systems: Cytokine-HPA-Axis Interactions

4.2.1 Immunophysiology

Although Niels Jerne remarked that the immune and nervous systems share important features such as threat detection and memory and "penetrate most other tissues of our body" (Jerne 1974, p. 387), he also indicated that "they seem to be kept separate from each other by the so-called blood-brain barrier" (Jerne 1974, p. 387) and that lymphocytes stimulated with antigen in vitro "will produce specific antibody molecules, in the absence of any nerve cells" (Jerne 1985, p. 852). Other immunologists, while embracing the idea of features common to the immune and nervous systems, proposed to move from an antigen-centered to a more organism-centered view of the immune system (Coutinho et al. 1984). These authors, in fact, agreed with Jerne's idea that the immune system is active in the absence of antigenic stimulation, but proposed a more physiological approach with a more important place for in vivo studies.

The term immunophysiology had started to become used more in the early 1980s to describe immune functions in a particular organ, such as the gut, or of relatively ignored cell types like Natural Killer cells (Dobbins 1982; Oldham 1983). But a broader vision of immunophysiology also emerged, which aimed to "bring the self-regulated immune system into conformity with other body systems ... based on the existence of afferent-efferent pathways between immune and neuroendocrine structures" (Besedovsky and Sorkin 1977, p. 1). Thus, Besedovsky affirmed that neurotransmitters and neuropeptides released by nerve endings and neuroendocrine hormone production and action can constitute "external immunoregulatory signals [super]imposed upon autoregulatory mechanisms" (Besedovsky et al. 1985, p. 750s). Moreover, lymphoid cells were not only shown to express receptors for glucocorticoids and catecholamines, but also synthesize "[c]lassic pituitary hormones, [such as] growth hormone and prolactin [that] appear to have distinct roles as immunomodulators" (Kelley 1988, p. 2095).

4.2.2 Stress and Immunity

In parallel with more physiological views of the immune system, a perspective emerged that acknowledged the influence of stress on immunity and proposed that this was mediated by neuroendocrine systems. Hans Selye had already shown before WWII that "[e]xposure to general (systemic) stress initiates a chain of physiologic reactions [resulting in thymic involution followed by adrenal enlargement], which are essentially similar, irrespective of the particular stress agent employed" (Selye

1948, p. 186). He next related that "the predominantly emotional stress caused by immobilization" also produced several of these responses (Selye and Fortier 1950, pp. 153–155). In addition this "neurogenic stress-situation" was found to result in "inhibition of inflammation" (Selye 1955, p. 124). Given that the anti-inflammatory effects of glucocorticoids have been widely known since the 1950s (Gordon and Katsh 1949; Glyn 1998), these findings gave rise to the idea that both physiological and psychological stressors could modulate inflammation through the release of corticosteroid hormones from the adrenals. Further work in the 1960s indicated that a shuttle box stressor to induce electric shock avoidance learning in rodents was also accompanied by hypertrophy of the adrenals and hypotrophy of the spleen and thymus as well as altered susceptibility to various viral infections (Rasmussen 1969). These animal findings, along with some clinical work, then gave rise to the more general hypothesis that "[s]tress and emotional distress may influence the function of the immunologic system via central nervous system and possibly endocrine mediation" (Solomon 1969, p. 335).

The stress response as a physiological concept was progressively linked to the HPA-axis by Hans Selye between the 1950s and 1970s (Selye 1950, 1976). During that time it had indeed become clear that acute and chronic stress were associated with increased corticosteroid production in animals and humans (Hale et al. 1957; Mason et al. 1961; Treiman et al. 1970; Weiss 1970; Arguelles et al. 1972; Bassett et al. 1973; Tache et al. 1976). Some authors therefore proposed that "psychosocial processes influence the susceptibility to some infections, to some neoplastic processes, and to some aspects of humoral and cell-mediated immune responses" (Stein et al. 1976, p. 439) and that "[t]hese psycho-social effects may be related to hypothalamic activity, the autonomic nervous system, and neuro-endocrine activity" (Miller 1977, p. 413). After some debate about how states of mind could give rise to changes in immune responses, the term psychoneuroimmunology was put forward in the early 1980s "to refer to studies of neuroendocrine mechanisms mediating the effects of behavior on immune function—and vice versa" (Ader and Cohen 1985, p. 103). It is nevertheless interesting to observe how from Selye's initial observations on the effects of systemic stressors on immune organs (even though at that time the thymus was not yet considered to be an immune organ) a progressive shift took place towards a perspective in which neurogenic or psychological stressors affect immune responses through activation of the HPA-axis.

4.2.3 Communication Between Immune and Neuroendocrine Systems

During the twentieth century, the idea of multicellular organisms displaying a division of labor in "a society of cells" that was orchestrated by intercellular communication (Reynolds 2017) was progressively rendered more concrete with the notions of hormones, neurotransmitters and later cytokines as soluble secreted mediators. In parallel, and regarding organ functions, the metaphoric image of the brain as "governor" in which it is considered "as the executive control center of the body" also gained traction (Fuller 2014, p. 100). This image was, in part, reinforced by the very idea of neuroendocrine systems, which attributes an important role to the hypothalamus in controlling pituitary-endocrine gland axes. From the 1970s onwards a perspective emerged in which peripheral immune responses activate hypothalamic neurons and "bring about changes in hormone levels," for example, increased corticosteroid concentrations, which, in turn, may "suppress[] a potentially harmful expansion of lymphoid tissue" and "prevent accumulation of macrophages in the delayed hypersensitivity reaction" (Besedovsky and Sorkin 1977, p. 4, 7, 9). In such a perspective, the hypothalamus would receive afferent information of ongoing peripheral immune responses and activate efferent pathways in the form of the HPA-axis resulting in corticosteroid-mediated inhibition of immune responses as some form of feedback loop.

4.3 How Cytokines and the HPA-Axis Interact

4.3.1 Cytokines Can Interact with the HPA-Axis at Different Levels

In 1977, Besedovsky argued that "[i]t is a rational assumption that the primary link between the immune and the neuroendocrine system is effected by one or another of the multiple events known to follow immunization" and proposed that "chemical mediators" produced by immune cells after stimulation "may influence the endocrine target glands either directly or more likely via hypothalamus-hypophysis" (Besedovsky and Sorkin 1977, p. 6). Interleukins or cytokines, as soluble polypeptides released by activated immune cells, were obvious candidates to constitute such mediators. Interestingly, it was shown in the 1980s that IL-1 can stimulate pituitary mRNA expression and secretion of adrenocorticotropin hormone (ACTH) as well as hypothalamic production of corticotropin-releasing hormone (CRH) (Berkenbosch et al. 1987; Bernton et al. 1987; Brown et al. 1987; Sapolsky et al. 1987). If other cytokines, such as IL-6 or tumor necrosis factor-alpha, produced after exposure of host cells to bacterial fragments, also increase corticosteroid release, IL-1beta seems to be the most potent (Besedovsky et al. 1991; Dunn 1992; Matta et al. 1992). Although it has been shown that IL-1 can induce corticosteroid secretion by acting directly on the adrenal (Andreis et al. 1991), this cannot explain why increases in plasma corticosterone after peripheral IL-1 administration require an intact pituitary and the action of CRH (Berkenbosch et al. 1987; Gwosdow et al. 1990; Van Der Meer et al. 1996). These findings indicate that although cytokines may act at different levels of the HPA-axis, activation at the level of the hypothalamus seems necessary for increasing the concentration of circulating corticosteroid after systemic IL-1 administration.

4.3.2 Mechanisms Mediating IL-1-Induced HPA-Axis Activation

But if one supposes that activation of the HPA-axis by peripheral IL-1 involves CRH neurons in the hypothalamus, then one would need to explain how such a response would circumvent the blood-brain barrier (BBB) that prevents water-soluble polypeptides like cytokines from passively leaving the blood and acting in the brain parenchyma. Given that in rodents 1) a functional blood-brain barrier is absent in the hypothalamic median eminence (Gross 1992), 2) IL-1 decreases the median eminence's content of CRH (Berkenbosch et al. 1987; Watanobe et al. 1991), 3) IL-1 type 1 receptor mRNA expression in the median eminence (Cunningham et al. 1992; Yabuuchi et al. 1994) and 4) intra-median eminence administration of IL-1ra attenuates IL-1-induced adrenocorticotropin hormone secretion (Matta et al. 1993), the action of circulating IL-1 on CRH-containing terminals in the median eminence would be one obvious mechanism by which IL-1-induced HPA-axis activation can be explained.

Another hypothesis emerged after it was shown that microinjection of the toxin 6-hydroxydopamine into the catecholaminergic fiber bundle connecting the brainstem to hypothalamus reduces IL-1-induced increases in hypothalamic CRH mRNA and plasma corticosterone levels (Chuluyan et al. 1992; Parsadaniantz et al. 1995). This, along with the repeated observation that peripheral IL-1 administration stimulates hypothalamic noradrenalin turnover, both in the paraventricular hypothalamus containing CRH cell bodies and in the median eminence in which CRH neurons terminate (Dunn 1988; Kabiersch et al. 1988; Mohankumar and Mohankumar 2005), indicated that brainstem catecholaminergic projections to the hypothalamus play an important role in IL-1-induced activation of the HPA-axis. In addition, prostaglandin synthesis was shown to mediate IL-1-induced activation of the HPA-axis, as administration of a cyclooxygenase inhibitor can attenuate increased hypothalamic noradrenalin turnover and plasma corticosterone after peripheral injection of this cytokine (Dunn and Chuluyan 1992). This pharmacological finding was soon followed by the demonstration of cyclooxygenase-2 induction associated with brain endothelial cells in response to peripheral IL-1 injection (Cao et al. 1996; Lacroix and Rivest 1998). Not surprisingly, these different findings gave rise to intense research aimed at testing various hypotheses of how peripheral IL-1 can activate hypothalamic CRH neurons, resulting in HPA-axis activation.

In these endeavors, some groups chose to use IL-1 as a stimulus, while others preferred to administer bacterial lipopolysaccharides (LPS) fragments as an inducer of IL-1 and other pro-inflammatory cytokines in mononuclear leukocytes. In doing so, the former used a robust, well-characterized, and progressively standardized single cytokine, while the latter allowed the possible production and interaction of different pro-inflammatory cytokines and preserved a potential role for monocytes and macrophages in immune-to-brain signaling. With respect to macrophages, it was shown that peripheral LPS administration induces IL-1 beta production, not only in liver Kupffer cells, but also in brain circumventricular organs such as the median eminence, lacking a functional blood-brain barrier (Chensue et al. 1991; Van Dam et al. 1992). Given that vagal nerve fibers in the liver can bind and react to IL-1

(Niijima 1996; Goehler et al. 1997), these findings raise the possibility that paracrine actions of IL-1beta, either on peripheral nerves or in the median eminence, could also play a role in activation of the HPA-axis.

The results of several intervention strategies, mostly lesion- and pharmacology-based, indicate that prostaglandin-dependent activation of catecholaminergic brainstem projections to hypothalamic CRH neurons underlies peripheral IL-1-induced HPA-axis activation (Ericsson et al. 1997). Interestingly, vagal afferent fibers in the brainstem terminate close to catecholaminergic neurons (Sumal et al. 1983) and subdiaphragmatic vagotomy attenuates adrenocorticotropin hormone responses after intraperitoneal, but not necessarily intravenous, administration of IL-1beta (Kapcala et al. 1996; Ericsson et al. 1997; Wieczorek and Dunn 2006). Regarding the role of mononuclear leukocytes, it was shown early on that elimination of macrophages by administration of dichloromethylene-diphosphonate-filled liposomes prevents increases of adrenocorticotropin hormone and corticosterone in response to low doses of bacterial LPS (Derijk et al. 1991). However, and more surprisingly, this same intervention increases hormonal responses after chronic peripheral IL-1beta infusion, suggesting that endogenous macrophage-derived mediators exert an inhibitory effect on IL-1-induced HPA-axis activation (Van Der Meer et al. 1996). The brain contains bona fide macrophages, in addition to immunocompetent microglia, in areas without a functional blood-brain barrier, such as the meninges and circumventricular organs as well as in perivascular spaces (Williams et al. 2001; Prinz et al. 2021). Interestingly, eliminating brain macrophages by intracerebroventricular administration of these dichloromethylene-diphosphonate-filled liposomes attenuates acute adrenocorticotropin hormone and corticosterone responses after peripheral IL-1beta injection, but increases them in response to bacterial LPS (Serrats et al. 2010). Taken together, these findings indicate that different mechanisms can underlie and modulate peripheral IL-1-induced activation of the HPA-axis and that the actual involvement of such mechanisms may depend on the conditions, both experimental and physiological.

4.4 Outstanding Questions and Perspectives

Most recently, intervention strategies have often involved the use of genetically modified organisms, for example, in experiments with knock-out mice or optogenetics. Many of the conclusions drawn from old intervention strategies have been confirmed with these newer techniques, including differences between the mechanisms mediating the effects of IL-1 and LPS on the HPA-axis. Thus, the lack of effect of administration of IL-1ra on bacterial LPS-induced increases in plasma ACTH and corticosterone was corroborated in mice genetically deficient in the IL-1 type 1 receptor (Dunn 2000; Matsuwaki et al. 2017). It is indeed important to consider the results of different lines of evidence when addressing a research question in the life sciences, as each single line or approach, and even the latest technology, is associated with particular limitations and biases (Munafo and Davey Smith 2018). In this context, it is also important to bear in mind that the use of genetically modified organisms may have less utility in addressing the role of metabolites, cells, or particular structural organizations. For example, when addressing the question of which cell types mediate the neuroendocrine or other effects of IL-1, most recent studies have used promotor-specific Cre-Lox recombinant mice. But one of the major bottlenecks of such approaches is the specificity of the promoters used, since specific promotors may not yet be widely available for all cell types and the specificity of promoters that have already been used may be questioned (Chaskiel et al. 2021).

Another important issue related to the interpretation of experimental findings is that many regulatory processes in biology are functionally redundant. Thus it is important to keep in mind that a regulatory response that appears to be top-down, in the sense that it involves entities at a perceived higher level of organization, typically the brain, influencing lower level entities, does not exclude other forms of regulation. So, even though it has been shown that in the case of systemic administration of IL-1, corticosterone release depends, to a large extent, on the hypothalamus (Berkenbosch et al. 1987; Gwosdow et al. 1990; Van Der Meer et al. 1996), this does not mean that every time plasma corticosterone concentrations increase that the full HPA-axis is active. Indeed, IL-1 can induce corticosteroid secretion by acting directly on the adrenal, where its signaling receptor is expressed (Andreis et al. 1991; Engstrom et al. 2008). Therefore, an increase in circulating glucocorticoid concentrations under certain inflammatory conditions or after administration of some molecule alone cannot be taken to reflect HPA-axis activation or indicate interactions between neuroendocrine and immune systems as such. In addition, the existence of functionally redundant regulatory mechanisms at different levels of perceived organization raises questions about how these interact and in which contexts they are mostly activated.

Beyond the progressively unravelling of mechanisms mediating the effects of cytokines on the HPA-axis and vice versa, it is important to point out that some conceptual shifts have occurred in the process. Back In 1977, Besedovsky and Sorkin wrote that "the possibility that the immune response itself could bring about changes in hormone levels has not been previously considered" (Besedovsky and Sorkin 1977, p. 4). Interestingly, some 15 years later, several paper titles referred to "infection as stressor" or "immune system-mediated stress response" (Eskay et al. 1990; Dunn 1993). Indeed, while Selye in the mid-twentieth century certainly did consider infection as a systemic or physiological stressor, subsequent research interests had progressively moved to studying the effects of so-called neurogenic, emotional, or psychological stressors on immune responses. Considering infection as a systemic, homeostatic, or physiological stressor led several groups in the 1990s to compare the brain circuits involved in activation of the HPA-axis by these two broad categories of stressors. Thus, systemic, homeostatic, or physiological stressors, including hypoglycemia and peripheral IL-1 injection, are thought to activate the HPA-axis through ascending catecholaminergic brainstem projections, whereas stimulation of the HPA-axis by neurogenic, emotional, or psychological stressors, such as restraint or foot shock, seem to involve forebrain circuits (Herman et al. 1996; Li et al. 1996; Herman and Cullinan 1997; Sawchenko et al. 2000).

Interestingly, most recently both homeostatic-physiological stressors and emotional-psychological stressors have been proposed to give rise to inflammation (Konsman 2019). In this respect, it is certainly going to be important in the future to specify subcategories of inflammation (Meizlish et al. 2021).

4.5 Key References

Besedovsky et al., *Journal of Immunology,* 1985. This review provides important insights into immune-neuroendocrine interactions.

Dinarello, *European Journal of Immunology,* 2007. This review, by one of the main actors in the field, gives a historical overview of the coming into being of cytokines as a class of intercellular messengers.

Herman & Cullinan, *Trends in Neurosciences,* 1997 and Sawchenko et al., *Progress in Brain Research,* 2000. These two review articles make the case that distinct categories of stressor mobilize different brain circuits that converge in the hypothalamus to activate the hypothalamo-pituitary-adrenal (HPA) axis.

References

Ader R, Cohen N (1985) High time for psychoimmunology. Nature 315:103–104

Andreis PG, Neri G, Belloni AS, Mazzocchi G, Kasprzak A, Nussdorfer GG (1991) Interleukin-1 beta enhances corticosterone secretion by acting directly on the rat adrenal gland. Endocrinology 129:53–57

Arguelles AE, Martinez MA, Hoffman C, Ortiz GA, Chekherdemian M (1972) Corticoadrenal and adrenergic overactivity and hyperlipidemia in prolonged emotional stress. Hormones 3:167–174

Aydin S (2015) A short history, principles, and types of ELISA, and our laboratory experience with peptide/protein analyses using ELISA. Peptides 72:4–15

Bassett JR, Cairncross KD, King MG (1973) Parameters of novelty, shock predictability and response contigency in corticosterone release in the rat. Physiol Behav 10:901–907

Berkenbosch F, Van Oers J, Del Rey A, Tilders F, Besedovsky H (1987) Corticotropin-releasing factor-producing neurons in the rat activated by interleukin-1. Science 238:524–526

Bernton EW, Beach JE, Holaday JW, Smallridge RC, Fein HG (1987) Release of multiple hormones by a direct action of interleukin-1 on pituitary cells. Science 238:519–521

Besedovsky H, Sorkin E (1977) Network of immune-neuroendocrine interactions. Clin Exp Immunol 27:1–12

Besedovsky HO, Del Rey A, Klusman I, Furukawa H, Monge Arditi G, Kabiersch A (1991) Cytokines as modulators of the hypothalamus-pituitary-adrenal axis. J Steroid Biochem Mol Biol 40:613–618

Besedovsky HO, Del Rey A, Sorkin E (1985) Immune-neuroendocrine interactions. J Immunol 135:S750–S754

Billiau A, Matthys P (2009) Interferon-gamma: a historical perspective. Cytokine Growth Factor Rev 20:97–113

Bonner WA, Hulett HR, Sweet RG, Herzenberg LA (1972) Fluorescence activated cell sorting. Rev Sci Instrum 43:404–409

Brazeau P, Vale W, Burgus R, Ling N, Butcher M, Rivier J, Guillemin R (1973) Hypothalamic polypeptide that inhibits the secretion of immunoreactive pituitary growth hormone. Science 179:77–79

Brown SL, Smith LR, Blalock JE (1987) Interleukin 1 and interleukin 2 enhance proopiomelanocortin gene expression in pituitary cells. J Immunol 139:3181–3183
Cao C, Matsumura K, Yamagata K, Watanabe Y (1996) Endothelial cells of the rat brain vasculature express cyclooxygenase-2 mRNA in response to systemic interleukin-1 beta: a possible site of prostaglandin synthesis responsible for fever. Brain Res 733:263–272
Chaskiel L, Dantzer R, Konsman JP (2021) Brain perivascular macrophages do not mediate Interleukin-1-induced sickness behavior in rats. Pharmaceuticals (Basel) 14:1030
Chensue SW, Terebuh PD, Remick DG, Scales WE, Kunkel SL (1991) In vivo biologic and immunohistochemical analysis of interleukin-1 alpha, beta and tumor necrosis factor during experimental endotoxemia. Kinetics, Kupffer cell expression, and glucocorticoid effects. Am J Pathol 138:395–402
Chuluyan HE, Saphier D, Rohn WM, Dunn AJ (1992) Noradrenergic innervation of the hypothalamus participates in adrenocortical responses to interleukin-1. Neuroendocrinology 56:106–111
Coons AH, Kaplan MH (1950) Localization of antigen in tissue cells; improvements in a method for the detection of antigen by means of fluorescent antibody. J Exp Med 91:1–13
Coons AH, Snyder JC, Sheever FS, Murray E (1950) Localization of antigen in tissue cells; antigens of Rickettsiae and mumps virus. J Exp Med 91:31–38
Coons AH, Leduc EH, Kaplan MH (1951) Localization of antigen in tissue cells. VI. The fate of injected foreign proteins in the mouse. J Exp Med 93:173–188
Coons AH, Leduc EH, Connolly JM (1955) Studies on antibody production. I. A method for the histochemical demonstration of specific antibody and its application to a study of the hyperimmune rabbit. J Exp Med 102:49–60
Coutinho A, Forni L, Holmberg D, Ivars F, Vaz N (1984) From an antigen-centered, clonal perspective of immune responses to an organism-centered, network perspective of autonomous activity in a self-referential immune system. Immunol Rev 79:151–168
Cunningham ETJ, Wada E, Carter DB, Tracey DE, Battey JF, De Souza EB (1992) In situ histochemical localization of type I interleukin-1 receptor messenger RNA in the central nervous system, pituitary, and adrenal gland of the mouse. J Neurosci 12:1101–1114
Derijk R, Van Rooijen N, Tilders FJ, Besedovsky HO, Del Rey A, Berkenbosch F (1991) Selective depletion of macrophages prevents pituitary-adrenal activation in response to subpyrogenic, but not to pyrogenic, doses of bacterial endotoxin in rats. Endocrinology 129:330–338
Dinarello CA (2007) Historical insights into cytokines. Eur J Immunol 37S1:S34–S45
Dobbins WO 3rd (1982) Gut immunophysiology: a gastroenterologist's view with emphasis on pathophysiology. Am J Physiol 242:G1–G8
Dunn AJ (1988) Systemic interleukin-1 administration stimulates hypothalamic norepinephrine metabolism paralleling the increased plasma corticosterone. Life Sci 43:429–435
Dunn AJ (1992) The role of interleukin-1 and tumor necrosis factor alpha in the neurochemical and neuroendocrine responses to endotoxin. Brain Res Bull 29:807–812
Dunn AJ (1993) Infection as a stressor: a cytokine-mediated activation of the hypothalamo-pituitary-adrenal axis? Ciba Found Symp 172:226–239
Dunn AJ (2000) Effects of the IL-1 receptor antagonist on the IL-1- and endotxoin-induced activation of the HPA-axis and cerebral biogenic amines in mice. Neuroimmunomodulation 7:36–45
Dunn AJ, Chuluyan HE (1992) The role of cyclo-oxygenase and lipoxygenase in the interleukin-1-induced activation of the HPA axis: dependence on the route of injection. Life Sci 51:219–225
Engstrom L, Rosen K, Angel A, Fyrberg A, Mackerlova L, Konsman JP, Engblom D, Blomqvist A (2008) Systemic immune challenge activates an intrinsically regulated local inflammatory circuit in the adrenal gland. Endocrinology 149:1436–1450
Ericsson A, Arias C, Sawchenko PE (1997) Evidence for an intramedullary prostaglandin-dependent mechanism in the activation of stress-related neuroendocrine circuitry by intravenous interleukin-1. J Neurosci 17:7166–7179
Eskay RL, Grino M, Chen HT (1990) Interleukins, signal transduction, and the immune system-mediated stress response. Adv Exp Med Biol 274:331–343

Felber JP (1963) ACTH antibodies and their use for a radio-immunoassay for ACTH. Experientia 19:227–229
Fuller S (2014) Neuroscience, Neurohistory, and the history of science: a tale of two brain images. Isis 105:100–109
Glyn J (1998) The discovery and early use of cortisone. J R Soc Med 91:513–717
Goehler LE, Relton JK, Dripps D, Kiechle R, Tartaglia N, Maier SF, Watkins LR (1997) Vagal paraganglia bind biotinylated interleukin-1 receptor antagonist: a possible mechanism for immune-to-brain communication. Brain Res Bull 43:357–364
Gordon AS, Katsh GF (1949) The relation of the adrenal cortex to the structure and phagocytic activity of the macrophagic system. Ann N Y Acad Sci 52:1–30
Gross PM (1992) Circumventricular organ capillaries. Prog Brain Res 91:219–233
Gwosdow AR, Kumar MS, Bode HH (1990) Interleukin 1 stimulation of the hypothalamic-pituitary-adrenal axis. Am J Physiol 258:E65–E70
Hale HB, Sayers G, Sydnor KL, Sweat ML, Van Fossan DD (1957) Blood adrenocorticotrophic hormone and plasma corticosteroids in men exposed to adverse environmental conditions. J Clin Investig 36:1642–1646
Harris GW (1955) Neural control of the pituitary gland. Edward Arnold, London
Hartley JL (2006) Cloning technologies for protein expression and purification. Curr Opin Biotechnol 17:359–366
Herman JP, Cullinan WE (1997) Neurocircuitry of stress: central control of the hypothalamo-pituitary-adrenocortical axis. Trends Neurosci 20:78–84
Herman JP, Prewitt CM, Cullinan WE (1996) Neuronal circuit regulation of the hypothalamo-pituitary-adrenocortical stress axis. Crit Rev Neurobiol 10:371–394
Jerne NK (1974) Towards a network theory of the immune system. Ann Immunol (Paris) 125C:373–389
Jerne NK (1985) The generative grammar of the immune system. EMBO J 4:847–852
Julius MH, Masuda T, Herzenberg LA (1972) Demonstration that antigen-binding cells are precursors of antibody-producing cells after purification with a fluorescence-activated cell sorter. Proc Natl Acad Sci U S A 69:1934–1938
Kabiersch A, Del Rey A, Honegger CG, Besedovsky HO (1988) Interleukin-1 induces changes in norepinephrine metabolism in the rat brain. Brain Behav Immun 2:267–274
Kapcala LP, He JR, Gao Y, Pieper JO, Detolla LJ (1996) Subdiaphragmatic vagotomy inhibits intra-abdominal interleukin-1 beta stimulation of adrenocorticotropin secretion. Brain Res 728: 247–254
Kelley KW (1988) Cross-talk between the immune and endocrine systems. J Anim Sci 66:2095–2108
Koehler G, Milstein C (1975) Continuous cultures of fused cells secreting antibody of predefined specificity. Nature 256:495–497
Konsman JP (2019) Inflammation and depression: a nervous plea for psychiatry to not become immune to interpretation. Pharmaceuticals (Basel) 12:29
Lacroix S, Rivest S (1998) Effect of acute systemic inflammatory response and cytokines on the transcription of the genes encoding cyclooxygenase enzymes (COX-1 and COX-2) in the rat brain. J Neurochem 70:452–466
Li HY, Ericsson A, Sawchenko PE (1996) Distinct mechanisms underlie activation of hypothalamic neurosecretory neurons and their medullary catecholaminergic afferents in categorically different stress paradigms. Proc Natl Acad Sci U S A 93:2359–2364
Mason JW, Brady JV, Polish E, Bauer JA, Robinson JA, Rose RM, Taylor ED (1961) Patterns of corticosteroid and pepsinogen change related to emotional stress in the monkey. Science 133: 1596–1598
Matsuwaki T, Shionoya K, Ihnatko R, Eskilsson A, Kakuta S, Dufour S, Schwaninger M, Waisman A, Muller W, Pinteaux E, Engblom D, Blomqvist A (2017) Involvement of interleukin-1 type 1 receptors in lipopolysaccharide-induced sickness responses. Brain Behav Immun 66:165–176

Matta SG, Weatherbee J, Sharp BM (1992) A central mechanism is involved in the secretion of ACTH in response to IL-6 in rats: comparison to and interaction with IL-1 beta. Neuroendocrinology 56:516–525
Matta SG, Linner KM, Sharp BM (1993) Interleukin-1 alpha and interleukin-1 beta stimulate adrenocorticotropin secretion in the rat through a similar hypothalamic receptor(s): effects of interleukin-1 receptor antagonist protein. Neuroendocrinology 57:14–22
Meizlish ML, Franklin RA, Zhou X, Medzhitov R (2021) Tissue homeostasis and inflammation. Annu Rev Immunol 39:557–581
Miller TR (1977) Psychophysiologic aspects of cancer: the James Ewing lecture. Cancer 39:413–418
Mohankumar SM, Mohankumar PS (2005) Systemic interleukin-1beta stimulates the simultaneous release of norepinephrine in the paraventricular nucleus and the median eminence. Brain Res Bull 65:451–456
Munafo MR, Davey Smith G (2018) Robust research needs many lines of evidence. Nature 553: 399–401
Nakane PK, Pierce GB Jr (1967) Enzyme-labeled antibodies for the light and electron microscopic localization of tissue antigens. J Cell Biol 33:307–318
Niijima A (1996) The afferent discharges from sensors for interleukin 1 beta in the hepatoportal system in the anesthetized rat. J Auton Nerv Syst 61:287–291
Oldham RK (1983) Natural killer cells: artifact to reality: an odyssey in biology. Cancer Metastasis Rev 2:323–336
Oppenheim JJ (2018) The future of the cytokine discipline. Cold Spring Harb Perspect Biol 10
Oppenheim JJ, Gery I (1993) From lymphodrek to interleukin 1 (IL-1). Immunol Today 14:232–234
Parsadaniantz SM, Gaillet S, Malaval F, Lenoir V, Batsche E, Barbanel G, Gardier A, Terlain B, Jacquot C, Szafarczyk A, Et A (1995) Lesions of the afferent catecholaminergic pathways inhibit the temporal activation of the CRH and POMC gene expression and ACTH release induced by human interleukin-1beta in the male rat. Neuroendocrinology 62:586–595
Porter JC (1973) Neuroendocrine systems: the need for precise identification and rigorous description of their operations. Prog Brain Res 39:1–6
Prinz M, Masuda T, Wheeler MA, Quintana FJ (2021) Microglia and central nervous system-associated macrophages-from origin to disease modulation. Annu Rev Immunol 39:251–277
Rasmussen AF Jr (1969) Emotions and immunity. Ann N Y Acad Sci 164:458–462
Reynolds AS (2017) Discovering the ties that bind: cell-cell communication and the development of cell sociology. In: Lidgard S, Nyhart LK (eds) Biological individuality: integrating scientific, philosophical, and historical perspectives. University of Chicago Press, Chicago
Rivier C, Vale W, Guillemin R (1973) An in vivo corticotropin-releasing factor (CRF) assay based on plasma levels of radioimmunoassayable ACTH. Proc Soc Exp Biol Med 142:842–845
Rodger NW, Beck JC, Burgus R, Guillemin R (1969) Variability of response in the bioassay for a hypothalamic somatotrophin releasing factor based on rat pituitary growth hormone content. Endocrinology 84:1373–1383
Sapolsky R, Rivier C, Yamamoto G, Plotsky P, Vale W (1987) Interleukin-1 stimulates the secretion of hypothalamic corticotropin-releasing factor. Science 238:522–524
Sawchenko PE, Li HY, Ericsson A (2000) Circuits and mechanisms governing hypothalamic responses to stress: a tale of two paradigms. Prog Brain Res 122:61–78
Selye H (1948) The Gordon Wilson lecture: on the general-adaptation-syndrome. Trans Am Clin Climatolol Assoc 60:184–193
Selye H (1950) Stress and the general adaptation syndrome. Br Med J 1:1383–1392
Selye H (1955) "Critical period" for inhibition of inflammation by a primarily neurogenic stress-situation. Psychosom Med 17:124–127
Selye H (1976) Forty years of stress research: principal remaining problems and misconceptions. Can Med Assoc J 115:53–56
Selye H, Fortier C (1950) Adaptive reaction to stress. Psychosom Med 12:149–157

Serrats J, Schiltz JC, Garcia-Bueno B, Van Rooijen N, Reyes TM, Sawchenko PE (2010) Dual roles for perivascular macrophages in immune-to-brain signaling. Neuron 65:94–106

Solomon GF (1969) Emotions, stress, the central nervous system, and immunity. Ann N Y Acad Sci 164:335–343

Spitzer R (1968) The clinical importance of polypeptide hormone assays. Clin Biochem 1:216–223

Stein M, Schiavi RC, Camerino M (1976) Influence of brain and behavior on the immune system. Science 191:435–440

Sumal KK, Blessing WW, Joh TH, Reis DJ, Pickel VM (1983) Synaptic interaction of vagal afferents and catecholaminergic neurons in the rat nucleus tractus solitarius. Brain Res 277:31–40

Tache Y, Du Ruisseau P, Tache J, Selye H, Collu R (1976) Shift in adenohypophyseal activity during chronic intermittent immobilization of rats. Neuroendocrinology 22:325–336

Treiman DM, Fulker DW, Levine S (1970) Interaction of genotype and environment as determinants of corticosteroid response to stress. Dev Psychobiol 3:131–140

Utiger RD, Parker ML, Daughaday WH (1962) Studies on human growth hormone. I. A radio-immunoassay for human growth hormone. J Clin Invest 41:254–261

Van Dam AM, Brouns M, Louisse S, Berkenbosch F (1992) Appearance of interleukin-1 in macrophages and in ramified microglia in the brain of endotoxin-treated rats: a pathway for the induction of non-specific symptoms of sickness? Brain Res 588:291–296

Van Der Meer MJ, Sweep CG, Pesman GJ, Tilders FJ, Hermus AR (1996) Chronic stimulation of the hypothalamus-pituitary-adrenal axis in rats by interleukin 1beta: central and peripheral mechanisms. Cytokine 8:910–919

Vilcek J (2006) Fifty years of interferon research: aiming at a moving target. Immunity 25:343–348

Watanobe H, Sasaki S, Takebe K (1991) Evidence that intravenous administration of interleukin-1 stimulates corticotropin releasing hormone secretion in the median eminence of freely moving rats: estimation by push-pull perfusion. Neurosci Lett 133:7–10

Weiss JM (1970) Somatic effects of predictable and unpredictable shock. Psychosom Med 32:397–408

Wieczorek M, Dunn AJ (2006) Effect of subdiaphragmatic vagotomy on the noradrenergic and HPA axis activation induced by intraperitoneal interleukin-1 administration in rats. Brain Res 1101:73–84

Williams K, Alvarez X, Lackner AA (2001) Central nervous system perivascular cells are immunoregulatory cells that connect the CNS with the peripheral immune system. Glia 36:156–164

Yabuuchi K, Minami M, Katsumata S, Satoh M (1994) Localization of type I interleukin-1 receptor mRNA in the rat brain. Brain Res Mol Brain Res 27:27–36

Yalow RS, Berson SA (1960) Immunoassay of endogenous plasma insulin in man. J Clin Invest 39:1157–1175

Catecholamines and Immunomodulation 5

Maren Claus and Silvia Capellino

Abstract

Dopamine, norepinephrine, and epinephrine are endogenous catecholamines, known to play many different roles in the body, such as increasing heart rate, blood pressure, and blood glucose, controlling movement, and acting on mood and behavior. Catecholamines were first described as neurotransmitters of the sympathetic nervous system, but they also act as hormones released by the adrenal medulla, and increasing evidence indicates their crucial role in the immune system. In this chapter, we will describe how catecholamines modulate immune cells in physiologic as well as pathologic conditions and provide evidence indicating that catecholamines are also to be considered cytokines, that is compounds synthesized by immune cells, which act as local intercellular signaling mediators. We will then describe how this knowledge will be helpful for new therapeutic strategies.

Keywords

Epinephrine · Norepinephrine · Dopamine · Immune · Catecholamine

5.1 Introduction

Dopamine, norepinephrine, and epinephrine are the endogenous catecholamines, traditionally described as neurotransmitters of the sympathetic nervous system, and are responsible for the fight-or-flight response, as well as for many other

M. Claus · S. Capellino (✉)
Department of Immunology, IfADo-Leibniz Research Centre for Working Environment and Human Factors, Research Group of Neuroimmunology, Dortmund, Germany
e-mail: capellino@ifado.de

J. P. Konsman, T. M. Reyes (eds.), *Neuroendocrine-Immune System Interactions*, Masterclass in Neuroendocrinology 13,
https://doi.org/10.1007/978-3-031-21358-8_5

physiological functions. Dopamine is mainly produced in neuronal cell bodies in two areas of the brain: the substantia nigra and the ventral tegmental area. Norepinephrine is mostly synthesized in the locus coeruleus, whereas epinephrine is produced in neurons near the solitary tract. In addition to their synthesis in the (central) nervous system and their classical role as neurotransmitters, norepinephrine and epinephrine can also be synthesized and released in the adrenal medulla of the adrenal glands, and catecholamine release was described in many peripheral organs, thus suggesting a "hormonal" role of catecholamines. The first mention of the crosstalk between the sympathetic nervous system and the immune system can be found in an article from 1903 (Meltzer and Meltzer 1903). Since then, a number of important articles have been published, showing the complex interaction between the sympathetic nervous system and the immune system. Here, we will depict the effects of catecholamines on immune cells and immune function. The aim of this chapter is not to itemize all available publications on this subject, but rather to give an overview of principal results and milestones, and to discuss how this knowledge could be useful for future clinical applications.

Furthermore, we will describe how immune cells synthesize and release catecholamines by themselves, thus using catecholamine-driven immune modulation without any involvement of the nervous system. Based on these results, we will introduce the idea that catecholamines should also be considered as cytokines and components of the immune system.

5.2 Catecholamine Synthesis, Release, and Inactivation

Catecholamines are chemical compounds with a catechol group and a side-chain amine. Catecholamine synthesis starts with the conversion of the amino acid L-tyrosine to l-3,4-dihydroxyphenylalanine (levodopa or L-DOPA) by the enzyme tyrosine hydroxylase (TH), which is the rate-limiting enzyme for catecholamine synthesis (Schulz et al. 2004).

Dopamine is the first catecholamine synthesized from L-DOPA by the enzyme DOPA decarboxylase. Depending on the cell type, dopamine can be then released as such or further converted to norepinephrine (also called noradrenaline) and finally to epinephrine (or adrenaline) (Fig. 5.1).

The Discovery of Dopamine as a Neurotransmitter

Dopamine was first synthesized in 1910 by George Barger and James Ewens. Shortly after, Henry Dale, from the same lab, found it to be an epinephrine-like compound. However, it took almost 50 years until Arvid Carlsson, Nils-Åke Hillarp, and Bengt Falck recognized that dopamine itself was an endogenous agonist. For his work on dopamine, Carlsson was awarded the Nobel Prize in Physiology or Medicine in 2000, together with Eric Kandel and Paul Greengard.

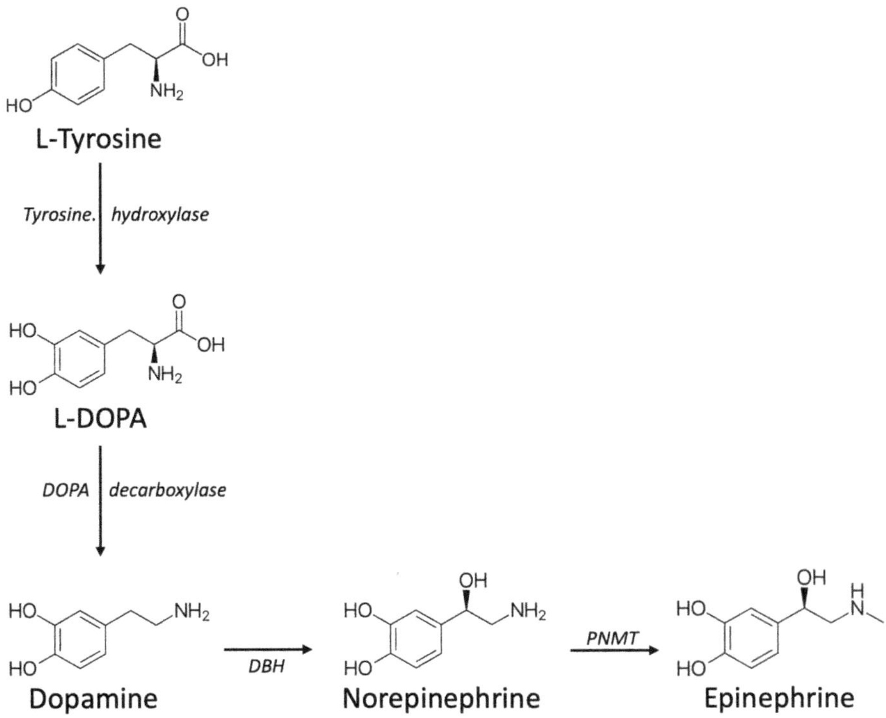

Fig. 5.1 Catecholamine synthesis. *DBH* dopamine β-hydroxylase, *PNMT* phenylethanolamine N-methyltransferase

After synthesis, catecholamines are stored in cytoplasmic vesicles. The mechanisms involved in catecholamine release have been studied mostly in chromaffin cells, but it is plausible that the same mechanisms are valid for other cell types as well. Many humoral factors, such as prostaglandins and a large number of peptides and other substances, are involved in catecholamine release, and additionally, catecholamines themselves may regulate their own release via presynaptic autoreceptors (Schulz et al. 2004; Langer 1980).

Metabolism and inactivation of catecholamines are very different among tissues and cell types. In most of the cases, catecholamine reuptake takes place in the same cells that secreted them, or in cells close by, but the extracellular breakdown of catecholamines is possible as well. Due to the instability of catecholamines, their application as pharmacological treatment would have only a low efficacy. Therefore, administration of precursors (L-DOPA) and blockade of reuptake and/or of degradation are used as a therapeutic strategy to increase their amount. For example, norepinephrine–dopamine reuptake inhibitors are used to boost catecholamines' effects and are useful to treat depression (Kintscher 2012), whereas catechol-O-methyltransferase inhibitors are used as supportive therapy to prolong the effect of L-DOPA in Parkinson's patients (Deane et al. 2004). As described below in this chapter, catecholaminergic pathways are also active in immune cells and catecholamine-modulating drugs can also affect immune function. Increasing our knowledge of the immune effects of these drugs is therefore crucial for the future development of catecholamine-induced immune modulation.

5.3 Catecholamine Receptors

Catecholamine receptors are transmembrane G-protein coupled receptors (GPCR). They are divided into dopaminergic receptors, binding dopamine, and adrenergic receptors, binding epinephrine and norepinephrine. Like many other neurotransmitter receptors, catecholamine receptors are expressed by many cells outside of the nervous systems. Here, we will focus on the expression and function of catecholamine receptors on immune cells.

5.3.1 Dopamine Receptors

Five different dopamine receptors (DR) were described in the late 1970s. These receptors are called D1-D5 DR and are encoded by different genes. An alternative splicing of D2-DR can lead to two variants of this receptor with distinct physiological, signaling, and pharmacological properties: $D2_S$ (D2-short) and $D2_L$ (D2-long) (Beaulieu et al. 2015). Splicing variants are also described for other DR, but with similar signaling properties. Depending on their biochemical properties and functional similarities, DR are grouped into two families: the D1-like and D2-like (Sibley et al. 1993). D1- and D5-DR belong to the D1-like DR family and are coupled to $G\alpha_s$, thus activating adenylate cyclase and increasing the cytosolic concentration of 3′-5′-cyclic adenosine monophosphate (cAMP), whereas D2-, D3-, and D4-DR belong to the D2-like DR family that are coupled to $G\alpha_i$ and inhibit adenylate cyclase and cAMP production (Beaulieu et al. 2015). Modulation of cAMP by DR regulates PKA. Among PKA substrates, phosphorylation of DARPP-32 and PPP1 in the dopamine receptor pathway has been intensively investigated (Svenningsson et al. 2004; Girault 2012; Beaulieu et al. 2015). Apart from the prevailing mechanism of cAMP regulation, DR can also regulate a variety of further signaling pathways, including alternate G-protein coupling but also non-G-protein mechanisms. A detailed description of cAMP-independent pathways activated by dopamine is presented by Beaulieu et al. (2015). To make things even more complicated, during chronic inflammation DR can switch the cytoplasmic subunit from $G\alpha_s$ to $G\alpha_i$ signaling during chronic inflammation (Jenei-Lanzl et al. 2015b). This phenomenon is probably responsible for the different effects of dopamine described in physiologic and in pathologic conditions (see below). Furthermore, DR can exist in oligomeric form (Perreault et al. 2014; Feng and Lu 2021). It is reported that DR can form heteromers with other DR but also with many other receptors, such as other GPCR and ionotropic receptors (Beaulieu et al. 2015; Feng and Lu 2021; Kamal and Jockers 2011; Perreault et al. 2014). Heteromerization is not a functional requirement, but a normal physiological function of GPCR. It is reported that heteromerization can occur early after receptor synthesis but can also take place in a very dynamic and transient way later on (Kasai and Kusumi 2014). Heteromerization confers a different signaling mechanism compared to that of single receptors. Therefore, a receptor agonist could induce the activation of one specific receptor but also of heteromers, thus causing different signaling mechanisms. This aspect is often not considered but is extremely important for the interpretation of results.

Of note, D1-like and D2-like DR have different binding affinity for dopamine. D2-like DR has high binding affinity and can bind already low dopamine concentrations, whereas D1-like DR has 10- to 100-fold lower binding affinity and can only bind at high concentration of dopamine (Klein et al. 2019). Therefore, dopamine can exert different intracellular effects depending on its concentration and thus on the activated DR.

5.3.2 Adrenergic Receptors

Norepinephrine and epinephrine exert their effect by binding to GPCR called "adrenergic receptors" or "adrenoceptors" (AR). AR are expressed within the central nervous system, as well as in almost all peripheral cells, and their activation has many consequences, including on blood pressure control, heart rate regulation, airway reactivity, and many areas of metabolism. Due to the wide spectrum of effects, AR modulation is currently used as a therapeutic target in many diseases, such as asthma, chronic obstructive pulmonary disease (COPD), and high blood pressure.

AR were firstly divided into two types, α-AR and β-AR, depending on their binding affinity for norepinephrine and epinephrine (Ahlquist 1948). Nowadays, AR are divided into three subtypes, based on their pharmacological properties (Bylund et al. 1994): α_1-AR, α_2-AR, and β-AR, each of them further divided into three subtypes, as described in Fig. 5.2.

For a summary of AR subtypes and their action, see Bylund et al. (1994), Hein (2006), Cosentino and Marino (2012), Scanzano and Cosentino (2015). In general, α-AR binds with stronger affinity to norepinephrine compared to epinephrine, and

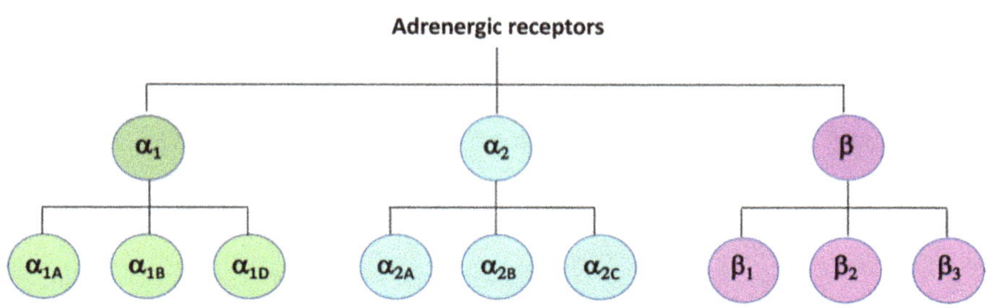

Fig. 5.2 Subdivision of adrenergic receptors

the opposite is true for β-AR. α_1-AR stimulation leads to activation of phospholipase C (PLC) and finally to Ca^{2+} release, but in vascular smooth muscle cells an influence of α_1-AR activation on calcium influx has also been described (Bylund et al. 1994). Also for α_2-AR many signaling pathways have been described. α_2-AR activation leads to inhibition of adenylate cyclase and reduction of cyclic adenosine monophosphate (cAMP) concentration, but α_2-AR were also shown to translocate extracellular calcium and to activate potassium channels (summarized in Bylund et al. 1994). β-AR are coupled to an intracellular $G\alpha_s$ subunit and lead to cAMP increase after stimulation, but other unconventional mechanisms are also described, with β_3-AR being responsible for protein kinase A (PKA) activation and Ca^{2+} entry in endomyocardial tissue (Gauthier et al. 1996).

Besides these classical signaling pathways, AR activation may act on many further non-canonical pathways, as described below with regard to immune cells. One reason is the presence of many ARs on the same cell and the different binding affinity of specific agonists for different AR subtypes. Also, prolonged or strong AR stimulation leads to reduced responsiveness to further stimuli, whereas low activation may lead to increased receptor signal. Moreover, AR can bind to other receptors and form heteromers, with different functions compared to the AR alone (Wnorowski and Jozwiak 2014; Hein 2006). The formation of heteromers can be very dynamic, as previously described (Kasai and Kusumi 2014), thus making the interpretation of results even trickier.

5.3.3 DR-AR Heteromers

As already described above, AR and DR can form heteromers with many other receptors, with a significant physiological impact on their function and intracellular signaling. Within the last few years, the formation of heteromers by DR and AR has been described. Rebois et al. (2012) demonstrated that D2-DR and β-AR can form heteromers in transfected HEK 293 cells. Gonzalez et al. (2012) showed that D4-DR heteromerized with α1b- and β1-AR in the pineal gland in rats. Stimulation of heteromers with dopamine inhibited adrenergic signaling, thus acting on melatonin synthesis. Interestingly, D4-DR is only expressed during the dark cycle in the pineal gland; therefore, this indirect effect of dopamine on the adrenergic signaling has a circadian rhythm. These results pinpoint the possible cross-reactivity of DR agonists on AR pathways (and *vice versa*) due to DR-AR heteromers, and also the dynamic changes between heteromeric and homomeric forms of DR and AR, which should be taken into consideration. Activation of the heteromers involves a signaling complex responsible for integrating the regulation of AC activity by $G\alpha_s$ and $G\alpha_i$.

How to Study GPCR Heteromers

It is well known that pharmacological features, trafficking, and signal transduction of GPCR are modulated by the formation of heteromers (Chandrasekera et al. 2013; Beaulieu et al. 2015). But whether heteromers are functional or not is not always clear, as it is difficult to clearly distinguish true interactions from random colocalization.

A growing number of methods have been developed to analyze heteromer formation. In early studies, potential candidates were co-expressed in HEK 293 cells and heteromers were subsequently detected by co-immunoprecipitation or assayed microscopically for colocalization (Chandrasekera et al. 2013). The latter approach could be refined by the use of the energy transfer-based techniques fluorescence resonance energy transfer (FRET) and bioluminescence resonance energy transfer (BRET) (Ayoub and Pfleger 2010). Further development of these methods (1) allows direct discrimination between heteromer-specific pharmacology and monomer/homomer-specific pharmacology by measuring BRET between one receptor and the ligand of the second receptor (Johnstone et al. 2021), (2) enables detection of β-arrestin recruitment to one receptor after stimulation of the second receptor (Mores et al. 2019), and 3) allows the examination of drug-induced changes in oligomer formation (Vidi et al. 2008). However, all these studies can only be performed in artificial systems using transfected cells.

Identification of GPCR heteromers in primary cells or tissues has become possible by methods such as the in situ proximity ligation assay (PLA) (Gomes et al. 2016) or the amplified luminescent proximity homogeneous assay (AlphaScreen) (Fernandez-Duenas et al. 2019). Here, labeled antibodies bind to the potential interaction partners. Labels located in close proximity—such as by direct protein–protein interaction in heteromers—can be detected by specific signal amplification and analyzed microscopically.

5.4 Involvement of Catecholamine Receptors in Immunity

5.4.1 Dopamine Receptors and Immunity

Dopamine is one of the best-studied neurotransmitters in the brain, due to its crucial role in several functions, such as reward-related behaviors and movement control. Despite a body of evidence suggesting a functional effect of dopamine on inflammation, its role as a humoral compound and immune transmitter was only described many years after the discovery of the immunomodulatory effect of norepinephrine and epinephrine. Nevertheless, a large number of studies have been published within the last few years confirming the crucial role of dopamine on immune responses (for reviews, see Thomas Broome et al. (2020), Feng and Lu (2021), Ugalde et al. (2021).

Early studies assessed the expression of DR in human peripheral blood mononuclear cells (PBMCs) by PCR or radioligand binding assays (Basu and Dasgupta 2000). In 2002, Mckenna et al. (2002) quantified DR expression on PBMCs via flow cytometry and reported that the highest expression of DR was in NK cells and B cells, whereas T cells and monocytes had lower DR levels, with great variance within donors.

Human NK cells express D2-D5 DR and lack D1-DR (Mckenna et al. 2002), but preliminary data from our group also suggest the expression of D1-DR on human NK cells (unpublished data). The literature on dopamine-mediated effects on human NK cells is limited at present, but current knowledge suggests that DR activation leads to an inhibitory function on human NK cells. In particular, upregulation of D5-DR in primary human NK cells pre-stimulated with interleukin-2 (IL-2) has been reported to suppress cell proliferation and NFκB-dependent IFNγ secretion (Mikulak et al. 2014). A recent publication supports the possible clinical relevance of dopaminergic modulation, as the treatment of patients with solid refractory tumors with a small molecule D2 antagonist in a phase II study led to more NK cell tumor infiltration and induction of cytokines (Stein et al. 2019). The results from mouse models also confirmed an effect of dopamine on NK cells, but the results seem contradictory. Zhao et al. (2013) reported that D1-like stimulation enhanced NK cytotoxicity in mice, whereas D2-like DR stimulation was responsible for NK cell inhibition. On the contrary, Nozaki et al. (1996) found that haloperidol, a D2-like DR antagonist, inhibited NK cell activity and paradoxically bromocriptine, a D2-like DR agonist, had the same effects. The authors hypothesized different pathways for the two drugs as an explanation for the unexpected results, but another reason could be the binding affinity of these drugs for other receptors, such as binding to serotoninergic receptors. Kavelaars et al. (2005) showed that knockout of the dopamine transporter (DAT) in mice led to a reduced activity of NK cells, but to higher cytokine production in LPS-induced macrophages. These effects could be due to a longer presence of dopamine in the extracellular space and thus to a prolonged dopaminergic effect. In rats with high susceptibility to dopamine, NK cell activity is lower than in hypodopaminergic rats (Teunis et al. 2004) and female hyperdopaminergic rats had lower NK cell numbers compared to hypodopaminergic female rats, thus suggesting an influence of sex hormones on dopamine-mediated effects on NK cells (Teunis et al. 2004). It is of interest that findings obtained in some *in vivo* stress models also suggest a role for dopamine in stress-mediated NK cell function. For example, restraint stress in mice caused an impairment in NK cell cytotoxicity and dopaminergic as well as adrenergic antagonists prior to stress induction could counteract this effect (Fiserova et al. 2002).

Despite the high DR expression detected on B cells (Mckenna et al. 2002), the functional role of dopamine in B cells is still poorly understood. In the 1990s, Bergquist et al. (1994, 1997) demonstrated that dopamine stimulation decreased IgG, IgA, and IgM secretion. Moreover, dopamine stimulation inhibited B cell proliferation. However, these described effects are apparently independent of canonical dopaminergic signaling and rather occur due to the formation of reactive oxygen species and consequent induction of apoptosis (Meredith et al. 2006). Stimulation of

D2-like DR by bromocriptine was reported to inhibit B cell proliferation as well as IgG secretion and expression of activation markers such as CD23, CD25, and CD71 (Morkawa et al. 1993). In rheumatoid arthritis, expression of D2-DR on B cells negatively correlated with disease activity, but no functional experiments were performed (Wei et al. 2015). Recently, Papa et al. demonstrated that stimulation of DR on germinal center and memory B cells leads to differentiation in plasma cells and to a rapid translocation of ICOSL to the cell membrane, thus enhancing T–B cell interaction (Papa et al. 2017). Of note, the authors also reported that these dopaminergic mechanisms are not conserved between mice and humans.

Most of the studies investigating the role of dopamine on the immune system have focused on T cells. DR expression has been well described in T cells (Arreola et al. 2016), and DR expression seems to be dynamic, as it changes during T cell maturation and increases after anti-CD3/CD28 stimulation (Kustrimovic et al. 2014). The effects of dopamine can be very different and seem to depend on T cell status. In general, dopamine activates resting T cells and inhibits activated T cells. However, dopamine can even lead to opposite effects on T cells, depending on its concentration, the activation state of the cells, and the specific DR bound (for summary, see Feng and Lu (2021), Thomas Broome et al. (2020), Levite (2016)). For example, D2-DR and D3-DR both belong to the D2-like DR family, but they have opposite effects on T cells, with D2-DR stimulating the release of the anti-inflammatory IL-10 and D3-DR promoting secretion of the pro-inflammatory TNFα (Besser et al. 2005). This example highlights the complexity of dopamine's action on T cells. It is of interest that only regulatory T cells (T_{reg}) were found to synthesize dopamine (Cosentino et al. 2007). Secreted dopamine can then act in an autocrine way and decrease secretion of IL-10 and transforming growth factor (TGF)-β, or in a paracrine way, thus leading to effector T cell (T_{eff}) suppression via D1-like DR (Cosentino et al. 2007).

Monocytes and macrophages are part of the immediate innate immune response against pathogens and shape adaptive immunity. Further, they are key players in tissue homeostasis. Monocytes circulate in peripheral blood and migrate to sites of inflammation after activation, which is paralleled by differentiation into macrophages. Macrophages are tissue-resident, with different phenotypic and functional features in different tissues. Human monocytes express high amounts of D2- and D3-DR and lower amounts of D4- and D5-DR (Mckenna et al. 2002), whereas a recent publication described D5-DR as the highest expressed DR in human macrophages (Nolan et al. 2019). Similarly to T cells, current knowledge suggests a multifaceted role for dopamine in monocyte function. D1-DR signaling blocked the production of pro-inflammatory cytokines from monocytes and LPS-activated macrophages (Bone et al. 2017), thus suggesting an anti-inflammatory effect. However, a study on HIV patients showed that dopamine had a pro-inflammatory role on macrophages, via D5-DR activation (Nolan et al. 2019). Another study suggested different effects of D2-like DR activation in physiologic and pathologic conditions (Gaskill et al. 2012). Indeed, the study showed that dopamine has a clear pro-inflammatory effect on human untreated macrophages and increased IL-6 and CCL-2 release, but it had less clear effects on macrophages pre-treated with LPS, where dopamine increased IL-6 and CCL-2 but decreased TNFα (Gaskill et al. 2012).

Dendritic cells (DCs) are specific antigen-presenting cells that stimulate and shape T cell responses during infection and in steady state. Immature DCs exhibit phagocytic features and present processed antigens to T cells. Dendritic cells are responsible for priming of T cells and for their differentiation into effector cells. Nakano et al. (2008) showed the presence of DR on monocyte-derived dendritic cells (Mo-DC) and suggested that D2-like DR blockade might lead to an increase in cAMP and consequently to dopamine release by Mo-DC, which then leads to D1-like DR activation on T cells (Nakano et al. 2009). Therefore, dopamine could be involved in DC-T cell interaction (Pacheco et al. 2009), but further studies are required.

Dopamine has also been reported to regulate granulocytes, as summarized by Pinoli et al. (2017). Granulocytes belong to the innate immune system and include neutrophils, eosinophils, basophils, and mast cells. Except for basophils, the influence of dopamine has been demonstrated for all granulocytes. In neutrophils, D1-like DR expression was reported in 1999 (Sookhai et al. 1999), and D2-like DR expression was reported shortly after (Pereira et al. 2003; Boneberg et al. 2006). Later it was also demonstrated that D1-like and D2-like DR are co-expressed on the same cells (Mckenna et al. 2002; Chen et al. 2014). Activation of DR usually inhibits neutrophil functions, for example their ability to adhere to the endothelium, as well as decreased cell migration, phagocytic activity, and inhibited superoxide anion production (for a review, see Pinoli et al. (2017)). It was also reported that dopamine increased neutrophil apoptosis under physiological conditions as well as during inflammation (Sookhai et al. 1999). Also, treatment with L-Dopa was shown to induce neutropenia in Parkinson's patients and caused an unbalanced DR expression on neutrophils (Cordano et al. 2015). In eosinophils, all five DR have been detected, but little is known about their functional effects. Podolec et al. showed a reduced amount of eosinophils in rats treated with high concentrations of L-Dopa or apomorphine, whereas an opposite effect was shown at low doses (Podolec et al. 1979). These findings could be of clinical relevance for transplantations, as described by Takkenberg et al. (2004), where administration of dopamine in patients waiting for heart transplantation reduced peripheral eosinophilia and eosinophilic myocarditis in the transplanted heart. In mast cells, effects of dopamine have been described only in mice and in a rat cell line. In the rat cell line RBL-2H3, many dopaminergic agents inhibited degranulation, but these effects were probably independent of DR (Seol et al. 2004). In contrast, dopamine induced degranulation of mast cells derived from mouse bone marrow in a D1-like DR-dependent way (Mori et al. 2013). To the best of our knowledge, expression of DR on human mast cells as well as possible functional effects has not been described so far.

In summary, the published findings are partially conflicting, probably due to different expression levels of DR in immune cell subtypes and to the fact that the dopaminergic compounds used have different binding affinities for DR, and so cannot be directly compared. Also, the presence of heteromers can alter the intracellular pathway activated by specific DR. Nevertheless, these results demonstrate the strong influence of dopamine on immune cell responses and open up a possible new way to modulate the immune response.

5.4.2 Adrenergic Receptors and Immunity

Adrenergic receptors (AR) have been described on almost all immune cells, and there is a growing body of evidence available for their role in the immune response. Here, we will therefore introduce only an overview of the main evidence regarding AR activation on B cells, T cells, NK cells, dendritic cells, and monocytes/macrophages.

In general, adrenergic signals seem to have an inhibitory effect on the functions of immune cells. Under specific conditions, however, AR agonists can also be activators. A pro- or anti-inflammatory environment and the various stages of cell differentiation and maturation may shape the expression pattern of AR or lead to preferential activation of different signaling pathways downstream of the AR.

Among lymphocytes, B cells express the greatest numbers of β2-AR per cell, which is considered their main receptor responding to epinephrine and norepinephrine (Karaszewski et al. 1990; Kohm and Sanders 2001). Chronic administration of adrenergic agonists in asthma seems to inhibit B cell function (Mansfield and Nelson 1982), while short-term adrenergic signals after psychological stress might have stimulating effects on B cells in humans (Matthews et al. 1995), which underlines the impact of the duration of adrenergic stimulation. A variety of mouse models have been developed to study the effects of adrenergic signaling on B cell function. Reduced production of IgM and IgG1 after depletion of norepinephrine prior to immunization *in vivo* as well as increased secretion of IgG1 by mouse B cells *in vitro* in the presence of β-AR stimulation suggests that an optimal B cell response to antigenic stimulation requires adrenergic signals. The enhancing effect on IgG1 secretion has been attributed to stimulation of cAMP/PKA-mediated signal pathways by β2-AR and direct upregulation of the costimulatory molecule CD86 through β2-AR signaling, which positively regulates IgG1 production (Kohm and Sanders 1999; Podojil and Sanders 2005). Production of IFNγ-dependent IgG2a is affected through β2-AR signaling in both B cells and T_H1 helper T cells. Chronically stressed mice showed decreased immune response to viral infection with a lower number of antibody-secreting cells and thus reduced amount of virus-specific IgM and IgG (Kennedy et al. 2005; Sheridan et al. 1998). Similarly, enhanced production of IgE was also dependent on adrenergic signaling (Pongratz et al. 2006), which underlines the clinical importance of stress and stress management in allergy and allergic asthma and questions the benefit of long-term administration of adrenergic agonists in the treatment of lung diseases.

Adrenergic signals shape T cell functions either directly, by regulating thymocyte differentiation (Leposavic and Pilipovic 2018) and interfering with activation, differentiation, and effector function (see below), or indirectly, by inhibiting the production of T cell activating cytokines by dendritic cells ((Takenaka et al. 2017) and below).

Expression levels of β-AR differ between T cells with greater receptor density on CD8 cytotoxic T cells than on CD4 helper T cells (Fan and Wang 2009). Functional effects of epinephrine and norepinephrine are mainly mediated by β2-AR (Borger et al. 1998). β2-AR numbers and responsiveness increase further with activation and differentiation (Korichneva and Tkachuk 1990; Carlson et al. 1994; Slota et al. 2015; Wahle et al. 2001; Fan and Wang 2009). There is only one study describing the induction of β3-AR expression on T cells by treatment with the mitogen concanavalin A (ConA). However, no evidence for functional consequences was found (Borger et al. 1998). Recently, the presence of β3-AR mRNA was shown in T_{reg} (Freier et al. 2010), and expression of β1-AR mRNA in T_{reg} cells was linked to the response of these cells to psychological stress (Cosentino et al. 2007; Freier et al. 2010). There is only little evidence for the presence of α1-AR and α2-AR on T cells (Jetschmann et al. 1997; Kavelaars 2002).

β2-AR is expressed on naive and T_H1, but not T_H2 CD4 helper T cells (Swanson et al. 2001). Consequently, adrenergic signals differentially impact the function of polarized CD4 T cells: β2-AR stimulation of naive T cells through norepinephrine or terbutaline promotes differentiation into T_H1 cells and increases production of IFNγ (Swanson et al. 2001). Yet, T_H1 cells can be suppressed by adrenergic signals and β2-AR signaling has been involved in driving T cell polarization toward a T_H2 phenotype (Heijink et al. 2003). A recent study showed that treatment of naive T cells with the β2-AR agonist terbutaline led to preferential differentiation of T_H17 cells over T_H1 cells (Carvajal Gonczi et al. 2017). Despite high expression of β2-AR, the inhibitory effects of the β2-AR agonist fenoterol on T_H1 cytokine production and proliferation are smaller in activated (polarized) than in resting naive T cells (Heijink et al. 2003). This has been attributed to altered signal transduction of $G\alpha_s$-coupled receptors in polarized cells (Heijink et al. 2005).

CD8 cytotoxic T cells are key players in the control of tumors and viral infection. Adrenergic signals limit antiviral function in humans and mice (Grebe et al. 2009; Estrada et al. 2016). Interestingly, moderate exercise promotes specific mobilization of CMV-specific memory T cells in a β2-AR-dependent manner. The authors speculate that regular moderate exercise might exert additional antiviral protection through frequent β2-AR-dependent mobilization of virus-specific T cells (Kunz et al. 2020). Moreover, norepinephrine modulates memory CD8 T cell function by enhancing inflammatory cytokine production and reducing proliferation in response to activation (Slota et al. 2015).

Adrenergic signaling increases inhibitory programmed cell death protein 1 (PD-1) on T cells, interferes with cellular metabolism, and leads to T cell exhaustion (Yang et al. 2019; Qiao et al. 2021). Thus, adrenergic signals interfere with CD8 T cell anti-tumor response. Beta-blockade resulted in a better tumor control in melanoma patients (Gandhi et al. 2021). Further, blocking β2-AR led to enhanced glycolysis and oxidative phosphorylation, promoting more effective cytokine production and a more potent cytolytic response of tumor-infiltrating lymphocytes (TIL) (Qiao et al. 2021). In chronically stressed mice, application of the beta-blocker propranolol could improve the outcome of PD-1 checkpoint blockade therapy (Bucsek et al. 2017), and treatment of mice with propranolol increased the number of CD8 TIL and enhanced cancer vaccine efficacy, which underscores the great importance of stress management in tumor patients. Interestingly, TIL were not affected by propranolol, while naive CD8 T cells were sensitive to β2-AR signaling, suggesting that β-AR blockade affects CD8 T cells mainly during the early priming phase (Daher et al. 2019).

Human NK cells express high levels of β2-AR but not β1-AR. Further, expression of α1- and α2-AR in CD16+ lymphocytes has been shown (Jetschmann et al. 1997; Xiao et al. 2010). Functional effects of epinephrine have mainly been attributed to β2-AR (reviewed in Ricon et al. (2019), Scanzano and Cosentino (2015)). However, epinephrine, but not norepinephrine, has also been shown to modulate the expression level of α1- and α2-AR on NK cells *in vivo* (Jetschmann et al. 1997). Elevated plasma levels of epinephrine through stress or exercise, as well as infusion of epinephrine, cause a rapid increase of NK cell numbers in peripheral blood, possibly by interfering with integrin-mediated adhesion to blood vessels (Benschop et al. 1997). Of note, increase of plasma epinephrine by moderate exercise induces the specific relocalization of distinct, highly differentiated NK cell subsets (Bigler et al. 2015; Graff et al. 2018). This response is impaired in latent CMV infection, which is associated with reduced expression of β2-AR and blunted isoproterenol-induced cAMP production in NK cells (Bigley et al. 2015).

NK cell cytotoxicity and cytokine production are inhibited by norepinephrine and epinephrine (Sun et al. 2018; Theorell et al. 2014; Ruiz-Medina et al. 2018), but there is evidence of enhanced NK cell function after treatment with sub-micromolar concentrations of epinephrine (Hellstrand et al. 1985). The stimulatory effect of low physiological concentrations of epinephrine was confirmed by the finding that induction of chronic stress through repeated social disruption had a "priming" effect on NK cell functions in mice (Tarr et al. 2012).

The negative influence of acute stress on NK cell function was already described in the 1980s: NK cell effector functions in patients undergoing upper abdominal surgery or elective coronary artery bypass grafting correlated with the patient's stress response during and after surgery (Tonnesen et al. 1984, 1987), and the decreased NK cell function after trauma or thermal injury was linked to adrenergic signaling (Blazar et al. 1986). Accordingly, a growing number of clinical studies have described beneficial effects of pre- and perioperative application of beta-blockers—usually in combination with COX2 inhibitors—in cancer surgery. Despite the heterogeneity of tumors, the data suggest a beneficial effect of β-AR blockade on NK cell-mediated tumor control (reviewed in Ricon et al. (2019)). Importantly, the extent of NK cell modulation and the resulting diminished tumor control due to surgical stress and β2-AR stimulation is modulated by sex and age (Ben-Eliyahu et al. 2000; Page et al. 2008). Chronic stress also negatively affects immune function (Segerstrom and Miller 2004). Reasons for chronic life stress can be complex and are difficult to define or control in humans. Therefore, the majority of studies on chronic stress are conducted in rodents (Patchev and Patchev 2006). Continuous administration of β2-AR agonists or chronic stress abrogates immunostimulatory effects of IL-12 on rat NK cells (Levi et al. 2011). In line with this, permanently elevated epinephrine levels might promote leukemia progression through reduced NK activity in chronically stressed rats (Inbar et al. 2011). Adrenergic signaling also interferes with NK cell effector function against viral infections. Mice receiving a β2-AR agonist were more susceptible to MCMV infection (Wieduwild et al. 2020). Similar findings have been described in humans. For example, daughters of breast cancer patients who experienced high levels of distress exhibited elevated plasma concentrations of epinephrine, which were paralleled by reduced NK cell activity (Cohen et al. 2002). Moderate physical exercise, psychological interventions, and other stress-reducing techniques have been shown to reduce epinephrine levels and thereby to counteract the negative effects of chronic stress (reviewed in Moraes et al. (2018)). In addition, there is growing evidence that stress-reducing activities such as forest bathing (Wen et al. 2019) or mindfulness-based stress reduction (MBSR) techniques (Fang et al. 2010) lead to a decrease in stress hormone levels and increased NK cell activity in healthy volunteers, and similarly in breast cancer patients and HIV-infected patients (Witek-Janusek et al. 2008; Kenne Sarenmalm et al. 2017; Robinson et al. 2003; Rao et al. 2017). Conversely, eustress induced by voluntary wheel running or an enriched environment leads to increased NK cell anti-tumor activity in mice. The authors of this work linked these effects to β-AR signals as they could be reversed by addition of the beta-blocker propranolol (Pedersen et al. 2016; Song et al. 2017).

NK cells lacking β2-AR display impaired NK cell expansion and memory formation in response to MCMV infection, which implies a role of intrinsic β2-AR signaling in optimal NK cell function (Diaz-Salazar et al. 2020). Hence, the effect of epinephrine and norepinephrine on NK cells depends on the duration of exposure, the dose, and also on the context, which may be influenced by other cytokines and factors.

Upon activation, DCs secrete large amounts of pro-inflammatory cytokines (Segura 2016), which in turn modulate T_H cell differentiation and function. Activation of β2-AR on human DCs reduces secretion of pro-inflammatory cytokines like IL-12, IL-23, TNFα, and IL-6, which results in inhibition of T_H1 differentiation of CD4 T cells and a shift toward T_H2 (Panina-Bordignon et al. 1997; Goyarts et al. 2008). Recently, inhibition of DC migration through epinephrine and β1-AR has been described (Mori et al. 2013). Mouse DCs express both β2-AR and α1-AR, with α1-AR stimulating (Maestroni 2000) and β2-AR inhibiting migration of DCs (Maestroni and Mazzola 2003). Norepinephrine enhances secretion of IL-33 through β2-AR, which enables DCs to directly promote T_H2 responses (Yanagawa et al. 2011). Further, short-term stimulation of α2-AR accelerates antigen uptake and consequently enhances immune responses (Yanagawa et al. 2010). The authors of this study suggest a possible contribution of adrenergic signals in DCs to stress-induced progression of T_H2-related allergic disorders.

Stimulation of murine DCs with Toll-like receptor 2 (TLR2) ligand in the presence of β2-AR agonist salbutamol inhibited IL-12 production, whereas IL-6 and IL-23 secretion was stimulated, resulting in a shift of cytokine pattern toward T_H17 priming. This finding involves adrenergic signals in the optimal control of bacterial or fungal infection (Happel et al. 2005), but also suggests a contribution of catecholamines to the development and progression of inflammatory diseases through IL-17 and T_H17 cells (Happel et al. 2005).

Peripheral blood monocytes express high numbers of β2-AR and treatment with the β2-AR agonist salmeterol interfering with the release of pro-inflammatory cytokines (Oddera et al. 1997; Guirao et al. 1997; Li et al. 2003), production and release of oxygen radicals (Schopf and Lemmel 1983), and phagocytosis (Borda et al. 1998). There is evidence that AR density is transiently modulated by physical exercise, but with variable kinetics and direction of modulation depending on the type of exercise (Ratge et al. 1988; Fragala et al. 2011).

The anti-inflammatory action of β2-AR agonists might contribute to increased reactivation of latent viral infection after highly stressful events, since β2-AR has been implicated in the direct stimulation of the human cytomegalovirus immediate early enhancer/promoter in monocytic cells (Fragala et al. 2011). Bone marrow monocytes that were differentiated in the presence of increased sympathetic signals during sepsis are functionally different regarding their cytokine responses (Cohen et al. 2004), and Mizuno et al. suggested the administration of β2-AR agonists to be beneficial in the treatment of sepsis by inhibiting LPS-induced production of IL-18 (Mizuno et al. 2005). Catecholamines skew monocyte differentiation toward immunosuppressive M2. Beta-blockade reduces the number of M2 monocytes and improves the control of opportunistic bacterial infections in severely burned children (Kobayashi et al. 2011).

Adrenergic signals can also exert pro-inflammatory effects under certain conditions. LPS-mediated IL-1β production is synergistically increased in monocytes and macrophages in the presence of a selective α1-AR agonist (Grisanti et al. 2011). Monocytes are involved in onset and progression of atherosclerosis (Chavez-Sanchez et al. 2014), and disease progression has been related to psychological stress (Stansfeld et al. 2002; Yusuf et al. 2004). Recently, adrenergic signaling during chronic stress has been shown to alter hematopoiesis in mice and humans, resulting in increased proliferation of inflammatory monocytes and accelerated atherosclerosis in mice (Heidt et al. 2014).

On the contrary, β2-AR expression is decreased on differentiated macrophages, and consequently these cells might be less susceptible to adrenergic signals (Joseph et al. 1981; Baker et al. 1994). β2-AR expression on human macrophages is further modulated by maturation state and cellular function (Radojcic et al. 1991; Gabanyi et al. 2016). In acutely stressed men, higher scores in anticipatory cognitive stress appraisal (ACSA) were related to lower macrophage microbicidal potential. This association was linked to norepinephrine released during the stress response (Kuebler et al. 2015).

Release of gut-derived noradrenaline in severe sepsis leads to enhanced production of pro-inflammatory cytokines by liver macrophages (or Kupffer cells) and finally promotes organ damage. This was associated with activation of α2-AR on Kupffer cells and led to the development of the concept of "sympathetic excitotoxicity in sepsis" (Miksa et al. 2005; Miksa et al. 2009). Importantly, macrophages are involved in the complex process of wound healing (Kloc et al. 2019). Administration of the β-adrenergic antagonist timolol improves wound healing in diabetic mice by shifting macrophage phenotype to anti-inflammatory M2 (Yang et al. 2020), and beta-blockade improves wound healing in chronically stressed mice (Romana-Souza et al. 2010). Moreover, acute stress affected macrophage function in wound healing in human individuals (Kuebler et al. 2013).

Among granulocytes, neutrophils express all AR except for α2b-AR (Scanzano et al. 2015) and stimulation of β2-AR inhibits the respiratory burst (Nielson 1987; Brunskole Hummel et al. 2013), which is a key function in the control of pathogens. The impact of adrenergic signals on the function of basophils, eosinophils, and mast cells has mainly been studied in the context of allergic diseases of the respiratory tract. There is still only limited knowledge about the effects of AR signaling on these cells, as is summarized in (Scanzano and Cosentino 2015).

What to Keep in Mind When Analyzing Catecholamine Receptors

Many factors affect the outcome of experiments studying the function of catecholamine receptors. Differentiation state of cells in vivo and during culture can change expression levels and downstream signaling of catecholamine receptors (Scanzano and Cosentino 2015; Heijink et al. 2003; Fan et al. 2018).

As mentioned earlier, GPCR can form heteromers with specific pharmacologic properties. In addition, many receptors are subject to circadian regulation (Gonzalez et al. 2012). Further, different types of agonists and antagonists induce variable downstream signals from the same receptor (Wingler and Lefkowitz 2020). Similarly, ligand concentration, duration of stimulus, or repeated application of stimuli induces changes in receptor structure or intracellular phosphorylation patterns, which in turn mediate coupling to different G proteins and altered receptor trafficking (Hilger et al. 2018). Finally, age- or disease-related changes in catecholamine secretion and expression of molecules downstream of catecholamine receptor activation lead to altered signaling properties (Van Gastel et al. 2021).

In this context, it is essential to consider the effects of e.g. maturation and differentiation state of cultured cells, but also the influence of housing conditions on the outcome of experimental mouse models or pre-existing life stress in human subjects, different kinds of stress applied in man and mice, etc., on the organism and subsequently on the function of the receptors studied.

5.5 Synthesis of Catecholamines in Immune Cells

Nowadays, it is well established that immune cells are not just able to respond to external catecholamines, but that many, if not all of them, can synthesize catecholamines as well. Bergquist et al. described in the 1990s the presence of catecholamines in immune cells, thus suggesting an autocrine loop of catecholamines on immune modulation (Bergquist et al. 1994). Musso et al. then confirmed the direct synthesis of catecholamines in lymphocytes (Musso et al. 1996). In their work, they added L-tyrosine and L-Dopa to lymphocyte cell culture and measured higher intracellular catecholamine levels, whereas no catecholamine increase was measured after addition of D-Dopa. In this work, Musso et al. also described different amounts of catecholamines in lymphocyte subpopulations, with T cells containing L-Dopa and norepinephrine, whereas B cells only contained L-Dopa (Musso et al. 1996). These crucial findings were confirmed only a few years later by Marino et al. (1999). In their work, HPLC was used to measure biosynthesis as well as degradation of catecholamines in PBMCs. Furthermore, Marino et al. nicely demonstrated that catecholamines can not only be stored in but also released from immune cells. Later on, catecholamine synthesis by immune cells was also demonstrated in peripheral organs. Qui et al. demonstrated the expression of TH in different lymphoid organs, with lymph nodes showing the highest and thymus the lowest concentration (Qiu et al. 2004). Also here, TH and catecholamine amounts were higher in isolated lymphocytes activated with ConA compared to unstimulated lymphocytes, thus suggesting a role for catecholaminergic pathway on immune activation. Catecholamine synthesis was also demonstrated in the synovial tissue of rheumatoid arthritis patients, where B cells, macrophages, neutrophils, mast cells, and fibroblasts, but not T cells, were found to be positive for TH (Capellino et al. 2010). The amount of TH+ cells has been shown to increase in the lymphoid organs during arthritis in a mouse model (Capellino et al. 2012), thus suggesting a direct involvement of catecholamine synthesis in arthritis. Papa et al. described dopamine synthesis in T follicular helper cells (T_{fh}) isolated from tonsils (Papa et al. 2017). After synthesis, dopamine was released by T_{fh} cells and used for strengthening T–B cell interaction. Synthesis of catecholamines by immune cells in physiologic as well as during pathologic conditions is nowadays confirmed and described very well in many articles, but it is not the goal of this chapter to itemize all of them.

Due to the crucial role of catecholamines in immune cell function, modulation of catecholamine synthesis could represent a good therapeutic target. Therefore, it is of relevance to understand the mechanisms responsible for catecholamine synthesis in immune cells. Cosentino and his group suggested pro-inflammatory pathways as key modulators in PBMCs (Cosentino et al. 2002; Ferrari et al. 2004). In their studies, catecholamine amounts were increased after treatment of PBMCs with phytohemagglutinin (Cosentino et al. 2002) and with the PKC activator TPA (Ferrari et al. 2004). Interestingly, D1-like DR blockade antagonized catecholamine production in TPA-treated PBMCs (Ferrari et al. 2004). On the contrary, an increase in cAMP is responsible for dopamine storage in human dendritic cells (Nakano et al. 2009), which led the authors to postulate an involvement of D1-like DR activation in dopamine synthesis (Nakano et al. 2009). These apparently contradictory results could be due to different activated pathways in resting and activated immune cells, or to different mechanisms involved in distinct leukocyte subtypes. A further study described hypoxic conditions as a possible trigger of TH expression and catecholamine synthesis (Jenei-Lanzl et al. 2015a). Considering the fact that hypoxic conditions often occur during chronic inflammation in the inflamed site, these results suggest a further mechanism involved in catecholamine synthesis during immune activation.

Despite the interesting results, further investigations are required to better understand the mechanisms responsible for catecholamine synthesis, storage, and release in different leukocyte subpopulations in order to use this knowledge for future therapeutic targets.

5.6 Conclusions and Perspectives

Historically, neurotransmitters, including catecholamines, have been described as compounds synthesized and released by the nervous system to transmit signals to other neurons. Similarly, cytokines are often described as small immunomodulating agents released by immune cells. Nowadays, it is well known that these descriptions are far too restrictive. Catecholamines are not just neurotransmitters but multifaceted compounds with many different functions in the body. At the beginning of the twentieth century, it became clear that catecholamines can affect the immune system; therefore, they do not just act as neurotransmitters but also as hormones, i.e., as signaling molecules used to communicate between organs and tissues. Immune cells express catecholamine receptors and can react to exogenous catecholamines under physiological conditions. Due to the multifaceted effects of catecholamines, they can inhibit as well as stimulate immune responses. It is therefore not possible to assign them a common immune function. Moreover, catecholamines play a crucial role during chronic inflammation and chronic stress, as described above. In pathologic conditions, the effects of catecholamines can differ from their physiological effects, thus suggesting that the catecholaminergic pathway can be used as a supportive mechanism along with the classical cytokine-mediated pathways or as an alternative mechanism, used when the cytokine-mediated pathways are exhausted due to chronic activation.

Furthermore, with the recent discovery of catecholamine production within the immune system, catecholamines could also be considered cytokines (Fig. 5.3). Therefore, one could assume that catecholamine-mediated immune modulation could also be a newly discovered yet "classical" immune response, rather than an "alternative, non-canonical" mechanism.

This insight has many possible clinical consequences. First of all, we are now aware that neuromodulating drugs can directly affect the immune response, as well as many other physiological body functions. These possible side effects were already described, but the direct link between neurological therapies and immune response was not taken into consideration. On the one hand, physicians should always be aware of these possible side effects in order to avoid unexpected dysregulation of immune function in patients undergoing treatment with catecholaminergic drugs. On the other hand, the unexpected effects of these drugs on the immune system could also be useful as a new therapeutic strategy in immune dysfunctions. The concept of drug repurposing has already been put forward (Cosentino 2020; Cosentino and Marino 2016) and will hopefully be further exploited in the future. In order to be able

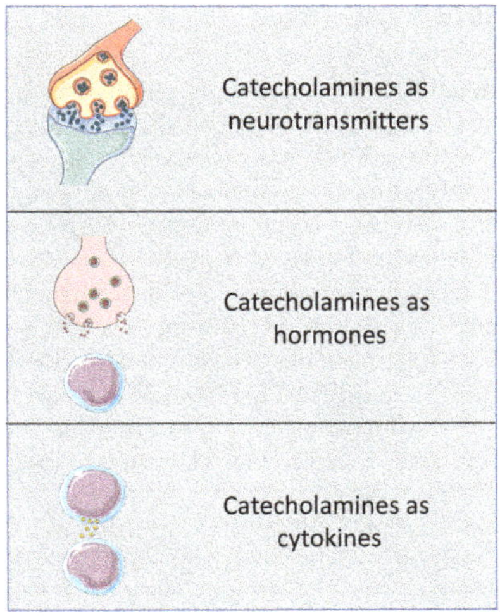

Fig. 5.3 Schematic representation of the multifaceted role of catecholamines in the body

to use catecholaminergic drugs as immunological therapy, the big issue will be to target only immune cells in order to avoid neurological side effects, and preferentially to couple catecholaminergic drugs to carriers such as functionalized nanoparticles specifically targeting only one type of immune cell. To the best of our knowledge, this strategy, to target dopaminergic drugs to the brain, is practicable (Guo et al. 2017; Malvindi et al. 2011), and it is in use to target immune cells for cancer therapy, but carriers able to deliver catecholaminergic drugs to specific immune cells are not available so far.

Future studies will therefore be necessary in order to understand how to use current knowledge of the immune functions of catecholamines for new therapeutic strategies and for a better knowledge of underlying mechanisms.

5.7 Key References

1. Basu & Dasgupta, *J Neuroimmunol,* 2000. One of the early studies assessing DR expression in human peripheral blood mononuclear cells (PBMCs).

2–3. Gonzalez et al., *PLoS Biol,* 2012 & Rebois et al., *Cell Signal*, 2012. These two publications are among the first to demonstrate that DR and AR can form heteromers.

4. Kunz et al., *Cell Stress Chaperones,* 2020. This study shows the positive effect of moderate physical stress on immune response.

5–6. Moraes et al., *Psychol Health Med,* 2018 & Ricon et al., *Cancer*, 2019. These reviews indicate the beneficial effects of β-AR modulation on immune responses and tumor control.

References

Ahlquist RP (1948) A study of the adrenotropic receptors. Am J Phys 153:586–600

Arreola R, Alvarez-Herrera S, Perez-Sanchez G, Becerril-Villanueva E, Cruz-Fuentes C, Flores-Gutierrez EO, Garces-Alvarez ME, De La Cruz-Aguilera DL, Medina-Rivero E, Hurtado-Alvarado G, Quintero-Fabian S, Pavon L (2016) Immunomodulatory effects mediated by dopamine. J Immunol Res 2016:3160486

Ayoub MA, Pfleger KD (2010) Recent advances in bioluminescence resonance energy transfer technologies to study GPCR heteromerization. Curr Opin Pharmacol 10:44–52

Baker AJ, Palmer J, Johnson M, Fuller RW (1994) Inhibitory actions of salmeterol on human airway macrophages and blood monocytes. Eur J Pharmacol 264:301–306

Basu S, Dasgupta PS (2000) Dopamine, a neurotransmitter, influences the immune system. J Neuroimmunol 102:113–124

Beaulieu JM, Espinoza S, Gainetdinov RR (2015) Dopamine receptors – IUPHAR Review 13. Br J Pharmacol 172:1–23

Ben-Eliyahu S, Shakhar G, Shakhar K, Melamed R (2000) Timing within the oestrous cycle modulates adrenergic suppression of NK activity and resistance to metastasis: possible clinical implications. Br J Cancer 83:1747–1754

Benschop RJ, Schedlowski M, Wienecke H, Jacobs R, Schmidt RE (1997) Adrenergic control of natural killer cell circulation and adhesion. Brain Behav Immun 11:321–332

Bergquist J, Tarkowski A, Ekman R, Ewing A (1994) Discovery of endogenous catecholamines in lymphocytes and evidence for catecholamine regulation of lymphocyte function via an autocrine loop. Proc Natl Acad Sci USA 91:12912–12916

Bergquist J, Josefsson E, Tarkowski A, Ekman R, Ewing A (1997) Measurements of catecholamine-mediated apoptosis of immunocompetent cells by capillary electrophoresis. Electrophoresis 18:1760–1766

Besser MJ, Ganor Y, Levite M (2005) Dopamine by itself activates either D2, D3 or D1/D5 dopaminergic receptors in normal human T-cells and triggers the selective secretion of either IL-10, TNFalpha or both. J Neuroimmunol 169:161–171

Bigler MB, Egli SB, Hysek CM, Hoenger G, Schmied L, Baldin FS, Marquardsen FA, Recher M, Liechti ME, Hess C, Berger CT (2015) Stress-Induced in vivo recruitment of human cytotoxic natural killer cells favors subsets with distinct receptor profiles and associates with increased epinephrine levels. PLoS One 10:e0145635

Bigley AB, Rezvani K, Pistillo M, Reed J, Agha N, Kunz H, O'connor DP, Sekine T, Bollard CM, Simpson RJ (2015) Acute exercise preferentially redeploys NK-cells with a highly-differentiated phenotype and augments cytotoxicity against lymphoma and multiple myeloma target cells. Part II: impact of latent cytomegalovirus infection and catecholamine sensitivity. Brain Behav Immun 49:59–65

Blazar BA, Rodrick ML, O'mahony JB, Wood JJ, Bessey PQ, Wilmore DW, Mannick JA (1986) Suppression of natural killer-cell function in humans following thermal and traumatic injury. J Clin Immunol 6:26–36

Bone NB, Liu Z, Pittet JF, Zmijewski JW (2017) Frontline Science: D1 dopaminergic receptor signaling activates the AMPK-bioenergetic pathway in macrophages and alveolar epithelial cells and reduces endotoxin-induced ALI. J Leukoc Biol 101:357–365

Boneberg EM, Von Seydlitz E, Propster K, Watzl H, Rockstroh B, Illges H (2006) D3 dopamine receptor mRNA is elevated in T cells of schizophrenic patients whereas D4 dopamine receptor mRNA is reduced in CD4+-T cells. J Neuroimmunol 173:180–187

Borda ES, Tenenbaum A, Sales ME, Rumi L, Sterin-Borda L (1998) Role of arachidonic acid metabolites in the action of a beta adrenergic agonist on human monocyte phagocytosis. Prostaglandins Leukot Essent Fatty Acids 58:85–90

Borger P, Hoekstra Y, Esselink MT, Postma DS, Zaagsma J, Vellenga E, Kauffman HF (1998) Beta-adrenoceptor-mediated inhibition of IFN-gamma, IL-3, and GM-CSF mRNA accumulation in activated human T lymphocytes is solely mediated by the beta2-adrenoceptor subtype. Am J Respir Cell Mol Biol 19:400–407

Brunskole Hummel I, Reinartz MT, Kalble S, Burhenne H, Schwede F, Buschauer A, Seifert R (2013) Dissociations in the effects of beta2-adrenergic receptor agonists on cAMP formation and superoxide production in human neutrophils: support for the concept of functional selectivity. PLoS One 8:e64556

Bucsek MJ, Qiao G, Macdonald CR, Giridharan T, Evans L, Niedzwecki B, Liu H, Kokolus KM, Eng JW, Messmer MN, Attwood K, Abrams SI, Hylander BL, Repasky EA (2017) beta-Adrenergic signaling in mice housed at standard temperatures suppresses an effector phenotype in CD8(+) T cells and undermines checkpoint inhibitor therapy. Cancer Res 77:5639–5651

Bylund DB, Eikenberg DC, Hieble JP, Langer SZ, Lefkowitz RJ, Minneman KP, Molinoff PB, Ruffolo RR Jr, Trendelenburg U (1994) International union of pharmacology nomenclature of adrenoceptors. Pharmacol Rev 46:121–136

Capellino S, Cosentino M, Wolff C, Schmidt M, Grifka J, Straub RH (2010) Catecholamine-producing cells in the synovial tissue during arthritis: modulation of sympathetic neurotransmitters as new therapeutic target. Ann Rheum Dis 69:1853–1860

Capellino S, Weber K, Gelder M, Haerle P, Straub RH (2012) First appearance and location of catecholaminergic cells during experimental arthritis and elimination by chemical sympathectomy. Arthritis Rheum 64:1110–1118

Carlson SL, Trauth K, Brooks WH, Roszman TL (1994) Enhancement of beta-adrenergic-induced cAMP accumulation in activated T-cells. J Cell Physiol 161:39–48

Carvajal Gonczi CM, Tabatabaei Shafiei M, East A, Martire E, Maurice-Ventouris MHI, Darlington PJ (2017) Reciprocal modulation of helper Th1 and Th17 cells by the beta2-adrenergic receptor agonist drug terbutaline. FEBS J 284:3018–3028

Chandrasekera PC, Wan TC, Gizewski ET, Auchampach JA, Lasley RD (2013) Adenosine A1 receptors heterodimerize with beta1- and beta2-adrenergic receptors creating novel receptor complexes with altered G protein coupling and signaling. Cell Signal 25:736–742

Chavez-Sanchez L, Espinosa-Luna JE, Chavez-Rueda K, Legorreta-Haquet MV, Montoya-Diaz E, Blanco-Favela F (2014) Innate immune system cells in atherosclerosis. Arch Med Res 45:1–14

Chen ML, Wu S, Tsai TC, Wang LK, Tsai FM (2014) Regulation of neutrophil phagocytosis of Escherichia coli by antipsychotic drugs. Int Immunopharmacol 23:550–557

Cohen M, Klein E, Kuten A, Fried G, Zinder O, Pollack S (2002) Increased emotional distress in daughters of breast cancer patients is associated with decreased natural cytotoxic activity, elevated levels of stress hormones and decreased secretion of Th1 cytokines. Int J Cancer 100:347–354

Cohen MJ, Shankar R, Stevenson J, Fernandez R, Gamelli RL, Jones SB (2004) Bone marrow norepinephrine mediates development of functionally different macrophages after thermal injury and sepsis. Ann Surg 240:132–141

Cordano C, Pardini M, Cellerino M, Schenone A, Marino F, Cosentino M (2015) Levodopa-induced neutropenia. Parkinsonism Relat Disord 21:423–425

Cosentino M (2020) Dopaminergic and adrenergic pathways as targets for drug repurposing in the neuroimmune network. J Neuroimmune Pharmacol 15:13–16

Cosentino M, Marino F (2012) Nerve Driven immunity: noradrenaline and adrenaline. Springer, Nerve-Driven Immunity

Cosentino M, Marino F (2016) The second insubria autumn school on neuroimmune pharmacology: repurposing established drugs for novel indications. J Neuroimmune Pharmacol 11:214–226

Cosentino M, Marino F, Bombelli R, Ferrari M, Rasini E, Lecchini S, Frigo G (2002) Stimulation with phytohaemagglutinin induces the synthesis of catecholamines in human peripheral blood mononuclear cells: role of protein kinase C and contribution of intracellular calcium. J Neuroimmunol 125:125–133

Cosentino M, Fietta AM, Ferrari M, Rasini E, Bombelli R, Carcano E, Saporiti F, Meloni F, Marino F, Lecchini S (2007) Human CD4+CD25+ regulatory T cells selectively express tyrosine hydroxylase and contain endogenous catecholamines subserving an autocrine/paracrine inhibitory functional loop. Blood 109:632–642

Daher C, Vimeux L, Stoeva R, Peranzoni E, Bismuth G, Wieduwild E, Lucas B, Donnadieu E, Bercovici N, Trautmann A, Feuillet V (2019) Blockade of beta-Adrenergic receptors improves CD8(+) T-cell priming and cancer vaccine efficacy. Cancer Immunol Res 7:1849–1863

Deane KH, Spieker S, Clarke CE (2004) Catechol-O-methyltransferase inhibitors versus active comparators for levodopa-induced complications in Parkinson's disease. Cochrane Database Syst Rev CD004553

Diaz-Salazar C, Bou-Puerto R, Mujal AM, Lau CM, Von Hoesslin M, Zehn D, Sun JC (2020) Cell-intrinsic adrenergic signaling controls the adaptive NK cell response to viral infection. J Exp Med 217

Estrada LD, Agac D, Farrar JD (2016) Sympathetic neural signaling via the beta2-adrenergic receptor suppresses T-cell receptor-mediated human and mouse CD8(+) T-cell effector function. Eur J Immunol 46:1948–1958

Fan X, Wang Y (2009) β2 Adrenergic receptor on T lymphocytes and its clinical implications. Prog Nat Sci 19:17–23

Fan Y, Chen Z, Pathak JL, Carneiro AMD, Chung CY (2018) Differential regulation of adhesion and phagocytosis of resting and activated microglia by dopamine. Front Cell Neurosci 12:309

Fang CY, Reibel DK, Longacre ML, Rosenzweig S, Campbell DE, Douglas SD (2010) Enhanced psychosocial well-being following participation in a mindfulness-based stress reduction program is associated with increased natural killer cell activity. J Altern Complement Med 16:531–538

Feng Y, Lu Y (2021) Immunomodulatory effects of dopamine in inflammatory diseases. Front Immunol 12:663102

Fernandez-Duenas V, Gomez-Soler M, Valle-Leon M, Watanabe M, Ferrer I, Ciruela F (2019) Revealing adenosine A2A-dopamine D2 receptor heteromers in Parkinson's disease post-mortem brain through a new AlphaScreen-based assay. Int J Mol Sci 20

Ferrari M, Cosentino M, Marino F, Bombelli R, Rasini E, Lecchini S, Frigo G (2004) Dopaminergic D1-like receptor-dependent inhibition of tyrosine hydroxylase mRNA expression and catecholamine production in human lymphocytes. Biochem Pharmacol 67:865–873

Fiserova A, Starec M, Kuldova M, Kovaru H, Pav M, Vannucci L, Pospisil M (2002) Effects of D2-dopamine and alpha-adrenoceptor antagonists in stress induced changes on immune responsiveness of mice. J Neuroimmunol 130:55–65

Fragala MS, Kraemer WJ, Mastro AM, Denegar CR, Volek JS, Hakkinen K, Anderson JM, Lee EC, Maresh CM (2011) Leukocyte beta2-adrenergic receptor expression in response to resistance exercise. Med Sci Sports Exerc 43:1422–1432

Freier E, Weber CS, Nowottne U, Horn C, Bartels K, Meyer S, Hildebrandt Y, Luetkens T, Cao Y, Pabst C, Muzzulini J, Schnee B, Brunner-Weinzierl MC, Marangolo M, Bokemeyer C, Deter HC, Atanackovic D (2010) Decrease of CD4(+)FOXP3(+) T regulatory cells in the peripheral blood of human subjects undergoing a mental stressor. Psychoneuroendocrinology 35:663–673

Gabanyi I, Muller PA, Feighery L, Oliveira TY, Costa-Pinto FA, Mucida D (2016) Neuro-immune interactions drive tissue programming in intestinal macrophages. Cell 164:378–391

Gandhi S, Pandey MR, Attwood K, Ji W, Witkiewicz AK, Knudsen ES, Allen C, Tario JD, Wallace PK, Cedeno CD, Levis M, Stack S, Funchain P, Drabick JJ, Bucsek MJ, Puzanov I, Mohammadpour H, Repasky EA, Ernstoff MS (2021) Phase I clinical trial of combination propranolol and Pembrolizumab in locally advanced and metastatic melanoma: safety, tolerability, and preliminary evidence of antitumor activity. Clin Cancer Res 27:87–95

Gaskill PJ, Carvallo L, Eugenin EA, Berman JW (2012) Characterization and function of the human macrophage dopaminergic system: implications for CNS disease and drug abuse. J Neuroinflammation 9:203

Gauthier C, Tavernier G, Charpentier F, Langin D, Le Marec H (1996) Functional beta3-adrenoceptor in the human heart. J Clin Invest 98:556–562

Girault JA (2012) Signaling in striatal neurons: the phosphoproteins of reward, addiction, and dyskinesia. Prog Mol Biol Transl Sci 106:33–62

Gomes I, Sierra S, Devi LA (2016) Detection of receptor heteromerization using in situ proximity ligation assay. Curr Protoc Pharmacol 75:2.16.1–2.16.31

Gonzalez S, Moreno-Delgado D, Moreno E, Perez-Capote K, Franco R, Mallol J, Cortes A, Casado V, Lluis C, Ortiz J, Ferre S, Canela E, Mccormick PJ (2012) Circadian-related heteromerization of adrenergic and dopamine D(4) receptors modulates melatonin synthesis and release in the pineal gland. PLoS Biol 10:e1001347

Goyarts E, Matsui M, Mammone T, Bender AM, Wagner JA, Maes D, Granstein RD (2008) Norepinephrine modulates human dendritic cell activation by altering cytokine release. Exp Dermatol 17:188–196

Graff RM, Kunz HE, Agha NH, Baker FL, Laughlin M, Bigley AB, Markofski MM, Lavoy EC, Katsanis E, Bond RA, Bollard CM, Simpson RJ (2018) beta2-Adrenergic receptor signaling mediates the preferential mobilization of differentiated subsets of CD8+ T-cells, NK-cells and non-classical monocytes in response to acute exercise in humans. Brain Behav Immun 74:143–153

Grebe KM, Hickman HD, Irvine KR, Takeda K, Bennink JR, Yewdell JW (2009) Sympathetic nervous system control of anti-influenza CD8+ T cell responses. Proc Natl Acad Sci USA 106: 5300–5305

Grisanti LA, Woster AP, Dahlman J, Sauter ER, Combs CK, Porter JE (2011) Alpha1-adrenergic receptors positively regulate Toll-like receptor cytokine production from human monocytes and macrophages. J Pharmacol Exp Ther 338:648–657

Guirao X, Kumar A, Katz J, Smith M, Lin E, Keogh C, Calvano SE, Lowry SF (1997) Catecholamines increase monocyte TNF receptors and inhibit TNF through beta 2-adrenoreceptor activation. Am J Phys 273:E1203–E1208

Guo Q, You H, Yang X, Lin B, Zhu Z, Lu Z, Li X, Zhao Y, Mao L, Shen S, Cheng H, Zhang J, Deng L, Fan J, Xi Z, Li R, Li CM (2017) Functional single-walled carbon nanotubes 'CAR' for targeting dopamine delivery into the brain of parkinsonian mice. Nanoscale 9:10832–10845

Happel KI, Dubin PJ, Zheng M, Ghilardi N, Lockhart C, Quinton LJ, Odden AR, Shellito JE, Bagby GJ, Nelson S, Kolls JK (2005) Divergent roles of IL-23 and IL-12 in host defense against Klebsiella pneumoniae. J Exp Med 202:761–769

Heidt T, Sager HB, Courties G, Dutta P, Iwamoto Y, Zaltsman A, Von Zur Muhlen C, Bode C, Fricchione GL, Denninger J, Lin CP, Vinegoni C, Libby P, Swirski FK, Weissleder R, Nahrendorf M (2014) Chronic variable stress activates hematopoietic stem cells. Nat Med 20: 754–758

Heijink IH, Vellenga E, Borger P, Postma DS, Monchy JG, Kauffman HF (2003) Polarized Th1 and Th2 cells are less responsive to negative feedback by receptors coupled to the AC/cAMP system compared to freshly isolated T cells. Br J Pharmacol 138:1441–1450

Heijink IH, Vellenga E, Oostendorp J, De Monchy JG, Postma DS, Kauffman HF (2005) Exposure to TARC alters beta2-adrenergic receptor signaling in human peripheral blood T lymphocytes. Am J Physiol Lung Cell Mol Physiol 289:L53–L59

Hein L (2006) Adrenoceptors and signal transduction in neurons. Cell Tissue Res 326:541–551

Hellstrand K, Hermodsson S, Strannegard O (1985) Evidence for a beta-adrenoceptor-mediated regulation of human natural killer cells. J Immunol 134:4095–4099

Hilger D, Masureel M, Kobilka BK (2018) Structure and dynamics of GPCR signaling complexes. Nat Struct Mol Biol 25:4–12

Inbar S, Neeman E, Avraham R, Benish M, Rosenne E, Ben-Eliyahu S (2011) Do stress responses promote leukemia progression? An animal study suggesting a role for epinephrine and prostaglandin-E2 through reduced NK activity. PLoS One 6:e19246

Jenei-Lanzl Z, Capellino S, Kees F, Fleck M, Lowin T, Straub RH (2015a) Anti-inflammatory effects of cell-based therapy with tyrosine hydroxylase-positive catecholaminergic cells in experimental arthritis. Ann Rheum Dis 74:444–451

Jenei-Lanzl Z, Zwingenberg J, Lowin T, Anders S, Straub RH (2015b) Proinflammatory receptor switch from Galphas to Galphai signaling by beta-arrestin-mediated PDE4 recruitment in mixed RA synovial cells. Brain Behav Immun 50:266–274

Jetschmann JU, Benschop RJ, Jacobs R, Kemper A, Oberbeck R, Schmidt RE, Schedlowski M (1997) Expression and in-vivo modulation of alpha- and beta-adrenoceptors on human natural killer (CD16+) cells. J Neuroimmunol 74:159–164

Johnstone EKM, See HB, Abhayawardana RS, Song A, Rosengren KJ, Hill SJ, Pfleger KDG (2021) Investigation of receptor heteromers using nanoBRET ligand binding. Int J Mol Sci 22

Joseph M, Tonnel AB, Capron A, Dessaint JP (1981) The interaction of IgE antibody with human alveolar macrophages and its participation in the inflammatory processes of lung allergy. Agents Actions 11:619–622

Kamal M, Jockers R (2011) Biological significance of GPCR heteromerization in the neuro-endocrine system. Front Endocrinol (Lausanne) 2:2

Karaszewski JW, Reder AT, Maselli R, Brown M, Arnason BG (1990) Sympathetic skin responses are decreased and lymphocyte beta-adrenergic receptors are increased in progressive multiple sclerosis. Ann Neurol 27:366–372

Kasai RS, Kusumi A (2014) Single-molecule imaging revealed dynamic GPCR dimerization. Curr Opin Cell Biol 27:78–86

Kavelaars A (2002) Regulated expression of alpha-1 adrenergic receptors in the immune system. Brain Behav Immun 16:799–807

Kavelaars A, Cobelens PM, Teunis MA, Heijnen CJ (2005) Changes in innate and acquired immune responses in mice with targeted deletion of the dopamine transporter gene. J Neuroimmunol 161:162–168

Kenne Sarenmalm E, Martensson LB, Andersson BA, Karlsson P, Bergh I (2017) Mindfulness and its efficacy for psychological and biological responses in women with breast cancer. Cancer Med 6:1108–1122

Kennedy SL, Nickerson M, Campisi J, Johnson JD, Smith TP, Sharkey C, Fleshner M (2005) Splenic norepinephrine depletion following acute stress suppresses in vivo antibody response. J Neuroimmunol 165:150–160

Kintscher U (2012) Reuptake inhibitors of dopamine, noradrenaline, and serotonin. Handb Exp Pharmacol:339–347

Klein MO, Battagello DS, Cardoso AR, Hauser DN, Bittencourt JC, Correa RG (2019) Dopamine: functions, signaling, and association with neurological diseases. Cell Mol Neurobiol 39:31–59

Kloc M, Ghobrial RM, Wosik J, Lewicka A, Lewicki S, Kubiak JZ (2019) Macrophage functions in wound healing. J Tissue Eng Regen Med 13:99–109

Kobayashi M, Jeschke MG, Asai A, Kogiso M, Yoshida S, Herndon DN, Suzuki F (2011) Propranolol as a modulator of M2b monocytes in severely burned patients. J Leukoc Biol 89: 797–803

Kohm AP, Sanders VM (1999) Suppression of antigen-specific Th2 cell-dependent IgM and IgG1 production following norepinephrine depletion in vivo. J Immunol 162:5299–5308

Kohm AP, Sanders VM (2001) Norepinephrine and beta 2-adrenergic receptor stimulation regulate CD4+ T and B lymphocyte function in vitro and in vivo. Pharmacol Rev 53:487–525

Korichneva IL, Tkachuk VA (1990) Alterations in beta-adrenoceptor density on T-lymphocytes upon activation with interleukin-2 and phytohaemagglutinin. Biomed Sci 1:84–88

Kuebler U, Wirtz PH, Sakai M, Stemmer A, Ehlert U (2013) Acute stress reduces wound-induced activation of microbicidal potential of ex vivo isolated human monocyte-derived macrophages. PLoS One 8:e55875

Kuebler U, Wirtz PH, Sakai M, Stemmer A, Meister RE, Ehlert U (2015) Anticipatory cognitive stress appraisal modulates suppression of wound-induced macrophage activation by acute psychosocial stress. Psychophysiology 52:499–508

Kunz HE, Agha NH, Hussain M, Lavoy EC, Smith KA, Mylabathula P, Diak D, Baker FL, O'connor DP, Bond RA, Katsanis E, Bollard CM, Simpson RJ (2020) The effects of beta1 and beta1+2 adrenergic receptor blockade on the exercise-induced mobilization and ex vivo expansion of virus-specific T cells: implications for cellular therapy and the anti-viral immune effects of exercise. Cell Stress Chaperones 25:993–1012

Kustrimovic N, Rasini E, Legnaro M, Marino F, Cosentino M (2014) Expression of dopaminergic receptors on human CD4+ T lymphocytes: flow cytometric analysis of naive and memory subsets and relevance for the neuroimmunology of neurodegenerative disease. J Neuroimmune Pharmacol 9:302–312

Langer SZ (1980) Presynaptic regulation of the release of catecholamines. Pharmacol Rev 32:337–362

Leposavic GM, Pilipovic IM (2018) Intrinsic and extrinsic thymic adrenergic networks: sex steroid-dependent plasticity. Front Endocrinol (Lausanne) 9:13

Levi B, Benish M, Goldfarb Y, Sorski L, Melamed R, Rosenne E, Ben-Eliyahu S (2011) Continuous stress disrupts immunostimulatory effects of IL-12. Brain Behav Immun 25:727–735

Levite M (2016) Dopamine and T cells: dopamine receptors and potent effects on T cells, dopamine production in T cells, and abnormalities in the dopaminergic system in T cells in autoimmune, neurological and psychiatric diseases. Acta Physiol (Oxf) 216:42–89

Li CY, Chou TC, Lee CH, Tsai CS, Loh SH, Wong CS (2003) Adrenaline inhibits lipopolysaccharide-induced macrophage inflammatory protein-1 alpha in human monocytes: the role of beta-adrenergic receptors. Anesth Analg 96:518–523, table of contents

Maestroni GJ (2000) Dendritic cell migration controlled by alpha 1b-adrenergic receptors. J Immunol 165:6743–6747

Maestroni GJ, Mazzola P (2003) Langerhans cells beta 2-adrenoceptors: role in migration, cytokine production, Th priming and contact hypersensitivity. J Neuroimmunol 144:91–99

Malvindi MA, Di Corato R, Curcio A, Melisi D, Rimoli MG, Tortiglione C, Tino A, George C, Brunetti V, Cingolani R, Pellegrino T, Ragusa A (2011) Multiple functionalization of fluorescent nanoparticles for specific biolabeling and drug delivery of dopamine. Nanoscale 3:5110–5119

Mansfield LE, Nelson HS (1982) Effect of beta-adrenergic agents on immunoglobulin G levels of asthmatic subjects. Int Arch Allergy Appl Immunol 68:13–16

Marino F, Cosentino M, Bombelli R, Ferrari M, Lecchini S, Frigo G (1999) Endogenous catecholamine synthesis, metabolism storage, and uptake in human peripheral blood mononuclear cells. Exp Hematol 27:489–495

Matthews KA, Caggiula AR, Mcallister CG, Berga SL, Owens JF, Flory JD, Miller AL (1995) Sympathetic reactivity to acute stress and immune response in women. Psychosom Med 57:564–571

Mckenna F, Mclaughlin PJ, Lewis BJ, Sibbring GC, Cummerson JA, Bowen-Jones D, Moots RJ (2002) Dopamine receptor expression on human T- and B-lymphocytes, monocytes, neutrophils, eosinophils and NK cells: a flow cytometric study. J Neuroimmunol 132:34–40

Meltzer SJ, Meltzer C (1903) On a difference in the influence upon inflammation between the section of the sympathetic nerve and the removal of the sympathetic Ganglion. J Med Res 10:135–141

Meredith EJ, Holder MJ, Rosen A, Lee AD, Dyer MJ, Barnes NM, Gordon J (2006) Dopamine targets cycling B cells independent of receptors/transporter for oxidative attack: implications for non-Hodgkin's lymphoma. Proc Natl Acad Sci USA 103:13485–13490

Miksa M, Wu R, Zhou M, Wang P (2005) Sympathetic excitotoxicity in sepsis: pro-inflammatory priming of macrophages by norepinephrine. Front Biosci 10:2217–2229

Miksa M, Das P, Zhou M, Wu R, Dong W, Ji Y, Goyert SM, Ravikumar TS, Wang P (2009) Pivotal role of the alpha(2A)-adrenoceptor in producing inflammation and organ injury in a rat model of sepsis. PLoS One 4:e5504

Mikulak J, Bozzo L, Roberto A, Pontarini E, Tentorio P, Hudspeth K, Lugli E, Mavilio D (2014) Dopamine inhibits the effector functions of activated NK cells via the upregulation of the D5 receptor. J Immunol 193:2792–2800

Mizuno K, Takahashi HK, Iwagaki H, Katsuno G, Kamurul HA, Ohtani S, Mori S, Yoshino T, Nishibori M, Tanaka N (2005) Beta2-adrenergic receptor stimulation inhibits LPS-induced IL-18 and IL-12 production in monocytes. Immunol Lett 101:168–172
Moraes LJ, Miranda MB, Loures LF, Mainieri AG, Marmora CHC (2018) A systematic review of psychoneuroimmunology-based interventions. Psychol Health Med 23:635–652
Mores KL, Cassell RJ, Van Rijn RM (2019) Arrestin recruitment and signaling by G protein-coupled receptor heteromers. Neuropharmacology 152:15–21
Mori T, Kabashima K, Fukamachi S, Kuroda E, Sakabe J, Kobayashi M, Nakajima S, Nakano K, Tanaka Y, Matsushita S, Nakamura M, Tokura Y (2013) D1-like dopamine receptors antagonist inhibits cutaneous immune reactions mediated by Th2 and mast cells. J Dermatol Sci 71:37–44
Morkawa K, Oseko F, Morikawa S (1993) Immunosuppressive property of bromocriptine on human B lymphocyte function in vitro. Clin Exp Immunol 93:200–205
Musso NR, Brenci S, Setti M, Indiveri F, Lotti G (1996) Catecholamine content and in vitro catecholamine synthesis in peripheral human lymphocytes. J Clin Endocrinol Metab 81:3553–3557
Nakano K, Higashi T, Hashimoto K, Takagi R, Tanaka Y, Matsushita S (2008) Antagonizing dopamine D1-like receptor inhibits Th17 cell differentiation: preventive and therapeutic effects on experimental autoimmune encephalomyelitis. Biochem Biophys Res Commun 373:286–291
Nakano K, Higashi T, Takagi R, Hashimoto K, Tanaka Y, Matsushita S (2009) Dopamine released by dendritic cells polarizes Th2 differentiation. Int Immunol 21:645–654
Nielson CP (1987) Beta-adrenergic modulation of the polymorphonuclear leukocyte respiratory burst is dependent upon the mechanism of cell activation. J Immunol 139:2392–2397
Nolan RA, Muir R, Runner K, Haddad EK, Gaskill PJ (2019) Role of macrophage dopamine receptors in mediating cytokine production: implications for neuroinflammation in the context of HIV-associated neurocognitive disorders. J Neuroimmune Pharmacol 14:134–156
Nozaki H, Hozumi K, Nishimura T, Habu S (1996) Regulation of NK activity by the administration of bromocriptine in haloperidol-treated mice. Brain Behav Immun 10:17–26
Oddera S, Silvestri M, Scarso L, Testi R, Rossi GA (1997) Salmeterol inhibits the allergen-induced mononuclear cell proliferation and downregulates GM-CSF release and HLA-DR expression by monocytes. Pulm Pharmacol Ther 10:43–49
Pacheco R, Prado CE, Barrientos MJ, Bernales S (2009) Role of dopamine in the physiology of T-cells and dendritic cells. J Neuroimmunol 216:8–19
Page GG, Fennelly AM, Littleton-Kearney MT, Ben-Eliyahu S (2008) Male–female differences in the impact of beta-adrenoceptor stimulation on resistance to experimental metastasis: exploring the effects of age and gonadal hormone involvement. J Neuroimmunol 193:113–119
Panina-Bordignon P, Mazzeo D, Lucia PD, D'ambrosio D, Lang R, Fabbri L, Self C, Sinigaglia F (1997) Beta2-agonists prevent Th1 development by selective inhibition of interleukin 12. J Clin Invest 100:1513–1519
Papa I, Saliba D, Ponzoni M, Bustamante S, Canete PF, Gonzalez-Figueroa P, Mcnamara HA, Valvo S, Grimbaldeston M, Sweet RA, Vohra H, Cockburn IA, Meyer-Hermann M, Dustin ML, Doglioni C, Vinuesa CG (2017) TFH-derived dopamine accelerates productive synapses in germinal centres. Nature 547:318–323
Patchev VK, Patchev AV (2006) Experimental models of stress. Dialogues Clin Neurosci 8:417–432
Pedersen L, Idorn M, Olofsson GH, Lauenborg B, Nookaew I, Hansen RH, Johannesen HH, Becker JC, Pedersen KS, Dethlefsen C, Nielsen J, Gehl J, Pedersen BK, Thor Straten P, Hojman P (2016) Voluntary running suppresses tumor growth through epinephrine- and IL-6-dependent NK cell mobilization and redistribution. Cell Metab 23:554–562
Pereira A, Mclaren A, Bell WR, Copolov D, Dean B (2003) Potential clozapine target sites on peripheral hematopoietic cells and stromal cells of the bone marrow. Pharmacogenomics J 3: 227–234

Perreault ML, Hasbi A, O'dowd BF, George SR (2014) Heteromeric dopamine receptor signaling complexes: emerging neurobiology and disease relevance. Neuropsychopharmacology 39:156–168

Pinoli M, Marino F, Cosentino M (2017) Dopaminergic regulation of innate immunity: a review. J Neuroimmune Pharmacol 12:602–623

Podojil JR, Sanders VM (2005) CD86 and beta2-adrenergic receptor stimulation regulate B-cell activity cooperatively. Trends Immunol 26:180–185

Podolec Z, Vetulani J, Bednarczyk B, Szczeklik A (1979) Central dopamine receptors regulate blood eosinophilia in the rat. Allergy 34:103–110

Pongratz G, Mcalees JW, Conrad DH, Erbe RS, Haas KM, Sanders VM (2006) The level of IgE produced by a B cell is regulated by norepinephrine in a p38 MAPK- and CD23-dependent manner. J Immunol 177:2926–2938

Qiao G, Chen M, Mohammadpour H, Macdonald CR, Bucsek MJ, Hylander BL, Barbi JJ, Repasky EA (2021) Chronic adrenergic stress contributes to metabolic dysfunction and an exhausted phenotype in T cells in the tumor microenvironment. Cancer Immunol Res

Qiu YH, Peng YP, Jiang JM, Wang JJ (2004) Expression of tyrosine hydroxylase in lymphocytes and effect of endogenous catecholamines on lymphocyte function. Neuroimmunomodulation 11:75–83

Radojcic T, Baird S, Darko D, Smith D, Bulloch K (1991) Changes in beta-adrenergic receptor distribution on immunocytes during differentiation: an analysis of T cells and macrophages. J Neurosci Res 30:328–335

Rao RM, Vadiraja HS, Nagaratna R, Gopinath KS, Patil S, Diwakar RB, Shahsidhara HP, Ajaikumar BS, Nagendra HR (2017) Effect of Yoga on sleep quality and neuroendocrine immune response in metastatic breast cancer patients. Indian J Palliat Care 23:253–260

Ratge D, Wiedemann A, Kohse KP, Wisser H (1988) Alterations of beta-adrenoceptors on human leukocyte subsets induced by dynamic exercise: effect of prednisone. Clin Exp Pharmacol Physiol 15:43–53

Rebois RV, Maki K, Meeks JA, Fishman PH, Hebert TE, Northup JK (2012) D2-like dopamine and beta-adrenergic receptors form a signaling complex that integrates Gs- and Gi-mediated regulation of adenylyl cyclase. Cell Signal 24:2051–2060

Ricon I, Hanalis-Miller T, Haldar R, Jacoby R, Ben-Eliyahu S (2019) Perioperative biobehavioral interventions to prevent cancer recurrence through combined inhibition of beta-adrenergic and cyclooxygenase 2 signaling. Cancer 125:45–56

Robinson FP, Mathews HL, Witek-Janusek L (2003) Psycho-endocrine-immune response to mindfulness-based stress reduction in individuals infected with the human immunodeficiency virus: a quasiexperimental study. J Altern Complement Med 9:683–694

Romana-Souza B, Porto LC, Monte-Alto-Costa A (2010) Cutaneous wound healing of chronically stressed mice is improved through catecholamines blockade. Exp Dermatol 19:821–829

Ruiz-Medina BE, Cadena-Medina DA, Esparza E, Arrieta AJ, Kirken RA (2018) Isoproterenol-induced beta-2 adrenergic receptor activation negatively regulates interleukin-2 signaling. Biochem J 475:2907–2923

Scanzano A, Cosentino M (2015) Adrenergic regulation of innate immunity: a review. Front Pharmacol 6:171

Scanzano A, Schembri L, Rasini E, Luini A, Dallatorre J, Legnaro M, Bombelli R, Congiu T, Cosentino M, Marino F (2015) Adrenergic modulation of migration, CD11b and CD18 expression, ROS and interleukin-8 production by human polymorphonuclear leukocytes. Inflamm Res 64:127–135

Schopf RE, Lemmel EM (1983) Control of the production of oxygen intermediates of human polymorphonuclear leukocytes and monocytes by beta-adrenergic receptors. J Immunopharmacol 5:203–216

Schulz C, Eisenhofer G, Lehnert H (2004) Principles of catecholamine biosynthesis, metabolism and release. Front Horm Res 31:1–25

Segerstrom SC, Miller GE (2004) Psychological stress and the human immune system: a meta-analytic study of 30 years of inquiry. Psychol Bull 130:601–630
Segura E (2016) Review of mouse and human dendritic cell subsets. Methods Mol Biol 1423:3–15
Seol IW, Kuo NY, Kim KM (2004) Effects of dopaminergic drugs on the mast cell degranulation and nitric oxide generation in RAW 264.7 cells. Arch Pharm Res 27:94–98
Sheridan JF, Dobbs C, Jung J, Chu X, Konstantinos A, Padgett D, Glaser R (1998) Stress-induced neuroendocrine modulation of viral pathogenesis and immunity. Ann NY Acad Sci 840:803–808
Sibley DR, Monsma FJ Jr, Shen Y (1993) Molecular neurobiology of dopaminergic receptors. Int Rev Neurobiol 35:391–415
Slota C, Shi A, Chen G, Bevans M, Weng NP (2015) Norepinephrine preferentially modulates memory CD8 T cell function inducing inflammatory cytokine production and reducing proliferation in response to activation. Brain Behav Immun 46:168–179
Song Y, Gan Y, Wang Q, Meng Z, Li G, Shen Y, Wu Y, Li P, Yao M, Gu J, Tu H (2017) Enriching the housing environment for mice enhances their NK cell antitumor immunity via sympathetic nerve-dependent regulation of NKG2D and CCR5. Cancer Res 77:1611–1622
Sookhai S, Wang JH, Mccourt M, O'connell D, Redmond HP (1999) Dopamine induces neutrophil apoptosis through a dopamine D-1 receptor-independent mechanism. Surgery 126:314–322
Stansfeld SA, Fuhrer R, Shipley MJ, Marmot MG (2002) Psychological distress as a risk factor for coronary heart disease in the Whitehall II Study. Int J Epidemiol 31:248–255
Stein MN, Malhotra J, Tarapore RS, Malhotra U, Silk AW, Chan N, Rodriguez L, Aisner J, Aiken RD, Mayer T, Haffty BG, Newman JH, Aspromonte SM, Bommareddy PK, Estupinian R, Chesson CB, Sadimin ET, Li S, Medina DJ, Saunders T, Frankel M, Kareddula A, Damare S, Wesolowsky E, Gabel C, El-Deiry WS, Prabhu VV, Allen JE, Stogniew M, Oster W, Bertino JR, Libutti SK, Mehnert JM, Zloza A (2019) Safety and enhanced immunostimulatory activity of the DRD2 antagonist ONC201 in advanced solid tumor patients with weekly oral administration. J Immunother Cancer 7:136
Sun Z, Hou D, Liu S, Fu W, Wang J, Liang Z (2018) Norepinephrine inhibits the cytotoxicity of NK92MI cells via the beta2adrenoceptor/cAMP/PKA/pCREB signaling pathway. Mol Med Rep 17:8530–8535
Svenningsson P, Nishi A, Fisone G, Girault JA, Nairn AC, Greengard P (2004) DARPP-32: an integrator of neurotransmission. Annu Rev Pharmacol Toxicol 44:269–296
Swanson MA, Lee WT, Sanders VM (2001) IFN-gamma production by Th1 cells generated from naive CD4+ T cells exposed to norepinephrine. J Immunol 166:232–240
Takenaka MC, Guereschi MG, Basso AS (2017) Neuroimmune interactions: dendritic cell modulation by the sympathetic nervous system. Semin Immunopathol 39:165–176
Takkenberg JJ, Czer LS, Fishbein MC, Luthringer DJ, Quartel AW, Mirocha J, Queral CA, Blanche C, Trento A (2004) Eosinophilic myocarditis in patients awaiting heart transplantation. Crit Care Med 32:714–721
Tarr AJ, Powell ND, Reader BF, Bhave NS, Roloson AL, Carson WE 3rd, Sheridan JF (2012) beta-Adrenergic receptor mediated increases in activation and function of natural killer cells following repeated social disruption. Brain Behav Immun 26:1226–1238
Teunis MA, Heijnen CJ, Cools AR, Kavelaars A (2004) Reduced splenic natural killer cell activity in rats with a hyperreactive dopaminergic system. Psychoneuroendocrinology 29:1058–1064
Theorell J, Gustavsson AL, Tesi B, Sigmundsson K, Ljunggren HG, Lundback T, Bryceson YT (2014) Immunomodulatory activity of commonly used drugs on Fc-receptor-mediated human natural killer cell activation. Cancer Immunol Immunother 63:627–641
Thomas Broome S, Louangaphay K, Keay KA, Leggio GM, Musumeci G, Castorina A (2020) Dopamine: an immune transmitter. Neural Regen Res 15:2173–2185
Tonnesen E, Huttel MS, Christensen NJ, Schmitz O (1984) Natural killer cell activity in patients undergoing upper abdominal surgery: relationship to the endocrine stress response. Acta Anaesthesiol Scand 28:654–660

Tonnesen E, Brinklov MM, Christensen NJ, Olesen AS, Madsen T (1987) Natural killer cell activity and lymphocyte function during and after coronary artery bypass grafting in relation to the endocrine stress response. Anesthesiology 67:526–533

Ugalde V, Contreras F, Prado C, Chovar O, Espinoza A, Pacheco R (2021) Dopaminergic signalling limits suppressive activity and gut homing of regulatory T cells upon intestinal inflammation. Mucosal Immunol 14:652–666

Van Gastel J, Leysen H, Boddaert J, Vangenechten L, Luttrell LM, Martin B, Maudsley S (2021) Aging-related modifications to G protein-coupled receptor signaling diversity. Pharmacol Ther 223:107793

Vidi PA, Chemel BR, Hu CD, Watts VJ (2008) Ligand-dependent oligomerization of dopamine D (2) and adenosine A(2A) receptors in living neuronal cells. Mol Pharmacol 74:544–551

Wahle M, Stachetzki U, Krause A, Pierer M, Hantzschel H, Baerwald CG (2001) Regulation of beta2-adrenergic receptors on CD4 and CD8 positive lymphocytes by cytokines in vitro. Cytokine 16:205–209

Wei L, Zhang C, Chen HY, Zhang ZJ, Ji ZF, Yue T, Dai XM, Zhu Q, Ma LL, He DY, Jiang LD (2015) Dopamine receptor DR2 expression in B cells is negatively correlated with disease activity in rheumatoid arthritis patients. Immunobiology 220:323–330

Wen Y, Yan Q, Pan Y, Gu X, Liu Y (2019) Medical empirical research on forest bathing (Shinrin-yoku): a systematic review. Environ Health Prev Med 24:70

Wieduwild E, Girard-Madoux MJ, Quatrini L, Laprie C, Chasson L, Rossignol R, Bernat C, Guia S, Ugolini S (2020) beta2-adrenergic signals downregulate the innate immune response and reduce host resistance to viral infection. J Exp Med 217

Wingler LM, Lefkowitz RJ (2020) Conformational basis of G protein-coupled receptor signaling versatility. Trends Cell Biol 30:736–747

Witek-Janusek L, Albuquerque K, Chroniak KR, Chroniak C, Durazo-Arvizu R, Mathews HL (2008) Effect of mindfulness based stress reduction on immune function, quality of life and coping in women newly diagnosed with early stage breast cancer. Brain Behav Immun 22:969–981

Wnorowski A, Jozwiak K (2014) Homo- and hetero-oligomerization of beta2-adrenergic receptor in receptor trafficking, signaling pathways and receptor pharmacology. Cell Signal 26:2259–2265

Xiao J, Huang HW, Peng YP, Bao JY, Huang Y, Qiu YH (2010) Modulation of natural killer cell function by alpha-adrenoreceptor-coupled signalling. Neuro Endocrinol Lett 31:635–644

Yanagawa Y, Matsumoto M, Togashi H (2010) Enhanced dendritic cell antigen uptake via alpha2 adrenoceptor-mediated PI3K activation following brief exposure to noradrenaline. J Immunol 185:5762–5768

Yanagawa Y, Matsumoto M, Togashi H (2011) Adrenoceptor-mediated enhancement of interleukin-33 production by dendritic cells. Brain Behav Immun 25:1427–1433

Yang Y, Ye Y, Chen C, Kong C, Su X, Zhang X, Bai W, He X (2019) Acute traumatic brain injury induces CD4+ and CD8+ T cell functional impairment by upregulating the expression of PD-1 via the activated sympathetic nervous system. Neuroimmunomodulation 26:43–57

Yang HY, Fierro F, So M, Yoon DJ, Nguyen AV, Gallegos A, Bagood MD, Rojo-Castro T, Alex A, Stewart H, Chigbrow M, Dasu MR, Peavy TR, Soulika AM, Nolta JA, Isseroff RR (2020) Combination product of dermal matrix, human mesenchymal stem cells, and timolol promotes diabetic wound healing in mice. Stem Cells Transl Med 9:1353–1364

Yusuf S, Hawken S, Ounpuu S, Dans T, Avezum A, Lanas F, Mcqueen M, Budaj A, Pais P, Varigos J, Lisheng L, Investigators IS (2004) Effect of potentially modifiable risk factors associated with myocardial infarction in 52 countries (the INTERHEART study): case-control study. Lancet 364:937–952

Zhao W, Huang Y, Liu Z, Cao BB, Peng YP, Qiu YH (2013) Dopamine receptors modulate cytotoxicity of natural killer cells via cAMP-PKA-CREB signaling pathway. PLoS One 8: e65860

Neuroendocrine-immune Interactions in Major Depressive Disorder: Glucocorticoids and Glucocorticoid Receptors

6

Frances Isabella Weston, Luca Sforzini, Annamaria Cattaneo, and Carmine Maria Pariante

Abstract

Major depressive disorder is a leading cause of disability worldwide; therefore, effective treatment options are crucial. However, due to the highly heterogeneous nature of depression, a comprehensive understanding of the disease is lacking and treatment options are limited. Whilst the pathology of depression is complex, neuroendocrine–immune interactions have consistently been linked to the disease. Hypothalamic–pituitary–adrenal (HPA) axis dysfunction has been identified as one of the main contributing factors, impacting 50–80% of patients with depression. The 'glucocorticoid resistance model' provided the first explanations of this dysfunction, suggesting reduced function of the glucocorticoid receptor; thus, glucocorticoid resistance, seen in some MDD patients, allows pro-inflammatory pathways to evade normal feedback inhibition by glucocorticoids. However, recent research has suggested alternative mechanisms, which identify cortisol as a pro-inflammatory mediator of stress reactions.

F. I. Weston (✉) · L. Sforzini
Department of Psychological Medicine, Institute of Psychiatry, Psychology and Neuroscience, King's College London, London, UK
e-mail: frances.i.weston@kcl.ac.uk

A. Cattaneo
Department of Pharmacological and Biomolecular Sciences, University of Milan, Milan, Italy

Laboratory of Biological Psychiatry, IRCCS Istituto Centro San Giovanni di Dio Fatebenefratelli, Brescia, Italy

C. M. Pariante
Department of Psychological Medicine, Institute of Psychiatry, Psychology and Neuroscience, King's College London, London, UK

National Institute for Health and Research Biomedical Research Centre at South London and Maudsley NHS Foundation Trust and King's College London, London, UK

J. P. Konsman, T. M. Reyes (eds.), *Neuroendocrine-Immune System Interactions*, Masterclass in Neuroendocrinology 13,
https://doi.org/10.1007/978-3-031-21358-8_6

Additional research into glucocorticoid dysfunction in MDD has also found single nucleotide polymorphisms in FKBP5, a key regulator of glucocorticoid receptor function, to play a role in HPA axis dysfunction and thus confer risk of depression. These effects are mediated by gene–environment interactions, specifically adverse early-life events. Whilst the underlying epigenetic mechanisms are not fully understood, increased FKBP5 mRNA expression and altered FKBP5 methylation are thought to play a role in impaired HPA axis function. An increased understanding of the interactions involving FKBP5 may in turn increase understanding of the pathophysiology of depression. This will allow identification of high-risk individuals who have past adverse early-life experiences. In turn, this may also impact the course of future antidepressant treatment and development.

Keywords

Major depressive disorder · Hypothalamic–pituitary–adrenal axis · Glucocorticoids · Glucocorticoid resistance · Inflammation · FKBP5 · Neuroendocrine immunology

6.1 Introduction

In recent decades, there has been a growing body of evidence for the association between the physiological functioning of the body and psychiatric state (Renoir et al. 2013). One of the best-established pathways involved in this mind–body interface is the neuroendocrine system, involving communication between the endocrine and nervous system (Toni 2004). This interaction is crucial for regulating homeostasis within the body; thus, a deviation in normal endocrine function can cause multiple pathological consequences.

One of the most researched neuroendocrine pathways involved in this mind–body interface is the hypothalamic–pituitary–adrenal (HPA) axis. HPA axis homeostasis is an integral component in maintaining a normal stress response, and directly influences the central nervous system, coordinating emotional, behavioural and physiological events (Seidman 2006). HPA axis activation triggers the release of the glucocorticoid hormone, cortisol (humans) or corticosterone (rodents), which plays a crucial role in anti-inflammatory and immunosuppressive processes required for maintaining homeostasis (Bellavance and Rivest 2014). Glucocorticoid receptors (GRs) are expressed on nearly all immune cell types; therefore, glucocorticoids have a wide range of immunomodulatory functions.

Extensive research has identified an association between HPA axis dysregulation and immune dysregulation, with psychiatric conditions such as major depressive disorder (MDD). Interestingly, the two most frequently reported physiological findings in MDD are HPA axis hyperactivation and increased inflammation (Pariante 2017). This is particularly important as MDD is currently the leading cause of disability around the world, increased in prevalence by 18.4% between

2005 and 2015 (Friedrich 2017), and was the largest contributor to nonfatal health loss in 2015 (Friedrich 2017). Furthermore, MDD is associated with the development of heart diseases, diabetes and stroke, and an increased risk of developing Alzheimer's disease (Lang and Borgwardt 2013). However, despite this impact, treatment options are still largely ineffective. Over 50% of patients do not respond to the first treatment prescribed, whilst 30% still do not respond following multiple different treatment attempts (Menke 2019). Individuals who do not respond or respond partially to treatments may be considered as having 'treatment-resistant' or 'partially responsive' depression, respectively (Sforzini et al. 2021). Therefore, MDD is currently a major mental and public health issue, and identifying its biomarkers is crucial to future successful treatment regimes (Mäntylä 2020).

Around 50–80% of depressed patients show hyperactivity of the HPA axis and glucocorticoid resistance (Anacker et al. 2011). However, to be able to stabilise this hyperactivity or identify patients most at risk to it, researchers must identify the causes. 'Glucocorticoid resistance' expresses the concept that glucocorticoid hormones are unable to exert their physiological actions, including the feedback inhibition on cortisol secretion and the anti-inflammatory action (see below). Thus, the role of GR resistance and its interplay with increased inflammation in MDD has been identified as an area of interest for research (Cattaneo et al. 2020).

In the search for molecular targets of HPA axis dysregulation in MDD, FK506 binding protein 51 (FKBP5), a key regulator of the GR, has been identified (Rao et al. 2016). Specifically, FKBP5 appears to mediate gene–environment interactions through altered genetic and epigenetic regulation. For example, FKBP5 genes confer sensitivity to adverse early-life events (AELEs), and gene variants appear to contribute to MDD development (Matosin et al. 2018). FKBP5 function in MDD will be further discussed in this chapter.

In this chapter, we will discuss the neuroendocrine–immune interactions in MDD by describing the systems individually, how they interact, and the significant role of the GR in MDD.

6.2 Physiology of the Hypothalamic–Pituitary–Adrenal Axis

The HPA axis is a neuroendocrine unit which incorporates a range of interactions between the hypothalamus, the pituitary gland and the adrenal glands (Fig. 6.1). The axis begins with corticotropin-releasing hormone (CRH), a hypothalamic hormone produced in the paraventricular nucleus (PVN), regulated by stress (Brook and Marshall 2001). Release of CRH into the hypophyseal portal system results in binding to CRH receptors in the anterior pituitary, initiating the release of adrenocorticotropic hormone (ACTH) (Chalmers et al. 1995). In turn, this stimulates glucocorticoid (cortisol) synthesis in the adrenal cortex. Cortisol circulates in the bloodstream and has a range of metabolic effects. To maintain regulation of the HPA axis, cortisol acts in a negative feedback loop, preventing further release of CRH or ACTH (Fig. 6.1), thus aiding in maintaining basal homeostasis (Brook and Marshall 2001).

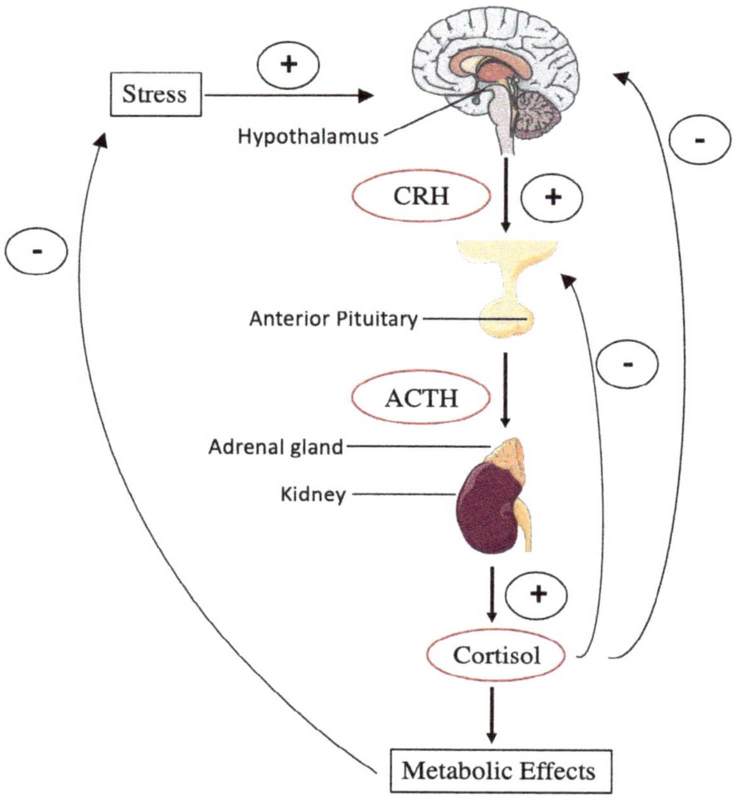

Fig. 6.1 The hypothalamic–pituitary–adrenal axis, representing the regulation of the production of cortisol from the adrenal cortex and of negative feedback regulation between the anterior pituitary and hypothalamus

This feedback regulation is mediated by cortisol binding to two cytoplasmic receptors: the lower affinity glucocorticoid receptor (GR) and higher affinity mineralocorticoid receptor (MR) (Arriza et al. 1988). Upon cortisol binding, the GR or MR will translocate to the nucleus and bind to glucocorticoid response elements (GREs), enhancing or suppressing gene transcription (Agler et al. 2007). Under normal conditions, MRs primarily help maintain cortisol levels, whilst under high stress conditions, GRs are progressively activated and bind cortisol to suppress the stress response (Kitchener et al. 2004). This GR activation inhibits CRH and ACTH expression, thus preventing HPA axis overactivation (Deng et al. 2015). However, if GR signalling is impaired, such as in chronic glucocorticoid stimulation, there will be a decrease in negative feedback on the hypothalamus and anterior pituitary, leading to increased glucocorticoid levels and dysregulation of the stress system. Reduced function of the GR is called glucocorticoid resistance, which can lead to HPA axis hyperactivity (Pariante and Lightman 2008). Therefore, normal GR activation is crucial in restoring homeostasis of the stress response (Pariante and Miller 2001).

One of the strongest regulating factors for GR sensitivity is FKBP5 (Wochnik et al. 2005; Davies et al. 2002), a 51 kDa immunophilin protein (Kang et al. 2008). Binding of FKBP5 to the GR reduces GR-binding affinity for cortisol, and inhibits normal GR translocation to the cell nucleus, thereby preventing transcription (Wochnik et al. 2005). FKBP5 transcription is induced by GR activation, providing a short-loop negative feedback regulatory mechanism which controls GR sensitivity (Vermeer et al. 2003). Thus, FKBP5 is a negative regulator of glucocorticoid action, terminating secretion of cortisol and the stress response. Increased chronic expression of FKBP5 has been shown to cause glucocorticoid resistance; therefore, correct functioning of FKBP5 is crucial in maintaining HPA axis homeostasis (Denny et al. 2000).

6.3 Impaired Endocrine Regulation in MDD

Impaired GR signalling leads to altered negative feedback regulation and has frequently been reported in MDD. The combined dexamethasone/CRH (dex/CRH) test is commonly used to determine feedback control of the HPA axis, and involves pre-treating individuals with 1.5 mg of dexamethasone, followed by stimulation with 100 μg of CRH the subsequent day. Post-administration, blood samples are collected to measure plasma cortisol and ACTH to test for a dysfunctional response. For example, using the combined dex/CRH test, a significantly increased cortisol response has been observed in acutely depressed male and female patients when compared with controls (Holsboer et al. 1987; Modell et al. 1997; Kunugi et al. 2004). Furthermore, within the depressed patient cohort, those with a history of attempted suicide had a significantly higher ACTH and cortisol response than those without a suicide attempt (Kunugi et al. 2004). In healthy individuals, dexamethasone binds to the GR and activates feedback inhibition, in turn lowering cortisol secretion. However, in depressed patients, cortisol secretion does not appear to be inhibited. This suggests that GR-mediated negative feedback is impaired in MDD. Furthermore, salivary cortisol, measured in the morning or evening, is significantly higher in MDD patients compared with controls (Vreeburg et al. 2009; Veen et al. 2011; Sorgdrager et al. 2017). Therefore, a number of clinical studies have shown that MDD is associated with an impaired HPA axis response to stressors throughout the circadian rhythm. However, the relationship between HPA axis dysregulation and depression is complex, as not all MDD patients have an abnormal cortisol response to the dex/CRH test. For example, patients with chronic MDD have shown no difference in salivary or serum cortisol levels following the dex/CRH test when compared with matched controls (Watson et al. 2002; Carpenter et al. 2009). This therefore suggests a slightly different pathophysiology in patients experiencing HPA axis abnormalities compared to those that are not, which may become more evident as depression lasts or becomes worse and the clinical picture resembles treatment resistance. For example, in another series of studies, Juruena and Pariante have shown that depressed patients show impaired feedback inhibition when challenged with dexamethasone, but normal response to prednisolone, a mixed GR and MR agonist, indicating a normal MR function (Juruena et al. 2006); however, severely depressed, treatment-resistant patients show impaired feedback inhibition even when challenged with prednisolone (Juruena et al. 2009, 2010, 2013).

HPA axis abnormalities have also been shown to influence and predict the development of MDD. When mean daytime cortisol was measured in patients at a mean age of 17.5 years, and upon follow-up at the age of 20, a previously high mean daytime cortisol level significantly predicted development of depression (Ellenbogen et al. 2011). Furthermore, an elevated cortisol response to the dex/CRH test in remitted patients has been correlated with a four-to-six-fold increased risk of relapse, compared to remitted patients with a normal cortisol response (Zobel et al. 2001). Similarly, a higher salivary cortisol response to low stress conditions in remitted MDD patients predicted a significant increase in depressive symptom score in the following 6 months (Morris et al. 2012). These studies suggest HPA axis abnormalities are closely related to the progression and long-term outcome of depression. As HPA dysregulation can precede a depressive episode, it suggests that abnormalities in the system arise prior to depression and may have a causal relationship.

A number of studies have shown differential HPA axis functioning between males and females (Kokras et al. 2019), suggesting that gender differences in MDD may be a variable that needs to be considered. This is particularly relevant, as following puberty girls are twice as likely to develop depression as adolescent boys (Salk et al. 2017). Additionally, depressed women have been found to have higher cortisol levels than depressed men, which is more prominent after a stressful or negative life event (Bangasser and Valentino 2014). Studies on rodents have shown that females have a larger stress-induced release of CRF, AVP, ACTH and cortisol compared to males, whilst negative feedback is also lower in females (Kokras et al. 2019). However, the directions of these changes are inconsistent between studies (Bangasser and Valentino 2014); thus, further research is required to understand the significance of these findings.

6.4 Activated Inflammatory Response in MDD

In addition to endocrine dysregulation seen in MDD, immune system alterations have also been associated with psychiatric disorders, with a well-established database of research linking immune system dysregulation with MDD. In fact, since the inflammation theory of depression was first introduced, a link between immune imbalance, depression and other psychiatric disorders has consistently been found (Stetler and Miller 2011; Pariante 2017). In a recent meta-analysis of 8887 depressed patients and matched controls, 58% of depressed patients had elevated serum C-reactive protein (CRP) levels (>1 mg/L), an acute-phase protein that rises in response to inflammation (Osimo et al. 2019). Additionally, a search of 11,813 depressed patients showed that 27% had low-grade inflammation (CRP >3 mg/L) (Osimo et al. 2019). In a similar meta-analysis of 5166 patients with depression, levels of CRP and inflammatory markers, such as interleukin (IL)-12, IL-6 and tumour necrosis factor (TNF)-α, were significantly higher in patients with depression than in controls (Osimo et al. 2020). These meta-analyses have been further corroborated by the UK Biobank study, the largest study to date including 26, 894 patients with depression and 59,001 controls (Pitharouli et al. 2021). Pitharouli et al. (2021) found significantly more patients with depression have CRP levels >3 mg/L when compared with controls. These results remained significant even after clinical and sociodemographic factors known to be associated with increased CRP, such as BMI, smoking, exposure to early-life adversity and low socio-economic circumstances, were adjusted for. Therefore, the aforementioned research demonstrates that increased circulating CRP is present in depression. Supporting this notion, studies have shown that both chronic and acute administration of pro-inflammatory challenges, such as interferon-alpha, produce a behavioural and emotional syndrome resembling depression (Nettis et al. 2020; Russell et al. 2019; Hepgul et al. 2016).

Additionally, in psychotropic-medication free MDD patients, plasma CRP is significantly correlated with inflammatory markers such as IL-6, TNF and soluble TNF receptor 2 (Felger et al. 2020). When comparing depressed patients with high versus low CRP (>3 mg/L vs. <3 mg/L), those with high CRP had a significant increase in these inflammatory markers, primarily driven by IL-6 and IL-1ra. Additionally, the cerebrospinal fluid concentrations of these cytokines in the high CRP group were also correlated with an increased Inventory of Depressive Symptomology Self Report (IDS-SR) score (Felger et al. 2020). Meta-analyses of cytokine-specific markers have found IL-1β, IL-6, TNF and CRP to be the most consistent biomarkers of inflammation in patients with MDD (Haapakoski et al. 2015).

Recent research has found evidence for differential relationships between inflammatory markers and different MDD symptoms, suggesting an 'inflammatory phenotype' of depression. For example, a higher polygenic risk score (PRS) for CRP has been associated with changes in appetite, fatigue and anhedonia, whilst a higher PRS for TNF-α is associated with fatigue (Kappelmann et al. 2021). Fried et al. (2019) similarly found CRP concentration to be associated with appetite and energy level whilst also reporting an association between IL-6, depression sum-score and 'aches and pains'. These findings suggest that specific inflammatory markers are associated with symptoms of depression, and thus may aid recruitment of patients with specific immune-related symptoms into clinical trials for immune-modulating drugs for MDD (Kappelmann et al. 2021). In terms of genetic regulation, it is also important to emphasise that the aforementioned study in the UK Biobank (Pitharouli et al. 2021) has found that the PRS for depression is associated with CRP levels, thus indicating a genetic contribution to inflammation, but only through metabolic and behavioural changes reflected by higher BMI and more frequent smoking behaviour in the patients, rather than a true immune-related genetic predisposition.

Interestingly, increased inflammatory markers in MDD patients have been associated with decreased treatment response, and therefore appear to predict antidepressant efficacy. For example, expression of mRNA for macrophage migration inhibitory factor and IL-1β in peripheral blood has been shown to predict decreased treatment response in MDD patients (Cattaneo et al. 2016). Additionally, increased BMI-corrected circulating CRP has been shown to be significantly elevated in treatment-resistant MDD patients (Chamberlain et al. 2019), whilst increased TNF-α, sTNF-R2 and IL-6 are associated with an increased number of failed treatments in MDD patients (Haroon et al. 2018). Furthermore, 6 mRNAs for P2RX7, IL-1β, IL-6, TNF-α, CXCL12 and GR have been found to differentiate treatment-resistant from treatment-responsive patients (Cattaneo et al. 2020). This research suggests MDD patients with pro-inflammatory biomarkers have a unique clinical profile that makes them more susceptible to treatment non-response, and thus may benefit from additional treatment strategies beyond first-line antidepressant regimes. It is also very interesting to note in this context that these authors find that resistance to antidepressants is associated with lower expression of GR mRNA together with high expression of pro-inflammatory mRNAs, thus supporting the notion that the increased inflammation in depression is associated with glucocorticoid resistance, at least as indicated by the low expression of GR (see below).

Considering this evidence, it is not surprising that conjunctive treatment of antidepressants with anti-inflammatories is a new treatment strategy currently in clinical trials, with the aim of targeting inflammation in those who exhibit increased concentrations of inflammatory biomarkers. This has recently successfully been shown with minocycline, an antibiotic that also has broad anti-inflammatory properties, and is able to penetrate the central nervous system through the blood–brain barrier (Nettis et al. 2021).

An additional consideration for research into inflammation in MDD is the impact of sex and sex hormones. For example, Moieni et al. (2015) administered endotoxin to both females and males to induce a pro-inflammatory cytokine response—increased IL-6 and TNF-a. Following endotoxin administration, females reported a significantly greater increase in depressed mood and social disconnectedness than males, suggesting females may be more susceptible to the effects of inflammation. Additionally, immune cells have sex hormone receptors, and thus respond directly to changes in sex hormone levels (Brundin et al. 2021). For example, two estrogen receptors (ER), ER-alpha and G-protein ER1, are associated with anti-inflammatory phenotypes, and expressed on human primary monocytes in peripheral blood. ER-alpha acts to inhibit IL-6 expression through NF-κB transcriptional inhibition, whilst G-protein ER1 is a vital co-regulator of ER-alpha and its aforementioned actions (Pelekanou et al. 2016). In addition, microglial cells contain estrogen receptors, and estrogens have an inhibitory effect on neuroinflammatory activity (Villa et al. 2016).

Studies have also shown fluctuating estradiol across the menstrual cycle impacts mood and neurological response to psychosocial stress (Albert et al. 2015), whilst during the different phases of the menstrual cycle, and thus times of hormone fluctuation, expression patterns of immune response genes differ (Brundin et al.

2021). For example, during the late luteal phase of the menstrual cycle, when estrogen levels are decreasing, high levels (>3 mg/L) of the inflammatory biomarker CRP are reported in women (Gold et al. 2016; Harding and Headon 2022). Furthermore, high CRP levels are significantly positively associated with premenstrual mood changes (Gold et al. 2016). Thus, the research discussed above suggests that hormone level fluctuations impact both the immune system and susceptibility to stress, and thus could influence the inflammatory component of depression.

6.5 Neuroendocrine–Immune System Interaction

A relationship between the two systems discussed above—neuroendocrine and immune—has also been extensively reported, with neuroendocrine–immune interactions impacting both the immune response and hormonal functioning to maintain homeostasis (Ashley and Demas 2017). Along with bidirectional crosstalk between the systems, neuroendocrine cells can also produce cytokines whilst immune cells produce low concentrations of hormones (Verburg-van Kemenade et al. 2017). Notably, HPA axis hyperactivity and increased inflammation are the two most consistently described pathophysiological findings in major depression (Pariante 2017). Thus, the contribution of the immune system and the consequent inflammation are important factors to consider when discussing neuroendocrine interactions with mood disorders.

As discussed, the HPA axis has been at the centre of the neuroendocrine–immune relationship with mood disorders, particularly in patients with more severe, treatment-resistant depression. A number of human and animal studies have provided evidence that the underlying physiological mechanisms of inflammation lie in stress-related pathways. In terms of stress-related pathways, it should be noted that so far, this chapter has mostly dealt with psychological stress. These pathways lead to increased circulating levels of monocytes and neutrophils, alongside activation of the HPA axis and thus increased cortisol levels. For example, following activation, immune cells such as monocytes are trafficked to the brain vasculature, triggering subsequent inflammatory signalling (Miller and Raison 2016). Multiple pro-inflammatory cytokines such as TNF-α, IL-1 and IL-6 act as activators of the HPA axis, whilst immune cells express both GR and adrenergic receptors (Ménard et al. 2017). Conversely, cortisol and corticosterone are potent anti-inflammatory steroid hormones and can regulate the expression of cytokines, adhesion molecules and chemoattractants (Eskandari et al. 2003). Therefore, the HPA axis is a central control centre for the inflammatory responses, both in the central nervous system and throughout the body (Otmishi et al. 2008).

Persistent HPA axis hyperactivation, such as that seen in some MDD patients, can lead to chronic glucocorticoid resistance. As the GR physiologically mediates HPA axis negative feedback, glucocorticoid resistance leads to both HPA axis hyperactivity and a decrease in GR-mediated anti-inflammatory action, and thus increased inflammatory markers (Pariante 2017). Lastly, the role of the kynurenine pathway, an alternate metabolic pathway for tryptophan degradation, is worth mentioning.

This pathway is frequently activated in both inflammation and depression (Sforzini et al. 2019; Savitz 2020), and may also be activated by an excess of cortisol (Menke 2019). These data suggest a potential role for the kynurenine pathway as a biological mechanism involved in neuroendocrine–immune interactions in MDD.

As discussed previously, sex differences have been seen in both the immune and endocrine systems; however, there is a limited body of research on neuroendocrine–immune crosstalk and sex. We will give only a brief summary, as an in-depth discussion is beyond the scope of this chapter (see Chap. 10). In a recent review, some sex differences in the medial preoptic area were suggested to involve endocrine–immune crosstalk (Arambula and McCarthy 2020). For example, Lenz et al. (2018) found a greater and more active number of mast cells in the medial preoptic area of male neonates than females, and these mast cells mediate some aspects of brain sexual differentiation. Therefore, it has been hypothesised that imbalances or changes to the endocrine–immune system may impact sex differences in the medial preoptic area, and thus differences in neurodevelopment (Arambula and McCarthy 2020). There have been no studies to date to confirm the above hypotheses; however, these suggestions may contribute to understanding sex differences in neuropsychiatric and neurodevelopmental conditions.

6.6 Neuroendocrine–Immune Models

6.6.1 The 'Glucocorticoid Resistance Model'

This coexistence of increased inflammation and HPA axis dysfunction in MDD has led to a debate regarding cause and effect: is HPA axis hyperactivity causing inflammation or *vice versa*? As the GR mediates negative feedback on the HPA axis, dysfunction of the GR leads to impaired HPA axis negative feedback and thus increased glucocorticoid levels. Increased glucocorticoids are known to inhibit immune function, as previously discussed; however, in depression high concentrations of pro-inflammatory cytokines co-exist with high levels of glucocorticoids (Zunszain et al. 2011). Since the 1990s, the 'glucocorticoid resistance model' has been the accepted explanation for this concurrence, due to the potential immune activating features of glucocorticoids in MDD (Munhoz et al. 2010). This model suggested a reduced function of the GR; thus the glucocorticoid resistance seen in some MDD patients allows pro-inflammatory pathways to evade normal feedback inhibition by glucocorticoids (Pariante 2017). The increased immune response is due to immune cells becoming resistant to the anti-inflammatory role of cortisol; thus, even if these patients have higher cortisol levels, there is less inhibition of inflammation. Therefore, according to the 'glucocorticoid resistance model', the increased HPA axis activation and consequential hypercortisolaemia are suggested to be the cause of the increased inflammatory response in depression, rather than a result of glucocorticoid resistance. Using models of depression-induced inflammation, recent research has confirmed the finding that glucocorticoids can potentiate pro-inflammatory processes (Horowitz et al. 2020). Studies showing a concomitant increase in inflammatory biomarkers and reduced GR expression or function support this model (Nikkheslat et al. 2015; Cattaneo et al. 2012, 2020; Mariani et al. 2021). However, in a recent meta-analysis testing the association between glucocorticoid resistance and increased inflammation in MDD, the original findings from the 'glucocorticoid resistance model' have not been upheld, and are thus being questioned (Perrin et al. 2019).

Perrin et al. (2019) analysed all 32 studies that have looked at both HPA axis and inflammation data in the same MDD patients (2087 patients), a surprisingly small number compared to the studies which have looked at one or the other factor. Combining the studies that reported dexamethasone suppression test results with those that recorded GR expression, or *in vitro* assays of GR function to quantify glucocorticoid resistance, the meta-analyses found limited evidence for a positive association between glucocorticoid resistance and inflammation. Therefore, despite being limited by the small number of studies, the analyses indicate that the 'glucocorticoid resistance model' may not be the only explanation for the complex relationship between glucocorticoid resistance and immune system escape and suggests a need for further research (Perrin et al. 2019). Additionally, despite the presence of increased inflammatory biomarkers (IL-6, TNF-alpha and MIF) and reduced GR mRNA expression in the 190 MDD patients analysed, the BIODEP study similarly found no clear evidence for a correlation between inflammatory biomarkers, GR mRNA expression and salivary cortisol levels (Cattaneo et al. 2020). Therefore, this research suggests that GR mRNA expression does not fully elucidate the mechanism behind increased inflammation in MDD.

Moreover, in recent research on rodents, different forms of stress resulted in different HPA axis function and inflammatory outcomes (Du Preez et al. 2020). For example, physical stress related to repeated injections induced increased corticosterone reactivity and decreased plasma TNF-α and IL-4, whilst the psychosocial stress of social isolation induced increased TNF-α and decreased corticosterone reactivity. Despite the presence of a depressive-like phenotype, neither circumstance found increased HPA axis activity and increased inflammatory biomarkers, which would have been expected in the 'glucocorticoid resistance model'. Therefore, if glucocorticoid resistance is not the cause of increased inflammation in depression, additional hypotheses are required to explain the relationship between high cortisol levels and inflammation in MDD.

6.6.2 The 'Pro-inflammatory Cortisol' Model as an Alternative to the Glucocorticoid Resistance Model

Despite the prominent anti-inflammatory effects of glucocorticoids, research over the years has indicated that during stress glucocorticoids may also have pro-inflammatory properties. Glucocorticoid secretion activated by chronic stress has been shown to increase NF-κB activation in the frontal cortex and hippocampus of rodents whilst also increasing pro-inflammatory gene expression, thus suggesting a pro-inflammatory role for glucocorticoids (Munhoz et al. 2010). NF-κB is a mediator of inflammatory responses, inducing expression of pro-inflammatory genes and regulating the survival of both innate immune cells and T cells; thus, increased levels result in a pro-inflammatory response (Liu et al. 2017). Additionally, when corticosterone signalling in mice is inhibited with metyrapone, a glucocorticoid synthesis inhibitor, inflammation decreased despite the introduction of social stress (Niraula et al. 2018).

Consistent with animal studies, research in humans has found that pre-treatment with hydrocortisone in healthy participants induces a systemic inflammatory response following high cortisol concentrations (Yeager et al. 2011). Yeager et al. (2016) later found that administration of cortisol, to concentrations which mimic those seen during systemic stress, induces an upregulation of monocytes, macrophages and neutrophils, and thus a pro-inflammatory response in participants (Yeager et al. 2016). Most recently, *in vitro* studies of human hippocampal progenitor cells have shown that administration of dexamethasone prior to an immune challenge enhances inflammatory effects in these neural cells and upregulates multiple innate immune genes (Horowitz et al. 2020). These effects were most potent when the hippocampal cells were exposed to cortisol 24 hours prior to immune challenge. Therefore, this research provides evidence for an alternative to the 'glucocorticoid resistance model', which proposes cortisol as a pro-inflammatory mediator of stress reactions, and thus potentially another mechanism for the coexistence of inflammation and HPA axis dysfunction in MDD. As this model has not been tested in MDD, further research is required for the application of the 'pro-inflammatory cortisol' model.

6.7 HPA Axis Dysregulation in MDD: The Role of FKBP5

6.7.1 FKBP in MDD

Research into the neuroendocrine-immune relationship would not be complete without analysing the role of the GR in HPA axis dysregulation in MDD. In fact, research into this pathophysiology has identified FKBP5, a regulator of GR sensitivity, as a potential molecular target at the HPA axis interface. In humans, single nucleotide polymorphisms (SNPs) in FKBP5 are associated with differential FKBP5 mRNA expression, leading to changes in GR sensitivity, and in turn HPA axis regulation (Binder et al. 2004). Due to the role of FKBP5 in HPA axis function, and the established link between HPA axis dysfunction and MDD, several studies investigated the relationship between FKBP5 and MDD (Klengel and Binder 2015).

When MDD patients were compared with healthy controls, FKBP5 SNPs (rs1360780, rs3800373, rs4713916, rs9296158, rs9394309, rs9470080) were found to be significantly associated with MDD status (Lekman et al. 2008; Szczepankiewicz et al. 2014) and with an increased recurrence of depressive episodes (Binder et al. 2004). Additionally, significantly higher FKBP5 mRNA levels in leukocytes have been reported in MDD patients, compared with controls (Cattaneo et al. 2012), and are seen *post-mortem* in both the hippocampus (Mamdani et al. 2015) and cortical regions of the brain (Tatro et al. 2009). This research provides evidence that certain FKBP5 SNP alleles are more frequent in MDD patients than controls and are associated with MDD risk. Thus, this suggests FKBP5 polymorphisms cause a baseline difference in GR sensitivity, increasing GR resistance and decreasing efficiency of negative feedback on the HPA axis, thereby increasing MDD risk. Some studies also show evidence of an effect of FKBP5 polymorphisms on response to antidepressants (Stamm et al. 2016; Fabbri et al. 2018). These variations may explain why HPA axis dysregulation has been seen prior to a depressive episode and can predict development of depressive symptoms or possibly the likelihood of response to antidepressants.

6.7.2 FKBP5: Gene–Environment Interactions

The strongest evidence for gene–environment interaction between glucocorticoid-related genes and a stressful environment is not offered by the GR or the MR but by FKBP5, and we are therefore discussing this more in depth. As with genetic variables, environmental stressors are also a major cause for the development of MDD. Adverse early-life events (AELEs) have been identified as a major environmental cause for the development of MDD (Kendler et al. 1999). Interestingly, AELEs have also been shown to impact HPA axis functioning (Mangold et al. 2010; Shapero et al. 2014), and FKBP5 has been identified as a key target (Klengel et al. 2013).

The relationship between different gene–environment interactions has been shown in a study evaluating FKBP5 polymorphisms (rs3800373, rs9296158, rs1360780, rs9470080) in preschool children (Scheuer et al. 2016). Minor allele carriers of the SNPs had significantly increased risk of developing MDD if exposed to adverse life events, compared with homozygous carriers of the major allele (Scheuer et al. 2016). It should be noted that a significantly increased risk of MDD was only seen in those exposed to mild-to-moderate adverse events, rather than severe. Further research corroborated these findings, comparing 148 adolescent patients with MDD to 143 typically developing controls (Piechaczek et al. 2019). Participants who had reported a higher number of early-life stressors (using the Munich Event List and modified Life Event Survey) and were also carriers of FKBP5 SNPs (rs3800373, rs1360780) had increased risk for being in the MDD group (Piechaczek et al. 2019). The researchers also showed that adolescent carriers of FKBP5 SNPs (rs3800373, rs9296158, rs9470080) who had experienced sociodemographic stressors, such as unemployment or migrant background of parents and lower secondary education of the participant, had an increased risk of MDD. This shows significant interactions between these SNPs and stress in predicting depression in adolescents. Kang et al. (2020) also reported significant interaction effects between childhood physical abuse and FKBP5 SNPs (rs3800373, rs1360780, rs4713916). Those carrying the minor alleles of the SNPs showed higher depression scores than non-carriers when exposed to abuse. Importantly, these SNPs alone had no significant effect on depression risk or symptoms, as exposure to early-life stress was vital for an increased risk of depression (Piechaczek et al. 2019; Kang et al. 2020). Therefore, the minor alleles of the FKBP5 polymorphisms act as risk alleles, increasing susceptibility to the pathogenic effects of childhood abuse in developing childhood and adolescent depression.

The research discussed provides evidence that genetic variation in FKBP5 impacts the risk of MDD by altering an individual's sensitivity to the effects of AELEs. Thus, genetically driven variability produces individual differences in the risk of MDD, and may explain why only some individuals who experience AELEs develop MDD. Importantly, whilst SNPs may act as a risk factor for MDD, these studies have shown that SNP interaction with AELEs is crucial in mediating significant risk of depression. Although previous studies discussed showed a relationship between FKBP5 and MDD, life history was not taken into account (Binder et al. 2004; Lekman et al. 2008; Tatro et al. 2009; Cattaneo et al. 2012; Szczepankiewicz et al. 2014; Mamdani et al. 2015). Therefore, it is possible that the cohort of patients used had in fact experienced early-life adverse events, which would explain the significant relationship observed between FKBP5 SNPs and MDD.

FKBP5 SNPs that interact with early-life adverse events are associated with increased FKBP5 mRNA expression, leading to changes in GR sensitivity and, in turn, HPA axis regulation (Binder et al. 2004). Reporter gene assays of SNPs rs1360780, rs9296158, rs9470080a and rs3800373 found that rs1360780 was situated closest to the GRE, which transcriptionally regulates FKPB5 (Klengel et al. 2013). The homozygous risk allele (AA), but not the protective (GG) genotype of rs1360780, was shown to mediate the interaction of intron 7 with the FKBP5 transcription start site via three-dimensional loop formation. AA also mediated intron 2 in a genotype-dependent interaction, again with the FKBP5 transcription start site. Consequently, this leads to enhanced FKBP5 gene transcription in response to GR (Klengel et al. 2013). FKBP5 response therefore differs between risk allele and protective allele carriers. The response caused by increased FKPB5 transcription is consistent with previous research that showed a genetic predisposition in FKBP5 can lead to a stronger cortisol reaction to stressors, and glucocorticoid resistance in healthy controls (Ising et al. 2008; Binder et al. 2008).

6.7.3 FKBP5: Epigenetic Mechanisms

As for gene–environment interaction, and even if some studies have shown some evidence of epigenetic regulation of the GR in response to stress (Witzmann et al. 2012; Mourtzi et al. 2021), the strongest evidence for epigenetic mechanisms operating in the regulation of glucocorticoid function comes from studies on FKBP5. Growing evidence suggests that epigenetic alterations are a key component by which environmental stressors interact with the genome, increasing the risk of depressive symptomology (Davies et al. 2019). It has been suggested that FKBP5 epigenetic and environmental mechanisms play a role in the pathophysiology of

HPA axis dysregulation, whereby FKBP5 epigenetic components bridge the genetic and environmental association and contribute to MDD pathophysiology (Lin and Tsai 2019). There has been a particular focus on AELEs, as adverse events in adulthood have not found a significant association with outcome (Lahti et al. 2016; Cristóbal-Narváez et al. 2017).

Early-life adverse events have been shown to impact allele-specific epigenetic modification of FKBP5. Epigenetics refers to the potentially heritable, though environmentally modifiable, regulation of gene expression and function mediated through non-DNA-encoded mechanisms (Sun et al. 2013). For example, common epigenetic modifications in MDD include histone acetylation and DNA methylation (Sun et al. 2013). Whilst the exact molecular epigenetic mechanisms behind the FKBP5 gene–environment interactions in MDD are not entirely clear, recent research has elucidated a few potential mechanisms.

Patients with the rs1360780 risk allele of FKBP5 who also experienced childhood abuse had significantly decreased DNA methylation in intron 7, compared with controls or protective genotype carriers (Klengel et al. 2013). Furthermore, greater childhood exposure to trauma correlated with a greater decrease in methylation. Multiple studies have further confirmed decreased intron 7 DNA methylation in MDD patients, and healthy carriers of the high-risk allele exposed to AELEs (Non et al. 2016; Tozzi et al. 2018; Klinger-König et al. 2019). Therefore, individuals with the high-risk FKBP5 allele appear to be more susceptible to epigenetic changes following childhood abuse.

In addition, this decrease in FKBP5 CpG methylation in patients with early-life stress and depressive phenotypes has been found to upregulate FKBP5 in peripheral blood. Importantly, this upregulation of FKBP5 in peripheral blood promoted NF-kB signalling in immune cells and showed positive correlation with pro-inflammatory genes such as interleukin and toll-like receptors (Zannas et al. 2019).

Therefore, these findings suggest that AELEs may chronically influence both the HPA axis and inflammatory regulation through epigenetic modification of the FKBP5 gene. As GR mediates the negative feedback response of the HPA axis, impaired function can lead to long-term HPA axis dysregulation. This may explain why certain patients—those with high-risk alleles—have a higher risk of MDD. Thus, this research shows how a genetic predisposition to react more strongly to environmental stress interacts with AELEs, and may increase the risk of MDD.

6.8 Conclusion

MDD is significantly linked to physiological modifications in our body, most notably neuroendocrine–immune interactions. Whilst research has shown there is a neuroendocrine and immune system interplay in MDD, our understanding of the exact mechanism is still limited. The 'glucocorticoid resistance model' provided the first possible explanation; however, it is clear that further modifications to this model are required following the discovery of the pro-inflammatory role of cortisol.

Therefore, whilst this research has further elucidated neuroendocrine–immune mechanisms in MDD, additional research in humans is still necessary to overcome the uncertainties that remain, and thus help us to understand the biological mechanism underpinning MDD. Research into GR dysfunction in MDD has found FKBP5, a key regulator in GR function, to be a potential biomarker for MDD, specifically in those with AELEs.

6.9 Key References

Binder et al. 2004: this was the first paper to find an association between FKBP5 polymorphisms, HPA axis dysregulation and increased susceptibility for depression.

Cattaneo et al. 2020: this research found six mRNAs, all associated with the immune response, differentiated treatment-resistant from treatment-responsive patients with depression.

Pariante and Miller 2001: the first review of the literature on the associations between depression, glucocorticoid resistance and the impact of antidepressants on the glucocorticoid receptor.

Pitharouli et al. 2021: this is the largest study to date demonstrating that increased CRP, and thus inflammation, is associated with depression.

Acknowledgments and Disclosures Dr. Pariante is supported by the Wellcome Trust strategy award to the Neuroimmunology of Mood Disorders and Alzheimer's Disease (NIMA) Consortium (104025), which is also funded by Janssen, GlaxoSmithKline, Lundbeck and Pfizer; by the NARSAD grant RE14032; by the U.K. Medical Research Council (MR/N015746/1); and by the National Institute for Health Research Biomedical Research Centre at South London and Maudsley National Health Service Foundation Trust and King's College London.

Dr. Pariante has received research and consultancy funding from pharmaceutical companies interested in the development of novel strategies for depression, such as Johnson & Johnson, Lundbeck and Boehringer Ingelheim.

Dr. Sforzini and Prof. Pariante have received research funding from the Innovative Medicines Initiative 2 Joint Undertaking under grant agreement No 853966–2, as part of the EU-PEARL project. This Joint Undertaking receives support from the European Union's Horizon 2020 research and innovation programme and EFPIA.

References

Agler M, Prack M, Zhu Y, Kolb J, Nowak K, Ryseck RP, Shen D, Cvijic ME, Somerville JM, Nadler SG, Chen T (2007) A high-content glucocorticoid receptor translocation assay for compound mechanism-of-action evaluation. J Biomol Screen 12:1029–1041

Albert K, Pruessner J, Newhouse P (2015) Estradiol levels modulate brain activity and negative responses to psychosocial stress across the menstrual cycle. Psychoneuroendocrinology 59:14–24

Anacker C, Zunszain PA, Carvalho L, Pariante CM (2011) The glucocorticoid receptor: pivot of depression and of antidepressant treatment? Psychoneuroendocrinology 36(3):415–425
Arambula SE, McCarthy MM (2020) Neuroendocrine-immune crosstalk shapes sex-specific brain development. Endocrinology 161(6):bqaa055
Arriza JL, Simerly RB, Swanson LW, Evans RM (1988) The neuronal mineralocorticoid receptor as a mediator of glucocorticoid response. Neuron 1:887–900
Ashley NT, Demas GE (2017) Neuroendocrine-immune circuits, phenotypes, and interactions. Horm Behav 87:25–34
Bangasser DA, Valentino RJ (2014) Sex differences in stress-related psychiatric disorders: neurobiological perspectives. Front Neuroendocrinol 35(3):303–319
Bellavance MA, Rivest S (2014) The HPA – immune axis and the immunomodulatory actions of glucocorticoids in the brain. Front Immunol 5:136
Brook CGD, Marshall NJ (2001) Essential endocrinology, 4th edn. Blackwell Science, Oxford
Binder EB, Salyakina D, Lichtner PC, Wochnik G, Ising M, Pütz B, Papiol S, Seaman SR, Lucae S, Kohli MA, Nickel T, Künzel HE, Fuchs B, Majer M, Pfennig A, Kern N, Brunner J, Modell S, Baghai TC, Deiml T, Zill P, Bondy B, Rupprecht R, Messer T, Köhnlein O, Dabitz H, Brückl TM, Müller N, Pfister H, Lieb R, Mueller J, Lõhmussaar E, Strom TM, Bettecken T, Meitinger TH, Uhr M, Rein T, Holsboer F, Muller-Myhsok B (2004) Polymorphisms in FKBP5 are associated with increased recurrence of depressive episodes and rapid response to antidepressant treatment. Nat Genet 36:1319–1325
Binder EB, Bradley RG, Liu W, Epstein MP, Deveau TC, Mercer KB, Tang Y, Gillespie CF, Heim CM, Nemeroff CB, Schwartz AC, Cubells JF, Ressler KJ (2008) Association of FKBP5 polymorphisms and childhood abuse with risk of posttraumatic stress disorder symptoms in adults. JAMA 299(11):1291–1305
Brundin P, Landgren BM, Fjällström P, Shamekh MM, Gustafsson JÅ, Johansson AF, Nalvarte I (2021) Expression of sex hormone receptor and immune response genes in peripheral blood mononuclear cells during the menstrual cycle. Front Endocrinol 12:721813
Carpenter LL, Ross NS, Tyrka AR, Anderson GM, Price LH (2009) Dex/CRH test cortisol response in outpatients with major depression and matched healthy controls. Psychoneuroendocrinology 34(8):1208–1213
Cristóbal-Narváez P, Sheinbaum T, Myin-Germeys I, Kwapil TR, Castro-Catala MD, Domínguez-Martínez T, Racioppi A, Monsonet M, HinojosaMarqués L, Winkel RV, Rosa A, Barrantes-Vidal N (2017) The role of stress-regulation genes in moderating the association of stress and daily-life psychotic experiences. Acta Psychiatr Scand 136(4)
Cattaneo A, Gennarelli M, Uher R, Breen GD, Farmer A, Aitchison KJ, Craig IW, Anacker C, Zunsztain PA, McGuffin P, Pariante CM (2012) Candidate genes expression profile associated with antidepressants response in the GENDEP Study: differentiating between baseline 'predictors' and longitudinal 'targets'. Neuropsychopharmacology 38:376–376
Cattaneo A, Ferrari C, Uher R, Bocchio-Chiavetto L, Riva MA, MRC ImmunoPsychiatry Consortium, Pariante CM (2016) Absolute measurements of macrophage migration inhibitory factor and interleukin-1-β mRNA levels accurately predict treatment response in depressed patients. Int J Neuropsychopharmacol 19(10):pyw045
Cattaneo A, Ferrari C, Turner L, Mariani N, Enache D, Hastings C, Kose M, Lombardo G, McLaughlin AP, Nettis MA, Nikkheslat N, Sforzini L, Worrell C, Zajkowska Z, Cattane N, Lopizzo N, Mazzelli M, Pointon L, Cowen PJ, Cavanagh J, Harrison NA, de Boer P, Jones D, Drevets WC, Mondelli V, Bullmore ET, Neuroimmunology of Mood Disorders and Alzheimer's Disease (NIMA) Consortium, Pariante CM (2020) Whole-blood expression of inflammasome- and glucocorticoid-related mRNAs correctly separates treatment-resistant depressed patients from drug-free and responsive patients in the BIODEP study. Transl Psychiatry 10(1):232
Chalmers DT, Lovenberg TW, Souza EB (1995) Localization of novel corticotropin-releasing factor receptor (CRF2) mRNA expression to specific subcortical nuclei in rat brain: comparison with CRF1 receptor mRNA expression. J Neurosci 15(10):6340–6350

Chamberlain SR, Cavanagh J, de Boer P, Mondelli V, Jones D, Drevets WC, Cowen PJ, Harrison NA, Pointon L, Pariante CM, Bullmore ET (2019) Treatment-resistant depression and peripheral C-reactive protein. Br J Psychiatry J Ment Sci 214(1):11–19
Davies TH, Ning Y, Sánchez ER (2002) A new first step in activation of steroid receptors: hormone-induced switching of FKBP51 and FKBP52 immunophilins. J Biol Chem 277(7):4597–4600
Davies MR, Kalsi G, Armour C, Jones IR, McIntosh AM, Smith DJ, Walters JT, Bradley JR, Kingston N, Ashford S, Beange I, Brailean A, Cleare AJ, Coleman JR, Curtis CJ, Curzons SC, Davis KA, Dowey LR, Gault VA, Goldsmith KA, Bennett MH, Hirose Y, Hotopf M, Hübel C, Kanz C, Leng J, Lyall DM, Mason BD, McAtarsney-Kovacs M, Monssen D, Moulton AA, Ovington NR, Palaiologou E, Pariante CM, Parikh S, Peel AJ, Price RK, Rimes KA, Rogers HC, Sambrook JG, Skelton M, Spaul A, Suarez EL, Sykes BL, Thomas KG, Young AH, Vassos E, Veale D, White KM, Wingrove J, Eley TC, Breen GD (2019) The Genetic Links to Anxiety and Depression (GLAD) Study: online recruitment into the largest recontactable study of depression and anxiety. Behav Res Ther 123
Deng Q, Riquelme D, Trinh LB, Low MJ, Tomić M, Stojilkovic SS, Aguilera G (2015) Rapid glucocorticoid feedback inhibition of ACTH secretion involves ligand-dependent membrane association of glucocorticoid receptors. Endocrinology 156(9):3215–3227
Denny WB, Valentine DL, Reynolds PD, Smith DF, Scammell JG (2000) Squirrel monkey immunophilin FKBP51 is a potent inhibitor of glucocorticoid receptor binding. Endocrinology 141(11):4107–4113
Du Preez A, Law T, Onorato D, Lim YM, Eiben P, Musaelyan K, Egeland M, Hye A, Zunszain PA, Thuret S, Pariante CM, Fernandes C (2020) The type of stress matters: repeated injection and permanent social isolation stress in male mice have a differential effect on anxiety- and depressive-like behaviours, and associated biological alterations. Transl Psychiatry 10(1):325
Ellenbogen MA, Hodgins S, Linnen A, Ostiguy CS (2011) Elevated daytime cortisol levels: a biomarker of subsequent major affective disorder? J Affect Disord 132(1–2):265–269
Eskandari F, Webster JI, Sternberg EM (2003) Neural immune pathways and their connection to inflammatory diseases. Arthritis Res Ther 5(6):251–265
Fabbri C, Corponi F, Albani D, Raimondi I, Forloni G, Schruers K, Kasper S, Kautzky A, Zohar J, Souery D, Montgomery S, Cristalli CP, Mantovani V, Mendlewicz J, Serretti A (2018) Pleiotropic genes in psychiatry: calcium channels and the stress-related FKBP5 gene in antidepressant resistance. Prog Neuro-Psychopharmacol Biol Psychiatry 81:203–210
Felger JC, Haroon E, Patel TA, Goldsmith DR, Wommack EC, Woolwine BJ, Le NA, Feinberg R, Tansey MG, Miller AH (2020) What does plasma CRP tell us about peripheral and central inflammation in depression? Mol Psychiatry 25(6):1301–1311
Fried E, Von Stockert S, Haslbeck J, Lamers F, Schoevers R, Penninx B (2019) Using network analysis to examine links between individual depressive symptoms, inflammatory markers, and covariates. Psychol Med 50(16):2682–2690
Friedrich MJ (2017) Depression is the leading cause of disability around the world. JAMA 317(15): 1517
Gold EB, Wells C, O'Neill Rasor M (2016) The association of inflammation with premenstrual symptoms. J Women's Health 25(9):865–874
Haapakoski R, Mathieu J, Kivimäki M (2015) Cumulative meta-analysis of interleukins 6 and 1 beta, tumour necrosis factor alpha and C-reactive protein in patients with major depressive disorder. Brain Behav Immun 49:205–215
Harding AT, Headon NS (2022) The impact of estrogens and their receptors on immunity and inflammation during infection. Cancers 14(4):909
Haroon E, Daguanno AW, Woolwine BJ, Goldsmith DR, Baer WM, Wommack EC, Felger JC, Miller AH (2018) Antidepressant treatment resistance is associated with increased inflammatory markers in patients with major depressive disorder. Psychoneuroendocrinology 95:43–49
Hepgul N, Cattaneo A, Agarwal K, Baraldi S, Borsini A, Bufalino C, Forton DM, Mondelli V, Nikkheslat N, Lopizzo N, Riva MA, Russell A, Hotopf M, Pariante CM (2016) Transcriptomics in interferon-α-treated patients identifies inflammation-, neuroplasticity- and oxidative stress-

related signatures as predictors and correlates of depression. Neuropsychopharmacology 41(10): 2502–2511
Holsboer F, Bardeleben UV, Wiedemann K, Müller OA, Stalla GK (1987) Serial assessment of corticotropin-releasing hormone response after dexamethasone in depression Implications for pathophysiology of DST nonsuppression. Biol Psychiatry 22:228–234
Horowitz MA, Cattaneo A, Cattane N, Lopizzo N, Tojo L, Bakunina N, Musaelyan K, Borsini A, Zunszain PA, Pariante CM (2020) Glucocorticoids prime the inflammatory response of human hippocampal cells through up-regulation of inflammatory pathways. Brain Behav Immun 87: 777–794
Ising M, Depping A, Siebertz A, Lucae S, Unschuld PG, Kloiber S, Horstmann S, Uhr M, Müller-Myhsok B, Holsboer F (2008) Polymorphisms in the FKBP5 gene region modulate recovery from psychosocial stress in healthy controls. Eur J Neurosci 28(2):389–398
Juruena MF, Cleare AJ, Papadopoulos AS, Poon L, Lightman S, Pariante CM (2006) Different responses to dexamethasone and prednisolone in the same depressed patients. Psychopharmacology 189(2):225–235
Juruena MF, Pariante CM, Papadopoulos AS, Poon L, Lightman S, Cleare AJ (2009) Prednisolone suppression test in depression: prospective study of the role of HPA axis dysfunction in treatment resistance. Br J Psychiatry J Ment Sci 194(4):342–349
Juruena MF, Cleare AJ, Papadopoulos AS, Poon L, Lightman S, Pariante CM (2010) The prednisolone suppression test in depression: dose-response and changes with antidepressant treatment. Psychoneuroendocrinology 35(10):1486–1491
Juruena MF, Pariante CM, Papadopoulos AS, Poon L, Lightman S, Cleare AJ (2013) The role of mineralocorticoid receptor function in treatment-resistant depression. J Psychopharmacol (Oxford, England) 27(12):1169–1179
Kang CB, Hong Y, Dhe-Paganon S, Yoon HS (2008) FKBP family proteins: immunophilins with versatile biological functions. Neurosignals 16(4):318–325
Kang CY, Shi J, Gong Y, Wei J, Zhang M, Ding H, Wang K, Yu Y, Wang S, Han J (2020) Interaction between FKBP5 polymorphisms and childhood trauma on depressive symptoms in Chinese adolescents: the moderating role of resilience. J Affect Disord 266:143–150
Kappelmann N, Czamara D, Rost N, Moser S, Schmoll V, Trastulla L, Stochl J, Lucae S, CHARGE inflammation working group, Binder EB, Khandaker GM, Arloth J (2021) Polygenic risk for immuno-metabolic markers and specific depressive symptoms: a multi-sample network analysis study. Brain Behav Immun 95:256–268
Kendler KS, Karkowski LM, Prescott CA (1999) Causal relationship between stressful life events and the onset of major depression. Am J Psychiatry 156(6):837–841
Kitchener P, Blasi FD, Borrelli E, Piazza PV (2004) Differences between brain structures in nuclear translocation and DNA binding of the glucocorticoid receptor during stress and the circadian cycle. Eur J Neurosci 19(7):1837–1846
Klengel T, Binder EB (2015) Epigenetics of stress-related psychiatric disorders and gene × environment interactions. Neuron 86(6):1343–1357
Klengel T, Mehta D, Anacker C, Rex-Haffner M, Pruessner JC, Pariante CM, Pace TW, Mercer KB, Mayberg HS, Bradley B, Nemeroff CB, Holsboer F, Heim CM, Ressler KJ, Rein T, Binder EB (2013) Allele-specific FKBP5 DNA demethylation mediates gene-childhood trauma interactions. Nat Neurosci 16(1):33–41
Klinger-König J, Hertel J, Van der Auwera S, Frenzel S, Pfeiffer L, Waldenberger M, Golchert J, Teumer A, Nauck M, Homuth G, Völzke H, Grabe HJ (2019) Methylation of the FKBP5 gene in association with FKBP5 genotypes, childhood maltreatment and depression. Neuropsychopharmacology 44(5):930–938
Kokras N, Hodes GE, Bangasser DA, Dalla C (2019) Sex differences in the hypothalamic-pituitary-adrenal axis: an obstacle to antidepressant drug development? Br J Pharmacol 176(21): 4090–4106
Kunugi H, Urushibara T, Nanko S (2004) Combined DEX/CRH test among Japanese patients with major depression. J Psychiatr Res 38(2):123–128

Lahti J, Ala-Mikkula H, Kajantie E, Haljas K, Eriksson JG, Räikkönen K (2016) Associations between self-reported and objectively recorded early life stress, FKBP5 polymorphisms, and depressive symptoms in midlife. Biol Psychiatry 80(11):869–877

Lang UE, Borgwardt S (2013) Molecular mechanisms of depression: perspectives on new treatment strategies. Cell Physiol Biochem 31:761–777

Lekman M, Laje G, Charney DS, Rush AJ, Paddock S (2008) The FKBP5-gene in depression and treatment response—an association study in the Sequenced Treatment Alternatives to Relieve Depression (STAR*D) cohort. Biol Psychiatry 63:1103–1110

Lenz KM, Pickett LA, Wright CL, Davis KT, Joshi A, McCarthy MM (2018) Mast cells in the developing brain determine adult sexual behavior. J Neurosci Off J Soc Neurosci 38(37): 8044–8059

Lin E, Tsai S (2019) Epigenetics and depression: an update. Psychiatry Investig 16(9):654–661

Liu T, Zhang L, Joo D, Sun SC (2017) NF-κB signaling in inflammation. Signal Transduct Target Ther 2:17023

Mamdani F, Rollins B, Morgan L, Myers RM, Barchas JD, Schatzberg AF, Watson SJ, Akil H, Potkin SG, Bunney WE, Vawter MP, Sequeira PA (2015) Variable telomere length across post-mortem human brain regions and specific reduction in the hippocampus of major depressive disorder. Transl Psychiatry 5(9):e636

Mangold DL, Wand G, Javors MA, Mintz JD (2010) Acculturation, childhood trauma and the cortisol awakening response in Mexican–American adults. Horm Behav 58(4):637–646

Mäntylä FL (2020) Major depressive disorder — a patient perspective to recovery. Inspire the Mind, November 20, 2020. https://www.inspirethemind.org/blog/major-depressive-disorder-a-patient-perspective-to-recovery

Mariani N, Cattane N, Pariante C, Cattaneo A (2021) Gene expression studies in Depression development and treatment: an overview of the underlying molecular mechanisms and biological processes to identify biomarkers. Transl Psychiatry 11(1):354

Matosin N, Halldorsdottir T, Binder EB (2018) Understanding the molecular mechanisms underpinning gene by environment interactions in psychiatric disorders: the FKBP5 model. Biol Psychiatry 83(10):821–830

Ménard C, Pfau ML, Hodes GE, Russo SJ (2017) Immune and neuroendocrine mechanisms of stress vulnerability and resilience. Neuropsychopharmacology 42(1):62–80

Menke A (2019) Is the HPA axis as target for depression outdated, or is there a new hope? Front Psych 10(101):1–8

Miller AH, Raison CL (2016) The role of inflammation in depression: from evolutionary imperative to modern treatment target. Nat Rev Immunol 16(1):22–34

Modell S, Yassouridis A, Huber J, Holsboer F (1997) Corticosteroid receptor function is decreased in depressed patients. Neuroendocrinology 65(3):216–222

Moieni M, Irwin MR, Jevtic I, Olmstead R, Breen EC, Eisenberger NI (2015) Sex differences in depressive and socioemotional responses to an inflammatory challenge: implications for sex differences in depression. Neuropsychopharmacology 40(7):1709–1716

Morris MC, Rao U, Garber J (2012) Cortisol responses to psychosocial stress predict depression trajectories: social-evaluative threat and prior depressive episodes as moderators. J Affect Disord 143(1–3):223–230

Mourtzi N, Sertedaki A, Charmandari E (2021) Glucocorticoid signaling and epigenetic alterations in stress-related disorders. Int J Mol Sci 22(11):5964

Munhoz CD, Sorrells SF, Caso JR, Scavone C, Sapolsky RM (2010) Glucocorticoids exacerbate lipopolysaccharide-induced signaling in the frontal cortex and hippocampus in a dose-dependent manner. J Neurosci Off J Soc Neurosci 30(41):13690–13698

Nettis MA, Veronese M, Nikkheslat N, Mariani N, Lombardo G, Sforzini L, Enache D, Harrison NA, Turkheimer FE, Mondelli V, Pariante CM (2020) PET imaging shows no changes in TSPO brain density after IFN-α immune challenge in healthy human volunteers. Transl Psychiatry 10(1):89

Nettis MA, Lombardo G, Hastings C, Zajkowska Z, Mariani N, Nikkheslat N, Worrell C, Enache D, McLaughlin A, Kose M, Sforzini L, Bogdanova A, Cleare A, Young AH, Pariante CM, Mondelli V (2021) Augmentation therapy with minocycline in treatment-resistant depression patients with low-grade peripheral inflammation: results from a double-blind randomised clinical trial. Neuropsychopharmacology 46(5):939–948

Nikkheslat N, Zunszain PA, Horowitz MA, Barbosa IG, Parker JA, Myint AM, Schwarz MJ, Tylee AT, Carvalho LA, Pariante CM (2015) Insufficient glucocorticoid signaling and elevated inflammation in coronary heart disease patients with comorbid depression. Brain Behav Immun 48:8–18

Niraula A, Wang Y, Godbout JP, Sheridan JF (2018) Corticosterone production during repeated social defeat causes monocyte mobilization from the bone marrow, glucocorticoid resistance, and neurovascular adhesion molecule expression. J Neurosci Off J Soc Neurosci 38(9): 2328–2340

Non AL, Hollister BM, Humphreys KL, Childebayeva A, Esteves KC, Zeanah CH, Fox NA, Nelson CA, Drury SS (2016) DNA methylation at stress-related genes is associated with exposure to early life institutionalization. Am J Phys Anthropol 161(1):84–93

Osimo EF, Baxter LJ, Lewis G, Jones PB, Khandaker GM (2019) Prevalence of low-grade inflammation in depression: a systematic review and meta-analysis of CRP levels. Psychol Med 49(12):1958–1970

Osimo EF, Pillinger T, Rodriguez IM, Khandaker GM, Pariante CM, Howes OD (2020) Inflammatory markers in depression: a meta-analysis of mean differences and variability in 5,166 patients and 5,083 controls. Brain Behav Immun 87:901–909

Otmishi P, Gordon J, El-Oshar S, Li H, Guardiola J, Saad M, Proctor M, Yu J (2008) Neuroimmune interaction in inflammatory diseases. Clin Med Circulat Respirat Pulmon Med 2:35–44

Pariante CM (2017) Why are depressed patients inflamed? A reflection on 20 years of research on depression, glucocorticoid resistance and inflammation. Eur Neuropsychopharmacol 27:554–559

Pariante CM, Lightman SL (2008) The HPA axis in major depression: classical theories and new developments. Trends Neurosci 31(9):464468

Pariante CM, Miller AH (2001) Glucocorticoid receptors in major depression: relevance to pathophysiology and treatment. Biol Psychiatry 49(5):391–404

Pelekanou V, Kampa M, Kiagiadaki F, Deli A, Theodoropoulos P, Agrogiannis G, Patsouris E, Tsapis A, Castanas E, Notas G (2016) Estrogen anti-inflammatory activity on human monocytes is mediated through cross-talk between estrogen receptor ERα36 and GPR30/GPER1. J Leukoc Biol 99(2):333–347

Perrin AJ, Horowitz MA, Roelofs J, Zunszain PA, Pariante CM (2019) Glucocorticoid resistance: is it a requisite for increased cytokine production in depression? A systematic review and meta-analysis. Front Psych 10:423

Piechaczek CE, Greimel E, Feldmann L, Pehl V, Schulte-Koerne G (2019) Interactions between FKBP5 variation and environmental stressors in adolescent Major Depression. Psychoneuroendocrinology 106:28–37

Pitharouli MC, Hagenaars SP, Glanville KP, Coleman J, Hotopf M, Lewis CM, Pariante CM (2021) Elevated C-reactive protein in patients with depression, independent of genetic, health, and psychosocial factors: results from the UK Biobank. Am J Psychiatry 178(6):522–529

Rao S, Yao Y, Ryan J, Li T, Wang DL, Zheng C, Xu Y, Xu Q (2016) Common variants in FKBP5 gene and major depressive disorder (MDD) susceptibility: a comprehensive meta-analysis. Sci Rep 6:32687

Renoir T, Hasebe K, Gray L (2013) Mind and body: how the health of the body impacts on neuropsychiatry. Front Pharmacol 4:158

Russell A, Hepgul N, Nikkheslat N, Borsini A, Zajkowska Z, Moll N, Forton D, Agarwal K, Chalder T, Mondelli V, Hotopf M, Cleare A, Murphy G, Foster G, Wong T, Schütze GA, Schwarz MJ, Harrison N, Zunszain PA, Pariante CM (2019) Persistent fatigue induced by interferon-alpha: a novel, inflammation-based, proxy model of chronic fatigue syndrome. Psychoneuroendocrinology 100:276–285

Salk RH, Hyde JS, Abramson LY (2017) Gender differences in depression in representative national samples: meta-analyses of diagnoses and symptoms. Psychol Bull 143:783–822

Savitz J (2020) The kynurenine pathway: a finger in every pie. Mol Psychiatry 25(1):131–147

Scheuer S, Ising M, Uhr M, Otto Y, Klitzing KV, Klein AM (2016) FKBP5 polymorphisms moderate the influence of adverse life events on the risk of anxiety and depressive disorders in preschool children. J Psychiatr Res 72:30–36

Seidman SN (2006) Psychoneuroendocrinology of mood disorders. In: Stein DJ, Kupfer DJ, Schatzberg AF (eds) The American psychiatric publishing textbook of mood disorders. American Psychiatric Association Publishing, p 117130

Sforzini L, Nettis MA, Mondelli V, Pariante CM (2019) Inflammation in cancer and depression: a starring role for the kynurenine pathway. Psychopharmacology 236(10):2997–3011

Sforzini L, Worrell C, Kose M, Anderson IM, Aouizerate B, Arolt V, Bauer M, Baune BT, Blier P, Cleare AJ, Cowen PJ, Dinan TG, Fagiolini A, Ferrier IN, Hegerl U, Krystal AD, Leboyer M, McAllister-Williams RH, McIntyre RS, Meyer-Lindenberg A, Meyer-Lindenberg A, Miller AH, Nemeroff CB, Normann C, Nutt D, Pallanti S, Pani L, Penninx BWJH, Schatzberg AF, Shelton RC, Yatham LN, Young AH, Zahn R, Aislaitner G, Butlen-Ducuing F, Fletcher C, Haberkamp M, Laughren T, Mäntylä F-L, Schruers K, Thomson A, Arteaga-Henríquez G, Benedetti F, Cash-Gibson L, Chae WR, De Smedt H, Gold SM, Hoogendijk WJG, Mondragón VJ, Maron E, Pariante CM (2021) A Delphi-method-based consensus guideline for definition of treatment-resistant depression for clinical trials. Mol Psychiatry. https://doi.org/10.1038/s41380-021-01381-x

Shapero BG, Black SK, Liu RT, Klugman J, Bender RE, Abramson LY, Alloy LB (2014) Stressful life events and depression symptoms: the effect of childhood emotional abuse on stress reactivity. J Clin Psychol 70(3):209–223

Sorgdrager FJ, Doornbos B, Penninx BW, Jonge PD, Kema IP (2017) The association between the hypothalamic pituitary adrenal axis and tryptophan metabolism in persons with recurrent major depressive disorder and healthy controls. J Affect Disord 222:32–29

Stamm TJ, Rampp C, Wiethoff K, Stingl J, Mössner R, O'Malley G, Ricken R, Seemüller F, Keck M, Fisher R, Gaebel W, Maier W, Möller HJ, Bauer M, Adli M (2016) The FKBP5 polymorphism rs1360780 influences the effect of an algorithm-based antidepressant treatment and is associated with remission in patients with major depression. J Psychopharmacol (Oxford, England) 30(1):40–47

Stetler C, Miller GE (2011) Depression and hypothalamic-pituitary-adrenal activation: a quantitative summary of four decades of research. Psychosom Med 73(2):114–126

Sun H, Kennedy PJ, Nestler EJ (2013) Epigenetics of the depressed brain: role of histone acetylation and methylation. Neuropsychopharmacology 38:124–137

Szczepankiewicz A, Leszczyńska-Rodziewicz A, Pawlak J, Narozna B, Rajewska-Rager A, Wilkosc M, Zaremba D, Maciukiewicz M, Twarowska-Hauser J (2014) FKBP5 polymorphism is associated with major depression but not with bipolar disorder. J Affect Disord 164:33–37

Tatro ET, Everall IP, Masliah E, Hult BJ, Lucero G, Chana G, Soontornniyomkij V, Achim CL, HIV Neurobehavioral Research Center (2009) Differential expression of immunophilins FKBP51 and FKBP52 in the frontal cortex of HIV-infected patients with major depressive disorder. J Neuroimmune Pharmacol 4(2):218–226

Toni R (2004) The neuroendocrine system: organization and homeostatic role. J Endocrinol Investig 27(6 Suppl):35–47

Tozzi L, Farrell C, Booij L, Doolin K, Nemoda Z, Szyf M, Pomares FB, Chiarella J, O'keane, V., & Frodl, T. (2018) Epigenetic changes of FKBP5 as a link connecting genetic and environmental risk factors with structural and functional brain changes in major depression. Neuropsychopharmacology 43:1138–1145

Veen G, Vliet IM, Derijk RH, Giltay EJ, Zitman FG (2011) Basal cortisol levels in relation to dimensions and DSM-IV categories of depression and anxiety. Psychiatry Res 185:121–128

Verburg-van Kemenade B, Cohen N, Chadzinska M (2017) Neuroendocrine-immune interaction: evolutionarily conserved mechanisms that maintain allostasis in an ever-changing environment. Dev Comp Immunol 66:2–23

Vermeer H, Hendriks-Stegeman BI, Burg BV, Buul-Offers SC, Jansen M (2003) Glucocorticoid-induced increase in lymphocytic FKBP51 messenger ribonucleic acid expression: a potential marker for glucocorticoid sensitivity, potency, and bioavailability. J Clin Endocrinol Metab 88(1):277–284

Villa A, Vegeto E, Poletti A, Maggi A (2016) Estrogens, neuroinflammation, and neurodegeneration. Endocr Rev 37(4):372–402

Vreeburg SA, Hoogendijk WJ, Pelt JV, Derijk RH, Verhagen JC, Dyck RV, Smit JH, Zitman FG, Penninx BW (2009) Major depressive disorder and hypothalamic-pituitary-adrenal axis activity: results from a large cohort study. Arch Gen Psychiatry 66(6):617–626

Watson SD, Gallagher P, Del-Estal D, Hearn A, Ferrier IN, Young AH (2002) Hypothalamic-pituitary-adrenal axis function in patients with chronic depression. Psychol Med 32(6): 1021–1028

Witzmann SR, Turner JD, Mériaux SB, Meijer OC, Muller CP (2012) Epigenetic regulation of the glucocorticoid receptor promoter 1(7) in adult rats. Epigenetics 7(11):1290–1301

Wochnik GM, Rüegg J, Abel GA, Schmidt U, Holsboer F, Rein T (2005) FK506-binding proteins 51 and 52 differentially regulate dynein interaction and nuclear translocation of the glucocorticoid receptor in mammalian cells. J Biol Chem 280(6):4609–4616

Yeager MP, Pioli PA, Guyre PM (2011) Cortisol exerts bi-phasic regulation of inflammation in humans. Dose-response: a publication of International Hormesis Society 9(3):332–347

Yeager MP, Pioli PA, Collins J, Barr F, Metzler S, Sites BD, Guyre PM (2016) Glucocorticoids enhance the in vivo migratory response of human monocytes. Brain Behav Immun 54:86–94

Zannas AS, Jia M, Hafner K, Baumert J, Wiechmann T, Pape JC, Arloth J, Ködel M, Martinelli S, Roitman M, Röh S, Haehle A, Emeny RT, Iurato S, Carrillo-Roa T, Lahti J, Räikkönen K, Eriksson JG, Drake AJ, Waldenberger M, Wahl S, Kunze S, Lucae S, Bradley B, Gieger C, Hausch F, Smith AK, Ressler KJ, Müller-Myhsok B, Ladwig KH, Rein T, Gassen NC, Binder EB (2019) Epigenetic upregulation of FKBP5 by aging and stress contributes to NF-κB-driven inflammation and cardiovascular risk. Proc Natl Acad Sci USA 116(23):11370–11379

Zobel AW, Nickel T, Sonntag A, Uhr M, Holsboer F, Ising M (2001) Cortisol response in the combined dexamethasone/CRH test as predictor of relapse in patients with remitted depression. a prospective study. J Psychiatr Res 35(2):83–94

Zunszain PA, Anacker C, Cattaneo A, Carvalho LA, Pariante CM (2011) Glucocorticoids, cytokines and brain abnormalities in depression. Prog Neuro-Psychopharmacol Biol Psychiatry 35(3):722–729

Brain Fluids, Blood–Brain Interfaces, and Their Involvement in Neuroimmune Regulation During Development and in Adulthood

7

Amel Amara and Jean-François Ghersi-Egea

Abstract

The homeostasis of brain fluids, which comprise the parenchymal interstitial fluid and the circulating cerebrospinal fluid, is crucial for proper brain development and brain function throughout life. The composition of brain fluids changes during development and is highly controlled by the blood–brain interfaces comprising the endothelium of the cerebral microvessels and pial vessels, the epithelium of the choroid plexuses, and the arachnoid membrane. This chapter discusses the sources and functions of the cerebrospinal fluid, which differ according to the developmental stages, and describes the organization of the brain fluid compartments. It gives an overview of the organization of brain interfaces and their functions, including neuroprotective and neuroendocrine functions, and explains how these interfaces complement each other and adapt as a function of brain developmental stages. The organization of brain fluid compartments in relation to the peripheral lymphatic system and the brain interfaces' differential permissiveness to immune cells is also described. They both contribute to the peculiar "immune privilege" that is attributed to the central nervous system while allowing the neuroimmune surveillance necessary to protect the brain.

Keywords

Cerebrospinal fluid · Interstitial fluid · Blood–brain barrier · Choroid plexus · Development · Neuroimmune regulation

A. Amara (✉) · J.-F. Ghersi-Egea
Fluid Team, Lyon Neuroscience Research Center, INSERM U1028, CNRS UMR5292, Lyon University, Bron, France
e-mail: amel.amara@univ-lyon1.fr; jean-francois.ghersi-egea@inserm.fr

J. P. Konsman, T. M. Reyes (eds.), *Neuroendocrine-Immune System Interactions*, Masterclass in Neuroendocrinology 13,
https://doi.org/10.1007/978-3-031-21358-8_7

Abbreviations

APC	Antigen-presenting cell
BBB	Blood–brain barrier
BCSFB	Blood–cerebrospinal fluid barrier
ChP	Choroid plexus
CNS	Central nervous system
CSF	Cerebrospinal fluid
DCLN	Deep cervical lymph node
EC	Endothelial cell
ECM	Extracellular matrix
GW	Gestational week
ISF	Interstitial fluid
NVU	Neurovascular unit
PVS	Perivascular space
Shh	Sonic hedgehog
TH	Thyroid hormone
TJ	Tight junction
TTR	Transthyretin
VEGF	Vascular endothelial growth factor

7.1 Introduction

Maintaining brain homeostasis is crucial to allow normal brain maturation during development and proper brain functioning throughout life. The composition of brain fluids is highly controlled by cellular interfaces that regulate the exchanges between the blood and the central nervous system (CNS) and protect the brain from potentially neurotoxic molecules circulating in the blood. The brain also benefits from an "immune privilege," and the brain fluids and blood–brain interfaces play an important part in the special interactions occurring between the immune system and the CNS. This chapter gives an overview of interstitial (ISF) and cerebrospinal fluid (CSF) system development and organization which is unique to the CNS. It then describes the different interfaces that complement each other to maintain brain homeostasis. It also discusses the concept of barrier immaturity during development and the controversies it generated. Finally, the chapter describes how the peculiar fluid organization and the different cellular barriers all contribute to the regulation of neuroimmune interactions in adults and during development.

7.2 The Organization of Brain Fluids During Development and in Adult

The brain is a peculiar organ in that it has a circulatory system of its own, the CSF system, which forms early during embryonic life. The development of the CNS occurs during the third and fourth gestational week (GW) in humans, on the embryonic day 9 (E9) and E7 in rats and mice, respectively, and results from a step-by-step process (DeSesso et al. 1999; Yuskaitis and Pomeroy 2017). It begins with the neurulation and the closure of the neural tube, which lead to the formation of the primitive vesicles giving birth to the different structures of the brain and the cerebellum, and of a fluid-filled internal space that will become the ventricular system (reviewed in Yuskaitis and Pomeroy 2017). Brain fluids both cushion and support brain cells and are distributed in different compartments of the CNS. The CSF fills the ventricular system and the subarachnoid spaces (SAS), whereas the ISF is located in the parenchyma between neural cells and extracellular matrix (ECM) components. These fluids differ according to their origin, their overall composition, and some of their properties, including hydrostatic pressure. The composition of both CSF and ISF is highly conserved across species in adults and during development. They both contribute to normal CNS development.

7.2.1 Cerebrospinal Fluid

The CSF is a colorless fluid that bathes the brain. It has a wide variety of functions in the CNS, from mechanical protection against shocks and vibration to the maintenance of brain homeostasis. It is secreted by different cells of the brain depending on the developmental stages. Our knowledge about embryonic and fetal CSF production, composition, and function has been gathered using relevant animal models (Box 7.1). The CSF fills the brain cavities forming the ventricular system before

Box 7.1. Animal models to study CSF during early development

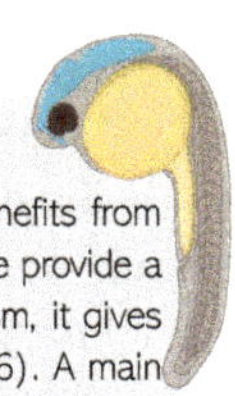

Our understanding of CSF space shaping and CSF functions at early developmental stages benefits from animal models adapted to these studies. External embryos of zebrafish (*Danio rerio*) for instance provide a large-scale access to the developing CSF and ventricles *in vivo*. As an entire transparent organism, it gives rise to excellent imaging allowing to follow fluid dynamics (Böhm et al. 2016; Fame et al. 2016). A main challenge to meet when studying brain development is the size of the model. The zebrafish embryos face out this problem since their size is important enough to easily perform ventricular manipulation and to sample CSF. In Australia, marsupials such as the opossum *Monodelphis domestica* is one of very few established laboratory models allowing easy access to the brain during development. The gestation is relatively short (14 days) and takes place externally, allowing easy and continuous observations and manipulations of littermates throughout development (Kiyonari et al. 2021). Other species have been used such as rodents which are also good models for their genetic similarities to humans, and because different manipulations can be performed during development (e.g., CSF collection, microdissection of the ventricular system, selected surgical procedures, etc.). Large mammalian systems such as sheep and pigs needs to be used to perform more complex surgical manipulations (reviewed in Fame et al. 2020).

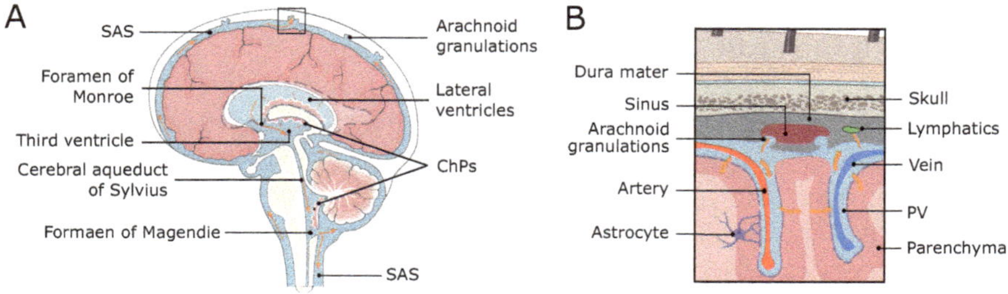

Fig. 7.1 Pathways of CSF flow. (**a**) The CSF (blue) freely circulates from the lateral ventricles to the third ventricle via the interventricular foramen of Monro. It then reaches the fourth ventricle through the cerebral aqueduct of Sylvius and flows through the foramen of Magendie, and laterally the foramina of Luschka (not shown) to fill the SAS surrounding both the brain and the spinal cord. From there it is reabsorbed into the blood and lymphatics. (**b**) Enlargement of square in (**a**) showing that CSF penetrates the PVS to mix with the ISF within the parenchyma (large orange arrows). It also reaches blood sinuses across arachnoid granulations, and dural lymphatic vessels (small orange arrows). Access to lymphatics also occurs through the cribriform plate and along cranial nerves (not shown here). ChPs: choroid plexuses; SAS: subarachnoid space; PVS: perivascular space (inspired by Ray and Heys 2019)

reaching the SAS that surround the entire organ and being reabsorbed into the blood or lymphatic circulation (Fig. 7.1). Its composition and its rate of renewal change throughout life and under pathological conditions.

7.2.1.1 Embryonic Cerebrospinal Fluid

The embryonic CSF refers to the fluid transitorily present at the earliest stages of brain development after closure of the neural tube. In humans, it starts to inflate the developing cerebral ventricles between GW3 and 4 (reviewed in Fame et al. 2020). This CSF is secreted by a homogeneous population of neuroepithelial precursors surrounding the cavity formed by the closure of the neural tube (reviewed in Gato et al. 2014; Yuskaitis and Pomeroy 2017; Fame et al. 2020). Within this primitive ventricular system, the CSF participates in the progressive formation of what will be the brain. It promotes the growth and differentiation of early progenitor cells of the brain anlage and therefore the beginning of the neurogenesis (reviewed in Gato and Desmond 2009; Gato et al. 2014). This function results from the positive hydrostatic pressure generated by the continuous secretion of the fluid, which is applied against the neuroepithelium lining the primitive cavity (Desmond and Jacobson 1977). Embryonic CSF also has a biological influence on the embryonic brain stem cell niche, by promoting survival, proliferation, and neurogenesis. The embryonic CSF composition is complex, and the factors responsible for the development and growth of neural stem cells are only beginning to be uncovered. Factors already known to be involved in mitogenic and neurogenic activity include fibroblast growth factors, insulin-like growth factors (IGFs), sonic hedgehog (Shh), bone morphogenic proteins, and Wnt glycoproteins (reviewed in Gato and Desmond 2009; Zappaterra and Lehtinen 2012; Gato et al. 2014). Besides inorganic anions and the previously mentioned factors, embryonic CSF contains numerous other components. Analyses in different species such as sheep and rats revealed that proteins including albumin, fetuin, alpha-fetoprotein, transferrin, and lipoproteins are the main components found in CSF in the early stages of development (Dziegielewska et al. 1980, 1981). In chickens, the level of these proteins is thirty times higher in embryos than in adults (Birge et al. 1974). Novel proteomic analyses have allowed for further refinement of the CSF composition and identified ECM components, enzymes, and cytokines within the embryonic CSF (reviewed in Gato et al. 2014).

This early period of brain development remains short and the neural tube cavity undergoes a rapid change in volume and morphology. The initial brain expansion is followed by the ventricular system formation in which the CSF circulates. The source of the CSF changes from a neuroepithelial to a choroid plexus secretion. Its composition changes over time, progressively giving way to the mature CSF.

7.2.1.2 Fetal and Adult Cerebrospinal Fluid

From Production to Resorption

The fetal and adult CSF is mostly secreted by specialized tissues called the choroid plexuses (ChPs). The ChPs originate from selected progenitor cells from the different primary brain vesicles and appear between GW 7 and 8 in humans (reviewed in Fame et al. 2020), and between E8.5 and E9.5 in mouse (reviewed in Lun et al. 2015, and see below 2.2.1). During this time, the ventricles expand and begin to form an interconnected network in which the fluid circulates throughout life. CSF volume progressively increases during brain development. Imaging studies and especially CT scanning studies estimate the CSF to be 50 mL in neonates, 140 mL in children, and up to 170 mL in adults and CSF volume continues to increase with age, especially in the SAS (Corns and Martin 2012). The ChPs produce at least 70% of the CSF, at a rate of 0.35 mL/min in human adults, and the majority of CSF is produced by the lateral ventricle ChPs (reviewed in Di Terlizzi and Platt 2006; Corns and Martin 2012). CSF production results from an active and controlled fluid secretion by the ChP epithelium. The mechanism involves various inorganic ion transporters and channels, resulting in a net flux of HCO_3^-, Na^+, and Cl^-, driving water movement into the ventricles. It is highly dependent on carbonic anhydrase and the Na^+/K^+-ATPase located in the ChP epithelium (reviewed in Ghersi-Egea et al. 2018). Extra choroidal sources of CSF are still debated. A secretion by the meninges surrounding the brain and a movement of ISF entering the ventricles across the leaky ependymal lining the ventricular walls have been hypothesized (Pollay and Curl 1967; reviewed in Di Terlizzi and Platt 2006, Fame et al. 2020). In human adults, CSF production is around 500 mL/day with a complete renewal of the fluid about three to four times per day (Zappaterra and Lehtinen 2012). CSF is distributed as follows: 25 mL in ventricles and 125 mL in cisterns, and both cranial and spinal SAS of the brain (Sakka et al. 2011).

The CSF circulates in distinct ways (Fig. 7.1a) (reviewed in Di Terlizzi and Platt 2006; Zappaterra and Lehtinen 2012). It flows from the lateral ventricles to the third ventricle via the interventricular foramen of Monro. It then reaches the fourth ventricle through the cerebral aqueduct of Sylvius before flowing through the foramen of Magendie and the foramina of Luschka to fill the SAS surrounding both the spinal cord and the brain. From there, it is reabsorbed in the blood and lymphatic system. It also penetrates the cerebral perivascular spaces (PVS) to mix with the ISF fluid (Fig. 7.1b). Finally, it also follows the velae to reach the internal cisterns directly from the ventricles.

The balance between CSF production and resorption is highly dynamic and allows the maintenance of a rather constant total volume of CSF. Changes in CSF dynamics in the brain may cause serious damage. Too much CSF accumulating in ventricles and SAS leads to hydrocephalus and neurodevelopmental diseases, whereas low CSF production leads to impaired brain growth (Corns and Martin 2012; Shen 2018). The ways in which the CSF is absorbed and distributed to the lymph nodes are still debated. The dominant view of CSF reabsorption is that the fluid can be reabsorbed directly into the meningeal venous network at the top of the brain through the arachnoid granulations (Fig. 7.1b). This seems to be a main route of CSF drainage in humans, but the presence of one-way pressure sensitive valves between CSF and venous blood within the granulations suggests that its function is the control of the intracranial pressure. Complementary studies showed that in many species the fluid also reaches the lymphatic network of the nasal mucosa throughout the cribriform plate of the ethmoid and along the cranial nerves (see 7.4.2). Finally, some studies support circadian cycles in CSF composition, production, and turnover (Myung et al. 2018).

Composition and Functions

The CSF is a complex fluid whose composition is finely tuned in healthy conditions. It is not an ultrafiltrate of plasma even though its inorganic ion composition is close (Abbott et al. 2010; Hladky and Barrand 2014). Modern analytical techniques show that it contains proteins, lipids, hormones, amino acids, glucose, and many other molecules and metabolites that affect a wide range of CNS functions (reviewed in Di Terlizzi and Platt 2006; Zappaterra and Lehtinen 2012).

Through exchanges with ISF, the CSF participates in the maintenance of brain homeostasis by buffering osmolarity, pH, and overall inorganic ion composition. Changes in this chemical environment can lead to neuronal dysfunctions. Na^+, for instance, is very important in transport and osmoregulation, and a decrease in its concentration can induce cellular swelling but also seizure and coma (Giuliani and Peri 2014). Regarding Ca^{2+}, changes in its concentration disturb neurotransmission (Odackal et al. 2015).

Besides homeostasis of inorganic ions, the CSF has other important buffering, delivery, and clearance functions. By filling the ventricles and the SAS surrounding the brain, the CSF acts as an absorber of physical shocks and regulates intracranial pressure (reviewed in Di Terlizzi and Platt 2006). This protection is, however, not sufficient to prevent traumatic brain injury in case of extreme force impacts. It also acts as an intracerebral transport system by distributing within the brain different hormones, growth factors, neurotransmitters, and neuropeptides (reviewed in Di Terlizzi and Platt 2006). Ventricular circulation of trophic factors and guidance molecules is especially important during fetal and postnatal development, to regulate stem cell behavior. CSF is also implicated in waste drainage by excreting toxic by-products of cellular metabolism and neurotoxic agents (Kaur et al. 2021).

Because these properties support brain development and function, the CSF can be considered a window allowing the assessment of brain health. As such it is a fluid in which biomarkers can be identified to categorize pathologies (e.g., inflammatory, neoplastic, metabolic) (Kansal and Irwin 2015). For example, in normal adult CSF, very low concentrations of proteins and cells are measured (Di Terlizzi and Platt 2006). A leak of blood-borne proteins, and especially albumin, into the CSF signals a damaged brain interface (Hühmer et al. 2006). Immune cells, mainly neutrophils and lymphocytes, can also enter the CSF following infection such as meningitis and induce serious and long-term neurological damage (Jaijakul et al. 2017) (see 3 below).

7.2.2 Brain Interstitial Fluid

The ISF is the fluid that occupies the space between neurons, glial cells, and ECM components that support the tissue. This dynamic and complex space is the compartment in which nutrients circulate and through which waste products are removed. ISF volume is more important during development than in adulthood. The interstitial space, which includes ISF and ECM, occupies 15–20% of the tissue volume in adults (reviewed in Lei et al. 2017) and up to 30–45% in newborns, as measured in rat (Simpson and Stephenson 1993).

The composition of the ISF is close but not identical to that of the CSF. Enzymes responsible for the degradation and remodeling of the ECM proteins are also found in ISF, as well as extracellular vesicles which are released by neural cells and contain signaling proteins implicated in cell–cell communication, synaptic function, and neuronal behavior (Ashok and Gabriele 2020).

There are few reports about ISF sources and mechanisms of production. Many correlation studies between CSF and ISF suggest that CSF is a source of ISF. Iliff and collaborators confirmed the link between both fluids by following fluorescent tracers injected into the cistern of the fourth ventricle (Iliff et al. 2012, 2015). Movements of inorganic ions, gaseous molecules, and organic molecules such as proteins, peptides, and enzymes occur between CSF and ISF through the ependymal lining of the ventricles, which does not display barrier properties as intercellular contacts between ependymal cells are made by adherents and gap junctions only. The extent of these exchanges is still being investigated and debated. Such exchanges also occur across the outer surfaces of the brain parenchyma, that is, across the pia mater and the glia limitans, these cell layers not being sealed by tight junctions (TJs) (Fig. 7.2c). There is also free movement of water and small and large solutes between ISF and CSF through PVS (reviewed in Hladky and Barrand 2014).

Additional sources of ISF are water generated by cell metabolism, and water transfer across brain capillaries (Iliff et al. 2012; Brinker et al. 2014; reviewed in Lei et al. 2017). Gaseous and lipid-soluble molecules present in the systemic circulation can passively cross the blood–brain barrier (BBB) and contribute to ISF

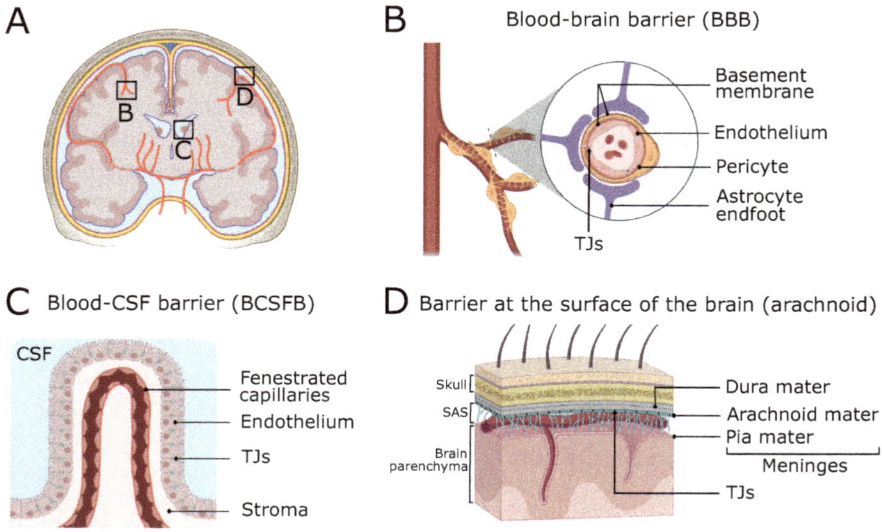

Fig. 7.2 Anatomical localization and organization of the cellular protective brain barriers. (**a**) Schematic representation illustrating the different sites of blood–brain exchanges detailed in (**b**), (**c**), and (**d**). (**b**) The blood–brain barrier is located at the endothelium of parenchymal capillaries and microvessels. Endothelial cells are joined by TJs and encircled by a basement membrane and by pericytes. Astrocytic processes surround these structures that form the neurovascular unit. (**c**) The blood–CSF barrier is situated at the ChPs. Blood vessels within the ChPs are fenestrated and do not form a restrictive barrier. The ChP epithelium is sealed by TJs restricting intercellular movements of molecules between the choroidal stroma and the CSF. (**d**) The barrier at the surface of the brain is located within the meninges that are composed of the dura, the arachnoid, and the pia mater. Blood vessels in the dura are fenestrated. The barrier between the CSF and the periphery is located at the arachnoid whose cells are sealed by TJs. The pia mater bordering the tissue has no barrier phenotype. Pial vessels circulate within the SAS before penetrating the brain parenchyma. The endothelium of these vessels already displays a tight phenotype that reduces exchanges between the blood and the subarachnoid CSF. ChP: choroid plexus; SAS: subarachnoid space; TJ: tight junction; CSF: cerebrospinal fluid (inspired by Engelhardt et al. 2017)

composition. Other blood-borne components such as ions, nutrients, and specific proteins can reach ISF across the BBB through channels, transporters, or via more complex pathways such as receptor-mediated endocytosis (see 2.1). The composition of ISF is also influenced by neural cell activities that are accompanied by exchanges of ions, proteins, and organic compounds between cells and the ISF. In astrocytes, iso-osmolarity between extra- and intracellular fluids is maintained by water movement through channel proteins such as Aquaporin 4 (Brinker et al. 2014; Lei et al. 2017; Ashok and Gabriele 2020).

In conclusion, both CSF and ISF participate in the maintenance of homeostasis and the proper functioning of the brain, and their composition is highly regulated by the cellular barriers located between the blood and these fluids.

7.3 The Blood–Brain Interfaces During Development and in Adult

Goldmann was one of the first scientists to formally demonstrate in 1909 the existence of blood–brain interfaces that make the CNS a "closed" compartment protected from systemic circulation. He performed peripheral injections of a dye (the trypan blue) in various species and noticed that all body regions were stained except the brain (Goldmann 1909, 1913). The notion of blood–brain barriers was first mentioned by Stern and Gautier in 1918: "The barrier that opposes the movement into the CSF of substances in the blood shows notable differences in different animal species" (translated from French). These specific cell layers tightly control or restrict the transport of blood-borne substances into the brain, thereby protecting this organ from neurotoxic molecules during development and throughout life. They also maintain brain homeostasis by regulating brain fluid composition and providing nutrients and other components the brain needs. It is now accepted that these cellular barriers also participate in neuroimmune surveillance, allowing an adapted response to injuries. The two main barriers in the CNS are the BBB located at the endothelium of the cerebral microvasculature (Fig. 7.2b) and the blood–CSF barrier (BCSFB) formed by the epithelium of the ChPs (Fig. 7.2c) and, at the surface of the brain, by the arachnoid (Fig. 7.2d).

7.3.1 The Blood–Brain Barrier

7.3.1.1 Organization and Function

The BBB is a major protective brain barrier that contributes to the normal development and integrity of the brain tissue. As the brain vasculature is the main route by which molecules reach the cerebral parenchyma, it is crucial for this interface to selectively control molecular transport into the brain. It is widely accepted that a dysfunction of this barrier contributes to the physiopathology of several diseases such as neurodegenerative disorders, cerebrovascular accidents, and neuroinflammatory diseases (reviewed in Abbott et al. 2010; Obermeier et al. 2013).

Different cell types confer its physical and biochemical properties on the BBB (Fig. 7.2b, reviewed in Kadry et al. 2020). First, the endothelial cells (ECs) lining the vessel walls are tightly sealed by junctional protein complexes. Claudin 5 and occludin are the most abundant proteins expressed in these TJs and prevent the paracellular diffusion even of small molecular weight molecules (Nitta et al. 2003). Other cell types are closely associated with ECs. They include pericytes and astrocytes whose end feet surround the endothelial monolayer. Pericytes regulate BBB-specific gene expression patterns in ECs, strengthen TJs, and repress non-specific transcytosis. They are therefore crucial to limiting the non-specific access of blood-borne molecules into the brain parenchyma across the cerebral endothelium (reviewed in Armulik et al. 2011). Perivascular astrocytes actively participate in the barrier phenotype maintenance by extending processes around ECs and secreting factors to modulate EC function (reviewed in Obermeier et al.

2013; Kadry et al. 2020). Collectively these different cell types form the neurovascular unit (NVU) and support the barrier properties.

Despite the tight nature of the cerebral endothelium, the BBB is also an active site of transport (reviewed in Abbott et al. 2010; Kadry et al. 2020). Simple diffusion of small molecules across the BBB is dependent on their lipophilicity, and many molecules cannot freely cross the plasma membrane of the BBB to enter the parenchyma. Glucose, amino acids, nucleotides, and vitamins are transported into the brain through carrier-mediated transport systems. Glucose, for instance, is carried by the Glut-1 transporter into the ISF (Qutub and Hunt 2005). Macromolecules such as proteins can enter the brain parenchyma by transcytosis through receptor-mediated transport systems. A well-described system is the insulin receptor-mediated transcytosis (Fishman et al. 1987). The transferrin receptor also induces an endocytotic process allowing iron bound to transferrin to reach the ISF, while the unbound transferrin recycles at the luminal, blood-facing membrane (Duffy and Pardridge 1987).

The barrier shows a highly restrictive transcellular permeability to numerous blood-borne toxic substances but also to therapeutics. This "multidrug resistance" mechanism mainly results from the action of several efflux transporters of the ATP-binding cassette (ABC) family, including P-glycoprotein (PGP/ABCB1), multidrug-resistant associated proteins (MRPs/ABCCs), and the human breast cancer resistance protein (BCRP/ABCG2) (Leslie et al. 2005; reviewed in Abbott et al. 2010). The BBB is also considered to be a metabolic barrier since it is associated with high expression of metabolizing enzymes such as peptidases and monoamine oxidases (Brownlees and Williams 1993).

7.3.1.2 Development of the Vascular Network and Blood–Brain Barrier

The BBB phenotype of the cerebral endothelium develops at early developmental stages. It is a step-by-step process that begins with angiogenesis within CNS parenchyma during embryonic development (E11 in mice, Abbott et al. 2010). Angiogenesis is guided by a concentration gradient of different signaling molecules produced by subventricular progenitor cells (e.g., vascular endothelial growth factor (VEGF), Wnt, Obermeier et al. 2013). New vessel sprouts rapidly exhibit several BBB properties including the appearance of short and primitive TJs and a lack of fenestrations that restrict the movement of proteins from blood to the brain. They also express nutrient transporters which provide the developing neural cells with their needs (Abbott et al. 2010; Obermeier et al. 2013). Non-specific pinocytosis across ECs remains an ongoing process in early angiogenesis, and new vessels also express leukocyte adhesion molecules such as intercellular adhesion molecule 1 (Icam-1), (Daneman et al. 2010). This angiogenesis stage is followed by a differentiation phase in which ECs of newborn vessels form interactions with pericytes. ECs release platelet-derived growth factor β, which signals to receptors expressed by pericytes (Hellström et al. 1999). Pericytes then start to assist angiogenesis and to participate in the development of the BBB phenotype, around GW12 in humans, and at E12 in rats (reviewed in Obermeier et al. 2013). They form long processes that fully cover the vessels. This almost full coverage is crucial for maintaining the barrier's integrity. Pericytes secrete several growth factors and morphogens contributing to the BBB formation (e.g., angiopoietin-1) (Armulik et al. 2011). Interactions with pericytes lead to the maturation of TJs, decreased transcytosis, and a downregulation of leukocyte adhesion molecules (Daneman et al. 2010). The vascular network extension and the maturation of the BBB phenotype largely continue after birth with the recruitment of perivascular astrocytes. They secrete for instance VEGF to stimulate angiogenesis (Alvarez et al. 2013) and Sonic hedgehog (Shh) to promote the expression of functional barrier proteins (Alvarez et al. 2011). To a much lesser extent, neurons on their own can directly contact ECs and astrocytes to participate in the maintenance of the BBB (Abbott et al. 2010; Obermeier et al. 2013).

Nutrient influx transporters display developmental expression profiles that vary from one to another to meet the brain needs which change during the different stages of development. The expression of efflux pumps, although essential to protect the brain from noxious endogenous or exogenous compounds, is more heterogeneous. For instance, P-glycoprotein expression is low during development and increases to adult levels only after birth under the probable influence of astrocytes (Strazielle and Ghersi-Egea 2015).

7.3.1.3 The History of an Immature Blood–Brain Barrier Concept

Historically, the BBB has long been considered immature during early developmental stages, making the blood–placenta barrier the main barrier protecting the fetal brain. For nearly a hundred years, a leaky BBB was considered necessary for the rapid growth of the brain, in order to facilitate the access of blood-borne nutrients. This idea has been supported by the weak expression of TJ proteins and of efflux transporters such as ABCB1 as well as by the absence of astrocytic end feet in the early stages of development (Strazielle and Ghersi-Egea 2015). Today this concept is challenged, as the entry of plasma proteins and other blood-borne molecules early in development reflects the importance of these compounds for brain maturation rather than a non-specific blood-to-brain leak. The greater entry of amino acids during early development, for instance, is due to a much higher expression of their transporters in brain ECs (reviewed in Saunders et al. 2014).

The maturity of the BBB during development is also supported by the early expression of TJs proteins such as occludin, claudin-5 (Virgintino et al. 2004), and ZO-1 (Nico et al. 1999) in the developing BBB. Even if these expressions appear weak, they are sufficient to allow the formation of TJs preventing paracellular diffusion. The permeability to endogenous albumin, for instance, is not different between development and adulthood in mice (Vorbrodt and Dobrogowska 1994). During later development in humans, TJs protein expression and function rapidly reach the levels observed in adult BBB (Virgintino et al. 2004).

7.3.2 The Blood–CSF Barrier

The BCSFB is located at two different sites of the brain: the ChPs and the arachnoid.

7.3.2.1 The Choroid Plexuses

Organization and Functions

The ChPs are secretory tissues that float in the CSF of each ventricle. The two largest ChPs are found in the lateral ventricles, a third one is located in the fourth ventricle, and a fourth small ChP is found in the third ventricle. They are all highly vascularized tissues constituting a cuboidal epithelial cell monolayer surrounding a core of connective tissue containing fenestrated blood capillaries (Fig. 7.2c). Next to being the main source of CSF, they also act as barriers between the blood and the CSF and participate in brain homeostasis by controlling the composition of CSF. They transport both endogenous and exogenous compounds between the two fluids. They also are a source of biologically active molecules that influence cerebral development and function.

Selective Barrier Properties

Unlike the BBB, the choroidal ECs are fenestrated and do not display a tight phenotype. The choroidal endothelium is therefore permeable not only to small molecules and water but also to larger solutes that cross this fenestrated endothelium by diffusion or vesicular transport to reach the stromal space. The selective barrier between blood and CSF is shifted at the level of the choroidal epithelium whose cells express a range of TJ proteins including integral membrane proteins (e.g., occludin, claudin-1, 2, 3, 19) and pericellular cytoplasmic proteins (e.g., ZO-1, ZO-2) (Kratzer et al. 2012). The BCSFB is, however, more permeable than the cerebral capillaries to water and selected inorganic ions, and this difference in permeability may be explained by a difference in TJ protein composition. The claudin-2 expressed in the ChP epithelium, for instance, is responsible for a decrease in the electrical resistance of the barrier by forming weak adhesion with other claudins (e.g., claudin-1) and aqueous pores with high conductance within the TJ strands (Furuse et al. 2001).

The leaky phenotype of choroidal ECs is maintained by products secreted by the epithelium, in particular VEGF, a factor well known for its potent angiogenic properties. Together with transforming growth factor β signaling, VEGF/VEGF-receptor signaling increases the vascular permeability of choroidal vessels by mediating the formation of endothelial fenestrations during development and throughout life (Maharaj et al. 2008).

The choroidal BCSFB is a highly active interface that, by dynamic transport mechanisms, on the one hand allows amino acid and glucose delivery to the CSF and on the other hand acts as a neuroprotective barrier preventing entry of noxious compounds into the CSF. In this latter respect, multidrug resistance-associated protein 1 (MRP1/ABCC1) and also MRP4 (ABCC4) and different organic anion transporters of the solute carrier SLC22 and SLC21/OATP families are the predominant choroidal efflux transporters (Ghersi-Egea et al. 2018). High levels of detoxifying enzymes (e.g., epoxide hydrolase, glutathione transferases, and peroxidases), associated with a machinery to transport or synthesize glutathione, cysteine, and metallothionein, also participate in protection of the brain from harmful compounds by favoring their elimination and ensuring protection against oxidative stress (Ghersi-Egea et al. 2018).

Finally, infiltrating immune cells can migrate from the systemic circulation across the fenestrated endothelium to reach the stroma and then the CSF across the epithelium, making the ChP a gateway for immune cell entry into the CNS (see 3.3.2) (Ransohoff and Engelhardt 2012).

Transport and Secretion of Biologically Active Molecules

The ChP-CSF system has an important role in neuroendocrine signaling. ChPs are a source of biologically active molecules that are important for the development of the brain and the maintenance of its integrity throughout life. ChPs either synthesize these factors or express transport proteins that allow the transport of vitamins, growth factors, and hormones (reviewed in Nilsson et al. 1992). The high expression of the sodium-dependent vitamin C transporter 2 within the ChP epithelium for instance allows the distribution of this essential vitamin to the brain (Ulloa et al. 2019). ChP-secreted factors can act directly on adult stem cells located along the lateral ventricles to regulate their behavior. During embryonic development, the ChP secretes Shh and IGF-2 that promote the proliferation of neural progenitors. Other factors that come from the blood and are transported across the BCSFB have also a role in the regulation of stem cell activity. Folates, for instance, are taken up by the epithelial cells and released within microvesicles into the CSF to reach the brain parenchyma and participate in brain development (Grapp et al. 2013). Other ChP-derived factors modulate global neural circuit formation, activity, and plasticity at postnatal stages and throughout life, such as the homeobox protein Otx2 (reviewed in Ghersi-Egea et al. 2018). The distribution of such active molecules to stem cells lining the cerebral ventricles and to cells located deeper in the brain parenchyma is facilitated by the absence of TJs between the cells forming the CSF–brain borders. However, the nature of these factors and their precise effects on adult stem cells need to be clarified in humans.

The ChP epithelium also expresses carrier proteins involved in the central distribution of circulating hormones. An example is transthyretin (TTR) which is a thyroid hormone (TH) carrier protein whose expression in the ChP epithelium has been first described in 1979 by Møllgård *and others*. THs are ubiquitous hormones synthesized in the thyroid gland that circulate in the blood in complex with TTR, released by the liver. During development, they regulate the transcription of specific genes within the brain. Insufficiency of TH during prenatal development in humans leads to cognitive retardation. In adults, it causes severe damage, including mental impairment and clinical depression (Richardson et al. 2015). TTR is the main protein synthesized and continuously secreted into the CSF by the ChP epithelium from early embryonic development and throughout life (reviewed in Richardson et al. 2015). Its choroidal expression occurs just prior to the period of brain growth when THs are required. Blood-borne THs reach the CSF by Oatp-14 (SLCO1C1)-dependent choroidal uptake followed by binding to choroidal TTR released in the CSF (reviewed in Richardson et al. 2015). THs can also be transported by Oatp14 across the BBB to reach the extracellular space (reviewed in Bernal et al. 2015).

The distribution of active molecules between blood and CSF is facilitated by the high blood flow in the ChP anastomosed capillaries. Measurements in rats suggest that the blood flow within the ChP increases during postnatal development, in line with the progressive increase of CSF production rate at the same time (Szmydynger-Chodobska et al. 1994).

Development of the Choroid Plexuses

Formation of the ChPs begins at early developmental stages, between GW7 and 8 in humans (Fame et al. 2020). The establishment of interspecies grafts in birds allowed the origin of each ChP to be followed. The different ChPs come from different primary brain vesicles. The wall of the fourth ventricle in the hindbrain gives rise to the first ChP. This is immediately followed by the formation of the ChP in each of the lateral ventricles originating from the prosencephalon. The ChP of the third ventricle, coming from the diencephalon, is the last to appear but matures more rapidly than the others (Lun et al. 2015; Ghersi-Egea et al. 2018). The molecular control of ChP development has not been completely elucidated. Genes and pathways implicated in the differentiation and proliferation of the choroidal epithelium include Notch, Shh, and neurogenin 2 signaling (reviewed in Lun et al. 2015).

The development of ChP tissue is linked to the development of its barrier properties. ChPs rapidly mature during development to fulfill their neuroprotective role. Choroidal TJs are present before GW8 in humans (Møllgård and Saunders 1986), and TJ-associated genes have been identified in ChP of early embryonic (E15) and postnatal (P2) stages in rats (Kratzer et al. 2013). The ChP epithelium of the developing brain also harbors various influx and efflux transport proteins. The ABC efflux transporters, for instance, are present in the early embryonic and fetal stages in rats, as well as in human neonates (Daood et al. 2008; Ek et al. 2010; Kratzer et al. 2013).

The efficiency of choroidal detoxification activities during fetal and perinatal stages suggests that a functional biochemical barrier supported by various enzymatic systems contributes to the protection of the developing brain. High levels of glutathione S-transferase (GST) mRNA expression and activity have been detected in ChPs of newborn rats and human fetuses (Ghersi-Egea et al. 2006; Kratzer et al. 2018). Glutathione peroxidase (GPX) genes were also highly expressed in ChP during the perinatal period (Kratzer et al. 2013; Saudrais et al. 2018). These perinatal expression levels are more important than in the liver, which remains immature at these stages with respect to its antioxidant and detoxifying functions.

The History of the Immature Blood–CSF Barrier Concept

As for the BBB, the BCSFB located in ChPs has long been considered immature during embryonic and postnatal development. This classical belief was supported by the high concentration of proteins present in the fetal CSF compared to that in adult CSF, as well as by the apparent developmental decrease in permeability toward lipid-insoluble diffusion markers such as sucrose or exogenous proteins (Dziegielewska and Saunders 1988). The morphology of embryonic and adult TJs, as well as their composition in TJ-associated proteins, is, however, similar, except for the pore-forming claudin 2 whose appearance is delayed during development (Johansson et al. 2008; Kratzer et al. 2013). The hypothesis of a transcellular pathway especially active during development therefore emerged. The currently prevailing concept is that a certain permeability to plasma proteins is not necessarily associated with immaturity. It is now accepted that CSF proteins are important for differentiation and migration of stem cells lining the ventricles that are in direct contact with the fluid. They also may transport growth factors, hormones, and vitamins. Finally, CSF proteins participate in ventricular expansion by exerting an oncotic pressure (also known as colloid osmotic pressure) (reviewed in Johansson et al. 2008; Saunders et al. 2012, 2018). In line with this, junctional complexes called "strap" junctions seal the neuroepithelial cells lining the ventricular system from GW8, when the ChP appears, up to GW22 in humans (Møllgård and Saunders 1986). These junctions have been reported in many other species and act to prevent proteins, but not smaller molecules, from escaping from CSF into the brain. The BCSFB is therefore selective but adapts its permeability to what the brain needs at this crucial period of development before the parenchymal vascular network fully develops.

7.3.2.2 The Arachnoid

The whole brain is covered by a network of membranes called the meninges. The meninges are organized into three layers, the dura mater against the skull, the arachnoid membrane, and the pia mater closely appended to the cerebral parenchyma (Fig. 7.2d). They provide support for the cerebral vessels before they enter the parenchyma and divide into the extensive microvascular network. The meninges also contain the subarachnoid CSF that protects the CNS from mechanical impacts. Because the arachnoid cells are sealed by TJs and express efflux ABC transporters such as PGP and BCRP (reviewed in Gomez-Zepeda et al. 2019), this cell layer forms the CSF–blood barrier at the external surface of the brain.

7.4 Neuroimmune Regulation

7.4.1 The CNS, an Immunoprivileged Site

The CNS constitutes a particular immune environment since no lymphatic circulatory system is present. The CNS is thus considered an immune privileged site, a concept already described 90 years ago in pioneer observations (Shirai 1921). These experiments showed that grafts implanted into the cerebral parenchyma were not accompanied by an immune response, in contrast to what is observed in other organs where polynuclear neutrophils as well as macrophages infiltrate the tissue. Additional studies also demonstrated that sources of antigens such as viruses, tumor cells, and bacterial agents administered into the cerebral parenchyma do not induce a stereotypic immune response (reviewed in Engelhardt et al. 2017).

This "immune privilege" makes sense as full-blown inflammatory reactions, leading to tissue edema, for example, would have deleterious neurological consequences. Therefore, the CNS needs effective but adapted immune responses to face infection and tissue damage. Failure in control of these immune responses leads to chronic immunopathological disorders (e.g., multiple sclerosis, neuromyelitis optica) (Engelhardt et al. 2017; Wells et al. 2018). The concept of neuroimmune privilege is now redefined to include the notion of immune surveillance.

The CNS immune surveillance allows detection of pathogens and injuries and determination of whether a response is required. Different populations of resident immune cells exist in the CNS. Microglial cells are the main innate immune cells residing in the healthy CNS parenchyma. In mice, they are generated at E8.5 from a yolk sac progenitor (Gomez Perdiguero et al. 2015; reviewed in Forrester et al. 2018). Some of these progenitors mature and become microglial precursors that invade the CNS at days E9.5–E10.5 (Gomez Perdiguero et al. 2015). The microglial cells proliferate locally within the brain and their pool is maintained throughout life without any renewal by bone marrow-derived immune cell precursors. They respond rapidly to pathogens and injuries by secreting pro-inflammatory cytokines. These cells are, however, unable to migrate into the peripheral lymphatic system to fulfill the function of antigen-presenting cells (APCs), in line with the lack of lymphatic vessels in the brain parenchyma. The situation is thus different from peripheral organs such as the skin or gut, where activated dendritic cells migrate via lymphatic vessels to lymph nodes.

In the last two decades, evidence has accumulated showing that immune surveillance of the brain occurs through CSF-filled spaces including ventricles, SAS, cisterns, and PVS. In line with this, tissue grafts transplanted directly into cerebral ventricles induce a reaction that leads to their rejection. The group of cell sentinels found in CSF and forming a line of defense against noxious immunogenic agents includes mainly (80%) central memory lymphocytes T cells (Svenningsson et al. 1995) but also dendritic cells and macrophages (Ousman and Kubes 2012; Lun et al. 2015). In the absence of a neuroinflammatory environment, these immune cells seem to be able to cross the brain barriers to infiltrate the fluid spaces. They play an important role in neuroimmune surveillance (reviewed in Ousman and Kubes 2012).

7.4.2 The Role of Brain Fluid Drainage Pathways in Neuroimmune Regulation

The CNS is not irrigated by lymphatic vessels; rather, distinct fluid drainage pathways allow soluble antigens or immune cells to reach lymph nodes.

7.4.2.1 The Cerebrospinal Fluid

The CSF is produced in ventricles and circulates freely, to reach the SAS but also large PVS (Fig. 7.1). All these spaces are immunologically competent since an adaptive immune system exists with the presence of APCs. This explains the rapid immune response that follows an infection located in a CSF compartment by opposition to the situation in the parenchyma, which is immune privileged. The number of immune cells in these CSF spaces remains low, however, in both humans and rodents (Engelhardt et al. 2017). They include, in addition to central memory T cells, myeloid cells, such as macrophages and dendritic cells, and Kolmer's epiplexus cells patrolling the apical surface of ChPs. Some mast cells are also found in meninges and PVS (reviewed in Ransohoff and Engelhardt 2012). All these resident cells promote host defense when in contact with pathogens. No other immune cells, such as neutrophils, granulocytes, or natural killer lymphocytes, are found in the healthy CSF. When APCs specifically interact with memory T cells, the latter are activated, and this leads to the setting up of an adaptive immune response to the detected danger such as an infection. Such neuroimmune surveillance needs a connection between the CSF and lymph nodes.

In parallel to being reabsorbed from the SAS through arachnoid granulations into the blood of dural venous sinuses (Fig. 7.1b), the CSF also drains into lymph nodes through different pathways (Fig. 7.3) (Ransohoff and Engelhardt 2012; Engelhardt et al. 2017). It first reaches the deep and superficial cervical lymph nodes by two main routes. It is reabsorbed through the cribriform plate following olfactory rootlets that reach the nasal mucosa which is rich in lymphatics draining mainly toward deep cervical lymph nodes (DCLNs) (Harling-berg et al. 1989). In mice, this pathway drains more than half of the CSF (Ma et al. 2017). This has been demonstrated by injecting labeled antigens into the CSF in ventricles. Half of these antigens were rapidly found in the DCLNs (Ransohoff and Engelhardt 2012). CSF also drains along the spinal nerves roots to reach the lumbar lymph nodes. Finally, CSF seems to reach nearby dural lymphatics that drain into the DCLNs along large vessels and cranial nerves (Aspelund et al. 2015). The dural lymphatics are highly plastic. For instance, their number and their size increase in response to a cerebrovascular accident. They are also involved in the development of neuroinflammatory diseases as observed in animal models of multiple sclerosis by facilitating the entry of immune cells into the brain (Louveau et al. 2018).

CNS-derived soluble antigens are transported via these different pathways into the lymphatic system and accumulate in the lymph nodes. As these pathways also allow the traffic of T cells and APCs, thus establishing efferent communication between the CNS and immune system, they compensate to some extent for the lack of lymphatic vessels in the CNS and are integrated in a functional neuro-immunological circuitry (Engelhardt et al. 2017).

7.4.2.2 The Interstitial Fluid

Studies using radioactive and fluorescent tracers have demonstrated that the ISF and its solutes drain out of the parenchyma along intramural perivascular pathways in the wall of the arterial network to DCLNs (Fig. 7.3) (reviewed in Engelhardt et al. 2017). The movement of molecules along basement membranes within these intramural perivascular ISF–drainage pathways is limited by their size. Larger proteins, for instance, aggregate and form immune complexes, thus impairing the drainage, as has been observed in Alzheimer's disease and ischemic stroke (Arbel-Ornath et al. 2013; reviewed in Engelhardt et al. 2017). This pathway is too narrow and therefore does not allow migration of APCs in lymph nodes.

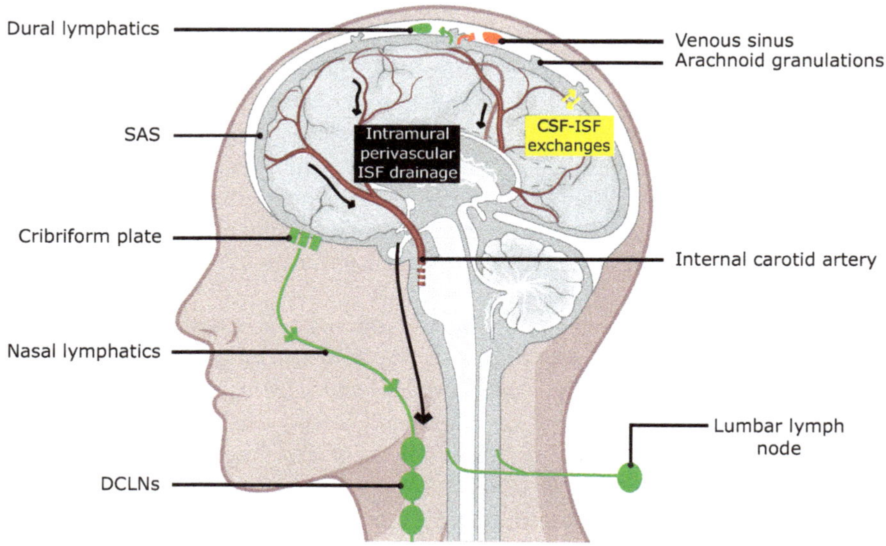

Fig. 7.3 Brain fluid drainage pathways. CSF drains into the blood across the arachnoid villi into the dural blood sinuses (red arrow) or into the lymphatic system (outlined in green). The CSF can reach either the dura lymphatic vessels or the nasal lymphatics through the cribriform plate to reach the DCLNs. CSF can also follow other cranial nerves to reach the lymphatic system (not shown). CSF from the lumbar SAS drains along spinal nerves to reach the lumbar lymph nodes. ISF drains along the basement membranes, in the wall of the cerebrovascular network, and then in the wall of the internal carotid artery to reach the DCLNs (black arrows). Exchanges between CSF and ISF mostly occur along arteries penetrating the brain parenchyma and across the external glia limitans (yellow arrows), or across the ependyma bordering the ventricles (not shown). CSF: cerebrospinal fluid; DCLNs: deep cervical lymph nodes; ISF: interstitial fluid; SAS: subarachnoid spaces (inspired by Ransohoff and Engelhardt 2012; Engelhardt et al. 2017)

7.4.2.3 Interconnection Between CSF and ISF

ISF and CSF drain to lymph nodes in distinct ways, but as mentioned previously, an interconnection exists between CSF and ISF and is located at two different sites (reviewed in Engelhardt et al. 2017). Diffusion of substances between ISF and CSF occurs across the ventricular ependyma and external glia limitans which lack a barrier phenotype (see 1.2). Both fluids also mix within the PVS. This gave rise to the notion of convective influx from CSF to ISF (Rennels et al. 1985). A proportion of tracers injected into the CSF can bypass the glia limitans surrounding cerebral vessels to reach the ISF. This led to the concept of glymphatic system (Iliff et al. 2012, reviewed in Forrester et al. 2018). This communication between PVS and ISF may be regulated by the water channel aquaporin 4 located in astrocyte end feet forming the glia limitans, although the rationale behind this regulation is not understood. Whether the PVS to ISF exchanges occur by simple diffusion or through convective flux, the latter being the basis of the glymphatic system theory, remains highly debated (Abbott et al. 2018). Tracers circulating in the ISF may drain back into the subarachnoid CSF to reach the regional lymph nodes. The second drainage pathway toward lymph nodes that occurs within the walls of cerebral arteries has been shown but has not yet been fully characterized.

As of today, the microanatomical basis of ISF-CSF–lymph connections as well as the contribution of these interconnections to the drainage of antigens and immune cell trafficking is not completely understood. The concept of a glymphatic system remains debated and is difficult to integrate with the immune privilege concept (Engelhardt et al. 2017; Abbott et al. 2018).

7.4.3 The Role of Brain Barriers in Neuroimmune Regulation

Brain barriers provide an anatomical basis for understanding CNS immune surveillance occurring through the CSF-filled ventricles and SAS. Due to their different properties, the barriers participate in the establishment of CNS compartmentalization with respect to communication with the immune system. Three potential immune cell entry sites into the CNS exist (Fig. 7.2a). Immune cells may gain access to the brain (1) by reaching the CNS parenchyma through the BBB, mainly across the postcapillary venules, (2) by crossing the BCSFB located at ChP epithelium to access the CSF, and (3) by reaching the SAS at the surface of the brain through the arachnoid barrier or across the wall of pial vessels.

7.4.3.1 The Blood–Brain Barrier

PVS formed along the cerebral vasculature and located between the basement membrane of ECs and the glia limitans harbor myeloid cells acting as APCs (reviewed in Ousman and Kubes 2012). That facilitates immune surveillance along the vascular tree. In order to migrate from blood to parenchyma, a leukocyte needs the expression of adhesion molecules at the surface of both the immune cell membrane and the vascular endothelium. These adhesion molecules trap the cells and initiate their transcellular migration from blood to the brain. In a healthy condition, the cerebral endothelium has a very low expression of these surface molecules. Therefore, migration of immune cells across the BBB remains a rare event, limited to activated T cells. These T cells are not able to cross the glia limitans if APCs that reside in the PVS do not recognize their cognate antigen. They undergo apoptosis or leave the CNS, probably by CSF drainage pathways (Engelhardt et al. 2017). BBB is therefore not a gate of entry into the CNS parenchyma for adaptive immune cells (T and B cells), and immune surveillance is limited to cerebral ventricles and SAS filled with CSF.

In neuroinflammatory diseases, the expression of adherent molecules such as P-selectin, Icam, and Vcam and the synthesis of chemokines are induced in the ECs, leading to the migration of immune cells within the brain. P-selectin, for instance, is rapidly expressed on the EC surface to fix P-selectin glycoprotein ligand 1 (PSGL-1) on T cells and initiate their interaction with the endothelium. These changes come from signals such as cytokines released by neural cells within the parenchyma. T cells can recognize specific antigens presented by APCs in the CSF-filled PVS. In multiple sclerosis, the interaction between T cells and APCs worsens the disease by facilitating the migration of lymphocytes and monocytes across the BBB as well as across the glia limitans to reach the nervous tissue (reviewed in Ransohoff and Engelhardt 2012). High amounts of the chemokine CXCL12 localized at the abluminal (brain-facing) surface of ECs are required to retain immune cells expressing its receptors CXCR4 and CXCR7 in PVS. In some inflammatory diseases, such as human multiple sclerosis and the animal model experimental autoimmune encephalomyelitis, CXCL12 is internalized within the ECs and relocates at the luminal (blood-facing) surface. This relocalization participates in the transendothelial migration of CXCR4- and CXCR7-expressing T cells from the blood to the PVS and parenchyma. The fine molecular mechanisms regulating both chemokine relocalization and immune cell migration in these inflammatory diseases remain to be deciphered (reviewed in Holman et al. 2011).

7.4.3.2 The Blood–CSF Barriers

Immune cell infiltration into the CSF is limited to lymphocytes and especially CD4+ and CD8+ T cells in healthy conditions (i.e., in the absence of neuroinflammation). Carrithers et al. demonstrated that the main routes of entry for these T cells are the ChP and the meninges (Carrithers et al. 2002).

The Choroid Plexuses

The ChPs are of particular interest in the context of immune surveillance. Their stroma harbors many resident immune cells such as short-lived macrophages, dendritic cells and their precursors, and T cells (Lun et al. 2015). ChPs are considered to be preferred sites for T cell migration from blood to ventricular CSF both under healthy conditions and during the early stages of neuroinflammation (reviewed in Ransohoff and Engelhardt 2012). Lymphocytes and especially T cells are found both in ventricular and subarachnoid CSF and in the choroidal stroma in healthy conditions. They first migrate from blood across the fenestrated ChP endothelium before reaching the CSF across the epithelial wall of the ChP. Migration across this barrier was confirmed with *in vitro* models (Engelhardt and Ransohoff 2012; Strazielle et al. 2016). The cellular and molecular mechanisms of this migration, which occurs without altering the barrier properties, still need to be investigated in detail, but chemokines and receptors implicated in these movements have been identified (Ousman and Kubes 2012; Engelhardt and Ransohoff 2012; Engelhardt et al. 2017). Extravasation across the ChP endothelium is facilitated by the constitutive expression of P-selectin, an adhesion protein at the endothelial surface that interacts with PSGL-1 expressed by T cells. The ChP epithelium is permissive to T-cell migration into the CSF, through a process that appears to involve the cell-surface proteoglycan syndecan-1 and selected chemokines (e.g., CCR6, CCL20) secreted by epithelial cells, which attract these infiltrating T cells and promote their entry into the CSF. The ChPs are also a gate of entry for the first T cells infiltrating the brain in the early stages of neuroinflammatory diseases (Reboldi et al. 2009). The epithelium also expresses cell-adhesion molecules (e.g., Icam-1, Vcam-1) capable of interaction with T cells, but these adhesion molecules are located at the apical side of epithelial cells, so invisible to immune cells present in the stroma. The precise cellular and molecular cues guiding immune cells from the choroidal stroma into the CSF without altering the epithelial barrier properties remain therefore to be identified.

ChPs are early sensors of both systemic and central injuries that generate inflammation (cerebral stroke, trauma, ischemia, multiple sclerosis, bacterial and viral infections). Even if ChPs are far from the injured site, these structures allow a rapid entry of innate immune cells in the brain, including monocytes and polynuclear neutrophils, or T lymphocytes, depending on the nature of the injury. ChP epithelium reacts to the injury by secreting pro-inflammatory molecules, including Interleukin-1 and various chemokines, in the ChP stroma and in the CSF (Szmydynger-Chodobska et al. 2012; Mottahedin et al. 2020). This could explain why the initial immune response in experimental autoimmune encephalomyelitis, a model of multiple sclerosis, occurs within the CSF-filled spaces before communicating with PVS and spreading within the parenchyma (Schmitt et al. 2012). Similarly, during postnatal development, systemic exposure to lipopeptides derived from gram-negative bacterial infection induces an infiltration of polynuclear neutrophils and monocytes across the BCSFB (Mottahedin et al. 2020).

The Arachnoid and Pial Vessels

T cells could also reach the CSF by crossing the superficial brain barriers. Subarachnoid CSF can be reached across the arachnoid, from the blood circulating in the dura, or across the vessel wall of pial vessels that lie within the SAS. The CNS parenchyma does not contain T cells under physiological conditions. The APCs residing in CSF decide whether T cells within the SAS-filled CSF enter the nearby parenchyma in response to receiving signals necessary for their activation and adhesion (reviewed in Engelhardt and Ransohoff 2012; Schläger et al. 2016). In the presence of such signals, T cells infiltrate the nervous tissue crossing the glia limitans to set up an inflammatory process. The pathways and mechanisms of T-cell migration across the glia limitans are, however, not clearly identified.

Recognition of antigens on macrophages by T cells in the SAS leads to the upregulation of adhesion molecules on ECs of pial vessels, in a way similar to that observed at the BBB in inflammatory conditions (see 3.3.1). This leads to a further recruitment of blood-circulating immune cells (reviewed in Engelhardt and Ransohoff 2012). Perturbation of immune cell trafficking through the meninges has been associated with inflammatory diseases. For example, in ischemia, polynuclear neutrophils coming from the nearby bone marrow of the skull are recruited, extravasate in the dura, and enter the brain, possibly across the arachnoid, to reach the injured tissue (Herisson et al. 2018).

7.5 Perspectives

Our understanding of how brain fluids and blood–brain barriers regulate brain homeostasis has been mainly established in adulthood. Although there is now a better insight into the development of the cerebroventricular system and blood–CNS interfaces in laboratory animal models (Box 7.1), the main challenge remains the access to data allowing a better understanding of such developmental processes in the human brain.

Investigating the CSF composition allows consideration of this fluid as a vector of active molecules that promote CNS homeostasis. Although the field is still at its beginning, developing strategies to selectively target active compounds of the CSF should open tracks to major therapeutic advances for neurodegenerative disorders, stroke, and neuroinflammatory diseases. The neurogenic properties of the embryonic CSF, for instance, have demonstrated their efficiency in adult brain repair (Carnicero et al. 2013).

Blood–CNS barriers have long been considered to be immature during early development, but this concept is challenged by several laboratories that emphasized the difference between a non-specific blood-to-brain leak and an adapted system for a growing tissue. Improved animal models, as well as *in vitro* model of brain barriers coupled with advanced imaging tools, will be useful in better understanding the dynamics of brain barriers and the interactions between cells forming the NVU, and between ECs and blood components (i.e., immune cells and soluble factors). The anatomy and functions of brain barriers are relatively similar between humans and rodents, but advances in *in vivo* imaging of blood vessels in the human CNS should be interesting to translate observations made in rodents. While there are now studies on the developmental specificities of brain barrier functions, including their neuro-endocrine functions, little is known about how aging and neurodegenerative disorders affect these functions. Drug delivery to the CNS is the major obstacle for treatment of central disorders since blood–brain barriers restrict access of numerous blood-borne therapeutic molecules to brain circuits. Advances in our understanding of the structure and function of the blood–CNS interfaces under physiological and pathological conditions including neurodegenerative diseases are important for the development of innovative therapeutic approaches to circumvent these cellular barriers. All the barrier components, and especially transporters, receptors, and their signaling pathways, are key targets for improving CNS drug entry as well as brain repair.

The CNS is characterized by its particular innate and adaptive immune response compared with other peripheral tissues. The cellular basis of immune privilege resides in the absence of a route to the lymph node for APCs from the parenchyma and PVS. However, this “privileged” concept is not associated with the complete absence of immune cells in the CNS and of neuroimmune surveillance but rather with an elaborate regulation of interactions between CNS-resident and blood-borne immune cells that take place in the fluid compartments. Adaptive immune cells, and especially lymphocytes, carry out immune surveillance of the CNS by patrolling CSF-filled spaces. Efferent and afferent connections between the CSF and the peripheral immune system form the functional equivalent of a lymphatic system. The blood–brain barriers provide an anatomical basis for immune surveillance by regulating blood-to-CNS extravasation of immune cells. A better understanding of the cellular and molecular mechanisms underlying immune surveillance remains, however, crucial to designing successful therapies targeting both arms of immune surveillance in neurological disorders such as multiple sclerosis, stroke, and Alzheimer's disease.

7.6 Key References

Goldmann (1909) One of the first studies using peripheral injection of a dye to demonstrate that the CNS is a closed compartment protected from systemic circulation by barriers.

Harling-Berg et al. (1989) One of the first studies to propose CSF drainage pathways into the DCLNs by injecting human serum albumin into the CSF of rats.

Shirai (1921) First demonstration of a central immune privilege concept, through the observation that transplantation of a rat sarcoma into the mouse parenchyma does not lead to rejection.

Ransohoff and Engelhardt (2012) This review describes the anatomical and cellular basis of immune surveillance in the CNS, by focusing on the anatomy of the blood–brain barriers, the molecular and cellular mechanisms involved in immune-cell trafficking, especially T cells, and the drainage pathways of the CSF and ISF toward lymphatics nodes.

Ghersi-Egea et al. (2018) This review describes the molecular anatomy and functions of the BCSFB in ChP in relation to neuroprotection, neuroimmune surveillance, and regulation of stem cell behavior.

Kadry et al. (2020) This review describes the development of the BBB, its structure, and functions. It also discusses the biomarkers used to assess barrier permeability in classical studies.

References

Abbott NJ, Patabendige AA, Dolman DE, Yusof SR, Begley DJ (2010) Structure and function of the blood-brain barrier. Neurobiol Dis 37(1):13–25

Abbott NJ, Pizzo ME, Preston JE, Janigro D, Thorne RG (2018) The role of brain barriers in fluid movement in the CNS: is there a 'glymphatic' system? Acta Neuropathol 135:387–407

Alvarez JI, Dodelet-Devillers A, Kebir H, Ifergan I, Fabre PJ, Terouz S, Sabbagh M, Wosik K, Bourbonnière L, Bernard M, van Horssen J, de Vries HE, Charron F, Prat A (2011) The hedgehog pathway promotes blood-brain barrier integrity and CNS immune quiescence. Science 334:1727–1731

Alvarez JI, Katayama T, Prat A (2013) Glial influence on the blood brain barrier. Glia 61:1939–1958

Arbel-Ornath M, Hudry E, Eikermann-Haerter K, Hou S, Gregory JL, Zhao L, Betensky RA, Frosch MP, Greenberg SM, Bacskai BJ (2013) Interstitial fluid drainage is impaired in ischemic stroke and Alzheimer's disease mouse models. Acta Neuropathol 126:353–364

Armulik A, Genové G, Betsholtz C (2011) Pericytes: developmental, physiological, and pathological perspectives, problems, and promises. Dev Cell 21:193–215

Ashok KS, Gabriele Z (2020) The interstitial system of the brain in health and disease. Aging Dis 11:200

Aspelund A, Antila S, Proulx ST, Karlsen TV, Karaman S, Detmar M, Wiig H, Alitalo K (2015) A dural lymphatic vascular system that drains brain interstitial fluid and macromolecules. J Exp Med 212:991–999
Bernal J, Guadaño-Ferraz A, Morte B (2015) Thyroid hormone transporters—functions and clinical implications. Nat Rev Endocrinol 11(7):406–417
Birge WJ, Rose AD, Haywood JR, Doolin PF (1974) Development of the blood-cerebrospinal fluid barrier to proteins and differentiation of cerebrospinal fluid in the chick embryo. Dev Biol 41: 245–254
Böhm UL, Prendergast A, Djenoune L, Nunes Figueiredo S, Gomez J, Stokes C, Kaiser S, Suster M, Kawakami K, Charpentier M, Concordet J-P, Rio J-P, Del Bene F, Wyart C (2016) CSF-contacting neurons regulate locomotion by relaying mechanical stimuli to spinal circuits. Nat Commun 7:10866
Brinker T, Stopa E, Morrison J, Klinge P (2014) A new look at cerebrospinal fluid circulation. Fluids Barriers CNS 11:10
Brownlees J, Williams CH (1993) Peptidases, peptides, and the mammalian blood? Brain barrier. J Neurochem 60:793–803
Carnicero E, Alonso MI, Carretero R, Lamus F, Moro JA, de la Mano A, Fernández JMF, Gato A (2013) Embryonic cerebrospinal fluid activates neurogenesis of neural precursors within the subventricular zone of the adult mouse brain. Cells Tissues Organs 198:398–404
Carrithers MD, Visintin I, Viret C, Janeway CA (2002) Role of genetic background in P selectin-dependent immune surveillance of the central nervous system. J Neuroimmunol 129:51–57
Corns R, Martin A (2012) Hydrocephalus Surgery (Oxford) 30:142–148
Daneman R, Zhou L, Kebede AA, Barres BA (2010) Pericytes are required for blood–brain barrier integrity during embryogenesis. Nature 468:562–566
Daood M, Tsai C, Ahdab-Barmada M, Watchko J (2008) ABC transporter (P-gp/ABCB1, MRP1/ ABCC1, BCRP/ABCG2) expression in the developing human CNS. Neuropediatrics 39:211–218
DeSesso JM, Scialli AR, Holson JF (1999) Apparent lability of neural tube closure in laboratory animals and humans. Am J Med Genet 87(2):143–162
Desmond ME, Jacobson AG (1977) Embryonic brain enlargement requires cerebrospinal fluid pressure. Dev Biol 57:188–198
Di Terlizzi R, Platt S (2006) The function, composition and analysis of cerebrospinal fluid in companion animals: Part I – Function and composition. Vet J 172:422–431
Duffy KR, Pardridge WM (1987) Blood-brain barrier transcytosis of insulin in developing rabbits. Brain Res 420:32–38
Dziegielewska KM, Saunders NR (1988) The development of the blood-brain barrier: proteins in fetal and neonatal CSF, their nature and origins. In: Meisami E, Timiras PJ (eds) Handbook of human growth and developmental biology, vol 1A. CRC, Boca Raton, FL, pp 169–191
Dziegielewska KM, Evans CAN, Fossan G, Lorscheider FL, Malinowska DH, Møllgård K, Reynolds ML, Saunders NR, Wilkinson S (1980) Proteins in cerebrospinal fluid and plasma of fetal sheep during development. J Physiol 300:441–455
Dziegielewska KM, Evans CAN, Lai PCW, Lorscheider FL, Malinowska DH, Møllgrd K, Saunders NR (1981) Proteins in cerebrospinal fluid and plasma of fetal rats during development. Dev Biol 83:193–200
Ek CJ, Wong A, Liddelow SA, Johansson PA, Dziegielewska KM, Saunders NR (2010) Efflux mechanisms at the developing brain barriers: ABC-transporters in the fetal and postnatal rat. Toxicol Lett 197:51–59
Engelhardt B, Ransohoff RM (2012) Capture, crawl, cross: the T cell code to breach the blood–brain barriers. Trends Immunol 33:579–589
Engelhardt B, Vajkoczy P, Weller RO (2017) The movers and shapers in immune privilege of the CNS. Nat Immunol 18:123–131
Fame RM, Chang JT, Hong A, Aponte-Santiago NA, Sive H (2016) Directional cerebrospinal fluid movement between brain ventricles in larval zebrafish. Fluids Barriers CNS 13:11

Fame RM, Cortés-Campos C, Sive HL (2020) Brain ventricular system and cerebrospinal fluid development and function: light at the end of the tube: a primer with latest insights. BioEssays 42:1900186

Fishman JB, Rubin JB, Handrahan JV, Connor JR, Fine RE (1987) Receptor-mediated transcytosis of transferrin across the blood-brain barrier. J Neurosci Res 18:299–304

Forrester JV, McMenamin PG, Dando SJ (2018) CNS infection and immune privilege. Nat Rev Neurosci 19:655–671

Furuse M, Furuse K, Sasaki H, Tsukita S (2001) Conversion of zonulae occludentes from tight to leaky strand type by introducing claudin-2 into Madin-Darby Canine kidney I cells. J Cell Biol 153:263–272

Gato A, Desmond ME (2009) Why the embryo still matters: CSF and the neuroepithelium as interdependent regulators of embryonic brain growth, morphogenesis and histogenesis. Dev Biol 327:263–272

Gato A, Alonso MI, Martín C, Carnicero E, Moro JA, De la Mano A, Fernández JMF, Lamus F, Desmond ME (2014) Embryonic cerebrospinal fluid in brain development: neural progenitor control. Croat Med J 55:299–305

Ghersi-Egea J-F, Strazielle N, Murat A, Jouvet A, Buénerd A, Belin M-F (2006) Brain protection at the blood–cerebrospinal fluid interface involves a glutathione-dependent metabolic barrier mechanism. J Cereb Blood Flow Metab 26:1165–1175

Ghersi-Egea J-F, Strazielle N, Catala M, Silva-Vargas V, Doetsch F, Engelhardt B (2018) Molecular anatomy and functions of the choroidal blood-cerebrospinal fluid barrier in health and disease. Acta Neuropathol 135:337–361

Giuliani C, Peri A (2014) Effects of hyponatremia on the brain. JCM 3:1163–1177

Goldmann EE (1909) Die äussere und innere Sekretion des gesunden und kranken Organismus im Lichte der 'vitalen Färbung' Beitr. Klin Chir 64:192–265

Goldmann EE (1913) Vitalfärbungen am Zentralnervensystem. Beitrag zur Physio-Pathologie des Plexus Chorioideus und der Hirnhäute. Abh Preuss Akad Wiss Physik -Math 1:1–60

Gomez Perdiguero E, Klapproth K, Schulz C, Busch K, Azzoni E, Crozet L, Garner H, Trouillet C, de Bruijn MF, Geissmann F, Rodewald H-R (2015) Tissue-resident macrophages originate from yolk-sac-derived erythro-myeloid progenitors. Nature 518:547–551

Gomez-Zepeda D, Taghi M, Scherrmann J-M, Decleves X, Menet M-C (2019) ABC transporters at the blood–brain interfaces, their study models, and drug delivery implications in gliomas. Pharmaceutics 12:20

Grapp M, Wrede A, Schweizer M, Hüwel S, Galla HJ, Snaidero N, Simons M, Bückers J, Low PS, Urlaub H, Gärtner J, Steinfeld R (2013) Choroid plexus transcytosis and exosome shuttling deliver folate into brain parenchyma. Nat Commun 4:2123

Harling-Berg C, Knopf PM, Merriam J, Cserr HF (1989) Role of cervical lymph nodes in the systemic humoral immune response to human serum albumin microinfused into rat cerebrospinal fluid. J Neuroimmunol Dec 25(2–3):185–193

Hellström M, Kalén M, Lindahl P, Abramsson A, Betsholtz C (1999) Role of PDGF-B and PDGFR-beta in recruitment of vascular smooth muscle cells and pericytes during embryonic blood vessel formation in the mouse. Development 126(14):3047–55

Herisson F, Frodermann V, Courties G, Rohde D, Sun Y, Vandoorne K, Wojtkiewicz GR, Masson GS, Vinegoni C, Kim J, Kim D-E, Weissleder R, Swirski FK, Moskowitz MA, Nahrendorf M (2018) Direct vascular channels connect skull bone marrow and the brain surface enabling myeloid cell migration. Nat Neurosci 21:1209–1217

Hladky SB, Barrand MA (2014) Mechanisms of fluid movement into, through and out of the brain: evaluation of the evidence. Fluids Barriers CNS 11:26

Holman DW, Klein RS, Ransohoff RM (2011) The blood–brain barrier, chemokines and multiple sclerosis. Biochim Biophys Acta (BBA) – Mol Basis Dis 1812:220–230

Hühmer AF, Biringer RG, Amato H, Fonteh AN, Harrington MG (2006) Protein analysis in human cerebrospinal fluid: physiological aspects, current progress and future challenges. Dis Markers 22:3–26

Iliff JJ, Wang M, Liao Y, Plogg BA, Peng W, Gundersen GA, Benveniste H, Vates GE, Deane R, Goldman SA, Nagelhus EA, Nedergaard M (2012) A paravascular pathway facilitates CSF flow through the brain parenchyma and the clearance of interstitial solutes, including amyloid β. Sci Transl Med 4:147ra111

Iliff JJ, Goldman SA, Nedergaard M (2015) Clearing the mind: Implications of dural lymphatic vessels for brain function. Lancet Neurol 14:977–979

Jaijakul S, Salazar L, Wootton SH, Aguilera E, Hasbun R (2017) The clinical significance of neutrophilic pleocytosis in cerebrospinal fluid in patients with viral central nervous system infections. Int J Infect Dis 59:77–81

Johansson PA, Dziegielewska KM, Liddelow SA, Saunders NR (2008) The blood–CSF barrier explained: when development is not immaturity. BioEssays 30:237–248

Kadry H, Noorani B, Cucullo L (2020) A blood–brain barrier overview on structure, function, impairment, and biomarkers of integrity. Fluids Barriers CNS 17:69

Kansal K, Irwin DJ (2015) The use of cerebrospinal fluid and neuropathologic studies in neuropsychiatry practice and research. Psychiatr Clin N Am 38:309–322

Kaur J, Fahmy LM, Davoodi-Bojd E, Zhang L, Ding G, Hu J, Zhang Z, Chopp M, Jiang Q (2021) Waste clearance in the brain. Front Neuroanat 15:665803

Kiyonari H, Kaneko M, Abe T, Shiraishi A, Yoshimi R, Inoue K, Furuta Y (2021) Targeted gene disruption in a marsupial, Monodelphis domestica, by CRISPR/Cas9 genome editing. Curr Biol 31(17):3956–3963.e4

Kratzer I, Vasiljevic A, Rey C, Fevre-Montange M, Saunders N, Strazielle N, Ghersi-Egea J-F (2012) Complexity and developmental changes in the expression pattern of claudins at the blood–CSF barrier. Histochem Cell Biol 138:861–879

Kratzer I, Liddelow SA, Saunders NR, Dziegielewska KM, Strazielle N, Ghersi-Egea J-F (2013) Developmental changes in the transcriptome of the rat choroid plexus in relation to neuroprotection. Fluids Barriers CNS 10:25

Kratzer I, Strazielle N, Saudrais E, Mönkkönen K, Malleval C, Blondel S, Ghersi-Egea J-F (2018) Glutathione conjugation at the blood–CSF barrier efficiently prevents exposure of the developing brain fluid environment to blood-borne reactive electrophilic substances. J Neurosci 38:3466–3479

Lei Y, Han H, Yuan F, Javeed A, Zhao Y (2017) The brain interstitial system: anatomy, modeling, in vivo measurement, and applications. Prog Neurobiol 157:230–246

Leslie EM, Deeley RG, Cole SPC (2005) Multidrug resistance proteins: role of P-glycoprotein, MRP1, MRP2, and BCRP (ABCG2) in tissue defense. Toxicol Appl Pharmacol 204:216–237

Louveau A, Herz J, Alme MN, Salvador AF, Dong MQ, Viar KE, Herod SG, Knopp J, Setliff JC, Lupi AL, Da Mesquita S, Frost EL, Gaultier A, Harris TH, Cao R, Hu S, Lukens JR, Smirnov I, Overall CC, Oliver G, Kipnis J (2018) CNS lymphatic drainage and neuroinflammation are regulated by meningeal lymphatic vasculature. Nat Neurosci 21(10):1380–1391

Lun MP, Monuki ES, Lehtinen MK (2015) Development and functions of the choroid plexus–cerebrospinal fluid system. Nat Rev Neurosci 16:445–457

Ma Q, Ineichen BV, Detmar M, Proulx ST (2017) Outflow of cerebrospinal fluid is predominantly through lymphatic vessels and is reduced in aged mice. Nat Commun 8:1434

Maharaj ASR, Walshe TE, Saint-Geniez M, Venkatesha S, Maldonado AE, Himes NC, Matharu KS, Karumanchi SA, D'Amore PA (2008) VEGF and TGF-β are required for the maintenance of the choroid plexus and ependyma. J Exp Med 205:491–501

Møllgård K, Saunders NR (1986) The development of the human blood-brain and blood-CSF barriers. Neuropathol Appl Neurobiol 12:337–358

Møllgård K, Jacobsen M, Jacobsen GK, Clausen PP, Saunders NR (1979) Immunohistochemical evidence for an intracellular localization of plasma proteins in human foetal choroid plexus and brain. Neurosci Lett 14(1):85–90

Mottahedin A, Blondel S, Ek J, Leverin A-L, Svedin P, Hagberg H, Mallard C, Ghersi-Egea J-F, Strazielle N (2020) N-acetylcysteine inhibits bacterial lipopeptide-mediated neutrophil transmigration through the choroid plexus in the developing brain. Acta Neuropathol Commun 8:4

Myung J, Wu D, Simonneaux V, Lane TJ (2018) Strong circadian rhythms in the choroid plexus: implications for sleep-independent brain metabolite clearance. J Exp Neurosci 12: 117906951878376
Nico B, Quondamatteo F, Herken R, Marzullo A, Corsi P, Bertossi M, Russo G, Ribatti D, Roncali L (1999) Developmental expression of ZO-1 antigen in the mouse blood–brain barrier. Dev Brain Res 114:161–169
Nilsson C, Lindvall-Axelsson M, Owman C (1992) Neuroendocrine regulatory mechanisms in the choroid plexus-cerebrospinal fluid system. Brain Res Rev 17:109–138
Nitta T, Hata M, Gotoh S, Seo Y, Sasaki H, Hashimoto N, Furuse M, Tsukita S (2003) Size-selective loosening of the blood-brain barrier in claudin-5–deficient mice. J Cell Biol 161:653–660
Obermeier B, Daneman R, Ransohoff RM (2013) Development, maintenance and disruption of the blood-brain barrier. Nat Med 19:1584–1596
Odackal J, Sherpa AD, Patel N, Colbourn R, Hrabetova S (2015) T-type calcium channels contribute to calcium disturbances in brain during hyponatremia. Exp Neurol 273:105–113
Ousman SS, Kubes P (2012) Immune surveillance in the central nervous system. Nat Neurosci 15: 1096–1101
Pollay M, Curl F (1967) Secretion of cerebrospinal fluid by the ventricular ependyma of the rabbit. Am J Physiol Legacy Content 213:1031–1038
Qutub AA, Hunt CA (2005) Glucose transport to the brain: a systems model. Brain Res Rev 49: 595–617
Ransohoff RM, Engelhardt B (2012) The anatomical and cellular basis of immune surveillance in the central nervous system. Nat Rev Immunol 12:623–635
Ray LA, Heys JJ (2019) Fluid flow and mass transport in brain tissue. Fluids 4:196
Reboldi A, Coisne C, Baumjohann D, Benvenuto F, Bottinelli D, Lira S, Uccelli A, Lanzavecchia A, Engelhardt B, Sallusto F (2009) C-C chemokine receptor 6–regulated entry of TH-17 cells into the CNS through the choroid plexus is required for the initiation of EAE. Nat Immunol 10:514–523
Rennels ML, Gregory TF, Blaumanis OR, Fujimoto K, Grady PA (1985) Evidence for a 'Paravascular' fluid circulation in the mammalian central nervous system, provided by the rapid distribution of tracer protein throughout the brain from the subarachnoid space. Brain Res 326:47–63
Richardson SJ, Wijayagunaratne RC, D'Souza DG, Darras VM, Van Herck SLJ (2015) Transport of thyroid hormones via the choroid plexus into the brain: the roles of transthyretin and thyroid hormone transmembrane transporters. Front Neurosci 9
Sakka L, Coll G, Chazal J (2011) Anatomy and physiology of cerebrospinal fluid. Eur Ann Otorhinolaryngol Head Neck Dis 128:309–316
Saudrais E, Strazielle N, Ghersi-Egea J-F (2018) Choroid plexus glutathione peroxidases are instrumental in protecting the brain fluid environment from hydroperoxides during postnatal development. Am J Phys Cell Phys 315:C445–C456
Saunders NR, Liddelow SA, Dziegielewska KM (2012) Barrier mechanisms in the developing brain. Front Pharmacol 3
Saunders NR, Dreifuss J-J, Dziegielewska KM, Johansson PA, Habgood MD, Møllgård K, Bauer H-C (2014) The rights and wrongs of blood-brain barrier permeability studies: a walk through 100 years of history. Front Neurosci 8:404
Saunders NR, Dziegielewska KM, Møllgård K, Habgood MD (2018) Physiology and molecular biology of barrier mechanisms in the fetal and neonatal brain: Barrier mechanisms in the fetal and neonatal brain. J Physiol 596:5723–5756
Schläger C, Körner H, Krueger M, Vidoli S, Haberl M, Mielke D, Brylla E, Issekutz T, Cabañas C, Nelson PJ, Ziemssen T, Rohde V, Bechmann I, Lodygin D, Odoardi F, Flügel A (2016) Effector T-cell trafficking between the leptomeninges and the cerebrospinal fluid. Nature 530:349–353
Schmitt C, Strazielle N, Ghersi-Egea J-F (2012) Brain leukocyte infiltration initiated by peripheral inflammation or experimental autoimmune encephalomyelitis occurs through pathways

connected to the CSF-filled compartments of the forebrain and midbrain. J Neuroinflammation 9:187
Shen MD (2018) Cerebrospinal fluid and the early brain development of autism. J Neurodevelop Disord 10:39
Shirai Y (1921) On the transplantation of the rat sarcoma in adult heterogeneous animals. Japan Med World 1:14–15
Simpson J, Stephenson T (1993) Regulation of extracellular fluid volume in neonates. Early Hum Dev 34:179–190
Stern L, Gautier R (1918) Le passage dans le liquide céphalo-rachidien de substances introduites dans la circulation et leur action sur le système nerveux central chez les différentes espèces animales. R C R d Ia Soc de Phys et d'hist natur de Genève 35:91–94
Strazielle N, Ghersi-Egea J-F (2015) Efflux transporters in blood-brain interfaces of the developing brain. Front Neurosci 9
Strazielle N, Creidy R, Malcus C, Boucraut J, Ghersi-Egea J-F (2016) T-lymphocytes traffic into the brain across the blood-CSF barrier: evidence using a reconstituted choroid plexus epithelium. PLoS One 11:e0150945
Svenningsson A, Andersen O, Edsbagge M, Stemme S (1995) Lymphocyte phenotype and subset distribution in normal cerebrospinal fluid. J Neuroimmunol 63:39–46
Szmydynger-Chodobska J, Chodobski A, Johanson CE (1994) Postnatal developmental changes in blood flow to choroid plexuses and cerebral cortex of the rat. Am J Phys Regul Integr Comp Phys 266:R1488–R1492
Szmydynger-Chodobska J, Strazielle N, Gandy JR, Keefe TH, Zink BJ, Ghersi-Egea J-F, Chodobski A (2012) Posttraumatic invasion of monocytes across the blood—cerebrospinal fluid barrier. J Cereb Blood Flow Metab 32:93–104
Ulloa V, Saldivia N, Ferrada L, Salazar K, Martínez F, Silva-Alvarez C, Magdalena R, Oviedo MJ, Montecinos H, Torres-Vergara P, Cifuentes M, Nualart F (2019) Basal sodium-dependent vitamin C transporter 2 polarization in choroid plexus explant cells in normal or scorbutic conditions. Sci Rep 9:14422
Virgintino D, Errede M, Robertson D, Capobianco C, Girolamo F, Vimercati A, Bertossi M, Roncali L (2004) Immunolocalization of tight junction proteins in the adult and developing human brain. Histochem Cell Biol 122
Vorbrodt AW, Dobrogowska DH (1994) Immunocytochemical evaluation of blood-brain barrier to endogenous albumin in adult, newborn and aged mice. Folia Histochem Cytobiol 32(2):63–70
Wells E, Hacohen Y, Waldman A, Tillema JM, Soldatos A, Ances B, Benseler S, Bielekova B, Dale RC, Dalmau J, Gaillard W, Gorman M, Greenberg B, Hyslop A, Pardo CA, Tasker RC, Yeh EA, Bar-Or A, Pittock S, Vanderver A, Banwell B (2018) Neuroimmune disorders of the central nervous system in children in the molecular era. Nat Rev Neurol 14:433–445
Yuskaitis CJ, Pomeroy SL (2017) Development of the nervous system. In: Fetal and neonatal physiology. Elsevier, p 1294–1313.e2
Zappaterra MW, Lehtinen MK (2012) The cerebrospinal fluid: regulator of neurogenesis, behavior, and beyond. Cell Mol Life Sci 69:2863–2878

Neuropeptide Binding Autoantibodies Regulating Neuroendocrine Communications

8

Serguei O. Fetissov and Mouna El Mehdi

Abstract

Immunoglobulin (Ig) molecules or antibodies binding neuropeptides or peptide hormones that regulate motivated behavior, stress, and emotion, such as α-MSH, ACTH, oxytocin, or ghrelin, are naturally present in humans and in rodents in both healthy and disease conditions. The functional role of such autoantibodies was analyzed, showing their constitutive participation as modulators of peptidergic signaling in the neuroendocrine system. In particular, IgG were shown to play a role of peptide carriers and allosteric modulators of peptide receptors. Moreover, the pathway of discovery of the origin of α-MSH-reactive autoantibodies leads to the identification of a specific bacterial protein, ClpB, as an antigen mimetic of α-MSH, a key anorexigenic peptide. This discovery contributed to a better understanding of the role of the immune system in neuroendocrine communication between gut microbiota and the host and was further used for the development of a new generation of probiotics to control appetite. Furthermore, experimental data showed that an autoimmune reaction against ClpB, leading to the production of cross-reactive pathogenic α-MSH-reactive autoantibodies, may underlie the origin of eating disorders. Thus, in this chapter, we show that the characteristics of autoantibodies reacting with peptide hormones may provide valuable data that help to better understand the molecular mechanisms of neuroendocrine communication and to create therapeutic approaches for altered motivated behavior, response to stress and emotions.

S. O. Fetissov (✉) · M. El Mehdi
Neuroendocrine, Endocrine and Germinal Differentiation and Communication Laboratory, Inserm UMR1239, University of Rouen Normandy, Mont-Saint-Aignan, France
e-mail: Serguei.Fetissov@univ-rouen.fr

J. P. Konsman, T. M. Reyes (eds.), *Neuroendocrine-Immune System Interactions*, Masterclass in Neuroendocrinology 13,
https://doi.org/10.1007/978-3-031-21358-8_8

Keywords

Neuropeptides · α-MSH · Oxytocin · ACTH, ghrelin · ClpB · Peptide hormones · Autoantibodies · Immunoglobulins · Motivated behavior · Stress · Feeding behavior · Appetite · Social behavior · Psychoneuroimmunology · Hypothalamo-pituitary-adrenal axis · Molecular mimicry · Gut microbiota-brain axis

Neuroendocrine system and immune system interactions, for the Masterclass in Neuroendocrinology Series

8.1 Introduction. Peptide Signaling in the Neuroendocrine System

Peptide hormones and neuropeptides are the main signaling molecules in the neuroendocrine system. Typically, small proteins of less than 100 amino acids are called peptides, but larger protein molecules may play roles similar to peptides in neuroendocrine signaling. Peptides are co-produced and co-released with a number of classical transmitters and combinations of both represent a chemical signature of a specific group of neurons or endocrine cells involved in a more or less specific function (Hökfelt et al. 2018). The distinction between peptide hormones and neuropeptides is more terminological than functional, since both can be produced and may act in neuronal and non-neuronal tissues; for example α-melanocyte stimulating hormone (α-MSH) is a neuropeptide produced in the brain and a hormone produced in the pituitary as well as a paracrine factor in some peripheral tissues, such as skin. In this chapter, the terms neuropeptide and peptide hormone will be used interchangeably.

The immune system has multiple ways to influence neuroendocrine signaling, and the present chapter is focused on the role of naturally occurring autoantibodies, i.e., immunoglobulins (Ig), reacting with some peptide hormones having well-known functions in the neuroendocrine system. IgG and other classes of Ig are most abundant in the blood, and therefore the discussion of their interactions with peptide hormones will primarily concern peripheral circulation. Indeed, IgG represent about 80% of all immunoglobulins in human blood, where they are at levels of 70 μM (10–20 mg/ml), second only to albumin (Wang et al. 2008; Hortin et al. 2008). IgG are also present in the central nervous system (CNS) at concentrations of 0.1–0.2 μM, i.e., approximately 500 times lower than in the systemic circulation and accounting for about 8% of the total protein in the cerebrospinal fluid (CSF) of adult humans (Davson and Segal 1996). A situation opposite to that of IgG occurs with neuropeptide hormones preferentially produced in the brain, such as oxytocin and vasopressin, with extremely low concentrations in the blood (~1–10 pg/ml) (Leng and Ludwig 2008). Nevertheless, systemic action of both central and peripheral neuropeptides is of physiological importance, and hence, the mechanisms regulating peripheral peptide signaling are of practical relevance. Moreover, since IgG are present in the CNS, we cannot exclude their central action in modulating neuropeptide signaling.

Beside their signaling properties, peptides are also a source of cellular nutrition, and undergo degradation within a few minutes by plasma and membrane-bound enzymes such as dipeptidyl peptidase (DPP). The only way for peptides to escape degradation is to be bound to a larger carrier molecule that will protect the peptide until it binds to its specific receptor. Only a few peptides, such as corticotropin-releasing hormone (CRH), have specific binding proteins (Ketchesin et al. 2017). Peptide stability is a key issue of their biological activity *in vivo,* both under natural conditions and when peptides are used as drugs. As will be discuss below, it is possible that IgG may improve peptide stability by playing the role of carrier molecules.

The biological action of peptides as messengers in the neuroendocrine systems occurs upon their binding to specific receptors, which are typically G-protein coupled receptors (GPCR). Activation of GPCR by peptides is modulated by endogenous allosteric modulators, e.g., activation of the melanocortin 4 receptor (MC4R) by α-MSH is enhanced by its co-binding of Ca^{2+} (Yu et al. 2020). As discussed below, IgG may also play a role as an allosteric modulator of MC4R, which is relevant to obesity and eating disorders.

8.2 Detection of Neuropeptide-Reactive Autoantibodies: Methodology

The term autoimmune refers to a self-reaction of the immune system. However, the term autoantibody does not automatically imply a pathological reaction leading to tissue damage by immune cells, which is present in autoimmune disorders such as diabetes type 1. In fact, many so-called natural autoantibodies are present in healthy individuals as a part of the total Ig pool (Lacroix-Desmazes et al. 1998). Table 8.1 summarizes the majority of studies showing the detection and analysis of autoantibodies reacting with various neuropeptides and peptide hormones in humans and rodents in health and disease. Autoantibodies of classes IgM, IgA and IgG have been detected by several techniques. For instance, the binding of plasma IgG to neuropeptides on brain sections has been visualized using immunohistochemistry. This technique is also useful for screening of human plasma for the presence of selective staining which can be further explored for the detection of binding antigens. For instance, it was used for the identification of IgG reactive with α-MSH in the plasma of patients with eating disorders (Fetissov et al. 2002) (Fig. 8.1).

Most commonly, plasma levels of neuropeptide-reactive autoantibodies of various Ig classes are assayed by enzyme-linked immunosorbent assay (ELISA). A technical protocol for neuropeptide autoantibody detection by ELISA is available (Fetissov 2011). The protocol allows the analysis of both the "free" and the "total" fraction of IgG autoantibodies, by incubating the plasma or serum samples in normal

Table 8.1 Occurrence of autoantibodies reactive with neuropeptides/peptide hormones

Neuropeptide/Peptide hormone-reactive autoantibodies in plasma	Species	Methods of detection	Clinical relevance	Reference
α-MSH, IgG, IgM, IgA	Hu, Rt, Mo	ELISA, IHC, SPR, RBI	Eating disorders	Fetissov et al. (2002, 2005, 2006, 2008), Hamze Sinno et al. (2009), Coquerel et al. (2012), Lucas et al. (2019), Tennoune et al. (2014, 2015), Molfino et al. (2020), Karaiskos et al. (2010), Wallenius et al. (2019), Roubalova et al. (2021)
Galanin and α-MSH IgG in both plasma & CSF	Hu, Rt	ELISA	Alzheimer's disease	Costa et al. (2011), Fetissov et al. (2008)
ACTH IgG, IgM	Hu, Rt	ELISA	Aggressive behavior	Fetissov et al. (2002, 2005, 2006, 2008), Værøy et al. (2018), Tennoune et al. (2015), Karaiskos et al. (2010), Schaefer et al. (2013)
Oxytocin IgG, IgM	Hu, Rt	ELISA	Depression, eating disorders	Fetissov et al. (2006), Karaiskos et al. (2010), Garcia et al. (2011)
Vasopressin IgG, IgM	Hu, Rt	ELISA	Depression, eating disorders	Fetissov et al. (2006), Karaiskos et al. (2010), Garcia et al. (2011)
NPY, IgG, IgM, IgA	Hu, Rt	ELISA	Depression	Karaiskos et al. (2010), Garcia et al. (2012)
CRH, IgG, IgA	Hu, Rt	ELISA	Psychopathological traits	Karaiskos et al. (2010), Fetissov et al. (2008)
Ghrelin	Hu, Mo	ELISA, SPR	Obesity, anorexia, stress, rheumatoid arthritis	Takagi et al. (2013), Terashi et al. (2011), François et al. (2015, 2016), Porchas-Quijada et al. (2019), Espinoza-Garcia et al. (2021)
Orexin/hypocretin IgG, IgM, IgA	Hu	ELISA, RBI	Narcolepsy	Deloumeau et al. (2010), Tanaka et al. (2006), Black et al. (2005), Wallenius et al. (2019)
Leptin, IgG, IgA	Hu, Rt	ELISA, SPR	Obesity, diabetes	Bouhajja et al. (2018), Fetissov et al. (2008), Espinoza-Garcia et al. (2021)

(continued)

Table 8.1 (continued)

Neuropeptide/ Peptide hormone-reactive autoantibodies in plasma	Species	Methods of detection	Clinical relevance	Reference
Peptide YY, IgG, IgA	Hu, Rt	ELISA	Not studied	Fetissov et al. (2008)
Agouti-related protein, IgG, IgA	Hu, Rt	ELISA	Not studied	Fetissov et al. (2008)
Melanin-concentrating hormone, IgG, IgA	Hu, RT	ELISA	Not studied	Fetissov et al. (2008)
Vasoactive intestinal peptide (VIP) IgG	Hu	ELISA	Asthma, muscular exercise	Paul et al. (1985, 1989), Paul and Said (1988)
Bradykinin, IgG	Hu	ELISA	Alzheimer's disease	Myagkova et al. (2003)

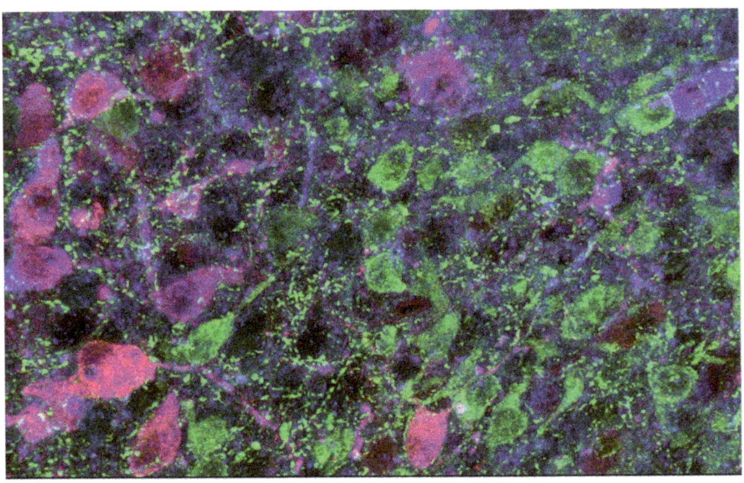

Fig. 8.1 Immunohistochemical visualization of plasma IgG from a patient with bulimia nervosa in the arcuate nucleus of the hypothalamus in a rat brain section. The patient's plasma contains α-MSH-reactive IgG, which binds to anorexigenic proopiomelanocortin neurons (pink). The neighboring population of orexigenic NPY/AgRP neurons was stained with commercial rabbit NPY antisera (green). Confocal fluorescence image by Serguei O. Fetissov

or dissociative buffers, respectively. Such an approach may provide complementary data on the relative levels of "free" and "complexed" antibodies, which may reflect altered peptidergic signaling in some pathological conditions. For instance, changes of plasma levels of free and total fractions of IgG autoantibodies reactive with adrenocorticotropic hormone (ACTH) in humans with various levels of aggressive behavior showed that their relative ratio was increased in the group of most

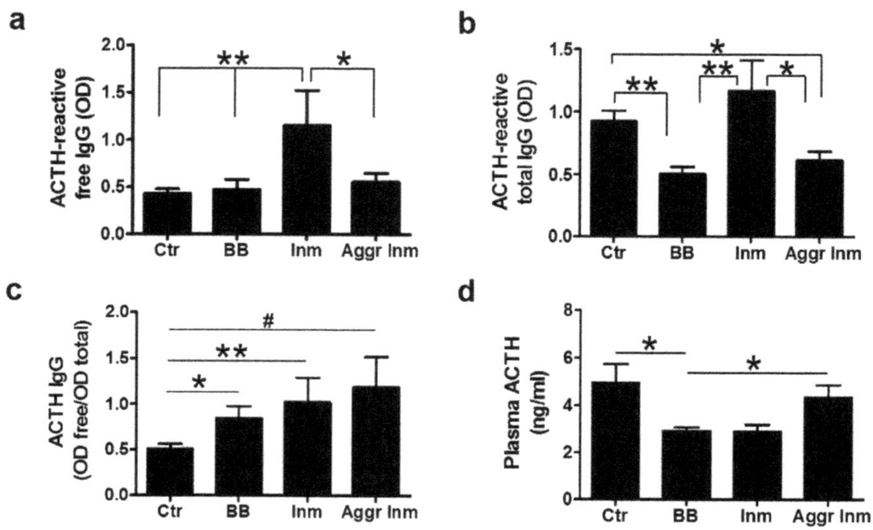

Fig. 8.2 Plasma levels of free (**a**) and total (**b**) ACTH-reactive IgG, their ratios (**c**) and plasma concentrations of ACTH peptide (**d**) in subjects with various degree of aggressive behavior. *Ctr* non-aggressive controls, *BB* body builders, *Inm* inmates, *Aggr inm* inmates committed violent aggression. Figure from Vaeroy et al. (2018)

aggressive subjects and was correlated negatively with the plasma concentration of ACTH peptide (Fig. 8.2). Such data indicate the role of plasma IgG as modulators of peptidergic signaling in the mechanism of stress-response and in stress-related disorders (Vaeroy et al. 2019). The role of ACTH-reactive IgG in modulating ACTH-induced cortisol secretion is discussed below.

Other techniques of peptide hormone reactive autoantibodies detection, such as the liquid-phase radiobinding assay (RBA), are of clinical relevance to classic autoimmune diseases (Sebriakova and Little 1973). For instance, it is used for analysis of the plasma levels of anti-insulin antibodies in type 1 diabetes, which is often performed in insulin-treated patients (Demeester et al. 2015). Because of the abundant literature on anti-insulin autoantibodies in the autoimmune disease context, we did not include them in Table 8.1. Nevertheless, insulin-reactive IgG can be present at various levels in non-insulin-treated subjects and in other species, which may also classify them as natural peptide-reactive autoantibodies (Sodoyez et al. 1990; Nishii et al. 2010). In fact, the recognition of insulin by germinal B-cells supports the presence of natural insulin-like autoimmunity (Wan et al. 2016). Naturally occurring insulin autoantibodies may possibly modulate glucose metabolism, for instance, it was shown that the administration of anti-insulin antibodies in the portal vein influenced feeding behavior (Surina-Baumgartner et al. 1995).

Because the IgG class of antibodies typically undergo affinity maturation, the affinity kinetics of IgG interaction with neuropeptides can be an important property that can be characterized. For instance, increased affinity of insulin autoantibody was shown to be associated with diabetes type 1 (Achenbach et al. 2007). The affinity

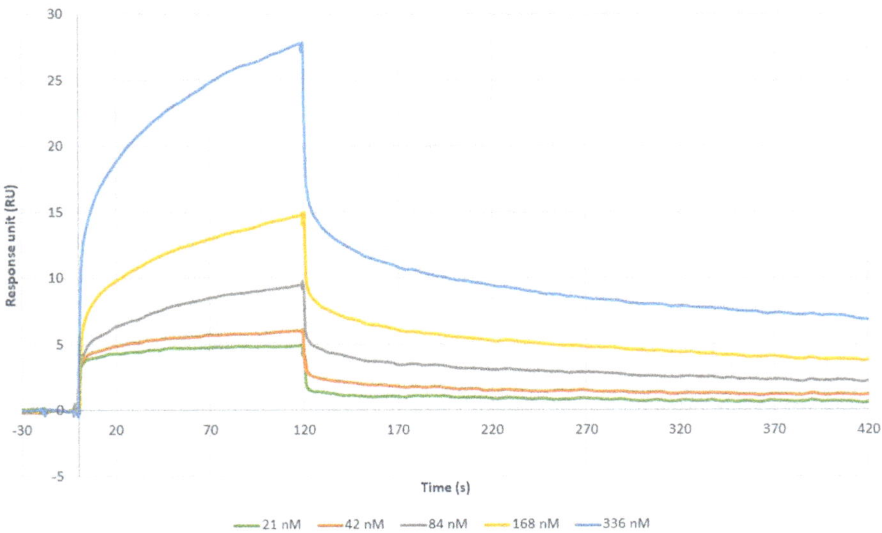

Fig. 8.3 SPR sensorgram obtained using a BIAcore T200 instrument for the interaction of the oxytocin peptide with serial dilutions of IgG (shown in different colors) purified from a healthy human plasma sample. Analysis of affinity kinetics shows that the equilibrium constant of this interaction (K_D) is 1.31×10^{-7} M. Image by Emilie Lahaye, Inserm UMR1239

kinetics can be analyzed using competitive RBA, typically used for the analysis of anti-insulin autoantibodies, or by the direct analysis of protein-protein interactions using the surface plasmon resonance (SPR) technique (Fig. 8.3).

8.3 Occurrence of Neuropeptide-Reactive Autoantibodies

Occurrence of autoantibodies reactive with several neuropeptides/peptide hormones in the plasma of humans and rodents has been reported in numerous studies (Table 8.1). It is possible that such occurrence may reflect a general phenomenon of immunoglobulin binding to peptidergic messengers involved in the neuroendocrine and other types of communication, e.g., in the immune system. In fact, autoantibodies reactive with cytokines have been detected in human plasma (Bendtzen et al. 1998). Like any Ig, neuropeptide-reactive autoantibodies are produced by B-cells and secreted into the systemic circulation, where they are usually detected in plasma or serum samples. For instance, the levels of IgG-reactive to ghrelin, the hormone of hunger produced in the stomach, is about 20-fold higher in plasma than in the hypothalamus or liver (Takagi et al. 2013).

The gastro-intestinal tract-associated lymphoid system is the largest part of the immune system in the body that may significantly contribute to the production of neuropeptide-reactive autoantibodies. In fact, gastric electrical stimulation was shown to activate c-fos marker in the gastric mucosal and submucosal layers and increase plasma levels of both ghrelin and ghrelin-reactive IgG (Gallas et al. 2011; Gallas and Fetissov 2011). The vicinity of the gut lymphoid system to the gut microbiota may serve as a natural antigenic stimulus for the production of neuropeptide-cross reactive autoantibodies. Therefore, the gut is probably another organ where neuropeptide-reactive autoantibodies can be readily detected. In the cerebrospinal fluid (CSF), the neuropeptide-reactive IgG can be detected at low levels, reflecting their limited passage across the blood–brain barrier (BBB). For instance, galanin-reactive IgG, possibly transporting galanin in immune complexes, were detected in human CSF, where their levels correlated negatively with the severity of cognitive impairment, suggesting a functional relevance of such IgG to Alzheimer's disease (Costa et al. 2011). The possible role of IgG as neuropeptide carriers to the brain is discussed below in more detail.

When analyzing plasma neuropeptide autoantibodies in some pathological conditions, it may be found that some patients display lower levels than healthy controls. Such results could be interpreted by some scientists as "negative," i.e., as showing the absence of such autoantibodies. However, considering the physiological role of neuropeptide-reactive IgG, such a decrease may reflect pathological changes relevant to the disease. For instance, lower levels of orexin-reactive IgG found in patients with narcolepsy (Black et al. 2005; Deloumeau et al. 2010) reflected only the free fraction of circulating autoantibody, while most of orexin-reactive-IgG in narcoleptic patients was present in the form of immune complexes (Deloumeau et al. 2010). Nevertheless, the relevance of IgG complexes with orexin to the pathophysiology of narcolepsy is still unknown and the search for the mechanism of orexin neuron-directed autoimmunity continues (Mahoney et al. 2019).

8.4 Role of Autoantibodies as Peptide Carriers

The function of immunoglobulins as carrier molecules of peptide hormones serves to improve peptide stability. Such a function was revealed by studying the peptide ghrelin, which is highly unstable in the circulation, where it is deacylated by plasma enzymes leading to the loss of ghrelin's orexigenic effect. Preincubation of ghrelin with IgG protected the hormone from deacylation, while the absence of IgG led to a complete loss of the acylated form. Importantly, the presence of ghrelin in the immune complex with IgG preserved its biological effect of stimulating food intake (Takagi et al. 2013). Furthermore, using chromatographic purification of IgG from human plasma, the same study showed that a large amount of the total pool of circulating ghrelin is present bound to IgG, further supporting the IgG carrier role for this peptide hormone. Moreover, while about half of the total ghrelin was found as the bound form in healthy subjects, in obese patients it was 4 times higher, which could be due to an increased affinity of IgG for ghrelin (Takagi et al. 2013). The phenomenon of increased carrier function of ghrelin by IgG may explain the enhanced orexigenic effects of exogenous ghrelin in obesity vs. non-obese subjects (Druce et al. 2005).

Another example, using a similar approach of peptide detection after IgG extraction from plasma, showed that IgG may serve as a carrier molecule for nerve growth factor in both healthy humans and in patients with autoimmune disorders (Dicou and Nerriere 1997). A protective effect of IgG for leptin was suggested by significant correlations between increased IgG binding to leptin and lower body mass index in healthy subjects, and inversely, decreased affinity of IgG for leptin in obesity (Bouhajja et al. 2018). As mentioned above, plasma ACTH correlates with ACTH-reactive IgG levels. Taken together, these data suggest that the main biological role of neuropeptide-reactive IgG autoantibodies is to serve as peptide carriers. This role can be exploited to improve peptide hormone stability after their pharmacological administration. Such a possibility was supported by the demonstration of the enhanced pharmacological effects of ghrelin combined with IgG in the alleviation of activity-based anorexia in mice (Legrand et al. 2016).

8.5 Role of Autoantibodies in Neuropeptide Receptor Activation (α-MSH, ACTH, Oxytocin)

Here we discuss only the effects of IgG in immune complexes with neuropeptides in activation of the corresponding neuropeptide receptors; this should not be confused with anti-receptor antibodies, which can be relevant to neuroendocrine signaling but constitute a different topic. As discussed above, the immune complex of IgG with ghrelin preserved the peptide's biological activity, suggesting that such a complex could bind and activate the corresponding peptide receptor. Indeed, an *in vitro* study indicated that ghrelin affinity of binding to its receptor GHS-R1a was increased when ghrelin was present as an immune complex with IgG from obese subjects (unpublished).

More detailed data of the IgG modulatory role in neuropeptide signaling were obtained for α-MSH binding to human MC4R (Lucas et al. 2019). Using a cAMP assay, it was found that IgG decreased the threshold of receptor activation and that this effect of IgG was not present in obesity. Moreover, using *in vitro* microscopy, it was shown that the α-MSH/IgG immune complex binds to the MC4R on the cell membrane and is then internalized, reflecting receptor activation. Of relevance to the critical role of MC4R activation in decreasing food intake, the internalization rate of α-MSH/IgG immune complexes was decreased in obesity but increased in anorexia nervosa. Analysis by epitope mapping of α-MSH binding to IgG suggested that the C-terminal of the α-MSH peptide needs to be available for MC4R activation, while its masking by IgG, for example in obesity, blocks the activation (Lucas et al. 2019).

The importance of binding by natural IgG to the different parts of a peptide hormone in modulating its biological activity has also been shown for the adrenocorticotropic hormone (ACTH) (Værøy et al. 2018). In fact, the predominant binding of IgG to the central part of 39 amino acid-long ACTH was associated with the preservation of ACTH-induced cortisol secretion by human cortical adrenal cells *in vitro*. In contrast, IgG binding to the N-terminal, which was more frequently observed in aggressive subjects, was associated with deficient cortisol secretion.

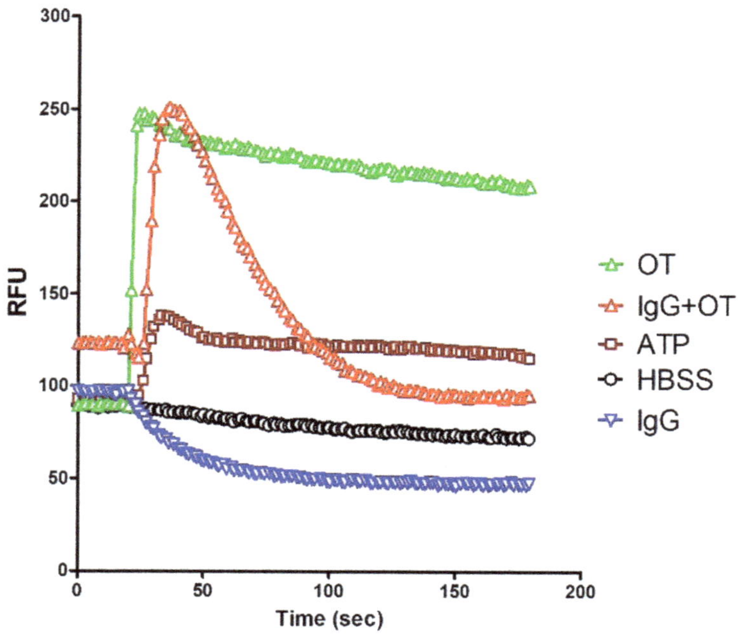

Fig. 8.4 IgG modulation of oxytocin-induced Ca^{2+} release by HEK293 cells expressing human oxytocin receptors. OT, oxytocin, ATP, adenosine triphosphate, HBSS, incubation buffer. IgG, example of IgG purified from plasma of a healthy subject. The area under the curve for oxytocin/IgG complex varies in different subjects (not shown), suggesting individual effects of IgG in complex with oxytocin in activation of oxytocin receptors. Data from the poster presentation of Lahaye et al. (2021)

A recent study showed that activation of the human oxytocin receptor, as assayed by Ca^{2+} release, can be modulated by plasma IgG in healthy subjects. As compared to oxytocin alone, oxytocin/IgG immune complex reduced the duration of Ca^{2+} release in all subjects. Furthermore, the amplitude of oxytocin/IgG complex-stimulated Ca^{2+} release varied in different subjects, sometimes being as high as after stimulation by oxytocin alone (Lahaye et al. 2021) (Fig. 8.4). Thus, the modulatory effects of IgG on neuropeptide-induced receptor activation suggests that naturally present IgG may contribute to the mechanism of individual variability in neuroendocrine signaling.

8.6 Catalytic Effect of Neuropeptide-Reactive IgG

It was found that IgG autoantibodies purified from a human plasma sample produced catalytic hydrolysis of vasoactive intestinal peptide (VIP), splitting it into two fragments (Paul et al. 1995). VIP, a 28-amino acid peptide from the glucagon/secretin family, stimulates vasodilation, suggesting a pathogenic role of neutralizing VIP IgG in some diseases such as asthma (Paul et al. 1989). The functional relevance of the proteolytic IgG activity is probably dependent on the unique IgG structure and is a part of a more general catalytic activity of some IgG, concerning not only proteins and peptides but also DNA and RNA (Nevinsky and Buneva 2002). In fact, natural autoantibodies reactive to VIP are present in both healthy subjects and in patients with various autoimmune disorders (Bangale et al. 2002). It cannot be excluded that some catalytic IgG may occur naturally and hydrolyze peptide hormones and neuropeptides other than VIP, contributing to individual biological activity of neuroendocrine signaling in health and disease.

8.7 Peptide-Autoantibody Complex Penetration Across the Blood–Brain Barrier

IgG passes across the blood–brain barrier by saturable transport (Zlokovic et al. 1990); however, several key questions remain regarding the mechanisms involved. For instance, the rapid perivascular transport called the "glymphatic system" has been suggested to play a role in the spread of IgG once it has crossed the blood–brain barrier (Iliff et al. 2012). Another study suggested that the majority of systemically derived IgG in the brain is sequestered within the endothelial cell compartment (St-Amour et al. 2013). IgG antibodies possess two antigen-binding domains (Fab) for binding unique epitopes and a single fragment crystallizable domain (Fc) for the binding of cell surface receptors and the recruitment of effector functions. Known Fc receptors include the neonatal Fc receptor (FcRn or Brambell receptor) and Fc receptors specific for Ig isoforms (e.g., FcγR for IgG). Several studies showed that FcRn mediated the transport of IgG across peripheral vascular cells in both directions, while in the CNS the transport occurs mainly in the brain-to-blood direction (Roopenian et al. 2003; Schlachetzki et al. 2002). Nevertheless, the role of Fc and FcRn in the transport of IgG in the CNS remain controversial, and other mechanisms involving antigen receptor binding at the brain endothelial blood vessels, for example to transferrin and insulin receptors, have been discussed (Kouhi et al. 2021). Moreover, low-affinity binding to transferrin receptors may be necessary for the transport of IgG into the CNS (Bien-Ly et al. 2014). Penetration of the brain by neuropeptides in immune complexes with IgG still needs to be clarified, a preliminary study showed the presence of IgG in several brain structures involved in the barrier function after peripheral injection in the rat of α-MSH/IgG immune complexes (Fig. 8.5).

8.8 Antigenic Origin and Discovery Pathway to New Drugs

The search for the antigenic origin of α-MSH-reactive autoantibodies led to the identification of a bacterial mimetic protein of α-MSH, represented by caseinolytic protease B (ClpB) (Tennoune et al. 2014). Indeed, either active immunization of mice with ClpB or their gavage with ClpB-producing *Escherichia coli* (*E. coli*) K12 bacteria resulted in increased levels of anti-ClpB α-MSH-cross-reactive antibodies.

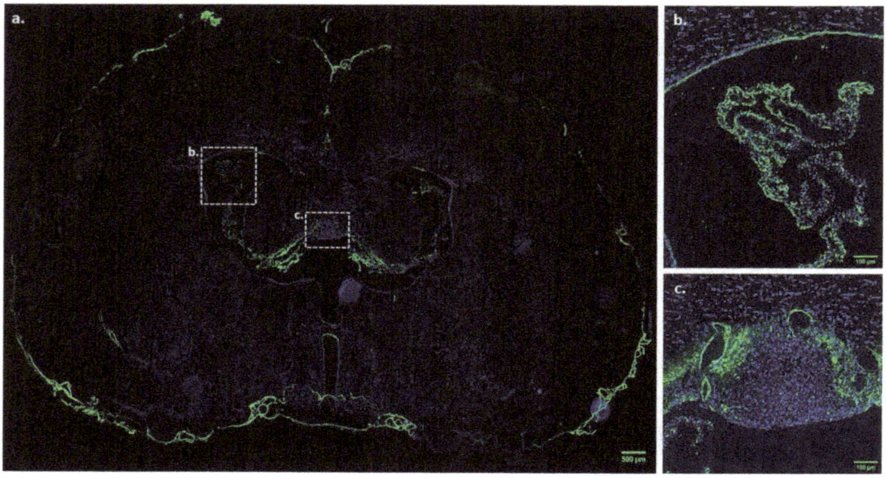

Fig. 8.5 Immunohistochemical detection of rat IgG in the rat brain 1 min after intracardiac injection of α-MSH/IgG immune complexes. (**a**) Coronal brain section at the level −0.8 mm from Bregma. Immunopositive structures (green) include: leptomeninges, choroid plexus, ventricular ependyma and the subcortical organ. Inserts (**b**) and (**c**) are shown with higher magnification, illustrating the choroid plexus and ventricular ependyma in (**b**) and the subfornical organ in (**c**) Nuclear counterstaining with DAPI (blue). Fluorescence image by Mouna El Mehdi

Importantly, when mice received a ClpB mutant *E. coli* strain, no increase in such antibodies was noticed. Indeed, the protein sequence of ClpB was found to contain an α-MSH-like fragment which is conserved in the ClpB molecule produced by a few bacterial taxa, including the whole family of Enterobacteriaceae. This finding suggests that a specific neuroendocrine signaling (ex. the melanocortin system) may be functionally connected to specific gut microorganisms (ex. Enterobacteriaceae), based on molecular mimicry between neuropeptides and bacterial proteins synthetized by such microorganisms (Fetissov 2019).

The discovery of bacteria producing an α-MSH-like anorexigenic protein led to a better understanding of the role of gut microbiota in the regulation of appetite and triggered a new line of research aimed at the development of ClpB-based probiotics for controlling appetite and body weight (Fetissov 2017). The proof of concept preclinical study was successfully accomplished in obese mice using a food-grade *Hafnia alvei* strain that expressed ClpB with an α-MSH-like epitope, similar to *E. coli* (Legrand et al. 2020). A follow-up study in overweight subjects confirmed the clinical efficacy of *H. alvei* by increasing a feeling of satiety and in reducing body weight and hip circumference (Déchelotte et al. 2021). Therefore, the discovery of α-MSH autoantibodies has served as a starting point for the development of a new generation of precision probiotics that can be further developed into biopharmaceutical drugs. Considering multiple examples of molecular mimicry between several neuropeptides and microbial proteins produced by commensal and pathogenic microorganisms, as well as the natural presence of neuropeptide-reactive Ig (Table 8.1), one may suggest that such autoantibodies have been induced by homologous antigens derived from gut microbiota. Several research projects are ongoing to identify such antigens. Increased prevalence of some neuropeptide-reactive autoantibodies have also been found in *Helicobacter pylori* and *Candida albicans*-positive children and their relevance to short stature has been discussed (Stawerska et al. 2015).

8.9 Pathogenic Role of α-MSH Autoantibodies

Since the initial discovery of α-MSH-reactive autoantibodies in patients with eating disorders in 2002 (Fetissov et al. 2002), several studies have pursued research on their origin, mechanism of action, and clinical relevance. Based on this research, a pathophysiological model of eating disorders has been proposed by Fetissov and Hökfelt in 2019, postulating that both anorexia nervosa and bulimia appear as an autoimmune reaction to the gut bacterial protein ClpB (Fetissov and Hökfelt 2019). Such a model can be contrasted with the pathophysiology of diabetes type 1, where an autoimmune reaction is directed against insulin, a protein hormone critically involved in the regulation of glucose metabolism. In the case of eating disorders, bacterial ClpB is naturally present in healthy gut microbiota and physiologically participates in signaling satiety. An autoimmune reaction against ClpB, leading to production of pathogenic α-MSH-cross reactive IgG, may lead to the chronic activation of the melanocortin system with a loss of appetite and increased anxiety in both anorexia nervosa and bulimia. Neutralization of ClpB may cause a bulimic attack, also explaining why both anorexia and bulimia typically occur in the same patient. If this pathophysiological model is correct, then therapeutic strategies aimed at the elimination of the antigen and/or of the pathogenic α-MSH-autoantibodies may prove effective.

8.10 Conclusions

Neuropeptide-reactive IgG are involved in neuroendocrine signaling as carriers of neuropeptides, protecting them from degradation and as modulators of neuropeptide receptor activation. It is possible that such a role of IgG has evolved as the immune response to the neuropeptide-like microbial antigens constitutively present in the microbiota of all animals with a gut. It is clear that further analysis of autoantibodies that are reactive with neuropeptides/peptide hormones will allow a better understanding of their role in neuroendocrine signaling as well as their involvement in both physiological and pathophysiological regulation of motivated behavior and metabolism.

8.11 Key References

Fetissov, S. O. & Hökfelt, T. 2019. On the origin of eating disorders: altered signaling between gut microbiota, adaptive immunity and the brain melanocortin system regulating feeding behavior. *Curr Opin Pharmacol,* 48, 82–91.

This paper takes you on a 20-year journey, from the finding of α-MSH-reactive antibodies to the formulation of the pathophysiological model of eating disorders.

Takagi, K., Legrand, R., Asakawa, A., Amitani, H., François, M., Tennoune, N., Coëffier, M., Claeyssens, S., Do Rego, J.-C., Déchelotte, P., Inui, A. & Fetissov, S. O. 2013. Anti-ghrelin immunoglobulins modulate ghrelin stability and its orexigenic effect in obese mice and humans. *Nat Commun,* 4:2685.

This paper provides the first evidence of plasma IgG playing the role of a functional peptide hormone carrier, protecting it from degradation.

Værøy, H., Adori, C., Legrand, R., Lucas, N., Breton, J., Cottard, C., Do Rego, J.-C., Duparc, C., Louiset, E., Lefebvre, H., Déchelotte, P., Western, E., Andersson, S., Hökfelt, T. & Fetissov, S. O. 2018. Autoantibodies reactive to adrenocorticotropic hormone can alter cortisol secretion in both aggressive and nonaggressive humans. *Proc Natl Acad Sci USA,* 115, E6576-E6584.

This paper reveals the role of IgG in modulating the activity of the hypothalamo-pituitary adrenal axis in humans, i.e., involvement in individual variability of the stress response.

Acknowledgments Studies of neuropeptide autoantibodies are currently funded by the EC as a part of research projects exploring the role of gut microbiota in autism (H2020 "GEMMA") and in anorexia nervosa (ERAnet, "MIGBAN"), as well as by the PTM program of Inserm, France.

References

Achenbach P, Schlosser M, Williams AJ, Yu L, Mueller PW, Bingley PJ, Bonifacio E (2007) Combined testing of antibody titer and affinity improves insulin autoantibody measurement: Diabetes Antibody Standardization Program. Clin Immunol 122:85–90

Bangale Y, Cavill D, Gordon T, Planque S, Taguchi H, Bhatia G, Nishiyama Y, Arnett F, Paul S (2002) Vasoactive intestinal peptide binding autoantibodies in autoimmune humans and mice. Peptides 23:2251–2257

Bendtzen K, Hansen MB, Ross C, Svenson M (1998) High-avidity autoantibodies to cytokines. Immunol Today 19:209–211

Bien-Ly N, Yu YJ, Bumbaca D, Elstrott J, Boswell CA, Zhang Y, Luk W, Lu Y, Dennis MS, Weimer RM, Chung I, Watts RJ (2014) Transferrin receptor (TfR) trafficking determines brain uptake of TfR antibody affinity variants. J Exp Med 211:233–244

Black JL 3rd, Silber MH, Krahn LE, Fredrickson PA, Pankratz VS, Avula R, Walker DL, Slocumb NL (2005) Analysis of hypocretin (orexin) antibodies in patients with narcolepsy. Sleep 28: 427–431

Bouhajja H, Bougacha-Elleuch N, Lucas N, Legrand R, Marrakchi R, Kaveri SV, Jamoussi K, Ayadi H, Abid M, Mnif-Feki M, Fetissov SO (2018) Affinity kinetics of leptin-reactive immunoglobulins are associated with plasma leptin and markers of obesity and diabetes. Nutr Diabetes 8:32

Coquerel Q, Hamze Sinno M, Boukhettala N, Coëffier M, Terashi M, Bole-Feysot C, Breuillé D, Déchelotte P, Fetissov SO (2012) Intestinal inflammation influences α-MSH reactive autoantibodies: relevance to food intake and body weight. Psychoneuroendocrinology 37:94–106

Costa A, Bini P, Hamze-Sinno M, Moglia A, Franciotta D, Sinforiani E, Ravaglia S, Bole-Feysot C, Hökfelt T, Déchelotte P, Fetissov SO (2011) Galanin and alpha-MSH autoantibodies in cerebrospinal fluid of patients with Alzheimer's disease. J Neuroimmunol 240–241:114–120

Davson H, Segal MB (1996) Physiology of the CSF and blood-brain barriers. CRC Press, Boca Raton

Déchelotte P, Breton J, Trotin-Picolo C, Grube B, Erlenbeck C, Bothe G, Fetissov SO, Lambert G (2021) The probiotic strain H. alvei HA4597® improves weight loss in overweight subjects under moderate hypocaloric diet: a proof-of-concept, multicenter randomized, double-blind placebo-controlled study. Nutrients 13:1902

Deloumeau A, Bayard S, Coquerel Q, Déchelotte P, Bole-Feysot C, Carlander B, Cochen De Cock V, Fetissov SO, Dauvilliers Y (2010) Increased immune complexes of hypocretin autoantibodies in narcolepsy. PLoS One 5:e13320

Demeester S, Keymeulen B, Kaufman L, Van Dalem A, Balti EV, Van De Velde U, Goubert P, Verhaeghen K, Davidson HW, Wenzlau JM, Weets I, Pipeleers DG, Gorus FK (2015) Preexisting insulin autoantibodies predict efficacy of otelixizumab in preserving residual β-cell function in recent-onset type 1 diabetes. Diabetes Care 38:644–651

Dicou E, Nerriere V (1997) Evidence that natural autoantibodies against the nerve growth factor (NGF) may be potential carriers of NGF. J Neuroimmunol 75:200–203

Druce MR, Wren AM, Park AJ, Milton JE, Patterson M, Frost G, Ghatei MA, Small C, Bloom SR (2005) Ghrelin increases food intake in obese as well as lean subjects. Int J Obes Relat Metab Disord 29:1130–1136

Espinoza-Garcia AS, Hunot-Alexander C, Martinez-Moreno AG, Vazquez-Solorzano R, Porchas-Quijada M, Reyes-Castillo Z (2021) IgG antibodies reacting with ghrelin and leptin are correlated with body composition and appetitive traits in young subjects. Appetite 168:105685

Fetissov SO (2011) Neuropeptide autoantibodies assay. Methods Mol Biol 789:295–302

Fetissov SO (2017) Role of the gut microbiota in host appetite control: bacterial growth to animal feeding behaviour. Nat Rev Endocrinol 13:11–25

Fetissov SO (2019) Neuropeptide-like signaling in the microbiota-gut-brain axis. Behav Brain Sci 42:e70

Fetissov SO, Hökfelt T (2019) On the origin of eating disorders: altered signaling between gut microbiota, adaptive immunity and the brain melanocortin system regulating feeding behavior. Curr Opin Pharmacol 48:82–91

Fetissov SO, Hallman J, Oreland L, Af Klinteberg B, Grenbäck E, Hulting AL, Hökfelt T (2002) Autoantibodies against α-MSH, ACTH, and LHRH in anorexia and bulimia nervosa patients. Proc Natl Acad Sci USA 99:17155–17160

Fetissov SO, Harro J, Jaanisk M, Järv A, Podar I, Allik J, Nilsson I, Sakthivel P, Lefvert AK, Hökfelt T (2005) Autoantibodies against neuropeptides are associated with psychological traits in eating disorders. Proc Natl Acad Sci USA 102:14865–14870

Fetissov SO, Hallman J, Nilsson I, Lefvert AK, Oreland L, Hökfelt T (2006) Aggressive behavior linked to corticotropin-reactive autoantibodies. Biol Psychiatry 60:799–802

Fetissov SO, Hamze Sinno M, Coëffier M, Bole-Feysot C, Ducrotté P, Hökfelt T, Déchelotte P (2008) Autoantibodies against appetite-regulating peptide hormones and neuropeptides: putative modulation by gut microflora. Nutrition 24:348–359

François M, Schaefer JM, Bole-Feysot C, Déchelotte P, Verhulst FC, Fetissov SO (2015) Ghrelin-reactive immunoglobulins and anxiety, depression and stress-induced cortisol response in adolescents. The TRAILS study. Prog Neuro-Psychoph 59:1–7

François M, Takagi K, Legrand R, Lucas N, Beutheu Youmba S, Bôle-Feysot C, Cravezic A, Tennoune N, Do Rego J-C, Coëffier M, Inui A, Déchelotte P, Fetissov SO (2016) Increased ghrelin but low ghrelin-reactive immunoglobulins in a rat model of methotrexate chemotherapy-induced anorexia. Front Nutr 3

Gallas S, Fetissov SO (2011) Ghrelin, appetite and gastric electrical stimulation. Peptides 32:2283–2289

Gallas S, Sinno MH, Boukhettala N, Coëffier M, Dourmap N, Gourcerol G, Ducrotté P, Déchelotte P, Leroi A-M, Fetissov SO (2011) Gastric electrical stimulation increases ghrelin production and inhibits catecholaminergic brainstem neurons in rats. Eur J Neurosci 33:276–284

Garcia FD, Coquerel Q, Kiive E, Déchelotte P, Harro J, Fetissov SO (2011) Autoantibodies reacting with vasopressin and oxytocin in relation to cortisol secretion in mild and moderate depression. Prog Neuro-Psychopharmacol Biol Psychiatry 35:118–125

Garcia FD, Coquerel Q, Do Rego J-C, Cravezic A, Bole-Feysot C, Kiive E, Déchelotte P, Harro J, Fetissov SO (2012) Anti-neuropeptide Y plasma immunoglobulins in relation to mood and appetite in depressive disorder. Psychoneuroendocrinology 37:1457–1467

Hamze Sinno M, Do Rego JC, Coëffier M, Bole-Feysot C, Ducrotte P, Gilbert D, Tron F, Costentin J, Hökfelt T, Déchelotte P, Fetissov SO (2009) Regulation of feeding and anxiety by α-MSH reactive autoantibodies. Psychoneuroendocrinology 34:140–149

Hökfelt T, Barde S, Xu ZD, Kuteeva E, Rüegg J, Le Maitre E, Risling M, Kehr J, Ihnatko R, Theodorsson E, Palkovits M, Deakin W, Bagdy G, Juhasz G, Prud'homme HJ, Mechawar N, Diaz-Heijtz R, Ögren SO (2018) Neuropeptide and small transmitter coexistence: fundamental studies and relevance to mental illness. Front Neural Circuits 12:106

Hortin GL, Sviridov D, Anderson NL (2008) High-abundance polypeptides of the human plasma proteome comprising the top 4 logs of polypeptide abundance. Clin Chem 54:1608–1616

Iliff JJ, Wang M, Liao Y, Plogg BA, Peng W, Gundersen GA, Benveniste H, Vates GE, Deane R, Goldman SA, Nagelhus EA, Nedergaard M (2012) A paravascular pathway facilitates CSF flow through the brain parenchyma and the vlearance of interstitial solutes, including amyloid β. Sci Transl Med 4:147ra111

Karaiskos D, Mavragani CP, Sinno MH, Déchelotte P, Zintzaras E, Skopouli FN, Fetissov SO, Moutsopoulos HM (2010) Psychopathological and personality features in primary Sjogren's syndrome—associations with autoantibodies to neuropeptides. Rheumatology 49:1762–1769

Ketchesin KD, Stinnett GS, Seasholtz AF (2017) Corticotropin-releasing hormone-binding protein and stress: from invertebrates to humans. Stress 20:449–464

Kouhi A, Pachipulusu V, Kapenstein T, Hu P, Epstein AL, Khawli LA (2021) Brain disposition of antibody-based therapeutics: dogma, approaches and perspectives. Int J Mol Sci 22:6442

Lacroix-Desmazes S, Kaveri SV, Mouthon L, Ayouba A, Malanchère E, Coutinho A, Kazatchkine MD (1998) Self-reactive antibodies (natural autoantibodies) in healthy individuals. J Immunol Methods 216:117–137

Lahaye E, Dubessy C, Værøy H, FetissovSO (2021) Oxytocin receptor activation in humans is modulated by oxytocin-reactive immunoglobulin. PNIRS annual meeting, May 20–21st 2021 virtual

Legrand R, Lucas N, Breton J, Azhar S, Do Rego J-C, Déchelotte P, Coëffier M, Fetissov SO (2016) Ghrelin treatment prevents development of activity based anorexia in mice. Eur Neuropsychopharmacol 26:948–958

Legrand R, Lucas N, Dominique M, Azhar S, Deroissart C, Le Solliec M-A, Rondeaux J, Nobis S, Guérin C, Léon F, Do Rego J-C, Pons N, Le Chatelier E, Ehrlich SD, Lambert G, Déchelotte P, Fetissov SO (2020) Commensal Hafnia alvei strain reduces food intake and fat mass in obese mice-a new potential probiotic for appetite and body weight management. Int J Obes 44:1041–1051

Leng G, Ludwig M (2008) Neurotransmitters and peptides: whispered secrets and public announcements. J Physiol 586:5625–5632

Lucas N, Legrand R, Bôle-Feysot C, Breton J, Coëffier M, Akkermann K, Järv A, Harro J, Déchelotte P, Fetissov SO (2019) Immunoglobulin G modulation of the melanocortin 4 receptor signaling in obesity and eating disorders. Transl Psychiatry 9:87

Mahoney CE, Cogswell A, Koralnik IJ, Scammell TE (2019) The neurobiological basis of narcolepsy. Nat Rev Neurosci 20:83–93

Molfino A, Amabile MI, Imbimbo G, Emiliani A, Ramaccini C, Lahaye E, Takagi K, Fetissov SO (2020) Plasma enterobacterial ClpB levels and ClpB- and α-MSH-reactive immunoglobulins in lung cancer patients with and without anorexia. Nutrition 78:110952

Myagkova MA, Gavrilova SI, Lermontova NN, Kalyn YB, Selezneva ND, Zharikov GA, Kolykhalov IV, Abramenko TV, Serkova TP, Bachurin SO (2003) Content of autoantibodies to bradykinin and beta-amyloid(1-42) as a criterion for biochemical differences between Alzheimer's dementias. Bull Exp Biol Med 136:49–52

Nevinsky GA, Buneva VN (2002) Human catalytic RNA- and DNA-hydrolyzing antibodies. J Immunol Methods 269:235–249

Nishii N, Takasu M, Kojima M, Hachisu T, Wakabayashi K, Iwasawa A, Maeda S, Ohba Y, Kitagawa H (2010) Presence of anti-insulin natural autoantibodies in healthy cats and its interference with immunoassay for serum insulin concentrations. Domest Anim Endocrinol 38:138–145

Paul S, Said SI (1988) Human autoantibody to vasoactive intestinal peptide: increased incidence in muscular exercise. Life Sci 43:1079–1084

Paul S, Heinz-Erian P, Said SI (1985) Autoantibody to vasoactive intestinal peptide in human circulation. Biochem Biophys Res Commun 130:479–485

Paul S, Said SI, Thompson AB, Volle DJ, Agrawal DK, Foda H, De La Rocha S (1989) Characterization of autoantibodies to vasoactive intestinal peptide in asthma. J Neuroimmunol 23:133–142

Paul S, Gao QS, Huang H, Sun M, Thompson A, Rennard S, Landers D (1995) Catalytic autoantibodies to vasoactive intestinal peptide. Chest 107:125S–126S

Porchas-Quijada M, Reyes-Castillo Z, Muñoz-Valle JF, Durán-Barragán S, Aguilera-Cervantes V, López-Espinoza A, Vázquez-Del Mercado M, Navarro-Meza M, López-Uriarte P (2019) IgG anti-ghrelin immune complexes are increased in rheumatoid arthritis patients under biologic therapy and are related to clinical and metabolic markers. Front Endocrinol (Lausanne) 10:252

Roopenian DC, Christianson GJ, Sproule TJ, Brown AC, Akilesh S, Jung N, Petkova S, Avanessian L, Choi EY, Shaffer DJ, Eden PA, Anderson CL (2003) The MHC class I-like IgG receptor controls perinatal IgG transport, IgG homeostasis, and fate of IgG-Fc-coupled drugs. J Immunol 170:3528–3533

Roubalova R, Prochazkova P, Dvorak J, Hill M, Papezova H, Kreisinger J, Bulant J, Lambertova A, Holanova P, Bilej M, Tlaskalova-Hogenova H (2021) Altered serum immunological and biochemical parameters and microbiota composition in patients with AN during realimentation. Front Nutr 8:680870

Schaefer JM, Fetissov SO, Legrand R, Claeyssens S, Hoekstra PJ, Verhulst FC, Van Oort FVA (2013) Corticotropin (ACTH)-reactive immunoglobulins in adolescents in relation to antisocial behavior and stress-induced cortisol response. The TRAILS study. Psychoneuroendocrinology 38:3039–3047

Schlachetzki F, Zhu C, Pardridge WM (2002) Expression of the neonatal Fc receptor (FcRn) at the blood–brain barrier. J Neurochem 81:203–206

Sebriakova M, Little JA (1973) A method for the determination of plasma insulin antibodies and its application in normal and diabetic subjects. Diabetes 22:30–40

Sodoyez JC, Sodoyez-Goffaux F, Koch M, Sondag D, Bouillenne C, François-Gérard C, Bosi E (1990) Clonally restricted insulin autoantibodies in a cohort of 2200 healthy blood donors. Diabetologia 33:719–725

St-Amour I, Paré I, Alata W, Coulombe K, Ringuette-Goulet C, Drouin-Ouellet J, Vandal M, Soulet D, Bazin R, Calon F (2013) Brain bioavailability of human intravenous immunoglobulin and its transport through the murine blood-brain barrier. J Cereb Blood Flow Metab 33:1983–1992

Stawerska R, Czkwianianc E, Matusiak A, Smyczyńska J, Hilczer M, Chmiela M, Lewiński A (2015) Prevalence of autoantibodies against some selected growth and appetite-regulating neuropeptides in serum of short children exposed to Candida albicans colonization and/or Helicobacter pylori infection: the molecular mimicry phenomenon. Neuro Endocrinol Lett 36: 458–464

Surina-Baumgartner DM, Langhans W, Geary N (1995) Hepatic portal insulin antibody infusion increases, but insulin does not alter, spontaneous meal size in rats. Am J Phys 269:R978–R982

Takagi K, Legrand R, Asakawa A, Amitani H, François M, Tennoune N, Coëffier M, Claeyssens S, Do Rego J-C, Déchelotte P, Inui A, Fetissov SO (2013) Anti-ghrelin immunoglobulins modulate ghrelin stability and its orexigenic effect in obese mice and humans. Nat Commun 4:2685

Tanaka S, Honda Y, Inoue Y, Honda M (2006) Detection of autoantibodies against hypocretin, hcrtrl, and hcrtr2 in narcolepsy: anti-Hcrt system antibody in narcolepsy. Sleep 29:633–638

Tennoune N, Chan P, Breton J, Legrand R, Chabane YN, Akkermann K, Jarv A, Ouelaa W, Takagi K, Ghouzali I, François M, Lucas N, Bole-Feysot C, Pestel-Caron M, Do Rego JC, Vaudry D, Harro J, Dé E, Déchelotte P, Fetissov SO (2014) Bacterial ClpB heat-shock protein, an antigen-mimetic of the anorexigenic peptide [alpha]-MSH, at the origin of eating disorders. Transl Psychiatry 4:e458

Tennoune N, Legrand R, Ouelaa W, Breton J, Lucas N, Bole-Feysot C, Rego J-CD, Déchelotte P, Fetissov SO (2015) Sex-related effects of nutritional supplementation of Escherichia coli: Relevance to eating disorders. Nutrition 31:498–507

Terashi M, Asakawa A, Harada T, Ushikai M, Coquerel Q, Sinno MH, Déchelotte P, Inui A, Fetissov SO (2011) Ghrelin reactive autoantibodies in restrictive anorexia nervosa. Nutrition 27: 407–413

Værøy H, Adori C, Legrand R, Lucas N, Breton J, Cottard C, Do Rego J-C, Duparc C, Louiset E, Lefebvre H, Déchelotte P, Western E, Andersson S, Hökfelt T, Fetissov SO (2018) Autoantibodies reactive to adrenocorticotropic hormone can alter cortisol secretion in both aggressive and nonaggressive humans. Proc Natl Acad Sci USA 115:E6576–E6584

Vaeroy H, Schneider F, Fetissov SO (2019) Neurobiology of aggressive behavior—role of autoantibodies reactive with stress-related peptide hormones. Front Psychiatry 10

Wallenius M, Lind A, Akel O, Karlsson E, Svensson M, Arvidsson E, Ramelius A, Törn C, Palm L, Lernmark Å, Elding Larsson H (2019) Autoantibodies in Pandemrix®-induced narcolepsy: nine candidate autoantigens fail the conformational autoantibody test. Autoimmunity 52:185–191

Wan X, Thomas JW, Unanue ER (2016) Class-switched anti-insulin antibodies originate from unconventional antigen presentation in multiple lymphoid sites. J Exp Med 213:967–978

Wang W, Wang EQ, Balthasar JP (2008) Monoclonal antibody pharmacokinetics and pharmacodynamics. Clin Pharmacol Ther 84:548–558

Yu J, Gimenez LE, Hernandez CC, Wu Y, Wein AH, Han GW, Mcclary K, Mittal SR, Burdsall K, Stauch B, Wu L, Stevens SN, Peisley A, Williams SY, Chen V, Millhauser GL, Zhao S, Cone RD, Stevens RC (2020) Determination of the melanocortin-4 receptor structure identifies Ca2+ as a cofactor for ligand binding. Science 368:428–433

Zlokovic BV, Skundric DS, Segal MB, Lipovac MN, Mackic JB, Davson H (1990) A saturable mechanism for transport of immunoglobulin G across the blood-brain barrier of the guinea pig. Exp Neurol 107:263–270

Part III

Modulation

Postnatal Development of Neuroimmune Responses

9

Sarah J. Spencer

Abstract

The early-life period represents a time of particular vulnerability to environmental and lifestyle stimuli that can shape our physiology long-term. The neuroimmune system is somewhat immature around the time of birth and a growing body of evidence has illustrated how immune stimuli experienced at this time can have a lasting influence on peripheral and central responses to further immune challenge. This chapter will detail current knowledge on the postnatal development of the neuroimmune system and how this is shaped by exposure to various stimuli. Further, field-relevant issues of experimental design, as well as novel techniques, technologies, and inter-disciplinary approaches that are expanding our capacity for understanding in this field will be discussed.

Keywords

Cytokines · Fever · Lipopolysaccharide · Microglia · Postnatal programming

Abbreviations

ACTH	Adrenocorticotropic hormone
BDNF	Brain-derived neurotrophic factor
BrdU	Bromodeoxyuridine

S. J. Spencer (✉)
School of Health and Biomedical Sciences, RMIT University, Melbourne, VIC, Australia

ARC Centre of Excellence for Nanoscale Biophotonics, RMIT University, Melbourne, VIC, Australia
e-mail: sarah.spencer@rmit.edu.au

J. P. Konsman, T. M. Reyes (eds.), *Neuroendocrine-Immune System Interactions*, Masterclass in Neuroendocrinology 13,
https://doi.org/10.1007/978-3-031-21358-8_9

C1q	Complement component q
C3	Complement 3
Cd11b	Cluster of differentiation molecule 11b
COX2	Cyclo-oxygenase-2
CSF1R	Colony-stimulating factor 1 receptor
Cx3cr1	C-X3-C motif chemokine receptor 1
GR	Glucocorticoid receptor
HPA axis	Hypothalamic-pituitary-adrenal axis
Iba-1	Ionized calcium-binding adaptor molecule 1
IκB	Inhibitory factor κB
IL-1β	Interleukin-1β
LPS	Lipopolysaccharide
NGFI-A	Nerve growth factor inducible factor A
P	Postnatal day
PAMPS	Pathogen-associated molecular patterns
poly i:c	Polyinosinic:polycytidylic acid
PSD-95	Postsynaptic density 95
SNAP-25	Synaptosomal-associated protein 25
SPF	Specific-pathogen-free
TLR	Toll-like receptor
TNBS	Trinitrobenzenesulfonic acid

9.1 Introduction

The mammalian-acquired immune system is crucial for adaptive resilience to pathogens encountered across the lifespan. It is also the basis upon which most vaccinations protect against future exposure. The innate immune system, on the other hand, responds to generalized and conserved microbial patterns, known as pathogen-associated molecular patterns (PAMPS), raising a fast, non-specific immune activation for effectiveness against a multitude of pathogens (Dantzer et al. 2008; Spencer et al. 2008). While necessarily broadly responsive, this innate immune system is still inherently adaptive, for instance becoming hyporesponsive during pregnancy (Barrientos et al. 2019; Brunton and Russell 2011; Spencer et al. 2008). It is also eminently programmable. The innate immune system is highly responsive to bacterial or viral infection, toxins or injury, and even to a high-fat diet, particularly when these stimuli are encountered during specific developmental windows of vulnerability. These programming effects can last a lifetime (Barrientos et al. 2019). Exposure to stressful or nutritional challenges early in life can lead to lasting changes in central and peripheral responsiveness to further challenge (Boisse et al. 2004; Spencer et al. 2006a). These exposures can also potentially program long-term changes in the central nervous system's key immune cells, microglia, priming these cells to hyper-respond to further challenges. This microglial priming can contribute to lasting vulnerability to neuroinflammation, cognitive decline, and aging-related diseases such as dementia and Alzheimer's disease (De Luca et al. 2016; Hoeijmakers et al. 2017; Ziko et al. 2014). Recent developments in genetic engineering and neuroimaging techniques have meant that our understanding of the innate immune system's capacity for malleability is rapidly expanding. This chapter will describe historical knowledge on the postnatal development of neuroimmune responses; will discuss latest advice on experimental procedures; will cover information about novel techniques, technologies, and inter-disciplinary approaches; and will detail future developments in research in neuroimmune neuroendocrinology.

9.2 What Murine Postnatal Development Means for Humans

Much of our understanding of human neuroimmune and neuroendocrine development comes from non-human animal models, chiefly rats and mice. While many of the mechanisms and developmental sequences are well conserved between mammals, the absolute timing of events obviously differs. Although rats, mice, and humans are considered altricial species (under-developed at birth), many of the developmental processes that occur prenatally in humans do not take place until the first few weeks of life in the rat and mouse (Semple et al. 2013). These differences can prove useful in scientific study, allowing us to interrogate the development of physiological functions that occur in utero in humans in individual, relatively independent, murine animal models, without the surgical access that would otherwise be required.

For example, rodents are unresponsive to the anorexigenic effects of the adipose-derived hormone, leptin, until approximately the second to third week of life, due to a lack of mature connectivity between the arcuate nucleus and other regions of the hypothalamus. This connectivity is stimulated to become fully mature by a leptin surge that occurs at around postnatal day (P)6-10 in these species (Bouret 2010; Bouret et al. 2004a, b; Sominsky et al. 2018b). In non-human primates, sheep, and likely humans, the hypothalamic satiety-signaling connections are robust at birth, probably consolidating in the third trimester of pregnancy (Grayson et al. 2006; Oyama et al. 1992). Likewise, hypoxia-ischemia white matter injury that occurs prenatally in humans is consistent with similar white matter injury in postnatal rats and mice at 3–7 days after birth (Jisa et al. 2018).

On the other hand, both rodents and humans display a postnatal stress hypo-responsive period that is directly related to maternal presence. The rat and mouse hypothalamic-pituitary-adrenal (HPA) axis, the apex of the neuroendocrine stress response, is immature at birth and hypo-responsive to stress throughout the first 14 days of the suckling period as shown in (Levine 1957; Levine et al. 1957; Suchecki 2018). It develops into full adrenal responsiveness to adrenocorticotropic hormone (ACTH) only once the dam starts leaving the nest for protracted periods or in the case of an earlier imposed maternal separation (Levine 1957; Levine et al. 1957; Suchecki 2018). The human HPA axis displays a similar profile of postnatal stress hypo-responsiveness, albeit with scalable differences in the broad timing. In

Table 9.1 Equating mouse, rat, and human ages at various stages of development. For instance, the average weaning age for humans is 6 months and for mice is 28 days, therefore during the suckling period 56 days is approximately equivalent to one human year of life, but age-equivalence changes as the animals mature

	Mice	Rats	Humans
Suckling period Mice weaned at ~28 days, rats weaned at ~21 days	56.8 days	42.4 days	1 year
Adolescence weaning to puberty	3.7 days	10.5 days	1 year
Adulthood	2.6 days	11.8 days	1 year
Reproductive senescence	8.8 days	17.1 days	1 year
Lifetime	9.1 days	13.8 days	1 year

Adapted from Dutta and Sengupta (2016); Sengupta (2013)

human babies, the cortisol response to vaccination stress is robust at 2 months of age but dampened in the period between 4 and 18 months, before maturing to adult-like levels after this time (Gunnar et al. 1996; Lewis and Ramsay 1995a, b). As with rats, disrupted maternal care in humans can accelerate the maturation of the axis, as is summarized in Suchecki (2018). In another example of convergent human and rodent postnatal development, human prefrontal cortex connections with the rest of the brain are thought to still be maturing even into late adolescence, meaning children and adolescents are emotionally more reactive and can be vulnerable to disruption of this maturation process by alcohol and other drugs of abuse until this time (Drzewiecki and Juraska 2020). In rodents, analogous connections are remodeling with a similar time frame (Drzewiecki and Juraska 2020).

Pallav Sengupta and their team have put forward one of the few considered attempts to equate human, rat, and mouse developmental ages based on life milestones, broadly likening 1 day in the life of an adult rat to 35 days in a human and 1 day in the life of a mouse to 40 human days (Dutta and Sengupta 2016; Sengupta 2013). Notably, this estimated time equivalence alters somewhat across the lifespan (Table 9.1) with a much greater proportion of the rat and mouse's life spent dependent upon their mother for nutrition, thermoregulation and safety than is seen in humans. Early neuroimmune and neuroendocrine trajectories clearly do not necessarily proceed at the same rate and these assumptions are based on lifespan and selected physical developmental milestones. However, we can very broadly assume that the first week of a rat/mouse's life is approximately equivalent to the third trimester of gestation in humans, the second week is equivalent to the human's first 4 months and the first 40 days is equivalent to the childhood/pre-teen phases, with due consideration of non-linear developmental processes. Other attempts to define equivalence between species have focused on neural development and have concluded a similar time structure (Workman et al. 2013). While rats and mice are the more commonly used experimental subjects in scientific research, emerging alternatives may provide a useful additional perspective. For example, the spiny mouse (*Acomys* species) is becoming commonly used in developmental research because of its precocial development, relatively long gestation period, and small litters, allowing different parallels with humans to be assessed (Box 9.1).

Box 9.1: Spiny Mouse

The spiny mouse (*Acomys* sp.; *A. cahirinus* pictured) has been used in developmental research since at least the 1980s and it continues to be a valuable tool for investigating neurodevelopmental processes with strong temporal parallels to those seen in humans. The desert-origin spiny mice have a gestation period of around 38–45 days, approximately twice as long as that of the rat (*Rattus norvegicus domestica*) or mouse (*Mus musculus*), and they deliver pups that are advanced in their development relative to newborn rats/mice. Organogenesis is complete at birth, as with term humans, but unlike rats and mice. Spiny mice also undergo most of their neurogenesis prior to birth. They thus tend to be more similar to humans around the time of birth than are rats and mice and are useful for investigating the neuronal origins of behavior, of in utero brain development, and of late-pregnancy developmental defects. Other notable advantages for research include model-specificity for near-term birth asphyxia (Hutton et al. 2009; Ireland et al. 2008), a propensity to develop type 2 diabetes (Shafrir et al. 2006), remarkable spontaneous non-scarring skin regeneration (Gaire et al. 2021), paternal investment in rearing the young, and the recent finding that females menstruate as humans do rather than resorb their uterine lining cyclically as in rats and mice (Bellofiore et al. 2017). All of these differences make the spiny mouse a valuable tool for neurodevelopmental and other research focuses (Dickinson et al. 2017; Haughton et al. 2016) and a useful potential model for future studies of postnatal neuroimmune development with relevance to humans.

Photograph courtesy of Prof. David Walker and Dr Bobbi Fleiss, RMIT University, 2021.

9.3 Postnatal Development of Neuroimmune Responses

9.3.1 The Postnatal Neuroimmune Environment Programs Neuroimmune Responses Long-Term

From birth to P14, the rat and mouse neuroimmune and neuroendocrine systems are relatively immature. The neuroendocrine stress axis, the HPA axis, is notably immature at this time. As such, the animal responds to both psychological stress and immune challenge with a dampened HPA axis response relative to the adult, leading to less corticosterone release than in adulthood (Levine 2002). Corticosterone usually acts to curtail nuclear factor (NF)κB-mediated transcription of pro-inflammatory cytokines and so to suppress the immune response (Spencer et al. 2011), reflective of a close relationship between these systems. For instance, a peripheral immune challenge of 100 μg/kg *Escherichia coli* lipopolysaccharide (LPS) is necessary to achieve the same febrile profile in neonates as 50 μg/kg is in adults (Boisse et al. 2004).

This relative immaturity leaves the animal differentially responsive to neuroimmune stimuli from the environment in this early period. As such, rats that are exposed to bacterial endotoxin, LPS or the viral mimetic, polyinosinic: polycytidylic acid (poly i:c), at P14 are hyporesponsive to a homeotypic challenge as adults, having reduced febrile, cyclo-oxygenase-2 (COX2) and cytokine responses (Boisse et al. 2004; Ellis et al. 2005). Early work suggested that an HPA axis mechanism may be at play. Early-life stress and poor maternal care had been shown to modulate later stress responses via modifications to HPA axis negative feedback (Champagne et al. 2008; Champagne and Meaney 2001; Hellstrom et al. 2012) and Hodgson et al. had shown that exposure to *Salmonella enteritidis* every second day for the first week of life also caused exacerbated corticosterone responses to stress in later life. This effect was associated with a two-fold increase in tumor colonization related to an impairment in natural killer cell activity (Hodgson et al. 2001). However, other studies examining adult responses to endotoxin after a neonatal endotoxin challenge were unable to find a direct effect on HPA axis function that might be responsible for the attenuated neuroimmune response (Walker et al. 2006). Evidence now suggests that these suppressed adult responses to a homeotypic immune challenge are likely to be due to the early-life endotoxin leading to a persistent up-regulation of the LPS-receptor, toll-like receptor (TLR)4 (or in the case of poly i:c, TLR3), and to the constitutive expression of COX2 in the liver. These changes represent essentially a priming of peripheral tissues that leads to heightened efficiency of the prostaglandin-mediated response to the LPS, faster activation of the HPA axis, and more robust corticosterone-mediated negative feedback on the transcription of pro-inflammatory cytokines (Mouihate et al. 2010). The basis for these long-term changes may also lie in alterations to the central nervous system's major immune cell population, microglia. Microglia are immature during the perinatal period in the rat and postnatal endotoxin has an enduring effect on these cells, increasing the number of new microglia in the hippocampus, parietal cortex and prefrontal cortex in both the neonatal and adult phases (as assessed with bromodeoxyuridine (BrdU) and ionized calcium-binding adaptor molecule 1 (Iba-1) co-labeling) (Bland et al. 2010). We discuss postnatal microglial development in more detail in Sect. 9.4.

Notably, such a programming effect of LPS on future immune responses is not seen if a single exposure to the LPS challenge is experienced as early as P7; an effect that may be related to the postnatal development of the stress-hyporesponsive period coming into effect after this time. Although multiple endotoxin exposures prior to P7 can stimulate a similar effect to that seen with a single dose at P14 (Walker et al. 2006), a single P7 LPS exposure leads to a lasting reduction in body weight, but no effect on febrile responses (Spencer et al. 2006b). P14 and P21 LPS reduce adult febrile responses to a second challenge of the same type, with corresponding reductions in COX2, and by P28 LPS yields adult-like responses and no lasting programming effect. Similarly, LPS at P14 but not P28 exacerbates later trinitrobenzenesulfonic acid (TNBS)-colitis (Spencer et al. 2007). It is also notable that in contrast to LPS, infection of rat pups on P4 with native *E. coli* sensitizes their febrile responses as adults to LPS and reduces, rather than increases, the corticosterone response (Bilbo et al. 2010). These findings with *E. coli* indicate that different neuroimmune stimuli can have contrasting outcomes for the individual, potentially dependent upon changes to the various TLRs involved.

9.3.2 The Postnatal Neuroimmune Environment Programs Stress and Anxiety

Additionally to a lasting neuroimmune phenotype, neonatal LPS can also lead to long-term anxiety-like behavior and an enhanced corticosterone response to psychological stress. This effect may be mediated by persistent microglial activation since hippocampal microglial soma density is lastingly increased in rats given LPS at P5 and this is associated with reduced open-arm exploration in the elevated plus maze (Sominsky et al. 2012). Notably, these effects are potentially even transmitted to the next generation, with the offspring of neonatally LPS-exposed males and females displaying increased anxiety-like behavior and those of neonatally LPS-exposed females also showing a potentiated corticosterone response to stress. It is likely that at least the anxiogenic effect is due to an impact on parenting behavior, since cross-fostering normalized the behavior (Walker et al. 2012). Neonatal immune challenge can also reprogram the adrenal and central catecholaminergic network, leading to an increase in tyrosine hydroxylase activity in the adrenal gland, locus coeruleus, ventral tegmental area and substantia nigra, with an increase in tyrosine hydroxylase protein and increased phosphorylation of all three serine residues in the N-terminal region of tyrosine hydroxylase, potentially facilitating differential activation of the sympathomedullary system (Ong et al. 2017; Sominsky et al. 2012).

9.3.3 Postnatal Diet and Its Neuroimmune Programming Effects

It is not only direct stimulation of the developing immune system with a bacterial or viral stimulus that can result in long-term changes to neuroimmune function. The environmental and dietary postnatal milieu can have a pronounced long-term impact. Childhood obesity is associated with increased rates of periodontal inflammation and respiratory or chemotherapy-related infections (Falagas and Kompoti 2006). Rats and mice that are overfed as neonates undergo accelerated weight gain that is reminiscent of postnatal catchup growth following in utero growth restriction. They have accelerated maturation of their HPA axes (Boullu-Ciocca et al. 2005) and as adults they have disrupted HPA axis function, including excessive HPA axis responses to restraint stress in females (Cai et al. 2016; Sominsky et al. 2017; Spencer and Tilbrook 2009). They also have over-active responses to challenge with LPS (but not with poly i:c) (Clarke et al. 2012). Mechanistically, these effects are likely to be related to the early overfeeding causing an increased expression of TLR4 in the inguinal fat, and hyper-phosphorylation of inhibitory factor (I)κB, which leads to an overexpression of pro-inflammatory cytokines to the stimulus (Clarke et al. 2012). Postnatally overfed rats also have long-term changes in microglial morphology and function that may contribute to exaggerated responses to immune challenge. As such, rats fed in small litters during lactation, with subsequent increased access to the dam's milk and consequent obesity, have more microglia in the hippocampus and hypothalamus. Furthermore, these microglia are primed to be hyper-reactive to LPS, with *E. coli* LPS triggering microglial proliferation in these regions 24 h later while having no proliferative effect in controls (Cai et al. 2015; Ziko et al. 2014).

9.4 Postnatal Development of Microglia

9.4.1 Microglia Express an Age-Dependent Morphological and Gene Signature Through Postnatal Development

Alongside the developmental programming of peripheral innate immune functionality, the central immune system also undergoes significant postnatal development in rodents. The major immune cell population of the brain, spinal cord, and retina is the microglial cell. Microglia start to develop early in prenatal life, first appearing from a yolk sac origin at around 8 days after conception in mice and 4.5–5 weeks into the pregnancy in humans (Ginhoux et al. 2010; Sominsky et al. 2018a). This proliferation is followed by a second wave of microglia infiltrating from hematopoietic-derived monocytes at around P3 in mice. This latter wave of cells likely contributes to microglia-dependent central nervous system development but then these cells undergo apoptosis such that principally those of yolk sac origin remain into maturity (Askew et al. 2017; Swinnen et al. 2013).

As they develop, the morphology, gene expression, and even primary function of these immune cells change substantially. Microglia in the embryonic rodent brain are ameboid (blob-like) in shape with either no processes or a single stout process and no evidence of ramified branches. On the day of birth in the rodent, the overall morphology is similar to that seen at early ages, but females have more microglia than males in some brain regions (cortex, hippocampal CA3, and amygdala). By P4, ramified microglia start to appear in both sexes, but the number of microglia is now higher in males in most brain regions. By P30 the brain microglia start to take on adult-like appearance and distribution, with more ramified than ameboid cells in most regions and a balance of more microglia in females than in males, a pattern that is largely consistent with that seen in young adulthood (P60; Box 9.2) (Hanamsagar et al. 2017; Matcovitch-Natan et al. 2016; Schwarz et al. 2012). In the adult animal, microglial morphology differs throughout the brain, depending upon the tissue type, region, age, sex, and the presence or absence of an immune challenge or injury (Arcuri et al. 2017; Masuda et al. 2019). For example, they are more elongated in white matter than in gray matter, more populous in females than in males, and more ameboid in response to pathogens or injury than in the healthy brain, as is comprehensively reviewed in (Sominsky et al. 2018a). This observation, that males have more cortical, hippocampal, and amygdala microglia than females in the early postnatal phase was elegantly derived by Schwarz and Bilbo and may explain why males are more vulnerable to immune challenge at this time than females and likely to overexpress central cytokines in response to such a challenge (Schwarz and Bilbo 2012). As microglia develop, they also express an age-dependent gene expression signature (Hanamsagar et al. 2017; Matcovitch-Natan et al. 2016). Microglia present in the young brain have relatively increased expression of genes relating to cell cycle and differentiation. Immediately prior to birth, the expression of neuronal development genes peaks and these are relatively reduced by adulthood (Matcovitch-Natan et al. 2016).

Box 9.2: Microglial Development

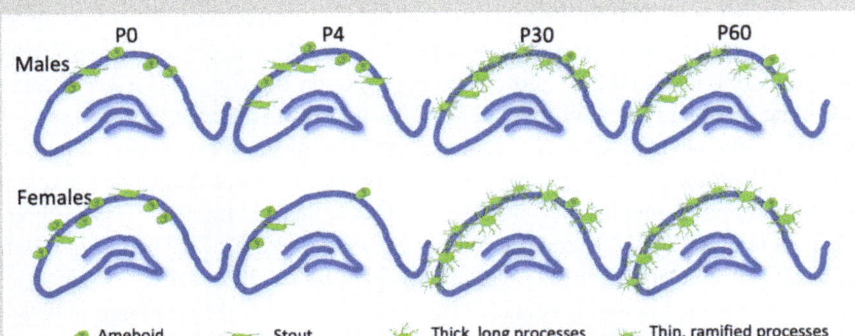

At parturition, microglial morphology is similar to that in embryonic development, but females have more microglia than males in some brain regions. By postnatal day (P)4, ramified microglia are seen in both males

(continued)

Box 9.2 (continued)
and females, but there are now more microglia in males than in females, potentially making males more vulnerable to immune challenge at this time. By P30 the microglia are similar to those seen in adults (P60), with more ramified than ameboid cells and more microglia in females than in males. Adapted from (Schwarz et al. 2012).

9.4.2 Microglia Support Functional Development of the Postnatal Brain

These morphological and gene changes as the microglia mature support different functions as the brain develops. For example, microglia refine neuronal networks by promoting programmed cell death and phagocytosis of apoptotic cells. As cortical neurogenesis proceeds, microglia in this region transition from being ameboid in morphology and phagocytic in activity prenatally to ramified in morphology and promoting of cell survival postnatally (Cunningham et al. 2013; Ueno et al. 2013). In the prenatal cortical neurogenesis phase, microglia actively phagocytose neural precursor cells; microglial suppression or depletion at this time leads to an aberrant increase in the numbers of these cells (Cunningham et al. 2013). Postnatally, on the other hand, microglia promote neuronal white matter survival, as is seen with a reduction in subcortical white matter development in mice deficient in functional microglial elements cluster of differentiation molecule (Cd)11b or C-X3-C motif chemokine receptor 1 (Cx3cr1) (Ueno et al. 2013).

As the brain develops, microglia are also important in eliminating supernumerary synapses and strengthening important synaptic connections, a process that can be influenced by exposure to immune and other challenges at this time (Paolicelli et al. 2011; Stevens et al. 2007). Electron- and stimulated emission depletion microscopy of presynaptic terminal proteins (synaptosomal-associated protein 25, SNAP-25) and excitatory post-synaptic proteins (post-synaptic density 95, PSD-95) located within the cytoplasm of microglial cells in the first postnatal week has revealed that microglia phagocytose synaptic elements, at least in the hippocampus (Paolicelli et al. 2011). In the absence of functional microglia, this pruning does not take place as it should and excitatory synapses are delayed in their maturation (Paolicelli et al. 2011; Zhan et al. 2014). Likewise, during the development of the retina in rodents, both eyes receive inputs from overlapping ganglion cells until a few days after eye-opening at the beginning of the second week after birth. These ganglion cells are then rearranged into eye-specific regions in a process of selective pruning. Microglia selectively prune those projections and synapses in a complement-dependent manner with those synapses that contain complement component q (C1q) and complement 3 (C3) being specifically targeted for elimination (Stephan et al. 2012; Stevens et al. 2007). Emerging work suggests that microglia may accomplish this perinatal synaptic refinement not by a direct phagocytic role, but by reshaping the synaptic material through a process of "nibbling" or "trogocytosis" (Weinhard et al. 2018). Microglia are also likely to be important in synaptic pruning or reshaping in humans. At least, similar timing of microglial gene changes during development and synaptic pruning in the cerebral cortex have been identified in non-human primates (Sasaki et al. 2014a, b). Furthermore, a microglial developmental index of gene expression derived from mice has been established as useful in identifying differences between healthy and diseased human brain samples, and between human males and females, suggesting similar developmental progression (Hanamsagar et al. 2017).

In addition to a programming role for microglia in eliminating unnecessary synaptic connections, these cells are also important in the postnatal phase for synapse formation and maturation, as is reviewed in Sominsky et al. (2018a). In the absence of microglia, the turnover of dendritic spines in the mouse motor cortex pyramidal neurons is disrupted, leading to fewer spines in this region (Parkhurst et al. 2013). This effect is principally driven by microglial brain-derived neurotrophic factor (BDNF), since specific elimination of this factor within microglia recapitulates this effect (Parkhurst et al. 2013). The end result is fewer new dendritic spines in the neonatal cortex, and impaired motor learning and fear-conditioning performance in adulthood (Parkhurst et al. 2013). Specific cytokines are also important in this microglia-dependent synaptic plasticity. For example, interleukin (IL)-1β is a microglia- (and other central nervous system cell-) derived pro-inflammatory cytokine that may contribute to deficits in synaptic plasticity in the hippocampus when microglia are absent, since antagonizing IL-1β signaling rescues these effects (Rogers et al. 2011).

9.4.3 Postnatal Microglial Perturbations Can Program Long-Term Brain Development

These mechanisms all speak to the importance of the neuroimmune milieu in pathways and processes that are not traditionally considered part of the immune system or an immune response. They also illustrate how these systems, already in place to carefully sculpt the developing brain, can be particularly vulnerable to early-life immune challenges. Notably, then, postnatal (and prenatal) immune challenge with LPS, *E. coli*, *S. enteritidis*, or an over-nutrition model, leads to short- and long-term microglial proliferation, pre-disposition to a less ramified morphology, hyper-reactivity to subsequent immune challenge and impairments in synaptic pruning (Bland et al. 2010; O'Loughlin et al. 2017). For example, P4 *E. coli* infection leads to an early increase in the expression of microglial surface antigens that are associated with microglial activation, including Cd11b, C3, and major histocompatibility complex (MHC)II; these are further increased in response to LPS in adulthood. Notably, there are no early-life or long-term alterations in the expression of the astrocyte marker glial fibrillary acidic protein (GFAP) (Bilbo et al. 2005, 2007). Likewise, postnatal LPS leads to persistent increases in microglial soma density in the hippocampus (Sominsky et al. 2012), while postnatal overfeeding leads to early and lasting microglial activation in the paraventricular nucleus of the hypothalamus and in the hippocampus and vulnerability of these microglia to stimulation with further immune challenge (De Luca et al. 2016; Ziko et al. 2014). These morphological and gene expression differences in microglia are associated with functional deficits, including anxiety-like behavior (Sominsky et al. 2012) and cognitive deficits (De Luca et al. 2016). Williamson and colleagues have illustrated that Cd11b-positive cells, i.e. microglia, release IL-1β during hippocampal-dependent learning and that this release is exaggerated in microglial cells from rats that were exposed to an immune challenge during the postnatal period. In vivo, LPS prior to a learning task leads to a similar exaggerated IL-1β response and a memory impairment (Williamson et al. 2011). Interestingly, a transient absence of microglia and their peripheral equivalent, circulating monocytes, in early postnatal life does not appear to impair long-term cognitive hippocampal or microglial development (Soch et al. 2020), although short-term plasticity may be affected (Paolicelli et al. 2011). This work suggests that developmental plasticity of the neuroimmune/neuroendocrine axes may be highly resilient.

9.5 The Positive and Negative of Postnatal Programming

Whether this neonatal programming of differences in the response to immune challenge is a positive or negative adaptation is still unclear. On the one hand, a more efficient immune response might appear to be a positive outcome, with the individual displaying less sickness behavior and less withdrawal from functional behaviors. On the other hand, we also know that a robust febrile response is highly adaptive in the removal of pathogens. Rabbits show increased survival associated with increased elevated temperatures (up to a point) in response to *Pasteurella multocida* challenge (Kluger et al. 1998). Likewise, when individuals cannot mount a febrile response, such as when exothermic animals are maintained in afebrile ambient temperatures, they present with higher pathogen loads and increased mortality (Hart 1988; Kluger et al. 1998). It is likely that fever is useful for creating an environment that is not conducive to pathogen survival (Jiang et al. 2000). Fever also enhances additional host defense strategies including neutrophil phagocytosis of pathogens. There also seem to be some clear detrimental effects of neonatal immune challenge in some cases. For instance, multiple immune challenges with *S. enteritidis* over several days lead to tumor colonization (Hodgson et al. 2001). A similar protocol leads to anxiety and HPA axis hyperactivity (Sominsky et al. 2012). Postnatal overfeeding, leading to lasting hyper-responsiveness to immune challenge with bacterial endotoxin, also appears to be clearly detrimental. These individuals have impaired cognition, fertility, and HPA axis responses, and their altered microglial profile suggests they will be vulnerable to further immune or stress challenges in adulthood (Soch and Spencer 2020).

In a series of experiments in the rat, we attempted to address the question of whether neonatal immune programming of the adult immune response is adaptive or detrimental, by giving adult animals a potentially life-threatening sepsis-like challenge (Spencer et al. 2010). In this work, male rats treated neonatally with *E. coli*

LPS had, respectively, reduced hypothermic and enhanced hyperthermic responses to 1 or 3 mg/kg of the same LPS in adulthood. Since the hypothermic component of the response to a septic dose of endotoxin is thought to be a dysregulated, maladaptive response (Blatteis 2006; Remick and Xioa 2006), it may be argued that the neonatal programming effect was beneficial in this case. However, the effect was not associated with alterations in other indices of health, such as weight loss, food intake or activity (Spencer et al. 2010). It can also be argued that our experimental animals routinely live in quite artificial surroundings, where specific-pathogen-free (SPF) conditions prevent their access to normal immune stimuli that would otherwise provide necessary training to both the acquired and innate immune systems (Pittman 2019). With this in mind, it would seem that neonatal immune challenge under these experimental conditions may be beneficial or normalizing, but this may depend upon the timing and type of neonatal challenge, the sex of the individual and the outcome measured later in life.

9.6 Experimental Considerations

9.6.1 The Maternal-Offspring Unit

One of the major experimental considerations associated with perinatal research is the maternal-offspring unit. In addition to the genetic impact, maternal behavior is a key determinant of lifelong behavior of the offspring. Mothers of all species are somewhat heterogeneous in their parenting style and such differences can manifest in their offspring. A well-known illustration of this is the high-intensity maternal licking-grooming behavior identified by Meaney and colleagues in the early 2000s, which can program HPA axis responsiveness to later stress. High-intensity parenting leads to an increase in the hippocampal expression of nerve growth factor inducible factor A (NGFI-A) during grooming, relative to low-intensity nursing (Hellstrom et al. 2012). This NGFI-A increases histone acetylation of the glucocorticoid receptor (GR), facilitating demethylation and therefore increased expression and activity of the GR and efficient negative feedback in response to HPA axis activation when the individual encounters stress (Champagne et al. 2008; Champagne and Meaney 2001). This postnatal variable of maternal behavior, coupled with maternal physiology (dictating milk composition, thermoregulatory capacity, etc.), antenatal impacts, and indeed the genetic component, means that littermates tend to be more similar than non-littermates by most measures (Zorrilla 1997). As such, littermates represent technical replicates or pseudo-replicates and should not be treated as fully independent experimental subjects.

Most rodents, including mice and rats, are multiparous species and can produce litters of ten or more offspring. Practically and ethically, it is clearly not desirable to discard nine pups in favor of using a single randomly selected representative from each litter (Chhor et al. 2017). Allocating littermates to different experimental outcomes within the same design is a useful strategy (e.g., one to an immunohistochemical analysis, one to analysis of gene expression, one to behavior, etc.).

Alternatively, statistical analysis could incorporate litter as a covariate that is nested under each treatment condition or could include a randomized complete block design to statistically define any effect of litter and account for the variation due to this effect (Lazic and Essioux 2013; Zorrilla 1997). In a between-litter design, it may also be ideal to include at least one "experimental" and one control animal within a litter to ensure that any influence of maternal care is spread evenly across the groups. Likewise, both males and females can be utilized as representatives of their separate experimental groups from the same litter, allowing additional valuable information to be obtained.

It is worth noting that this problem is not only a consideration for work investigating perinatal determinants of lifelong physiology. It is likely that animal suppliers deliver to the consumer groups of animals from the same litters, without necessarily reporting this, since pups are usually weaned into sibling groups and introducing adult rats and mice to new conspecifics is stressful for them, leading to fighting or dominance displays. The litter of origin is rarely reported in studies of adult animals, yet is just as likely to confound these experiments as in perinatal studies, a factor that should be considered going forward. These considerations are discussed in detail in Spencer and Meyer (2017). Consideration of the role of maternal-care and other maternally derived factors should be carefully made when designing studies into early-life programming of developmental factors and these considerations need to be carefully documented in reporting.

9.6.2 Litter Sizes

As described, the postnatal nutritional environment is important in programming long-term neuroimmune function. Postnatal overfeeding can lead to microglial proliferation and priming that lasts into adulthood, as well as hyper-responsiveness to an endotoxin challenge (Clarke et al. 2012; Ziko et al. 2014). Alongside these directly-neuroimmune-related considerations, postnatal overfeeding can influence fertility, cognitive function, satiety signaling, HPA axis reactivity and many other outcomes (Boullu-Ciocca et al. 2005; Morris et al. 2005; Plagemann 2006; Soch and Spencer 2020; Spencer and Tilbrook 2009). While studies on the postnatal nutritional impact of various physiological outcomes have been conducted by deliberately imposing a nutritional stimulus, such as manipulating litter sizes to create small or large litters (reflecting over- or under-nutrition respectively), these conditions can arise spontaneously in a laboratory breeding program and should be considered. Litters that are spontaneously abnormally large or small are likely to impose different nutritional challenges with long-term outcomes that could influence experimental findings. Unusually small litters can also reflect pregnancy issues that may be reflected in differences later on. Appropriately designed studies will use a standardized litter size and preferably report these considerations (e.g., Dinel et al. 2016).

9.6.3 Sex Differences

Despite a historical preference for animal experimentation being conducted on male subjects, there is now an expanding base of evidence that postnatal programming effects on male offspring can be quite different from those on females, and some funding agencies are now mandating that experimenters utilize both sexes (Chap. 10). For example, females show a remarkable resilience to early-life immune challenges in most studies, compared with males. Female microglia mature faster than male microglia do, and their maturation rate (unlike in males) is not affected by immune challenge with LPS (Hanamsagar et al. 2017). Female play behavior is not affected by pre- or postnatal exposure to LPS (unlike in males) (Hoffman et al. 2016; Taylor et al. 2012) and female HPA axis responses to stress are not hyper-activated after prenatal stress exposure (unlike in males) (Bronson and Bale 2014; Mueller and Bale 2008). On the other hand, females show an exacerbated inflammatory response, relative to males, in adulthood following a neonatal inflammatory injury (LaPrairie and Murphy 2007). These differences may be, in part, due to differences in the way responses to inflammatory challenge are mediated between the sexes. For example, in females, the response to mechanical pain hypersensitivity is thought to be of T cell origin, whereas in males it is driven by microglia (Sorge et al. 2015). It is currently recommended and becoming more commonplace that all discovery studies incorporate both males and females and even in cases where phenomenology is similar between the sexes, mechanistic explanations established in one should be verified in the other, as is discussed in detail in (Barrientos et al. 2019).

9.6.4 Gut Microbiome

In an extension of these points, it is only now becoming evident that the gut microbiome has an important role (Chap. 12), both in sculpting postnatal behavior and physiology and in possibly reversing the effects of early-life interventions. The perinatal microbiome can alter microglial maturation (Erny et al. 2015; Matcovitch-Natan et al. 2016; Thion et al. 2018) and can also influence peripheral macrophage development, including the postnatal replacement of tissue-resident macrophages from a bone-marrow-derived population (Bain et al. 2014). The perinatal role of the microbiome in programming the developing immune system is comprehensively reviewed in (Henneke et al. 2021). Rodents are coprophagic, and feces from controls and experimental animals may alter the gut microbiome of cage-mates when consumed. Standard experimental design is currently to house controls and experimental animals together in the same space, if not the same cage after weaning, a strategy that allows the animals to experience identical postnatal husbandry conditions. However, the shared microbiome that eventuates could actually partially mitigate the effects of the early-life interventions (e.g., Doenni et al. 2017). These considerations make it very important that the postnatal housing environment be considered and clearly defined to enable replication.

With these considerations in mind, it has been noted that our current approach to animal experimentation in removing extraneous variables actually leaves us with models that no longer closely represent wild or free-living animals (Pittman 2019). Most experimental animals are raised under SPF conditions, meaning they are not exposed to key developmental pathogens that may stimulate and train their immune systems. Likewise, most experimental animals are raised under uniquely stress-free and stimulation-free conditions, eliminating opportunities for training the HPA axis and limiting cognitive development. These artificial housing conditions are highly likely to chronically influence the animals' development and given the evident importance of the postnatal period in programming long-term neuroimmune function, the absence of such stimulation at this time may limit the applicability of these models to humans. For example, Bilbo and colleagues have demonstrated that wild-caught rats display very different short- and long-term responses to immune challenge and that colonization of pregnant rats with parasitic worms *Hymenolepsis diminuta* ameliorates the neonatal cytokine response to immune challenge with *E. coli* and prevents the microglial priming and cognitive dysfunction that otherwise occurs in adulthood after postnatal immune challenge (Williamson et al. 2016).

9.7 Latest Developments in the Field

9.7.1 Genetic and Dietary Supplements for Modifying Immune Cells

Recent years have seen an explosion of new experimental technologies that are advancing our understanding of how the postnatal environment can influence long-term neuroimmune development. Early genetic modifications to knock out (or in) specific genes of interest were immensely useful in our emerging understanding of their roles. For example, Paolicelli and colleagues used a cx3cr1 knockout mouse model to essentially remove the capacity for microglial activity (Paolicelli et al. 2011). With this model, they demonstrated that the lack of brain cx3cr1 leads to a deficit in synaptic pruning and an increase in long-term depression in the neonatal brain that is restored to normal by adulthood. These findings are indicative that microglial disruption during the developmental period (in utero and in the postnatal period) causes a delay but not a permanent disruption in the maturation of brain circuits (Paolicelli et al. 2011). Yet, Rogers and colleagues have shown long-term deficits in motor learning and hippocampal-dependent learning and memory in the same model, with some long-term deficits in cytokine production as well (Rogers et al. 2011). However, for the study of postnatal development, such knockout tools came with a significant limitation that any role for the gene in earlier developmental trajectories would obscure its involvement in postnatal outcomes. Thus, in the example of the Paolicelli study, the role of microglia in the postnatal period could not be distinguished from that in the antenatal.

This limitation has been overcome for study of adult animals with the advent of conditional knockout models. These models have been used to good effect in examining refined time periods across the lifespan. In a pioneering study, Parkhurst and team used conditional transgenic models to create a sophisticated solution for investigating the role of microglia in brain development without the confounding influence of circulating monocytes (Parkhurst et al. 2013). They initially removed both microglia and monocytes from the system by activating a tamoxifen-inducible cre-mediated recombinase that drove diphtheria toxin receptor (DTR) expression in both microglia and monocytes (*CX3CR1CreER/+:R26DsRed/+*). Since monocytes turn over rapidly and microglia were thought to be relatively more long-lived, the premise here is that the recombination is short-lived in monocytes but retained in microglia. This then allows microglia to be exclusively targeted with diphtheria toxin once sufficient time has elapsed (Parkhurst et al. 2013). With this model, Parkhurst et al demonstrated that microglial depletion from P19 leads to long-term deficits in conditioned fear, spatial memory, and motor learning-dependent synapse formation (Parkhurst et al. 2013). While again useful for study in adults, this work needed at least 18 days allowed for turnover of the monocyte population before microglia could be targeted, limiting the model's usefulness in mouse studies of postnatal development. It is also worth noting that recent data have suggested that microglial turnover, of at least some of the population, may be much faster than previously anticipated. Askew and colleagues have estimated that the entire microglial population could be replaced in a matter of 96 days in rodents, or that there are subpopulations of rapidly replacing microglia (Askew et al. 2017). Parkhurst and colleagues did report ablation of up to 99.1% of microglia in their model based on the presence of CD11b and Iba1 in whole brain and motor cortex, however (Parkhurst et al. 2013). Conditional knockouts targeting Cx3cr1 in the rat and Cx3cr1 or Cd11b in the mouse are also increasingly being utilized, albeit that they influence multiple immune cell types (De Luca et al. 2019).

Dietary models of immune cell manipulation have likewise been limited to the late postnatal phase or later, for the simple reason that mice and rats do not feed independently until this time. The Plexxikon "PLX" colony-stimulating factor 1 receptor (CSF1R)-inhibitors have been well-used in recent years to investigate microglial function in adults (Elmore et al. 2014; O'Neil et al. 2018). However, these have so far been nearly exclusively administered in the diet and so have not been used in studies of perinatal programming. Recent examples where these compounds have been delivered successfully via an intraperitoneal injection and showing that they are as effective in rats as in mice, albeit at different doses (Riquier and Sollars 2020), prove encouraging for the postnatal programming field.

9.7.2 Precision Imaging of Immune Cells

In addition to emerging tools for manipulating immune cells and neurons to determine their functions in postnatal programming, technologies for imaging changes in these cells in real time are emerging (Hutchinson 2020). Photo-stable nanoparticles can now be used to tag antigens of interest. For instance, nanodiamond can be imaged over minutes without deterioration of the signal and can stably emit a light signal for years (unlike traditional fluorescent labels) (Reineck et al. 2016; Reineck and Gibson 2017). Multiplexing can be achieved by utilizing multiple nanoparticle types with different wavelengths as well as different emission decay properties to separate the signals spectrally and temporally.

For cell culture applications the OnCELISA is proving useful in marrying cellular identity, morphological and functional characteristics to their activity, with respect to cytokine release. Here, the principle of the sandwich ELISA is employed, where the cell surface is functionalized with capture antibodies. When the cells are stimulated to release pro-inflammatory cytokines, these cytokines are captured on the cell. The preparation can then be visualized using a detection antibody tagged with a fluorescent label, allowing the identification of which cells are releasing cytokines (Liu et al. 2019). An in vivo analogue of this technology has also been developed in fiber-based sensing and imaging. The same principle applies with a fiber-based sensor bearing capture antibodies, allowing repeated sampling of cytokine release events. Using this technology, Zhang and colleagues were able to establish that IL-1β is released in the hippocampus within one hour of a peripheral LPS exposure and this response is resolved within four hours (Zhang et al. 2018). Fiber-optics approaches are also being developed for imaging specific regions of the brain with additional applications such as brain temperature measurements in the context of a pyrogenic stimulus (Musolino et al. 2016, 2019).

9.7.3 'Omics-Level Understanding of Postnatal Programming Changes

Recent years have seen unprecedented capacity for the generation of "big data" in the emergence of various 'omics technologies. Essentially detailing the entire spectrum of possible outputs, genomics refers to defining the entire genome; transcriptomics, the entire transcriptome; proteomics, all proteins expressed; metabolomics, the metabolic status, etc. While the data sets derived from these techniques are huge, requiring strong bioinformatics strategies, so too is the amount of detail we can obtain. Hanamsagar and colleagues have demonstrated this in the context of postnatal neuroimmune development by using transcriptomic profiling to derive a microglial developmental gene expression program from which they characterized a simplified microglial development index. An initial transcriptomic analysis was obtained from isolated mouse hippocampal microglia using Next-Generation RNA Sequencing. This analysis yielded thousands of genes with expression levels that increased or decreased across development. They then developed a microglia-specific developmental index based on global gene expression patterns by grouping significantly up- or down-regulated genes and scaling. This microglial development index can then be used within species to identify differences between sexes and between naïve and immune-challenged mice, as well as across species, to reveal differences between healthy brain samples and those with diseases such as Alzheimer's or autism (Hanamsagar et al. 2017). Thus, analysis of the whole transcriptome of a cell type is a powerful technique that can yield insight into how these cells develop and how they may be perturbed by external stimuli. Ultimately, this level of detail will allow us to build personalized medicine strategies based around an individual's own profile and responses to stimuli (Pintus et al. 2017). For example, such data are being used practically to help identify specific profiles associated with protective immunogenicity in response to vaccines (Cotugno et al. 2019).

9.8 Perspectives

The historical understanding that the innate immune system is established early in life and develops little throughout maturity is now being overturned by data demonstrating the continued growth and malleability of this system throughout life. Despite this, the perinatal period (in utero and immediately postnatally) remains a significant time of substantial change to this system. As such, it also represents a period of vulnerability when external influences, from immune to dietary, to environmental and psychological, can have lasting effects on how we respond to immune challenge and disease for the rest of our lives. This vulnerability encompasses both peripheral and central aspects of the neuroimmune response with the potential for prolonged changes to both depending upon early life experience. Indeed, our current understanding has broadened from a system-centric view of the body's immune responses to conceptualization of the immune system as being crucially interconnected with other body systems including the central and enteric nervous systems. This multisystem, multicellular interplay layers onto early life experience as the building blocks of the adult individual. Current advances in genetic manipulation, imaging, and computational technologies are giving us insight into how all these systems interact and how they are impacted by programming challenges during the early-life postnatal period. Future directions in the field will undoubtedly lead to personalized understanding and therapeutic interventions based on our biomarker signatures that reflect our early life experience. To achieve this end point it is crucial that we understand the steps in between for both biological sexes and across the lifespan, i.e., how early-life immunological, dietary, environmental, and psychological experience change our immune systems; how these changes persist, and how, if necessary, they can be reversed.

9.9 Key References

Barrientos et al., Brain, Behavior and Immunity, 2019. This article discusses why it is important to include both males and females in all discovery and mechanistic studies.

Schwarz & Bilbo, Hormones and Behavior, 2012. This publication describes how males have more cortical, hippocampal, and amygdala microglia than females in the early postnatal phase, which, in turn, may explain why the former are more vulnerable to immune challenges than the latter during this time period.

Sengupta, International Journal of Preventive Medicine, 2013. This article contains an attempt to equate human, rat, and mouse developmental ages based on life milestones, broadly likening 1 day in the life of an adult rat to 35 days in a human and 1 day in the life of a mouse to 40 human days.

Sominsky et al., The International Journal of Biochemistry and Cell Biology, 2018. This review briefly summarizes that microglia are more elongated in white matter than gray matter, more populous in females than in males, and more ameboid in response to pathogens or injury than in the healthy brain.

Spencer & Meyer, Brain Behavior and Immunity, 2017. This introductory review discusses important considerations relevant to the investigation of perinatal determinants of lifelong physiology in experimental animals.

Suchecki, Journal of Neuroendocrinology, 2018. This review discusses the hypo-responsiveness of the HPA axis to stress throughout the initial suckling period and how disrupted maternal care can accelerate the maturation of the axis in humans.

Acknowledgments This project was supported by funding from a National Health and Medical Research Council Career Development Fellowship II (APP1128646). Spiny mouse photograph, *Acomys cahirinis* courtesy of Prof. David Walker and Dr. Bobbi Fleiss, RMIT University, 2021.

References

Arcuri C, Mecca C, Bianchi R, Giambanco I, Donato R (2017) The pathophysiological role of microglia in dynamic surveillance, phagocytosis and structural remodeling of the developing CNS. Front Mol Neurosci 10:191

Askew K, Li K, Olmos-Alonso A, Garcia-Moreno F, Liang Y, Richardson P, Tipton T, Chapman MA, Riecken K, Beccari S et al (2017) Coupled proliferation and apoptosis maintain the rapid turnover of microglia in the adult brain. Cell Rep 18:391–405

Bain CC, Bravo-Blas A, Scott CL, Perdiguero EG, Geissmann F, Henri S, Malissen B, Osborne LC, Artis D, Mowat AM (2014) Constant replenishment from circulating monocytes maintains the macrophage pool in the intestine of adult mice. Nat Immunol 15:929–937

Barrientos RM, Brunton PJ, Lenz KM, Pyter L, Spencer SJ (2019) Neuroimmunology of the female brain across the lifespan: plasticity to psychopathology. Brain Behav Immun 79:39–55

Bellofiore N, Ellery SJ, Mamrot J, Walker DW, Temple-Smith P, Dickinson H (2017) First evidence of a menstruating rodent: the spiny mouse (Acomys cahirinus). Am J Obstet Gynecol 216:40 e41–40 e11

Bilbo SD, Biedenkapp JC, Der-Avakian A, Watkins LR, Rudy JW, Maier SF (2005) Neonatal infection-induced memory impairment after lipopolysaccharide in adulthood is prevented via caspase-1 inhibition. J Neurosci 25:8000–8009

Bilbo SD, Newsum NJ, Sprunger DB, Watkins LR, Rudy JW, Maier SF (2007) Differential effects of neonatal handling on early life infection-induced alterations in cognition in adulthood. Brain Behav Immun 21:332–342

Bilbo SD, Wieseler JL, Barrientos RM, Tsang V, Watkins LR, Maier SF (2010) Neonatal bacterial infection alters fever to live and simulated infections in adulthood. Psychoneuroendocrinology 35:369–381

Bland ST, Beckley JT, Young S, Tsang V, Watkins LR, Maier SF, Bilbo SD (2010) Enduring consequences of early-life infection on glial and neural cell genesis within cognitive regions of the brain. Brain Behav Immun 24:329–338

Blatteis CM (2006) Endotoxic fever: new concepts of its regulation suggest new approaches to its management. Pharmacol Ther 111:194–223

Boisse L, Mouihate A, Ellis S, Pittman QJ (2004) Long-term alterations in neuroimmune responses after neonatal exposure to lipopolysaccharide. J Neurosci 24:4928–4934

Boullu-Ciocca S, Dutour A, Guillaume V, Achard V, Oliver C, Grino M (2005) Postnatal diet-induced obesity in rats upregulates systemic and adipose tissue glucocorticoid metabolism during development and in adulthood: its relationship with the metabolic syndrome. Diabetes 54:197–203

Bouret SG (2010) Leptin, nutrition, and the programming of hypothalamic feeding circuits. Nestle Nutr Workshop Ser Pediatr Program 65:25–35. Discussion 35–29

Bouret SG, Draper SJ, Simerly RB (2004a) Formation of projection pathways from the arcuate nucleus of the hypothalamus to hypothalamic regions implicated in the neural control of feeding behavior in mice. J Neurosci 24:2797–2805

Bouret SG, Draper SJ, Simerly RB (2004b) Trophic action of leptin on hypothalamic neurons that regulate feeding. Science 304:108–110

Bronson SL, Bale TL (2014) Prenatal stress-induced increases in placental inflammation and offspring hyperactivity are male-specific and ameliorated by maternal antiinflammatory treatment. Endocrinology 155:2635–2646

Brunton PJ, Russell JA (2011) Allopregnanolone and suppressed hypothalamo-pituitary-adrenal axis stress responses in late pregnancy in the rat. Stress (Amsterdam, Netherlands) 14(1):6–12

Cai G, Dinan T, Barwood JM, De Luca SN, Soch A, Ziko I, Chan SMH, Zheng XY, Li S, Molero J et al (2015) Neonatal overfeeding attenuates acute central pro-inflammatory effects of short-term high fat diet. Front Neurosci 8:1–13

Cai G, Ziko I, Barwood J, Soch A, Sominsky L, Molero JC, Spencer SJ (2016) Overfeeding during a critical postnatal period exacerbates hypothalamic-pituitary-adrenal axis responses to immune challenge: a role for adrenal melanocortin 2 receptors. Sci Rep 6:21097

Champagne F, Meaney MJ (2001) Like mother, like daughter: evidence for non-genomic transmission of parental behavior and stress responsivity. Prog Brain Res 133:287–302

Champagne DL, Bagot RC, van Hasselt F, Ramakers G, Meaney MJ, de Kloet ER, Joels M, Krugers H (2008) Maternal care and hippocampal plasticity: evidence for experience-dependent structural plasticity, altered synaptic functioning, and differential responsiveness to glucocorticoids and stress. J Neurosci 28:6037–6045

Chhor V, Moretti R, Le Charpentier T, Sigaut S, Lebon S, Schwendimann L, Ore MV, Zuiani C, Milan V, Josserand J et al (2017) Role of microglia in a mouse model of paediatric traumatic brain injury. Brain Behav Immun 63:197–209

Clarke MA, Stefanidis A, Spencer SJ (2012) Postnatal overfeeding leads to obesity and exacerbated febrile responses to lipopolysaccharide throughout life. J Neuroendocrinol 24:511–524

Cotugno N, Ruggiero A, Santilli V, Manno EC, Rocca S, Zicari S, Amodio D, Colucci M, Rossi P, Levy O et al (2019) OMIC technologies and vaccine development: from the identification of vulnerable individuals to the formulation of invulnerable vaccines. J Immunol Res 2019: 8732191

Cunningham CL, Martinez-Cerdeno V, Noctor SC (2013) Microglia regulate the number of neural precursor cells in the developing cerebral cortex. J Neurosci 33:4216–4233

Dantzer R, O'Connor JC, Freund GG, Johnson RW, Kelley KW (2008) From inflammation to sickness and depression: when the immune system subjugates the brain. Nat Rev Neurosci 9: 46–56

De Luca SN, Ziko I, Sominsky L, Nguyen JC, Dinan T, Miller AA, Jenkins TA, Spencer SJ (2016) Early life overfeeding impairs spatial memory performance by reducing microglial sensitivity to learning. J Neuroinflamm 13:112

De Luca SN, Sominsky L, Soch A, Wang H, Ziko I, Rank MM, Spencer SJ (2019) Conditional microglial depletion in rats leads to reversible anorexia and weight loss by disrupting gustatory circuitry. Brain Behav Immun 77:77–91

Dickinson H, Ellery S, Davies-Tuck M, Tolcos M, Nitsos I, Walker DW, Miller SL (2017) Description of a method for inducing fetal growth restriction in the spiny mouse. J Dev Orig Health Dis 8:550–555

Dinel AL, Rey C, Bonhomme C, Le Ruyet P, Joffre C, Laye S (2016) Dairy fat blend improves brain DHA and neuroplasticity and regulates corticosterone in mice. Prostaglandins Leukot Essent Fatty Acids 109:29–38

Doenni VM, Song CM, Hill MN, Pittman QJ (2017) Early-life inflammation with LPS delays fear extinction in adult rodents. Brain Behav Immun 63:176–185

Drzewiecki CM, Juraska JM (2020) The structural reorganization of the prefrontal cortex during adolescence as a framework for vulnerability to the environment. Pharmacol Biochem Behav 199:173044

Dutta S, Sengupta P (2016) Men and mice: relating their ages. Life sciences 152:244–248

Ellis S, Mouihate A, Pittman QJ (2005) Early life immune challenge alters innate immune responses to lipopolysaccharide: implications for host defense as adults. FASEB J. 19:1519–1521

Elmore MR, Najafi AR, Koike MA, Dagher NN, Spangenberg EE, Rice RA, Kitazawa M, Matusow B, Nguyen H, West BL et al (2014) Colony-stimulating factor 1 receptor signaling is necessary for microglia viability, unmasking a microglia progenitor cell in the adult brain. Neuron 82:380–397

Erny D, Hrabe de Angelis AL, Jaitin D, Wieghofer P, Staszewski O, David E, Keren-Shaul H, Mahlakoiv T, Jakobshagen K, Buch T et al (2015) Host microbiota constantly control maturation and function of microglia in the CNS. Nature neuroscience 18:965–977

Falagas ME, Kompoti M (2006) Obesity and infection. Lancet Infect Dis 6:438–446

Gaire J, Varholick JA, Rana S, Sunshine MD, Dore S, Barbazuk WB, Fuller DD, Maden M, Simmons CS (2021) Spiny mouse (Acomys): an emerging research organism for regenerative medicine with applications beyond the skin. NPJ Regen Med 6:1

Ginhoux F, Greter M, Leboeuf M, Nandi S, See P, Gokhan S, Mehler MF, Conway SJ, Ng LG, Stanley ER et al (2010) Fate mapping analysis reveals that adult microglia derive from primitive macrophages. Science 330:841–845

Grayson BE, Allen SE, Billes SK, Williams SM, Smith MS, Grove KL (2006) Prenatal development of hypothalamic neuropeptide systems in the nonhuman primate. Neuroscience 143:975–986

Gunnar MR, Brodersen L, Krueger K, Rigatuso J (1996) Dampening of adrenocortical responses during infancy: normative changes and individual differences. Child Dev 67:877–889

Hanamsagar R, Alter MD, Block CS, Sullivan H, Bolton JL, Bilbo SD (2017) Generation of a microglial developmental index in mice and in humans reveals a sex difference in maturation and immune reactivity. Glia 65:1504–1520

Hart BL (1988) Biological basis of the behavior of sick animals. Neurosci Biobehav Rev 12:123–137

Haughton CL, Gawriluk TR, Seifert AW (2016) The biology and husbandry of the African spiny mouse (Acomys cahirinus) and the research uses of a laboratory colony. J Am Assoc Lab Anim Sci 55:9–17

Hellstrom IC, Dhir SK, Diorio JC, Meaney MJ (2012) Maternal licking regulates hippocampal glucocorticoid receptor transcription through a thyroid hormone-serotonin-NGFI-A signalling cascade. Philos Trans R Soc Lond B Biol Sci 367:2495–2510

Henneke P, Kierdorf K, Hall LJ, Sperandio M, Hornef M (2021) Perinatal development of innate immune topology. Elife 10

Hodgson DM, Knott B, Walker FR (2001) Neonatal endotoxin exposure influences HPA responsivity and impairs tumor immunity in Fischer 344 rats in adulthood. Pediatr Res 50: 750–755

Hoeijmakers L, Ruigrok SR, Amelianchik A, Ivan D, van Dam AM, Lucassen PJ, Korosi A (2017) Early-life stress lastingly alters the neuroinflammatory response to amyloid pathology in an Alzheimer's disease mouse model. Brain Behav Immun 63:160–175

Hoffman JF, Wright CL, McCarthy MM (2016) A critical period in Purkinje cell development is mediated by local estradiol synthesis, disrupted by inflammation, and has enduring consequences only for males. J Neurosci 36:10039–10049
Hutchinson MR (2020) Science convergence applied to psychoneuroimmunology: the future of measurement and imaging. Brain Behav Immun 88:262–269
Hutton LC, Ratnayake U, Shields A, Walker DW (2009) Neuropathology and functional deficits in a model of birth asphyxia in the precocial spiny mouse (Acomys cahirinus). Dev Neurosci 31: 523–535
Ireland Z, Dickinson H, Snow R, Walker DW (2008) Maternal creatine: does it reach the fetus and improve survival after an acute hypoxic episode in the spiny mouse (Acomys cahirinus)? Am J Obstet Gynecol 198(431):e431–e436
Jiang Q, Cross AS, Singh IS, Chen TT, Viscardi RM, Hasday JD (2000) Febrile core temperature is essential for optimal host defense in bacterial peritonitis. Infect Immun 68:1265–1270
Jisa KA, Clarey DD, Peeples ES (2018) Magnetic resonance imaging findings of term and preterm hypoxic-ischemic encephalopathy: a review of relevant animal models and correlation to human imaging. Open Neuroimag J 12:55–65
Kluger MJ, Kozak W, Conn CA, Leon LR, Soszynski D (1998) Role of fever in disease. Ann N Y Acad Sci 856:224–233
LaPrairie JL, Murphy AZ (2007) Female rats are more vulnerable to the long-term consequences of neonatal inflammatory injury. Pain 132(Suppl 1):S124–S133
Lazic SE, Essioux L (2013) Improving basic and translational science by accounting for litter-to-litter variation in animal models. BMC Neurosci 14:37
Levine S (1957) Infantile experience and resistance to physiological stress. Science 126:405
Levine S (2002) Regulation of the hypothalamic-pituitary-adrenal axis in the neonatal rat: the role of maternal behavior. Neurotox Res 4:557–564
Levine S, Alpert M, Lewis GW (1957) Infantile experience and the maturation of the pituitary adrenal axis. Science 126:1347
Lewis M, Ramsay D (1995a) Stability and change in cortisol and behavioral response to stress during the first 18 months of life. Dev Psychobiol 28:419–428
Lewis M, Ramsay DS (1995b) Developmental change in infants' responses to stress. Child Dev 66: 657–670
Liu G, Bursill C, Cartland SP, Anwer AG, Parker LM, Zhang K, Feng S, He M, Inglis DW, Kavurma MM et al (2019) A nanoparticle-based affinity sensor that identifies and selects highly cytokine-secreting cells. iScience 20:137–147
Masuda T, Sankowski R, Staszewski O, Bottcher C, Amann L, Sagar, Scheiwe C, Nessler S, Kunz P, van Loo G et al (2019) Spatial and temporal heterogeneity of mouse and human microglia at single-cell resolution. Nature 566:388–392
Matcovitch-Natan O, Winter DR, Giladi A, Vargas Aguilar S, Spinrad A, Sarrazin S, Ben-Yehuda H, David E, Zelada Gonzalez F, Perrin P et al (2016) Microglia development follows a stepwise program to regulate brain homeostasis. Science 353:aad8670
Morris MJ, Velkoska E, Cole TJ (2005) Central and peripheral contributions to obesity-associated hypertension: impact of early overnourishment. Exp Physiol 90:697–702
Mouihate A, Galic MA, Ellis SL, Spencer SJ, Tsutsui S, Pittman QJ (2010) Early life activation of toll-like receptor 4 reprograms neural anti-inflammatory pathways. J Neurosci 30:7975–7983
Mueller BR, Bale TL (2008) Sex-specific programming of offspring emotionality after stress early in pregnancy. J Neurosci 28:9055–9065
Musolino S, Schartner EP, Tsiminis G, Salem A, Monro TM, Hutchinson MR (2016) Portable optical fiber probe for in vivo brain temperature measurements. Biomed Opt Express 7:3069–3077
Musolino ST, Schartner EP, Hutchinson MR, Salem A (2019) Minocycline attenuates 3,4-methylenedioxymethamphetamine-induced hyperthermia in the rat brain. Eur J Pharmacol 858:172495

O'Loughlin E, Pakan JMP, Yilmazer-Hanke D, McDermott KW (2017) Acute in utero exposure to lipopolysaccharide induces inflammation in the pre- and postnatal brain and alters the glial cytoarchitecture in the developing amygdala. J Neuroinflamm 14:212
O'Neil SM, Witcher KG, McKim DB, Godbout JP (2018) Forced turnover of aged microglia induces an intermediate phenotype but does not rebalance CNS environmental cues driving priming to immune challenge. Acta Neuropathol Commun 6:129
Ong LK, Fuller EA, Sominsky L, Hodgson DM, Dunkley PR, Dickson PW (2017) Early life peripheral lipopolysaccharide challenge reprograms catecholaminergic neurons. Sci Rep 7: 40475
Oyama K, Padbury J, Martinez A, Chappell B, Stein H, Humme J (1992) Effects of fetal growth retardation on the development of central and peripheral catecholaminergic pathways in the sheep. J Dev Physiol 18:217–222
Paolicelli RC, Bolasco G, Pagani F, Maggi L, Scianni M, Panzanelli P, Giustetto M, Ferreira TA, Guiducci E, Dumas L et al (2011) Synaptic pruning by microglia is necessary for normal brain development. Science 333:1456–1458
Parkhurst CN, Yang G, Ninan I, Savas JN, Yates JR 3rd, Lafaille JJ, Hempstead BL, Littman DR, Gan WB (2013) Microglia promote learning-dependent synapse formation through brain-derived neurotrophic factor. Cell 155:1596–1609
Pintus R, Bassareo PP, Dessi A, Deidda M, Mercuro G, Fanos V (2017) Metabolomics and cardiology: toward the path of perinatal programming and personalized medicine. BioMed Res Int 2017:6970631
Pittman QJ (2019) How to make a better mouse for brain behavior and immunity. Brain Behav Immun 76:1–2
Plagemann A (2006) Perinatal nutrition and hormone-dependent programming of food intake. Horm Res 65(Suppl 3):83–89
Reineck P, Gibson BC (2017) Near-infrared fluorescent nanomaterials for bioimaging and sensing. Adv Opt Mater 5
Reineck P, Francis A, Orth A, Lau DWM, Nixon-Luke RDV, Rastogi ID, Razali WAW, Cordina NM, Parker LM, Sreenivasan VKA et al (2016) Brightness and photostability of emerging red and near-IR fluorescent nanomaterials for bioimaging. Adv Opt Mater 4:1549–1557
Remick DG, Xioa H (2006) Hypothermia and sepsis. Front Biosci 11:1006–1013
Riquier AJ, Sollars SI (2020) Astrocytic response to neural injury is larger during development than in adulthood and is not predicated upon the presence of microglia. Brain Behav Immun Health 1
Rogers JT, Morganti JM, Bachstetter AD, Hudson CE, Peters MM, Grimmig BA, Weeber EJ, Bickford PC, Gemma C (2011) CX3CR1 deficiency leads to impairment of hippocampal cognitive function and synaptic plasticity. J Neurosci 31:16241–16250
Sasaki T, Oga T, Nakagaki K, Sakai K, Sumida K, Hoshino K, Miyawaki I, Saito K, Suto F, Ichinohe N (2014a) Developmental expression profiles of axon guidance signaling and the immune system in the marmoset cortex: potential molecular mechanisms of pruning of dendritic spines during primate synapse formation in late infancy and prepuberty (I). Biochem Biophys Res Commun 444:302–306
Sasaki T, Oga T, Nakagaki K, Sakai K, Sumida K, Hoshino K, Miyawaki I, Saito K, Suto F, Ichinohe N (2014b) Developmental genetic profiles of glutamate receptor system, neuromodulator system, protector of normal tissue and mitochondria, and reelin in marmoset cortex: potential molecular mechanisms of pruning phase of spines in primate synaptic formation process during the end of infancy and prepuberty (II). Biochem Biophys Res Commun 444: 307–310
Schwarz JM, Bilbo SD (2012) Sex, glia, and development: interactions in health and disease. Horm Behav 62:243–253
Schwarz JM, Sholar PW, Bilbo SD (2012) Sex differences in microglial colonization of the developing rat brain. J Neurochem 120:948–963

Semple BD, Blomgren K, Gimlin K, Ferriero DM, Noble-Haeusslein LJ (2013) Brain development in rodents and humans: identifying benchmarks of maturation and vulnerability to injury across species. Prog Neurobiol 106–107:1–16

Sengupta P (2013) The laboratory rat: relating its age with human's. Int J Prev Med 4:624–630

Shafrir E, Ziv E, Kalman R (2006) Nutritionally induced diabetes in desert rodents as models of type 2 diabetes: Acomys cahirinus (spiny mice) and Psammomys obesus (desert gerbil). ILAR J 47:212–224

Soch A, Spencer SJ (2020) Consequences of early life overfeeding for microglia—perspectives from rodent models. Brain Behav Immun 88:256–261

Soch A, Sominsky L, Younesi S, De Luca SN, Gunasekara M, Bozinovski S, Spencer SJ (2020) The role of microglia in the second and third postnatal weeks of life in rat hippocampal development and memory. Brain Behav Immun 88:675–687

Sominsky L, Walker AK, Ong LK, Tynan RJ, Walker FR, Hodgson DM (2012) Increased microglial activation in the rat brain following neonatal exposure to a bacterial mimetic. Behav Brain Res 226:351–356

Sominsky L, Ziko I, Spencer SJ (2017) Neonatal overfeeding disrupts pituitary ghrelin signalling in female rats long-term; Implications for the stress response. PLoS One 12:e0173498

Sominsky L, De Luca S, Spencer SJ (2018a) Microglia: key players in neurodevelopment and neuronal plasticity. Int J Biochem Cell Biol 94:56–60

Sominsky L, Jasoni CL, Twigg HR, Spencer SJ (2018b) Hormonal and nutritional regulation of postnatal hypothalamic development. J Endocrinol 237:R47–R64

Sorge RE, Mapplebeck JC, Rosen S, Beggs S, Taves S, Alexander JK, Martin LJ, Austin JS, Sotocinal SG, Chen D et al (2015) Different immune cells mediate mechanical pain hypersensitivity in male and female mice. Nat Neurosci 18:1081–1083

Spencer SJ, Meyer U (2017) Perinatal programming by inflammation. Brain Behav Immun 63:1–7

Spencer SJ, Tilbrook A (2009) Neonatal overfeeding alters adult anxiety and stress responsiveness. Psychoneuroendocrinology 34:1133–1143

Spencer SJ, Boisse L, Mouihate A, Pittman QJ (2006a) Long term alterations in neuroimmune responses of female rats after neonatal exposure to lipopolysaccharide. Brain Behav Immun 20: 325–330

Spencer SJ, Martin S, Mouihate A, Pittman QJ (2006b) Early-life immune challenge: defining a critical window for effects on adult responses to immune challenge. Neuropsychopharmacology 31:1910–1918

Spencer SJ, Hyland NP, Sharkey KA, Pittman QJ (2007) Neonatal immune challenge exacerbates experimental colitis in adult rats: potential role for TNF-alpha. Am J Physiol 292:R308–R315

Spencer SJ, Mouihate A, Galic MA, Pittman QJ (2008) Central and peripheral neuroimmune responses: hyporesponsiveness during pregnancy. J Physiol 586:399–406

Spencer SJ, Field E, Pittman QJ (2010) Neonatal programming by neuroimmune challenge: effects on responses and tolerance to septic doses of lipopolysaccharide in adult male and female rats. J Neuroendocrinol 22:272–281

Spencer SJ, Galic MA, Pittman QJ (2011) Neonatal programming of innate immune function. Am J Physiol Endocrinol Metab 300:E11–E18

Stephan AH, Barres BA, Stevens B (2012) The complement system: an unexpected role in synaptic pruning during development and disease. Annu Rev Neurosci 35:369–389

Stevens B, Allen NJ, Vazquez LE, Howell GR, Christopherson KS, Nouri N, Micheva KD, Mehalow AK, Huberman AD, Stafford B et al (2007) The classical complement cascade mediates CNS synapse elimination. Cell 131:1164–1178

Suchecki D (2018) Maternal regulation of the infant's hypothalamic-pituitary-adrenal axis stress response: Seymour 'Gig' Levine's legacy to neuroendocrinology. J Neuroendocrinol 30:e12610

Swinnen N, Smolders S, Avila A, Notelaers K, Paesen R, Ameloot M, Brone B, Legendre P, Rigo JM (2013) Complex invasion pattern of the cerebral cortex bymicroglial cells during development of the mouse embryo. Glia 61:150–163

Taylor PV, Veenema AH, Paul MJ, Bredewold R, Isaacs S, de Vries GJ (2012) Sexually dimorphic effects of a prenatal immune challenge on social play and vasopressin expression in juvenile rats. Biol Sex Differ 3:15

Thion MS, Low D, Silvin A, Chen J, Grisel P, Schulte-Schrepping J, Blecher R, Ulas T, Squarzoni P, Hoeffel G et al (2018) Microbiome influences prenatal and adult microglia in a sex-specific manner. Cell 172:500–516 e516

Ueno M, Fujita Y, Tanaka T, Nakamura Y, Kikuta J, Ishii M, Yamashita T (2013) Layer V cortical neurons require microglial support for survival during postnatal development. Nat Neurosci 16: 543–551

Walker FR, Hodyl NA, Krivanek KM, Hodgson DM (2006) Early life host-bacteria relations and development: long-term individual differences in neuroimmune function following neonatal endotoxin challenge. Physiol Behav 87:126–134

Walker AK, Hawkins G, Sominsky L, Hodgson DM (2012) Transgenerational transmission of anxiety induced by neonatal exposure to lipopolysaccharide: implications for male and female germ lines. Psychoneuroendocrinology 37:1320–1335

Weinhard L, di Bartolomei G, Bolasco G, Machado P, Schieber NL, Neniskyte U, Exiga M, Vadisiute A, Raggioli A, Schertel A et al (2018) Microglia remodel synapses by presynaptic trogocytosis and spine head filopodia induction. Nat Commun 9:1228

Williamson LL, Sholar PW, Mistry RS, Smith SH, Bilbo SD (2011) Microglia and memory: modulation by early-life infection. J Neurosci 31:15511–15521

Williamson LL, McKenney EA, Holzknecht ZE, Belliveau C, Rawls JF, Poulton S, Parker W, Bilbo SD (2016) Got worms? Perinatal exposure to helminths prevents persistent immune sensitization and cognitive dysfunction induced by early-life infection. Brain Behav Immun 51:14–28

Workman AD, Charvet CJ, Clancy B, Darlington RB, Finlay BL (2013) Modeling transformations of neurodevelopmental sequences across mammalian species. J Neurosci 33:7368–7383

Zhan Y, Paolicelli RC, Sforazzini F, Weinhard L, Bolasco G, Pagani F, Vyssotski AL, Bifone A, Gozzi A, Ragozzino D et al (2014) Deficient neuron-microglia signaling results in impaired functional brain connectivity and social behavior. Nat Neurosci 17:400–406

Zhang K, Baratta MV, Liu G, Frank MG, Leslie NR, Watkins LR, Maier SF, Hutchinson MR, Goldys EM (2018) A novel platform for in vivo detection of cytokine release within discrete brain regions. Brain Behav Immun 71:18–22

Ziko I, De Luca S, Dinan T, Barwood JM, Sominsky L, Cai G, Kenny R, Stokes L, Jenkins TA, Spencer SJ (2014) Neonatal overfeeding alters hypothalamic microglial profiles and central responses to immune challenge long-term. Brain Behav Immun 41:32–43

Zorrilla EP (1997) Multiparous species present problems (and possibilities) to developmentalists. Dev Psychobiol 30:141–150

Sex Differences in Neuroendocrine-Immune Interactions

10

MiJin Cho, Gladys Shaw, Archana Venkatesan,
and Gretchen N. Neigh

Abstract

The contribution of sex to differences in immune function is integral for our understanding of diseases and disorders; however, systematic study of the contribution of sex to diseases and disorders and assessment of the contributing mechanisms did not develop as an area of study until the early 2000s. Our collective understanding of the influence of sex on immune function and the role of the endocrine system in these interactions is a burgeoning area of study promoted by the 2015 policy changes from the National Institutes of Health (NIH). Females are prone to neuroinflammatory disorders such as Alzheimer's disease, whereas males are more susceptible to peripheral inflammatory diseases such as cardiovascular disease. This chapter provides essential information on the known points of interactions between the immune and endocrine systems and aims to further understand the means by which sex influences these interactions. In addition, this chapter highlights critical aspects of design and interpretation in order to promote rigorous experimental design of future studies aimed at furthering our understanding of sex differences in immune-endocrine interactions.

Keywords

Immune · HPA axis · Sex differences HPG axis · Hormones

M. Cho · G. Shaw · A. Venkatesan · G. N. Neigh (✉)
Department of Anatomy & Neurobiology, Virginia Commonwealth University, Richmond, VA, USA
e-mail: chom6@vcu.edu; shawga@vcu.edu; venkatesana@vcu.edu; gnmccandless@vcu.edu; Gretchen.McCandless@vcuhealth.org

J. P. Konsman, T. M. Reyes (eds.), *Neuroendocrine-Immune System Interactions*,
Masterclass in Neuroendocrinology 13,
https://doi.org/10.1007/978-3-031-21358-8_10

10.1 Introduction

The contribution of sex to differences in immune function permeates our understanding of diseases and disorders; however, systematic study of the contribution of sex to diseases and disorders and assessment of the contributing mechanisms did not develop as an area of study until the early 2000s. Our collective understanding of the influence of sex on immune function and the role of the endocrine system in these interactions is a burgeoning area of study promoted by the 2015 policy changes from the National Institutes of Health (NIH) (Box 10.1). Similar initiatives are evident outside of the USA. In 2010, the Canada Institutes of Health Research (CIHR) implemented a portfolio policy that requires biomedical researchers to address their choice to include or exclude both sexes in all grant proposal types (Johnson et al. 2014), while the European Commission has issued a similar policy statement (European Commission 2020)

Box 10.1: NIH's Mandate on the Study of Sex as a Biological Variable (SABV)

The landmark mandate issued in 2015 presents "The Four Cs of Studying Sex to Strengthen Science" as a way of promoting the inclusion of male and female subjects in preclinical biomedical studies (Clayton and Collins 2014). The four Cs, or the four major pillars of studying **sex as a biological variable (SABV)**, are to consider the male and female sex, collection of data within each sex, characterization of sex differences, and to communicate sex-stratified data (Arnegard et al. 2020). It is critical to creating experimental designs that are consistent with these guidelines to fully encompass any and all sex-specific differences. Up-to-date information on the NIH position regarding the study of SABV and recommended resources can be found at: https://orwh.od.nih.gov/sex-gender/nih-policy-sex-biological-variable

Although the purposeful study of sex differences in diseases and disorders is a relatively new subfield, physicians, scientists, and the general public have long appreciated that women are more likely than men to develop autoimmune disorders such as Multiple Sclerosis, while men are more likely than women to succumb to disorders and diseases associated with inflammation, including heart disease (Shames 2002; Da Silva 1995). Specific to the central nervous system, females are more prone to neuroinflammatory disorders such as Alzheimer's disease and neuropathic pain, with gut microbiome-driven changes in autoimmunity acting as possible modulators (Loram et al. 2012; Markle et al. 2013), whereas males are more likely to develop inflammatory diseases (Mirza et al. 2015). Sex differences in immune responses are even demonstrated in our developing understanding of the sequalae that follow infection with SARS-CoV2, such that men are reported to be more likely to experience acute COVID-related complications due to a generalized uncontrolled cytokine response, also known as a ***cytokine storm*** (Shcherbak et al. 2021), whereas women are more likely to develop prolonged residual symptoms following COVID, perhaps due to residual immune activation (Förster et al. 2022). This chapter provides essential information to understand the known points at which sex influences interactions between the immune and endocrine systems. In addition, we discuss critical aspects of experimental design in order to promote rigorous experimental inquiry for future studies aimed at furthering our understanding of sex differences in neuroendocrine-immune interactions.

10.2 Primer on the Immune System

The immune system protects the organism from a wide range of perturbations including challenges external to the body like viruses and challenges internal to the body such as cancer cells. Like all systems, a delicate balance is required to ensure that the immune system eradicates threats to the organism without excessive collateral damage to the organism. Dysfunction of the immune system is bivalent between immunosuppression which can result in an inability to protect the organism and overactivation of the immune system which can damage the organism through excessive inflammation or autoimmune activation. There are two major branches to the immune system: innate and adaptive (Box 10.2). These systems together protect against both new challenges and recurrent exposures.

Box 10.2: Basic Composition of the Immune System

The two branches of the immune system are: innate and adaptive. The **innate immune system** shows a rapid response at the expense of specificity. The major cells of the innate immune system are epithelial cells, phagocytes (monocytes/macrophages, dendritic cells, neutrophils, microglia) and natural killer cells. The innate immune system responds to both pathogens and tissue damage. The innate response is characterized by an inflammatory response and this arm of the immune system has recently been proposed to also be engaged by non-pathogenic stimuli, including psychological stressors (Barrett et al. 2021; Bekhbat et al. 2019; Neigh and Ali 2016; Priyadarshini and Aich 2012). The **adaptive (acquired) immune system** is the slower portion of the immune response and is highly specific. Adaptive immune responses are facilitated by the actions of T cells and B cells and include both cytotoxic action and antibody production. The adaptive immune system is engaged by vaccination and holds a "memory" of prior immune challenges which enables an accelerated and highly specific response to a repetitive challenge.

The immune system, like the endocrine system, is distributed throughout the body and is capable of autocrine, paracrine, and endocrine communication. Cytokines and chemokines are two major classes of immune molecules which, like hormones, are capable of engaging most cells in all systems of the body. The complex biology of the immune system extends beyond the scope of this chapter. Additional resources to assist with developing an understanding of the immune system are provided in Sect. 10.7.

10.3 Primer on the Hypothalamic-Pituitary-Immune System (HPA) and Glucocorticoid Receptor (GR)

The HPA axis is named from its three major components: hypothalamus, pituitary and adrenals. Exposure to a pharmacological, physical, or physiological stressor causes release of corticotropin-releasing hormone (CRH) from the paraventricular nucleus of the hypothalamus. CRH-receptor binding in the anterior pituitary leads to adrenocorticotropic hormone (ACTH) production and release, which then binds to melanocortin type-2 receptors to promote glucocorticoid production and secretion from the adrenals. The primary glucocorticoid in humans is cortisol, while in most rodents the primary glucocorticoid is corticosterone. Glucocorticoids are transported throughout the body bound to corticosterone-binding globulin, and upon dissociation, the lipophilic nature of glucocorticoids allows them to readily access the intracellular space. Glucocorticoids function through two primary receptors: mineralocorticoid receptors and glucocorticoid receptors (GR). The culmination of the HPA axis response and the impact of HPA axis activation on immune system function is primarily mediated by actions of the GR. GRs are present on nearly every cell of the body and primarily function as transcription factors. The HPA axis response is terminated through negative feedback mediated by GRs at both the cellular level across organ systems and at the system level within the central nervous system. Dysfunction of the HPA axis can occur at the system level through either increased or prolonged activation, stemming from either increased activation of the system or deficits in negative feedback (Kennedy et al. 1988). HPA axis perturbations can also occur at the level of the cell across organ systems through a phenomenon known as **glucocorticoid resistance** (Binder 2009). In short, despite ample presence of circulating glucocorticoids, the receptor can become resistant to activation, frequently through alterations in co-chaperone protein function. An additional area of Further Reading recommended below delves into the critical nature of co-chaperones and glucocorticoid resistance (Bourke et al. 2012).

Glucocorticoids, through binding to GR, can mediate the body's anti-inflammatory and immunosuppressive responses via inhibition of hormones and hormone-like compounds, including prostaglandins and leukotrienes. In contrast, another transcription factor, NFkB, mediates pro-inflammatory responses through increasing the expression of genes that control pro-inflammatory cytokines and chemokines (Liu et al. 2017). Together, the GR and NFkB are capable of **transactivation and transrepression** such that the action of one of these transcription factors can regulate action of the other (McKay and Cidlowski 1999). For instance, the common clinical practice of prescribing corticosteroids to suppress inflammation is a result of the ability of GR to transrepress NFkB and thereby reduce inflammation. This critical crosstalk is also illustrated by the laboratory observation that a lab-based psychological stressor, the **Trier Social Stress Test (TSST)**, a method commonly used to assess reactivity to stress in humans (Heim et al. 2000), is capable of engaging activity of NFkB and promoting the production of systemic cytokines (Bierhaus et al. 2003). This response is typically kept in check, in part through the interactions of NFkB and GR, which prevent the inflammatory response from becoming excessive.

10.4 Sex and the Immune System

10.4.1 Tripartite Relationship Among HPA Axis, Hypothalamic-Pituitary-Gonadal (HPG) Axis, and Immune System

Interactions among key signaling molecules in the HPA axis and those within the HPG axis modify the function of the immune system and underlie a component of sex differences in immune function. Although commonly referred to as sex steroids and implied to be specific to each sex, the primary sex steroids, estrogens, progestins, and androgens, are present and functional in both sexes. Sex steroids interact with the HPA axis from the system level to molecular interactions. For instance, in the proestrus phase, in which serum progesterone and estradiol concentrations are higher, female rats exhibit prolonged CRH expression after exposure to an acute stressor, in comparison to females in diestrus (lower hormone levels) or males (Iwasaki-Sekino et al. 2009). In contrast, using an anxiety-related behavior prone strain of Wistar rats, female animals (in proestrus/estrus) exhibited a more rapid reduction in ACTH and cortisol post-CRH stimulation in comparison to male counterparts (Keck et al. 2002), indicating an important role for strain/genetics in the interactions of sex and HPA responsivity. Additional influence of sex on the HPA axis response occurs through actions of estradiol. Estradiol can impair glucocorticoid negative feedback, as dexamethasone (a synthetic glucocorticoid) was less effective in blocking diurnal and stress-induced HPA activation in ovariectomized females. Further, it was shown that estrogen receptor α acting in the hypothalamus was responsible for the ability of estradiol to prolong the HPA axis response (Weiser and Handa 2009). Other isoforms of estrogen receptors have also been shown to impact HPA regulation, such as estrogen receptor β (Frye et al. 2008; Shaw 2021) and the membrane-bound G-protein coupled estrogen receptor (Zheng et al. 2020). Given the intricate relationship between the HPA axis and the immune system, neuroendocrine effects of sex on the HPA axis can indirectly cause sex-related alterations in the immune response.

This complex multi-system interaction can also be shaped by early life experiences to fuel aberrant stress-related immune responses. Previous life experience shapes HPA axis function and sex is a critical regulator of this process (Bourke et al. 2012). Use of the TSST demonstrates an interaction between sex and stressful life history on the function of the HPA axis. Women with a history of early life stress and depression respond to the TSST with an ACTH and cortisol response six times greater than the control participants, due to HPA axis hyperactivity (Heim et al. 2000). Men with a prior history of depression showed a normal cortisol response to TSST, while women with a prior history of depression demonstrated an atypical response (Bagley et al. 2011). Research also found that prior trauma correlated negatively with basal ACTH concentrations in men but correlated positively in women (DeSantis et al. 2011). Importantly, as mentioned above, the TSST also activates NFkB and an associated cytokine response in peripheral blood mononuclear cells (Bierhaus et al. 2003) and this is dependent on sex, such that after the TSST, men demonstrate greater glucocorticoid sensitivity and reduced cytokine production compared to women (Rohleder et al. 2001). In addition, the effects of the TSST on the inflammatory response are modified by a life history of stress (Pace et al. 2006). Given that the HPG axis is known to interact with the HPA axis, and vice versa, stress-induced alterations in the HPA axis can interact with the function of the HPG axis and collectively, this may impact the immune system. These studies highlight the potential for neuroendocrine remodeling due to early life stress to exert recurrent system-level influences on immune responses to acute stressors throughout the lifespan.

Examination of the influence of adolescent stress on neural and immune endpoints has highlighted the sex-specific implications of environmental perturbations during puberty when the HPA and HPG axes are maturing. Exposure to adolescent stress interacts with sex to alter the adult response to an immune challenge of exposure to **lipopolysaccharide** (LPS). Female rats exposed to adolescent stress manifest an exaggerated neuroinflammatory response in the hippocampus following adult challenge with LPS, whereas male rats exposed to adult LPS following adolescent stress manifest a suppressed HPA axis response to LPS and an exaggerated peripheral inflammatory response (Bekhbat et al. 2019, 2021; Rowson et al. 2019). Further assessment of the pathways integral to these sex differences suggests that alterations in the response of estrogen receptor-alpha are key to the adult effects of adolescent stress in both sexes, despite the divergent phenotypic manifestations (Bekhbat et al. 2019, 2021) (Fig. 10.1).

10.4.2 Sex and the Immune System: Beyond the HPA Axis Connection

The interrelationship of the HPG axis, HPA axis, and immune system are fundamentally important to health and disease, but the HPA axis is not critical to all aspects of sex-immune interactions. Importantly, the immune system plays a key role in mediating brain sexual differentiation (Arambula and McCarthy 2020). For instance, the preoptic area (POA) is highly sexually dimorphic, in part due to differences in the immune system. Male POAs are exposed to higher concentrations of estrogens and androgens during the perinatal period. Estradiol promotes production of cyclooxygenase enzyme (COX2), the rate-limiting enzyme in synthesis of prostaglandin E2 (PGE2) in males. PGE2, an eicosanoid known for cytokine stimulation, subsequently activates a signal transduction pathway that stabilizes

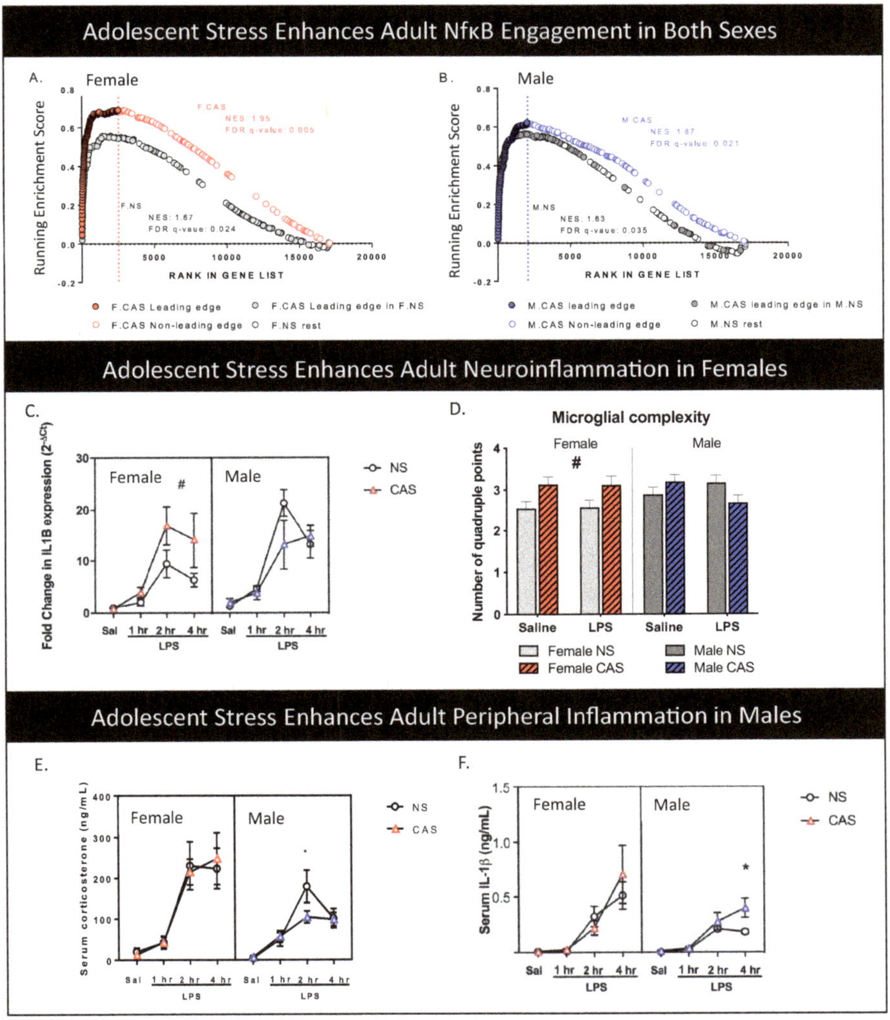

Fig. 10.1 Sex differences in neuroendocrine influences on the immune system are multifaceted, but one multimodal demonstration of the sustained effects of neuroendocrine engagement on immune-related outcomes comes from the work of Bekhbat and colleagues. A series of studies demonstrated that exposure to adolescent stress caused sustained changes in immune-related responses that differed by sex. Rats were exposed to chronic stress during adolescence and given an LPS challenge in adulthood. Panels **a** and **b** demonstrate the effects of a history of adolescent stress on the NFkB response to LPS in adulthood. RNA-Seq was used to assess the impact of an LPS trigger on gene expression in the hippocampus and leading-edge analysis demonstrated that rats with a history of adolescent stress (females in red and males in blue) manifested an exaggerated engagement of NFkB target genes as compared to the response in rats of the same sex but that lacked a chronic stress history. Although both sexes demonstrate an increased engagement of NFkB, the response diverges such that females exhibit enhanced neuroinflammatory markers (panel **c** shows IL-1B gene expression in the hippocampus and panel **d** shows the impact of stress history on microglial complexity) and males demonstrate an enhanced peripheral response (panel **e** shows a suppressed glucocorticoid response to LPS in males with a history of stress that cooccurs with an exaggerated peripheral IL-1B response shown in panel **f**). Collectively this work demonstrates that a neuroendocrine challenge such as chronic stress during development manifests

dendritic spine synapses on POA neurons. Thus, the activity of PGE2 in males allows the spine synapse density per dendrite unit to be almost twice that of females, which can then affect the neuroendocrine capacity of males. Cerebellar development also relies on such neuroendocrine-immune interactions. Male neonates have microglia that are more numerous and in a more activated state than in females (Arambula and McCarthy 2020). In contrast, females possess more numerous and activated microglia during early puberty (Schwarz et al. 2012) and exhibit a more sustained effect of adolescent stress on adult microglial morphology (Bekhbat et al. 2021). In addition, glial cells from male and female organisms respond differently to sex steroids: for example, stimulation with estradiol suppressed LPS-induced microglial IL-1β expression in neonatal males, while the opposite occurs in microglia of female origin (Loram et al. 2012).

Sex steroids can also directly impact immune cells. Lower cytokine production, particularly in male and ovariectomized female rodents, is associated with impaired immune function and decreased survival rates following hemorrhage and sepsis (Choudhry et al. 2005), while exogenous estradiol can combat this immunosuppression in males and ovariectomized females (Choudhry et al. 2007). The influence of estrogen on immune function is at least in part mediated by the ability of estrogen to promote antibody production and inhibit T cell proliferation (McMurray et al. 2001). In contrast, progesterone has an immune influence that differs by sex, such that progesterone mediates inhibition of dendritic cell function to a greater degree in females than in males (Butts et al. 2008). Furthermore, while testosterone and estradiol promote glucocorticoid-induced apoptosis of thymocytes, progesterone inhibits the process.

Much of the developing work related to the interactions among neuroendocrine and immune factors within the brain has been in the context of human immunodeficiency virus (HIV). HIV can be well controlled with antiretroviral therapy, but the chronic presence of foreign viral proteins precipitates chronic inflammation in the absence of viral replication or immunosuppression (Stadtler et al. 2021). Examination of people living with HIV (PLWH) has provided growing insight into the interactions among the neuroendocrine and immune systems in the context of sex. To date, it has been demonstrated that the chronic inflammatory nature of HIV has differential cognitive consequences for men versus women (Rubin et al. 2019), and the mediators of this difference may be a combination of alterations in glucocorticoid function (Bekhbat et al. 2018; Rubin et al. 2020) and sex steroids (Das et al. 2018; Devadas et al. 2018; Maki and Martin-Thormeyer 2009) and include cross-talk between steroid receptors and NfκB, which are capable of both transactivation and transrepression (Arambula and McCarthy 2020; Bekhbat et al. 2021). To illustrate this complexity, both sex and HIV status were found to impact the cytokine and HPA response to an acute stressor. HIV-positive men and women both showed a

Fig. 10.1 (continued) as differential immune outcomes in adulthood as a function of sex (panels **a** and **b** generated from data previously published in Bekhbat et al. 2021; panels **c** and **d** previously published in Bekhbat et al. 2021; panels **e** and **f** previously published in Bekhbat et al. 2019)

diminished TNF-α response to an acute stressor when compared to HIV controls. However, stressor-evoked cortisol increases over time were only observed in HIV+ men, a response that was absent in HIV males and all females regardless of HIV status (Hantsoo et al. 2019). Further study of interactions among glucocorticoids, sex steroids, and immune factors will be needed to understand sex differences in cognitive outcomes in the presence of chronic inflammation. Insight gained from examination of these relationships in the context of HIV will garner benefits for PLWH as well as provide a framework for understanding other conditions of chronic inflammation including aging, obesity, and autoimmune disorders.

10.5 Key Experimental Factors in Consideration of Sex Differences

Understanding similarities and differences between the sexes is of critical importance for the maintenance of health and defense against disease for both sexes. In order to effectively include SABV in scientific studies, an appreciation of the types of **sex differences** (Box 10.3) and awareness of the dimensions in which sex can influence outcomes are essential. Explaining sex differences requires careful consideration of origin of the difference. Sex differences can be precipitated by differences in sex chromosomes. In contrast, sex differences can be mediated by sex steroids either via exposure during development, as **organizational effects**, or due to steroid hormones in adulthood, as **activational effects**. Early sensitive periods, such as perinatal and puberty, determine the long-term effects of hormones and establish the organizational background on which activational effects of sex steroids function in adulthood (Sisk and Zehr 2005). Additional considerations such as sample source are important as peripheral and central nervous system concentrations of sex steroids are not necessarily equivalent or even reflective of one another.

Box 10.3: Classifications of Sex Differences

Sex differences in neuroendocrine-immune (NEI) interactions, particularly in response to stress, can be categorized into one of three classifications: sexual dimorphism, convergence, or quantitative/continuous. Sexual dimorphism refers to a behavioral, physiological, or morphological trait that has two different, exclusive forms: one in males and one in females. This type of trait is present in one sex and absent in the other, including male-exclusive courtship displays and female-specific postpartum aggression (McCarthy et al. 2012). Convergent, or compensatory, sex differences occur when both males and females exhibit similar behaviors or characteristics, but the contributing etiologies and underlying neurophysiology differ. An example of a convergent trait includes the biparental role of prairie voles, which is attributed to the increase in vasopressin in males and hormonal influences of pregnancy in

(continued)

Box 10.3 (continued)
females that lead to caregiver tendencies (Bangasser and Wicks 2017). In quantitative sexual differences, the disparate trait exists along a continuum of which the average male or female response falls separately, with varying degrees of overlap. Examples of quantitative/continuous sex differences include male and female responses to fear and stress, food preferences, and pain sensitivity (McCarthy et al. 2012).

10.5.1 Sex Chromosome Influences on Design

The effect of genetic factors on sex differences is a relatively new area of study, facilitated by the pivotal work of Art Arnold and the development of the **Four-Core-Genotypes** (Arnold 2020). The Four-Core-Genotypes model is a mouse model in which mice have been genetically modified such that the *Sry* gene, which is responsible for testicular development, is moved from the Y sex chromosome to another chromosome. The resulting XX mice develop testes, while XY mice, now lacking *Sry*, develop ovaries. Comparing these Four-Core-Genotypes mice with mice lacking the genetic modification reveals a distinction between sex differences related to sex chromosomes versus those related to gonadal hormones. In other words, if an endpoint differs by gonadal phenotype, then it is driven by sex steroids. However, if it differs by chromosome complement, it is driven by genetic variation. The two modulators may also interact to affect trait expression. Use of this model has begun to illuminate the sex chromosome complement influence versus the sex steroid influence on immune function. The origin of the sex difference of pathogenicity of some viruses has been linked to sex chromosome complement (Robinson et al. 2011) while critical components of immune system development have instead been linked to sex steroids (Ghosh et al. 2021). Continued work in this area will be foundational in developing a full understanding of the origin of sex differences in immune function and understanding the limits of the role of neuroendocrine mediators.

10.5.2 Organizational Influences of Sex Steroids

The organizational impact of steroids during development has a profound influence on sex differences observed in adulthood. Steroid biology changes markedly over the developmental window, including changes in available binding globulins and the dual roles of many steroids as both primary ligands for receptors and metabolic precursors for other steroids. For example, while testosterone has masculinizing effects on the brain and spinal cord of rats and mice, it is the conversion of testosterone to estradiol by aromatization that is responsible for many of these effects (McCarthy et al. 2012).

10.5.3 Activational Influences of Sex Steroid Influences on Experimental Design

Activational effects of sex steroids are those that are generated by the acute presence of the steroid hormone and are largely reversible by the removal of the steroid hormone. For instance, secondary sex characteristics (e.g., facial hair in males) are enhanced by the acute presence of sex steroids but decrease or removal of those sex steroids will modify the characteristic, and therefore the impact of sex steroid variation must be considered in experimental design at least in terms of interpretation of data, even if not fully incorporated into the design of the groups. Any trait that varies with the stages of the female reproductive cycle is necessarily classified as sexually dimorphic, as males lack this long-term cyclic variation. Consequently, the use of an experimental design that does not consider sex as a factor may be incomplete. A two-group design, including complete sample sizes for males and females, can be used to identify similarities and differences between males and females, but it will not account for the changing levels of female sex steroids (estrogen and progesterone) that vary across the female estrus cycle. Although a two-group design provides initial insight into the influence of SABV, interpretations of the data should acknowledge the limitations of a composite female group which includes subjects in varying stages of the estrous or menstrual cycle. A more comprehensive design could use a three-group or five-group design comparing males and females on days two or four of the reproductive cycle, which allows a nuanced understanding of potential sex differences (Becker et al. 2005). Realistically, a robust design of this nature is challenging to fund without demonstration of a sex difference, and the two-group design, with measurement of female sex steroid concentrations and histological assessment of cycle stage is an accepted approach for initial framework building work. Design strategies have also been proposed that allow for the assessment of effect size and power in smaller groups that can be built upon to inform design of more comprehensive studies of sex differences (Diester et al. 2019).

Study of gonadally intact males and females assesses sex-specific traits and behaviors, including endocrine feedback mechanisms and gonadal secretions, in the most translationally relevant manner. However, removal of gonads can facilitate the identification of sex differences as either sex hormone-related or sex chromosome-related. For example, if sex differences continue post-gonadectomy, the source can be narrowed down to organizational hormonal exposure, defined as steroid exposure during development, or variation in sex chromosome complement. Another commonly used method is to provide hormones to gonadectomized animals. To test for a male- or female-biased sex difference, researchers may administer testosterone or estradiol and/or progesterone to both sexes respectively. If the sex differences diminish, then these variations were likely due to adult gonadal steroid hormone levels. If sex differences persist, then the role of developmental steroid exposure and sex chromosome complements should be further explored. It is critical to emphasize that gonadectomy is a contrived experimental condition and interpretations should be limited (e.g., it is not appropriate to equate an ovariectomized female rodent to a menopausal human).

10.5.4 Additional Considerations: Aging and Gender

The global number of people aged 65 and older is expected to double between 2012 and 2050 (United Nations 2019). Sex, defined as the biological differences between men and women, and gender, defined as the sociocultural standards of masculinity and femininity, have both influenced the difference in disease rates between males and females. Adult women are more likely to develop autoimmune disorders than men, while men are more likely to develop inflammatory diseases such as cardiovascular disease. These differences have a plethora of causes, including genetic and microbiome differences between males and females (Markle et al. 2013; Arnold 2020; Bekhbat et al. 2021). The hormone milieu of females, in particular, changes significantly with increasing age. Concentrations of steroid hormones such as estrogen/progesterone and testosterone decrease in females and males, respectively (Neal-Perry et al. 2010). While some studies have highlighted differences between pre- and post-menopausal women, very little is known about changes in hormonal concentrations as they relate to age-related immune changes in males. Possible interventions, such as using testosterone replacement therapy to modulate immune response in aged males, have not been thoroughly explored. There is limited research on the efficacy of immunotherapies for therapeutic use in older men and women to improve cancer and autoimmunity interventions (Bupp et al. 2018) and these will be important areas of continued study.

10.6 Perspectives

Attention to the role of sex in the interactions between the neuroendocrine and immune systems will be critical in the development of robust and reliable interventions for diseases and disorders that engage the immune system. This chapter has identified some of the ways that sex can influence neuroendocrine and immune interactions and highlighted the critical need for consideration of sex in the design and interpretation of research within this field.

10.7 Key References

Bekhbat et al. (2019, 2021) demonstrate the long-standing effect of stress exposure during adolescence on the brain and provide a foundational characterization of the sex differences in manifestation of altered neuroendocrine-immune interactions.

Bourke et al. (2012) reviews the interactions among sex steroids and the glucocorticoid receptor and the critical role of co-chaperones as mediators of these interactions.

Dhabhar (2018) assesses the impact of short-term activation of the endocrine system in the enhancement of the immune system.

McCarthy et al. (2012) provides an accessible framework to categorize differences between the sexes into three categories: sexual dimorphism, sexual convergence, or quantitative/continuous sex differences.

Rohleder et al. (2001) was the first paper to directly explore the intersection of sex, acute stress, and immune reactivity in human subjects.

Kennedy et al. (1988) reviewed the influence that interpersonal relationships have on the impact of short- and long-term stressors on immune reactivity and susceptibility to disease onset.

Bierhaus et al. (2003) was the first to propose a mechanism to connect psychosocial stress to immune reactivity via the NFκB adrenergic signaling pathway.

References

Arambula SE, McCarthy MM (2020) Neuroendocrine-immune crosstalk shapes sex-specific brain development. Endocrinology 161:bqaa055

Arnegard ME, Whitten LA, Hunter C, Clayton JA (2020) Sex as a biological variable: a 5-year progress report and call to action. J Women's Heal 29:858–864

Arnold AP (2020) Four core genotypes and XY* mouse models: update on impact on SABV research. Neurosci Biobehav Rev 119

Bagley SL, Weaver TL, Buchanan TW (2011) Sex differences in physiological and affective responses to stress in remitted depression. Physiol Behav 104:180–186

Bangasser DA, Wicks B (2017) Sex-specific mechanisms for responding to stress. J Neurosci Res 95:75–82

Barrett TJ, Corr EM, van Solingen C, Schlamp F, Brown EJ, Koelwyn GJ, Lee AH, Shanley LC, Spruill TM, Bozal F et al (2021) Chronic stress primes innate immune responses in mice and humans. Cell Rep 36:109595

Becker JB, Arnold AP, Berkley KJ, Blaustein JD, Eckel LA, Hampson E, Herman JP, Marts S, Sadee W, Steiner M et al (2005) Strategies and methods for research on sex differences in brain and behavior. Endocrinology 146:1650–1673

Bekhbat M, Mehta CC, Kelly SD, Vester A, Ofotokun I, Felger J, Wingood G, Anastos K, Gustafson DR, Kassaye S et al (2018) HIV and symptoms of depression are independently associated with impaired glucocorticoid signaling. Psychoneuroendocrinology 96

Bekhbat M, Howell PA, Rowson SA, Kelly SD, Tansey MG, Neigh GN (2019) Chronic adolescent stress sex-specifically alters central and peripheral neuro-immune reactivity in rats. Brain Behav Immun 76:248–257

Bekhbat M, Mukhara D, Dozmorov MG, Stansfield JC, Benusa SD, Hyer MM, Rowson SA, Kelly SD, Qin Z, Dupree JL et al (2021) Adolescent stress sensitizes the adult neuroimmune transcriptome and leads to sex-specific microglial and behavioral phenotypes. Neuropsychopharmacology 46

Bierhaus A, Wolf J, Andrassy M, Rohleder N, Humpert PM, Petrov D, Ferstl R, von Eynatten M, Wendt T, Rudofsky G et al (2003) A mechanism converting psychosocial stress into mononuclear cell activation. Proc Natl Acad Sci USA 100:1920–1925

Binder EB (2009) The role of FKBP5, a co-chaperone of the glucocorticoid receptor in the pathogenesis and therapy of affective and anxiety disorders. Psychoneuroendocrinology 34

Bourke CH, Harrell CS, Neigh GN (2012) Stress-induced sex differences: Adaptations mediated by the glucocorticoid receptor. Horm Behav 62:210–218
Bupp MRG, Potluri T, Fink AL, Klein SL (2018) The confluence of sex hormones and aging on immunity. Front Immunol 9:1269
Butts CL, Bowers E, Horn JC, Shukair SA, Belyavskaya E, Tonelli L, Sternberg EM (2008) Inhibitory effects of progesterone differ in dendritic cells from female and male rodents. Gend Med 5:434–447
Choudhry MA, Schwacha MG, Hubbard WJ, Kerby JD, Rue LW, Bland KI, Chaudry IH (2005) Gender differences in acute response to trauma-hemorrhage. Shock 24:101–106
Choudhry MA, Bland KI, Chaudry IH (2007) Trauma and immune response—Effect of gender differences. Injury 38:1382–1391
Clayton JA, Collins FS (2014) Policy: NIH to balance sex in cell and animal studies. Nature 509: 282–283
Das B, Dobrowolski C, Luttge B, Valadkhan S, Chomont N, Johnston R, Bacchetti P, Hoh R, Gandhi M, Deeks SG et al (2018) Estrogen receptor-1 is a key regulator of HIV-1 latency that imparts gender-specific restrictions on the latent reservoir. Proc Natl Acad Sci USA 115:E7795–E7804
Da Silva JA (1995) Sex hormones, glucocorticoids and autoimmunity: facts and hypotheses. Ann Rheum Dis 54:6
DeSantis SM, Baker NL, Back SE, Spratt E, Ciolino JD, Maria MM, Dipankar B, Brady KT (2011) Gender differences in the effect of early life trauma on hypothalamic–pituitary–adrenal axis functioning. Depress Anxiety 28:383–392
Devadas K, Biswas S, Ragupathy V, Lee S, Dayton A, Hewlett I (2018) Modulation of HIV replication in monocyte derived macrophages (MDM) by steroid hormones. PLoS One 13: e0191916
Dhabhar FS (2018) The short-term stress response—mother nature's mechanism for enhancing protection and performance under conditions of threat, challenge, and opportunity. Front Neuroendocrinol 49:175–192
Diester CM, Banks ML, Neigh GN, Negus SS (2019) Experimental design and analysis for consideration of sex as a biological variable. Neuropsychopharmacology 44
European Commission, Directorate-General for Research and Innovation (2020) Gendered innovations 2: how inclusive analysis contributes to research and innovation: policy review. Publications Office
Förster C, Colombo MG, Wetzel A-J, Martus P, Joos S (2022) Persisting symptoms after COVID-19-prevalence and risk factors in a population-based cohort. Dtsch Arztebl Int 119(10):167–174
Frye CA, Koonce CJ, Edinger KL, Osborne DM, Walf AA (2008) Androgens with activity at estrogen receptor beta have anxiolytic and cognitive-enhancing effects in male rats and mice. Horm Behav 54:726–734
Ghosh MK, Chen KHE, Dill-Garlow R, Ma LJ, Yonezawa T, Itoh Y, Rivera L, Radecki KC, Wu QP, Arnold AP et al (2021) Sex differences in the immune system become evident in the perinatal period in the four core genotypes mouse. Front Endocrinol (Lausanne) 12
Hantsoo L, Kornfield S, Iannelli C, Podcasy J, Metzger D, Sammel MD, Epperson CN (2019) Glucocorticoid–immune response to acute stress in women and men living with HIV. J Behav Med 42:1153–1158
Heim C, Newport DJ, Heit S, Graham YP, Wilcox M, Bonsall R, Miller AH, Nemeroff CB (2000) Pituitary-adrenal and autonomic responses to stress in women after sexual and physical abuse in childhood. JAMA 284:592–597
Iwasaki-Sekino A, Mano-Otagiri A, Ohata H, Yamauchi N, Shibasaki T (2009) Gender differences in corticotropin and corticosterone secretion and corticotropin-releasing factor mRNA expression in the paraventricular nucleus of the hypothalamus and the central nucleus of the amygdala in response to footshock stress or psychological. Psychoneuroendocrinology 34:226–237
Johnson J, Sharman Z, Vissandjée B, Stewart DE (2014) Does a change in health research funding policy related to the integration of sex and gender have an impact? PLoS One 9

Keck ME, Wigger A, Welt T, Muller MB, Gesing A, Reul JMHM, Holsboer F, Landgraf R, Neumann ID (2002) Vasopressin Mediates the response of the combined dexamethasone/CRH test in hyper-anxious rats: implications for pathogenesis of affective disorders. Neuropsychopharmacology 26:94–105

Kennedy S, Kiecolt-Glaser JK, Glaser R (1988) Immunological consequences of acute and chronic stressors: mediating role of interpersonal relationships. Br J Med Psychol 61(Pt 1):77–85

Liu T, Zhang L, Joo D, Sun SC (2017) NF-κB signaling in inflammation. Signal Transduct Target Ther 2

Loram LC, Sholar PW, Taylor FR, Wiesler JL, Babb JA, Strand KA, Berkelhammer D, Day HEW, Maier SF, Watkins LR (2012) Sex and estradiol influence glial pro-inflammatory responses to lipopolysaccharide in rats. Psychoneuroendocrinology 37:1688–1699

Maki PM, Martin-Thormeyer E (2009) HIV, cognition and women. Neuropsychol Rev 19

Markle JGM, Frank DN, Mortin-Toth S, Robertson CE, Feazel LM, Rolle-Kampczyk U, von Bergen M, McCoy KD, Macpherson AJ, Danska JS (2013) Sex differences in the gut microbiome drive hormone-dependent regulation of autoimmunity. Science 339:1084–1088

McCarthy MM, Arnold AP, Ball GF, Blaustein JD, De Vries GJ (2012) Sex differences in the brain: the not so inconvenient truth. J Neurosci 32:2241–2247

McKay LI, Cidlowski JA (1999) Molecular control of immune/inflammatory responses: Interactions between nuclear factor-κB and steroid receptor-signaling pathways. Endocr Rev 20

McMurray RW, Suwannaroj S, Ndebele K, Jenkins JK (2001) Differential effects of sex steroids on T and B cells: modulation of cell cycle phase distribution, apoptosis and bcl-2 protein levels. Pathobiology 69:44–58

Mirza MA, Ritzel R, Xu Y, McCullough LD, Liu F (2015) Sexually dimorphic outcomes and inflammatory responses in hypoxic-ischemic encephalopathy. J. Neuroinflamm 12:32

Neal-Perry G, Nejat E, Dicken C (2010) The neuroendocrine physiology of female reproductive aging: An update. Maturitas 67:34–38

Neigh GN, Ali FF (2016) Co-morbidity of PTSD and immune system dysfunction: opportunities for treatment. Curr Opin Pharmacol 29:104–110

Pace TWW, Mletzko TC, Alagbe O, Musselman DL, Nemeroff CB, Miller AH, Heim CM (2006) Increased stress-induced inflammatory responses in male patients with major depression and increased early life stress. Am J Psychiatry 163

Priyadarshini S, Aich P (2012) Effects of psychological stress on innate immunity and metabolism in humans: a systematic analysis. PLoS One 7:e43232

Robinson DP, Huber SA, Moussawi M, Roberts B, Teuscher C, Watkins R, Arnold AP, Klein SL (2011) Sex chromosome complement contributes to sex differences in coxsackievirus B3 but not influenza A virus pathogenesis. Biol Sex Differ 2

Rohleder N, Schommer NC, Hellhammer DH, Engel R, Kirschbaum C (2001) Sex differences in glucocorticoid sensitivity of proinflammatory cytokine production after psychosocial stress. Psychosom Med 63:966–972

Rowson SA, Bekhbat M, Kelly SD, Binder EB, Hyer MM, Shaw G, Bent MA, Hodes G, Tharp G, Weinshenker D et al (2019) Chronic adolescent stress sex-specifically alters the hippocampal transcriptome in adulthood. Neuropsychopharmacology 44:1207–1215

Rubin LH, Neigh GN, Sundermann EE, Xu Y, Scully EP, Maki PM (2019) Sex differences in neurocognitive function in adults with HIV: patterns, predictors, and mechanisms. Curr Psychiatry Rep 21

Rubin LH, Langenecker SA, Phan KL, Keating SM, Neigh GN, Weber KM, Maki PM (2020) Remitted depression and cognition in HIV: The role of cortisol and inflammation. Psychoneuroendocrinology 114

Schwarz JM, Sholar PW, Bilbo SD (2012) Sex differences in microglial colonization of the developing rat brain. J Neurochem 120:948–963

Shames RS (2002) Gender differences in the development and function of the immune system. J Adolesc Health 30(4 Suppl):59–70

Shaw GA (2021) Mitochondria as the target for disease related hormonal dysregulation. Brain Behav Immun Heal 18:100350
Shcherbak SG, Anisenkova AY, Mosenko SV, Glotov OS, Chernov AN, Apalko SV, Urazov SP, Garbuzov EY, Khobotnikov DN, Klitsenko OA et al (2021) Basic predictive risk factors for cytokine storms in COVID-19 patients. Front Immunol 12
Sisk CL, Zehr JL (2005) Pubertal hormones organize the adolescent brain and behavior. Front Neuroendocrinol 26:163–174
Stadtler H, Shaw G, Neigh GN (2021) Mini-review: elucidating the psychological, physical, and sex-based interactions between HIV infection and stress. Neurosci Lett 747:135698
United Nations, Department of Economic and Social Affairs, Population Division (2019) World population ageing 2019: highlights (ST/ESA/SER.A/430)
Weiser MJ, Handa RJ (2009) Estrogen impairs glucocorticoid dependent negative feedback on the hypothalamic–pituitary–adrenal axis via estrogen receptor alpha within the hypothalamus. Neuroscience 159:883–895
Zheng Y, Wu M, Gao T, Meng L, Ding X, Meng Y, Jiao Y, Luo P, He Z, Sun T et al (2020) GPER-deficient rats exhibit lower serum corticosterone level and increased anxiety-like behavior. Neural Plast 2020:1–22

Further Recommended Reading

Arnold AP (2020) Four core genotypes and XY* mouse models: Update on impact on SABV research. Neurosci Biobehav Rev 119
Binder EB (2009) The role of FKBP5, a co-chaperone of the glucocorticoid receptor in the pathogenesis and therapy of affective and anxiety disorders. Psychoneuroendocrinology 34
Dhabhar FS (2018) The short-term stress response—mother nature's mechanism for enhancing protection and performance under conditions of threat, challenge, and opportunity. Front Neuroendocrinol 49:175–192
Fajgenbaum DC, June CH (2020) Cytokine storm. N Engl J Med 383
Kennedy S, Kiecolt-Glaser JK, Glaser R (1988) Immunological consequences of acute and chronic stressors: mediating role of interpersonal relationships. Br J Med Psychol 61(Pt 1):77–85
McKay LI, Cidlowski JA (1999) Molecular control of immune/inflammatory responses: Interactions between nuclear factor-κB and steroid receptor-signaling pathways. Endocr Rev 20
Rohleder N, Schommer NC, Hellhammer DH, Engel R, Kirschbaum C (2001) Sex differences in glucocorticoid sensitivity of proinflammatory cytokine production after psychosocial stress. Psychosom Med 63:966–972

Biological Clocks and Immune Function 11

William H. Walker II, O. Hecmarie Meléndez-Fernández, Darius D. Becker-Krail, and Randy J. Nelson

Abstract

Circadian rhythms are internal manifestations of the 24-h solar day that allow synchronization of biological and behavioral processes to the external temporal environment. In mammals, circadian rhythms are generated and sustained by the hypothalamic master clock, or suprachiasmatic nucleus (SCN). Within the SCN, a highly coupled network of rhythmically active neuronal and glial oscillators function to maintain rhythms of approximately 24 h. These circadian rhythms are set to precisely 24 h by exposure to the light-dark cycle or other external synchronizers (*zeitgeber*s). Virtually all aspects of physiology display circadian variation, including immune function. This chapter provides a brief review of the circadian system. We detail the neuroendocrine and autonomic nervous mechanisms by which circadian clocks communicate time of day signals to the immune system and how the immune system can feedback to alter circadian clocks. Next, we describe daily rhythms in immune function and circadian regulation of immune cell trafficking. We illustrate the functional relevance of circadian regulation of immune function via the effects of circadian disruption on disease. Finally, potential pitfalls and future directions in the field are discussed.

Keywords

Circadian rhythms · Circadian clocks · Immune function · Adaptive immune system · Innate immune system · Immune cell trafficking · Circadian disruption

W. H. Walker II (✉) · O. Hecmarie Meléndez-Fernández · D. D. Becker-Krail · R. J. Nelson
Department of Neuroscience, Rockefeller Neuroscience Institute, West Virginia University, Morgantown, WV, USA
e-mail: william.walker2@hsc.wvu.edu

J. P. Konsman, T. M. Reyes (eds.), *Neuroendocrine-Immune System Interactions*, Masterclass in Neuroendocrinology 13,
https://doi.org/10.1007/978-3-031-21358-8_11

11.1 Introduction

As a result of the Earth's 24-h periodic cycling of sunlight and temperature, circadian rhythms are thought to have convergently evolved to internalize this predictable light schedule and temporally regulate individuals' physiology and behavior. By synchronizing physiology with the Earth's solar days, organisms can better anticipate, adapt, and optimize the timing of biological processes and behaviors, potentially reducing energy expenditure and increasing fitness. Given this evolutionary advantage, circadian rhythms are highly conserved and exist across all kingdoms of life - animals (both vertebrates and invertebrates), plants, fungi, protists, archaea, and even bacteria. From redox reactions and gene transcription to sleep-wake and feeding behaviors, 24-h rhythms are essential to orchestrating life on Earth (Bhadra et al. 2017; Edgar et al. 2012; Patke et al. 2020).

In mammals, cellular and system-level rhythms are generated and sustained by an anterior hypothalamic region of the brain named the suprachiasmatic nucleus (SCN). Within the SCN, a highly coupled network of rhythmically active neuronal and glial oscillators function to maintain rhythms of approximately 24-h in the absence of environmental cues; however, for these rhythms to remain aligned with the Earth's periodic cycling of light days and dark nights, the SCN must be entrained through *zeitgebers*, or "time givers" (Welsh et al. 2010). Notably, environmental light serves as the most potent *zeitgeber* for the SCN. Photic information is transduced by the activity of intrinsically light-sensitive retinal ganglion cells in the eye that directly project to the SCN via the retinohypothalamic tract (Berson et al. 2002; Do and Yau 2010; Hastings et al. 2018). This environmental photic information adjusts or entrains the phase of the SCN's rhythmic activity (i.e., when it peaks and troughs within a 24-h period relative to the solar day). The SCN then relays this temporal information throughout the organism by synchronizing and organizing the rhythmic activity of a hierarchy of subsidiary oscillators (e.g., both extra-SCN central oscillators and peripheral oscillators) via hormonal, metabolic and autonomic signaling (Fig. 11.1a) (Mohawk et al. 2012). Although there are many mechanisms in place to facilitate circadian organization of physiology, the primary molecular mechanism of entrainment is thought to occur by phosphorylation-mediated activation of the transcription factor CREB and its subsequent phase shifting of circadian molecular clock gene transcription (see the following studies for a more in-depth overview of this mechanism: Albrecht et al. 1997; Gau et al. 2002; Ginty et al. 1993; Lee et al. 2010; von Gall et al. 1998).

At the molecular level, circadian rhythms are generated and maintained by a complex series of interlocking transcription-translation feedback loops (TTFLs) called the "circadian molecular clock." The molecular clock primarily works to drive 24-h rhythms in gene and protein expression through transcriptional, post-transcriptional, translational, and post-translational mechanisms, in turn temporally organizing nearly all aspects of cellular function (Chaix et al. 2016; Partch et al. 2014; Takahashi 2017). In mammals, transcriptome-wide sequencing studies have revealed that 43% of the rodent genome exhibits circadian rhythmicity, and approximately 80% of genes are thought to be rhythmic in primates (Mure et al. 2018; Zhang

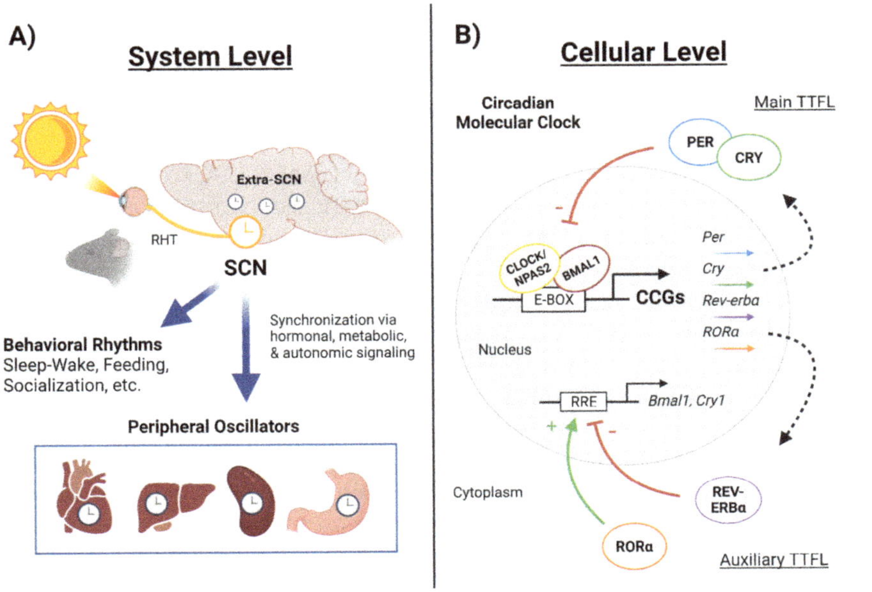

Fig. 11.1 Circadian rhythms are generated and maintained at both the system and cellular levels. (**a**) In mammals, the suprachiasmatic nucleus (SCN) is the principal rhythm-generating endogenous "clock", located in the anterior hypothalamus of the brain. Environmental light activates intrinsically light-sensitive retinal ganglion cells in the eye and this photic information is then relayed directly to the SCN via the retinohypothalamic tract (RHT). Entrained by this light, the SCN then coordinates and synchronizes the rhythmic activity of many central (extra-SCN) and peripheral oscillators (e.g., heart, liver, spleen, GI tract, etc.) through various endocrine, metabolic, and autonomic signaling mechanisms. Together, this hierarchical circadian system temporally regulates nearly all physiology and behavior, from sleep-wake and feeding behaviors down to metabolism and immune function. This temporal organization allows the organism to anticipate, adapt to and survive in the Earth's discrete periods of light and dark. (**b**) In mammals, cellular rhythms are generated by the circadian molecular clock, a complex series of transcription-translation feedback loops (TTFLs). At its core, the transcription factors CLOCK (or its paralogue NPAS2) and BMAL1 heterodimerize and bind to E-Box elements to promote the transcription and expression of hundreds of clock-controlled genes (CCGs). Among these CCGs, CLOCK/NPAS2:BMAL1 drive the expression of PERs and CRYs, which accumulate in the cytoplasm throughout the day. Into the night, PER and CRY will eventually heterodimerize and shuttle back into the nucleus to inhibit their own expression through repressing CLOCK/NPAS2:BMAL1 activity – thus completing the main TTFL. Alongside the main TTFL, CLOCK/NPAS2:BMAL1 also drive the expression of REV-ERBα and RORα as one of the auxiliary TTFLs that help sustain and strengthen the molecular clock. These two nuclear receptors compete at ROR response elements (RRE's) to competitively regulate the transcription and expression of BMAL1, CRY1, and many other CCGs – where RORα promotes expression and REV-ERBα inhibits expression. Taken together, this complex series of interlocking TTFLs cycles roughly every 24-h and temporally controls nearly all aspects of cellular physiology. Figure created with BioRender.com

et al. 2014). Highly conserved across species, many homologous transcription factors have been identified and characterized to serve as the basis of molecular rhythm generation (Dunlap 1999), all sharing a similar function of maintaining oscillating feedback loops that regulate downstream clock-controlled genes (CCGs). Although the molecular clock was first extensively characterized in *Drosophila* (Rosato et al. 2006), seminal work that was awarded the Nobel Prize in Physiology or Medicine in 2017 (Sehgal 2017), here we will primarily focus on the mammalian circadian molecular clock.

In mammals, *circadian locomotor output cycles kaput* (CLOCK), or the CLOCK paralogue *neuronal PAS domain protein 2* (NPAS2), and *brain and muscle ARNT-like protein 1* (BMAL1; encoded by *Arntl*) serve as integral transcription factors that drive the positive arm of the core TTFL (i.e., activators of the system, promoting the transcription of both the negative arm core clock genes, as well as the expression of clock-controlled genes (Fig. 11.1b) (Hogenesch et al. 1997; Ikeda and Nomura 1997; King et al. 1997; Zhou et al. 1997). Notably, CLOCK/NPAS2 and BMAL1 contain basic helix-loop-helix (bHLH) domains that allow for binding to regulatory E-box elements to drive transcription of hundreds of clock-controlled genes (CCGs), genes that are directly or indirectly regulated by the positive arm but do not play a role in the core clock complex. *Period* and *Cryptochrome* serve as the key repressors in the negative arm of the core TTFL (i.e., repressors of the system, genes when translated feedback to inhibit transcription of the positive arm core clock genes.) (Griffin et al. 1999; Shearman et al. 1997). Throughout the day, CLOCK/NPAS2: BMAL1 heterodimerize in the nucleus to drive the transcription and translation of *Per* and *Cry* genes (*Per1, Per2, Per3* & *Cry1, Cry2*), yielding an accumulation of PERs and CRYs in the cytoplasm. Into the night, PERs and CRYs heterodimerize, translocate back into the nucleus, and together repress their own transcription through inhibition of the CLOCK/NPAS2:BMAL1 complex – completing the feedback loop (Shearman et al. 2000). This feedback loop resets upon targeted ubiquitylation-dependent proteasomal degradation of the PERs and CRYs, relieving the CLOCK/NPAS2:BMAL1 complex to resume activity again (Gallego and Virshup 2007). Notably, this self-sustaining TTFL takes roughly 24 h to cycle and serves to bottom-up coordinate rhythmicity in physiology (Partch et al. 2014; Takahashi 2017).

Although CLOCK/NPAS2, BMAL1, PER, and CRY are indeed integral to driving the molecular clock, there are many auxiliary TTFLs that contribute to the stability and robustness of the molecular clock's rhythmicity. The most prominent of these auxiliary TTFLs is the regulation of BMAL1 and CRY1 by *reverse-ErbA alpha* (REV-ERBα; encoded by *Nr1d1*) and *RAR-related orphan receptor alpha* (RORα; encoded by *Nr1f1*). These clock-controlled nuclear receptors compete at retinoic acid-related orphan receptor response elements (RRE) to regulate the expression of BMAL1, CRY1, and many other CCGs – where RORα promotes expression and REV-ERBα represses expression, opposite of each other (Akashi and Takumi 2005; Preitner et al. 2002). Interlaced with the core TTFL, the RORα and REV-ERBα feedback loop runs in antiphase with PER expression and has been shown to both sustain and strengthen the rhythmicity of the circadian molecular clock through regulating BMAL1 activity and subsequently delaying CRY1 expression (Bugge et al. 2012; Cho et al. 2012; Pett et al. 2016; Relógio et al. 2011; Ukai-Tadenuma et al. 2011). This auxiliary TTFL, as well as others, creates a system of redundancy to not only reinforce the rhythmicity of the molecular clock, but also protect its function, such that mutation or loss of clock proteins can be compensated for to minimize interruption to rhythmicity. Importantly, the only clock protein that *cannot* be compensated for is BMAL1; loss of BMAL1 in mice (*Bmal1-/-* knockouts) entirely abolishes rhythms across the molecular, cellular, and behavioral levels (McDearmon et al. 2006). Disruption to circadian rhythms by loss of BMAL1 and subsequent molecular clock function results in decreased activity, disrupted metabolism, impaired immune function, and early mortality (Kondratov et al. 2006; McDearmon et al. 2006; Sun et al. 2006; Yang et al. 2016). These results further underscore the critical importance of circadian rhythmicity of physiology for survival.

Taken together, circadian rhythms are integral to life on Earth and are regulated by a dynamic, highly coordinated system of rhythmic signaling entrained to the environmental light-dark cycle. The SCN is the central endogenous clock in mammals that serves to transduce environmental light information to generate and synchronize rhythms in physiology at the system level all the way down to the cellular level via circadian molecular clock function. While there is extensive literature on the many physiological processes regulated by this pervasive endogenous time-keeping system, here we will cover (1) the current understanding of how the circadian system temporally controls immune function in mammals through neuroendocrine and autonomic nervous system mechanisms and (2) the consequences of disrupting this circadian regulation, as demonstrated in both clinical and preclinical studies.

11.2 Rhythms in Immune Function

More than 60 years ago studies began to demonstrate diurnal variations in host response to lethal immune challenges (Halberg et al. 1960; Shackelford and Feigin 1973). More recently, however, the mechanisms underlying this response have been uncovered. To date, there is evidence for rhythms in virtually every immune cell type and aspect of immune function (Labrecque and Cermakian 2015; Man et al. 2016; Scheiermann et al. 2018; Waggoner 2020). Studies report circadian oscillations in immune cell trafficking, phagocytosis, secretion of complement, production of chemokines/cytokines and histamine, and expression of pattern recognition receptors (He et al. 2018; Keller et al. 2009; Man et al. 2016; Nakamura et al. 2017; Scheiermann et al. 2012, 2018; Silver et al. 2012). Furthermore, the functionality of these oscillations has become apparent as there are clear time-of-day effects in susceptibility to bacterial infections, vulnerability to viral infections, response to vaccination, and propensity to tumor development (Cermakian et al. 2021; Halberg et al. 1960; He et al. 2018; Hrushesky et al. 1999; Long et al. 2016; Shackelford and Feigin 1973; Zhuang et al. 2017).

Circadian rhythms in immune function are well described in both the innate and adaptive immune systems. For example, the circadian clock regulates the inflammatory immune response in macrophages (Keller et al. 2009). Also, isolated splenic macrophages demonstrate oscillations in TNF-α and IL-6 secretion following *ex vivo* treatment of LPS. Peak expression occurred following isolation during the subjective day ~ CT8-CT12 (i.e., 8–12 h into the inactive phase) with a ~twofold change from peak to nadir (Keller et al. 2009). Additionally, isolated splenic macrophages from adrenalectomized mice display oscillations in TNF-α and IL-6 secretion demonstrating that glucocorticoid secretion was not responsible for the daily variation in cytokine production. Rather, the authors concluded that the cell-intrinsic clock within splenic macrophages likely governs these oscillations. The authors also reported that ~8% of the macrophage transcriptome oscillates across the day, including genes that regulate pattern recognition receptors and cytokine secretion (Keller et al. 2009).

Additional studies have demonstrated the importance of the cell-intrinsic clock within macrophages. For instance, BMAL-1 regulates macrophage IL-1β production (Early et al. 2018). Macrophage-specific BMAL-1 knockout prevents the oscillations in endotoxin-induced cytokine response in vivo and in cultured cells (Gibbs et al. 2012). Administration of synthetic nuclear receptor subfamily 1 group D member 1 (NR1D1 a.k.a REV-ERBα) ligand to cultured macrophages suppresses IL-6 secretion following LPS treatment (Gibbs et al. 2012). The proportion of macrophages that produce the pro-inflammatory cytokine IL-12 in response to an LPS challenge is dependent on the phase of the circadian clock (Allen et al. 2019). Specifically, Allen and colleagues have demonstrated that the relative expression of *Dbp* and *Nfil3* determines the cells that produce IL-12 following LPS challenge (Allen et al. 2019). The importance of clock genes in regulating immune function is not a macrophage-specific trait. Indeed, *Per 1* knockout mice display significantly altered rhythms of IFN-γ and cytolytic factor (e.g., granzyme B/perforin) secretion in splenic natural killer (NK) cells. Furthermore, BMAL-1 is necessary for optimal dendritic cell (DC) function as DC-specific BMAL-1 knockout mice lose time-of-day differences in gut mucosa parasite clearance (Hopwood et al. 2018). The circadian clock is also necessary for optimal mast cell function. Loss of *CLOCK* within bone marrow-derived mast cells abolishes rhythms in histamine production across the day (Nakamura et al. 2017). Additionally, desynchronization of the clock via aberrant light/dark cycles or chronic restraint stress prevents daily oscillations in plasma histamine levels (Nakamura et al. 2017).

For an adaptive immune response to be generated optimal interaction between antigen-presenting cells and lymphocytes must occur within lymph nodes, the timing of which has demonstrated functional relevance. For example, intravenous injections of ovalbumin (OVA) peptide-loaded DCs during the day increased the number of OVA-specific CD8+ T cells relative to nighttime injections in mice (Fortier et al. 2011). In addition, immunizing mice with a toll-like receptor 9 ligand (TLR9) at a time of enhanced TLR9 responsiveness (i.e., when peritoneal macrophages expressed the highest amount of TLR9) results in an improved adaptive immune response as measured by lymphocyte proliferation within the lymph nodes (Silver et al. 2012). The cell-intrinsic clock is also necessary for optimal T cell function. Indeed, *CLOCK* mutant mice display blunted T cell proliferation following stimulation of the T cell receptor (Fortier et al. 2011). CD8+ T cell specific BMAL-1 knockout abolishes time of day differences in CD8+ T cell response following vaccination with OVA peptide-loaded DCs (Nobis et al. 2019). Additionally, in a mouse model of experimental autoimmune encephalomyelitis (EAE), T cell specific BMAL-1 deletion prevented the previously reported time-of-day differences in disease severity, which is dependent on the timing immunization in mice (i.e., mice immunized during the late light phase have accelerated disease progression) (Druzd et al. 2017). Rhythms in B cell function are less studied and remain an area of active investigation. However, studies have demonstrated time-of-day differences in B cell trafficking to lymph nodes. B cell-specific BMAL-1 knockout abolishes daily oscillations in B cell trafficking to lymph nodes (Druzd et al. 2017). Furthermore, similarly to T cells, timing of lymph node trafficking has functional significance, as immunization during the period of lymphocyte accumulation in LNs enhances antibody responses (Suzuki et al. 2016).

11.3 Clock Signaling to the Immune System (Autonomic and Endocrine Systems)

As detailed above, the mammalian circadian clock (SCN) integrates signals from photic and other visual cues along with neural and environmental inputs to coordinate both central and peripheral downstream outputs in physiology and behavior. As a result, the SCN communicates locally by secreting neuropeptides, neurotransmitters (Kalsbeek and Buijs 2002; Reghunandanan and Reghunandanan 2006), and via direct axonal projections primarily within other hypothalamic nuclei that consequently relay molecular signals (Gamble et al. 2014; Hastings et al. 2007; Tonsfeldt and Chappell 2012) to specific brain regions that then facilitate the circadian outputs in physiology and behavior (Inouye and Kawamura 1979; Meyer-Bernstein et al. 1999; Silver et al. 1996). However, despite the vast influence of the SCN on multiple physiological rhythms, its immediate outputs are few (Vujovic et al. 2015). The SCN primarily communicates within the hypothalamus, through direct innervation of the subparaventricular zone (SPZ) (Abrahamson and Moore 2001; Kriegsfeld et al. 2004; Swanson and Cowan 1975), and both direct and indirect innervation of the dorsomedial (DMH) nuclei (Vujovic et al. 2015). From these regions, specialized effector neurons relay the cellular message to other brain regions controlling rhythmic behaviors. The dorsal SPZ projects to the medial preoptic area which regulates rhythms in body temperature, whereas the ventral SPZ projects to the DMH, along with GABAergic inputs from the SCN, triggering humoral and neurotransmitter release to regions governing rhythms in corticosteroid production (medial paraventricular nucleus; PVN), arousal states (lateral hypothalamic area) (Berthoud and Münzberg 2011; Leibowitz 1970), and sleep (ventrolateral preoptic area) (Arrigoni et al. 2019; Asala et al. 1990; Lu et al. 2001; Mondino et al. 2021). Thus, lesions to the DMH impair all of these biological rhythms (Chou et al. 2003). Altogether, these hypothalamic nuclei work in concert to maintain organism-wide energetic homeostasis.

The SCN also mediates neuroendocrine (Nader et al. 2010) and autonomic control (Dibner et al. 2010), primarily via GABAergic (Kalsbeek et al. 2006; Wang et al. 2003), and vasoactive intestinal peptide-mediated signaling to the PVN (Jones et al. 2021). These neurons project to corticotropin-releasing hormone (CRH)-containing neurons of the PVN (Vrang et al. 1995), which in turn, extend to the median eminence, and secrete CRH to the hypophyseal portal system. These facilitate the release of adenocorticotropic hormone (ACTH) from the anterior pituitary. Circulating ACTH then stimulates the adrenal secretion of corticosterone (Buijs et al. 2003b). Corticosterone, a "stress" hormone, can modulate the immune response by reducing inflammation; however, sustained or dramatically elevated corticosterone levels may impair both innate (Gao et al. 2017) and inducible (Stier et al. 2009) immune responses. Further, glucocorticoids entrain peripheral oscillators by binding to the promoter region, the glucocorticoid response element (GRE), resulting in shifted clock gene expression in these tissues (Ramakrishnan et al. 2002). Immune cells are one such example; cytokine production, leukocyte distribution, cell proliferation, and apoptosis have been proved to be regulated by circadian variation in glucocorticoid concentrations (Fu and Lee 2003).

Other targets of SCN projections include oxytocin-containing neurons of the PVN. These, along with CRH-producing neurons, mediate autonomic inputs to the periphery through direct projections to sympathetic preganglionic neurons of the intermediolateral column of the spinal cord and in the dorsal motor nucleus of the vagus (Buijs et al. 2003a; Dibner et al. 2010). Sympathetic innervation of the spleen and bone marrow elicits a rhythmic norepinephrine (NE) release, which triggers a similarly rhythmic immune cell response (Bellinger et al. 1993; Cano et al. 2001; Dokur et al. 2004; Elenkov et al. 2000; Pick et al. 2019). Indeed, various studies have demonstrated that severing or otherwise losing splenic or bone marrow sympathetic tracts impair the rhythmic immune response, in both rodents (Logan et al. 2011; Rosas-Ballina et al. 2008; Pick et al. 2019) and humans (Hoover et al. 2017), further supporting the understanding that sympathetic innervation mediates the NE-dependent circadian immune cell activity.

Perhaps the most distinct and well-known SCN-mediated autonomic output is pineal melatonin release. Melatonin is a sleep-promoting hormone, primarily by signaling night or "dark" information. During daytime, SCN release of GABA to the PVN inhibits PVN neuronal activity; however, during the nighttime, a multisynaptic pathway (SCN → PVN → intermediolateral cell column of the spinal cord → superior cervical ganglion → pineal gland) stimulates melatonin release (Kalsbeek et al. 2000). Removal of the GABAergic 'brakes' on the PVN allows the PVN to signal to the intermediolateral cell column of the spinal cord. Next, preganglionic NEergic neurons project to the superior cervical ganglion, which subsequently innervate the pineal gland. NE from these fibers then promote the biosynthesis of melatonin (Fig. 11.2). As it is synthesized by the pinealocytes, melatonin is released into the blood and CSF (Reiter et al. 2014). Melatonin serves as the endogenous ligand for two G-protein coupled receptors, MT1 and MT2, and binds to these receptors both peripherally and centrally to carry out its diverse actions. Specifically relevant to this chapter, MT2 receptors are present in the spleen and MT1 receptors

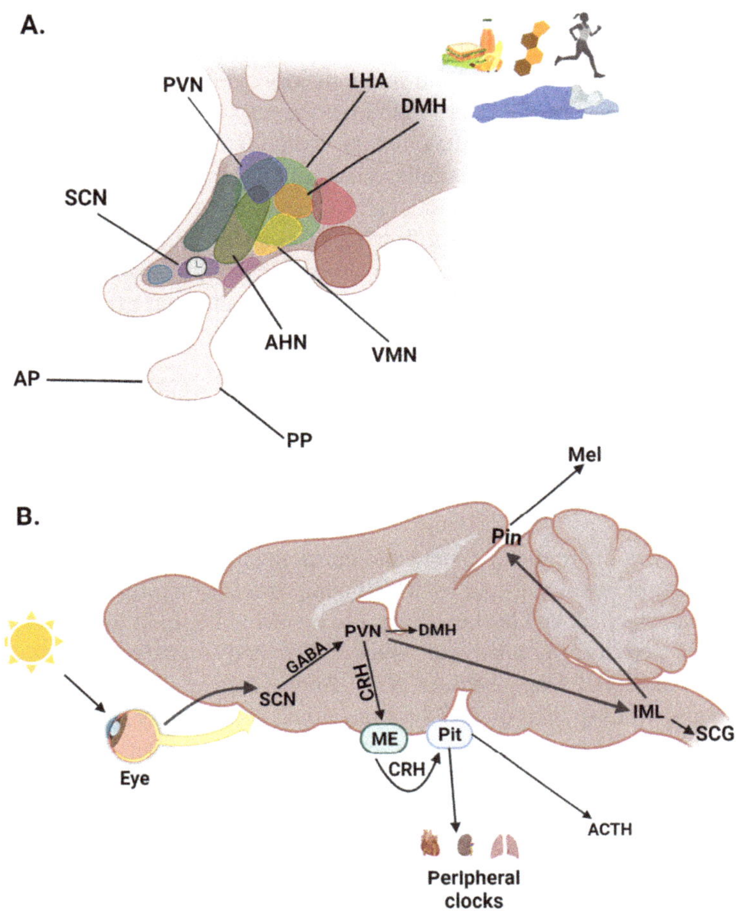

Fig. 11.2 Hypothalamic nuclei, primary neuroendocrine pathways, and output rhythms. (**a**) The suprachiasmatic nucleus (SCN) is the central circadian clock regulator, which projects to other hypothalamic nuclei to initiate the signaling of autonomic and neuroendocrine cascades, concluding in rhythms in glucocorticoid production, temperature, activity, sleep-wake, and feeding. (**b** The paraventricular nucleus (PVN) projects to the intermediolateral cell column (IML) of the spinal cord, from where preganglionic NEergic neurons project to the superior cervical ganglion (SCG), which subsequently innervate the pineal gland, causing the release of melatonin. The suprachiasmatic nuclei (SCN) send GABAergic projections to corticotropin-releasing hormone (CRH)-containing neurons of the paraventricular nucleus (PVN). These extend to the median eminence (ME), where they secrete CRH to the hypophyseal portal system. These facilitate the release of adrenocorticotropic hormone (ACTH) from the anterior pituitary (AP). Circulating ACTH then stimulates the adrenal secretion of corticosterone. Other abbreviations: DMH: dorsomedial hypothalamus; VMN: ventromedial nucleus; AHN: anterior hypothalamic area; AP: anterior pituitary: PP: posterior pituitary; Pin: pineal; Mel: melatonin. Figure created with BioRender.com

are found in the thymus, spleen, as well as in CD4 T cells, CD8 T cells, B cells, and monocytes (Slominski et al. 2012; Ren et al. 2017). Melatonin has immunomodulatory effects that can act as pro- or anti-inflammatory agents and has been posited to be an 'immune buffer' acting to promote immune response under immunosuppressive conditions, and as an anti-inflammatory agent under excess immunoreactivity (Carrillo-Vico et al. 2013). For example, melatonin increases survival and improves outcomes in rodent models of septic shock, reduces mortality following viral infections, increases antibody titers following vaccination, and can prevent rejection in experimental models of transplant (for a detailed review on melatonin and the immune system see Carrillo-Vico et al. 2013). Thus, melatonin exerts a wide range of actions on the immune system which are dependent on the immune environment at the time of signaling.

11.4 Crosstalk Between the Immune System and Circadian Clock

Signaling between the immune system and the circadian clock is not unidirectional. There is a clear bidirectional link by which components of the immune system can alter the circadian clock and lead to changes in behavioral output. Indeed, inflammation reduces activity rhythms and alters daily body temperature oscillations (Cavadini et al. 2007; Gegout-Pottie et al. 1999; Laste et al. 2013). Continuous subcutaneous administration of TNFα for three days significantly reduces locomotor activity and prolongs rest time (Cavadini et al. 2007). In addition, administration of lipopolysaccharide (LPS; a cell wall component of gram-negative bacteria), tumor necrosis factor (TNF), or interleukin-1 beta (IL-1β) phase-shifts activity rhythms (M Leone et al. 2012; Marpegán et al. 2005). However, this only occurs when animals are treated during the early active phase (Marpegán et al. 2005), and in the case of LPS is toll-like receptor-4 dependent and can be blocked via pretreatment with a TNF-α soluble receptor I antagonist (M Leone et al. 2012; Paladino et al. 2010).

Immune signaling can alter the circadian system at both the level of the SCN or peripheral clocks. Weekly intraperitoneal (i.p.) LPS administration for 60 days dampened SCN activation following photic stimulation (Palomba and Bentivoglio 2008). In addition, six days of subcutaneous injections of interferon-α (IFN-α; 2 MIU/kg) prevented photic induction of *Per* mRNA in the SCN (Ohdo et al. 2001). Photic induction of *Per* mRNA in the SCN was only abolished when IFN-α was administered at ZT12 and not ZT0 (Ohdo et al. 2001). Furthermore, independent of photic induction, subcutaneous administration of IFN-α altered the rhythms of *Bmal1*, *Clock*, *Per 1*, *Per 2* and *Per 3* mRNA within the SCN. A similar suppression of *Per 1* and *DBP* mRNA within the SCN was described following a single intravenous injection of LPS (1mg/kg) (Okada et al. 2008). In vivo, intravenous LPS administration (100 μg/kg) at ZT14, but not ZT2, or i.p. injection of LPS (5mg/kg) at ZT5 significantly increased SCN neuronal activation (measured via c-Fos) (Beynon and Coogan 2010; Guerrero-Vargas et al. 2014). A similar increase in c-Fos expression within the SCN was reported following a single active phase ICV injection of interferon-γ (IFN-γ) + TNF-α (Sadki et al. 2007). Co-culture of SCN slice preparations with LPS + IFN-γ + TNF-α reduced the frequency, but not amplitude, of the excitatory postsynaptic events (Lundkvist et al. 2002). Similarly, SCN co-cultured with IFN-γ alone lowered the average spike frequency and significantly reduced the Per1-Luc rhythm amplitude in individual SCN neurons (Kwak et al. 2008). The effects of immune signaling are not restricted to SCN neurons as other brain structures demonstrate time of day differences in c-Fos expression following an immune stimulus (Mul Fedele et al. 2020).

Importantly, astrocytes within the SCN respond to immune signals and may serve as an interface for immune-circadian signaling (Leone et al. 2006). TNF-α applied to cultures of SCN astrocytes altered both the phase and amplitude of PER2-Luc expression rhythms (Duhart et al. 2013). ICV injection of conditioned media from TNF-α treated SCN astrocytes at CT15 induced phase delays in activity rhythms and SCN activation in control mice, but not in tumor necrosis factor receptor 1 (TNFR-1) mutant mice (Duhart et al. 2013). Microglia within the SCN also respond to cytokine signaling. ICV injections of IFN-γ + TNF-α increased the number of microglia (measured by F4/80+ and CD11b+) and altered the morphology of microglia within the SCN (Bentivoglio et al. 2006; Deng et al. 2010). The studies examining the effects of immune signals on astrocytes and microglia within the SCN were obtained under rather artificial conditions (i.e., *in vitro* or ICV). Therefore, the physiological relevance remains to be examined, due to the high dose of cytokines or whether SCN glia are altered under basal conditions.

Immune signaling can also alter peripheral clocks. A single i.v. injection of LPS (0.045 μg/kg) is sufficient to transiently synchronize clock gene expression in equine peripheral blood (Murphy et al. 2007). Co-administration of phenylbutazone, a non-steroidal anti-inflammatory drug that inhibits prostaglandin PGE_2 synthesis, and LPS prevented synchronized clock gene expression. *Ex vivo* treatment of peripheral blood mononuclear cells (PBMC) with LPS or PGE_2 did not result in synchronized clock gene expression (Murphy et al. 2007). This suggests that the effects of LPS synchronization of clock genes in peripheral blood may be due to indirect actions on PBMCs. In humans, a single i.v. bolus of LPS (2 ng/kg) substantially (80–90%) reduced clock gene expression (*Clock*, *Cry1-2*, *Per2-3*, *CSNK1ε*, *Rora*, and *Rev-erb*) in peripheral blood leukocytes, neutrophils, and monocytes for up to 17 h post-infusion, irrespective of the time of administration (Haimovich et al. 2010). Similar disruptive effects on peripheral clock genes have been reported in rodent studies. Indeed, a single i.v. injection of LPS (1mg/kg) at ZT1 transiently suppressed clock gene expression (*Per1*, *Per2*, *DBP*) within the liver (Okada et al. 2008). In addition, reduced *Per1* and *Per2* expression in the heart following LPS administration has been reported (Yamamura et al. 2010). Furthermore, subcutaneous administration of IFN-α significantly altered *Per* expression within the adrenal glands and liver at ZT12, but not ZT0 (Ohdo et al. 2001).

Together, these studies suggest that the SCN and peripheral clocks are responsive to immune signaling. However, there are some caveats that should be considered. First, the current data demonstrate that the SCN and peripheral clocks are altered following an immune challenge/cytokine treatment; whether there is neuroimmune modulation of the circadian system under basal conditions remains to be determined. There is clear evidence that the SCN expresses receptors for several cytokines in a resting immune state, and these receptors display daily oscillations in expression (Beynon and Coogan 2010; Sadki et al. 2007). However, much work is needed to discover the role, if any, that these systems may play in maintaining and setting the circadian phase and output. For instance, mice lacking IL-1 or type I IFN receptors demonstrate unaltered body temperature and activity rhythms (Bohnet et al. 2004; Furuzawa et al. 2002). Also, the data demonstrate that the timing of the immune challenge plays an important role in the response of the circadian system, particularly the SCN. The SCN is primarily responsive to immune challenges during the early active phase of nocturnal animals and is less responsive or non-responsive to immune challenges during the rest phase; there are some exceptions to this phenomenon (Boggio et al. 2003; Guerrero-Vargas et al. 2014). Future studies should ascertain whether diurnal animals display altered daily sensitivity to immune challenges and determine the phase relationship relative to nocturnal animals. The role of immune signaling altering the biological clock is not yet fully understood. However, the putative role described by Cermakian and colleagues is likely. The authors argue that similarly to locomotion, temperature, and melatonin secretion, which have been demonstrated to exert feedback onto the SCN, immune humoral factors might interact with the SCN and subtly adjust entrainment and adaptation to environmental cues to aid in disease recovery (Cermakian et al. 2013). There are also data supporting this idea, as ablation of the SCN following an immune challenge leads to an exacerbated inflammatory response (Guerrero-Vargas et al. 2014).

11.5 Circadian Control of Immune Cell Trafficking

Immune cells display daily variations in both blood circulation and tissue surveillance. Broadly, blood leukocyte counts peak during the behavioral rest phase of animals (Pick et al. 2019). This effect has been demonstrated across multiple nocturnal and diurnal species including humans, mice, rats, and hamsters (Ackermann et al. 2012; Born et al. 1997; Dimitrov et al. 2009; He et al. 2018; Pelegrí et al. 2003; Prendergast et al. 2013). The number of circulating blood leukocytes is controlled via two mechanisms. First, the egress of leukocytes from hematopoietic organs (e.g., the bone marrow), which increases counts within the blood, and second, the migration of leukocytes from the blood into tissues, which decreases cell counts. Egress from the bone marrow primarily occurs during the rest phase of an organism (Lucas et al. 2008; Méndez-Ferrer et al. 2008). The migration of leukocytes into peripheral organs generally occurs during the active phase of individuals (He et al. 2018; Scheiermann et al. 2012). Egress and migration together are responsible for the peak of blood leukocyte counts occurring during the behavioral rest phase of animals (Fig. 11.3).

The mechanisms governing daily oscillations in egress of hematopoietic stem cells (HSCs) and their progenitors from the bone marrow have been well described

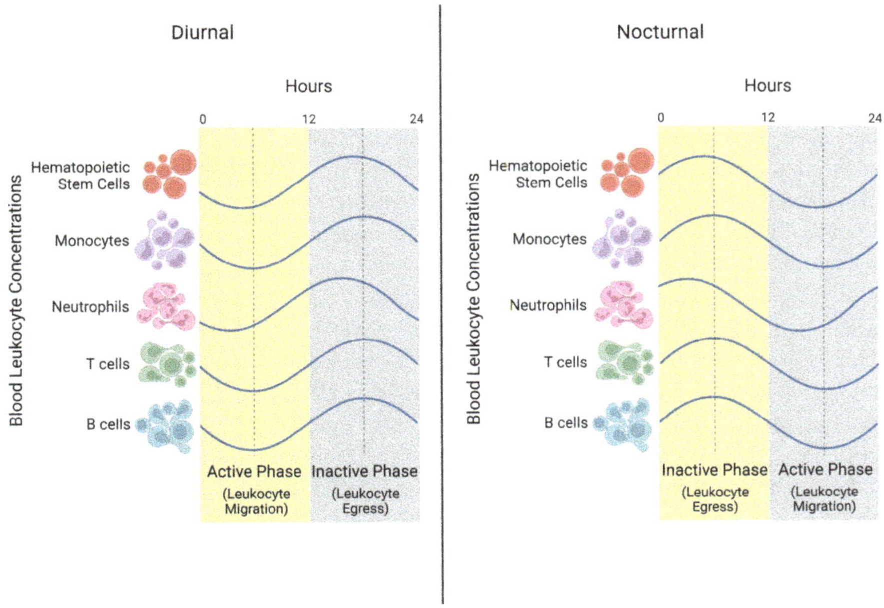

Fig. 11.3 Daily Rhythms in Blood Leukocyte Concentrations. The number of circulating blood leukocytes is controlled via the egress of leukocytes from hematopoietic organs (e.g., the bone marrow), which increases counts within the blood, and the migration of leukocytes from the blood into tissues, which decreases cell counts. Egress from the bone marrow primarily occurs during the rest phase, whereas emigration into tissues occurs primarily during the active phase. This holds true for both nocturnal (e.g., most mice and rats) and diurnal (e.g., humans) species. Figure created with BioRender.com

(Lucas et al. 2008; Méndez-Ferrer et al. 2008). Circulating HSC and their progenitors oscillate antiphase with the expression of CXCL12 within the bone marrow; HSCs peak in the general circulation about 5 h into the rest phase and reach their nadir approximately 5 h into the active phase (Méndez-Ferrer et al. 2008). CXCL12 expression is controlled via daily oscillations in norepinephrine secretion by the sympathetic nervous system and is clock-gene dependent as Bmal-1$^{-/-}$ mice fail to demonstrate oscillations in CXCL12 expression (Méndez-Ferrer et al. 2008). Specifically, secretion of norepinephrine from sympathetic nerves that innervate the bone marrow signal to stromal cells via the β3-adrenergic receptors resulting in decreased nuclear Sp1 transcription factor and the rapid downregulation of CXCL12 (Méndez-Ferrer et al. 2008). In addition to CXCL12, its receptor CXCR4 is also under circadian control (Lucas et al. 2008). CXCR4 expression on HSCs is synchronized with CXCL12 secretion from bone marrow stromal cells. Furthermore, similarly to CXCL12, oscillations in CXCR4 are clock-gene dependent, as Bmal-1$^{-/-}$ mice fail to demonstrate oscillations in CXCR4 expression on HSCs and are responsive to alterations in the circadian cycle via continuous light or conditions of jet lag (Lucas et al. 2008). As expected, circadian oscillations in CXCR4 expression are antiphase in humans and mice. In addition to sympathetic nervous system regulation of HSC bone marrow egress, recent studies have demonstrated a role for parasympathetic signaling as well (García-García et al. 2019). Specifically, García-García et al. (2019) reported that at night, which corresponds to the active phase of mice, parasympathetic cholinergic signals dampen sympathetic noradrenergic tone and decrease bone marrow egress of HSCs and leukocytes in mice. Additionally, parasympathetic activity triggered by light onset inhibits bone marrow vascular cell adhesion and homing, thus further promoting the release of cells within the blood during the rest phase of mice (García-García et al. 2019).

Circadian control of bone marrow egress is not specific to HSCs (García-García et al. 2019; Scheiermann et al. 2012). Indeed, monocytes also display daily variations in CXCR4 expression which control their release from the bone marrow (Chong et al. 2016). Alternatively, CXCR4 expression regulates bone marrow homing in neutrophils and some T cell subsets (Casanova-Acebes et al. 2013; Dimitrov et al. 2009; Zhang et al. 2015). Blockade of CXCR4 expression on neutrophils abolishes oscillations of neutrophils within the blood (Casanova-Acebes et al. 2013). In humans, naïve CD4+ and CD8+ T cell subsets in blood demonstrate a daytime nadir (Dimitrov et al. 2009). The numbers of naïve CD4+ and CD8+ T cells negatively correlate with cortisol rhythms (i.e., antiphase) and decrease after cortisol administration. Notably, naïve CD4+ and CD8+ T cells express the highest amounts of CXCR4, and CXCR4 expression is upregulated following cortisol administration. Thus, it is likely that daytime increases in cortisol act via CXCR4 to redistribute naïve T cells to the bone marrow linking the HPA-axis with daily rhythms in leukocyte concentrations within the circulation (Dimitrov et al. 2009).

In addition to examining daily alterations in egress from and homing to the bone marrow, much research has been conducted examining circadian oscillations in leukocyte migration to lymph nodes and other tissues (Druzd et al. 2017; He et al. 2018; Scheiermann et al. 2012; Shimba et al. 2018; Suzuki et al. 2016). As mentioned, leukocyte migration into peripheral tissues primarily occurs during the active phase. Similarly to the egress of leukocytes from the bone marrow, local sympathetic signals play major roles in the recruitment of leukocytes in tissues (Scheiermann et al. 2012). Perivascular sympathetic nerves act on β-adrenoreceptors expressed by endothelial cells and lead to tissue-specific oscillations in endothelial cell adhesion molecules and chemokine expression. Specifically, endothelial cell expression of CCL2 and ICAM-1 within skeletal muscle and endothelial cell P- and E-selectins, CXCL12 and VCAM-1 within the bone marrow are modulated via sympathetic signaling (Scheiermann et al. 2012). These effects are clock-gene dependent as circadian rhythms in leukocytes counts within the blood and tissues are absent in Bmal-1$^{-/-}$ mice (Scheiermann et al. 2012). Subjecting mice to an experimental jet lag paradigm abolishes circadian oscillations in the expression of ICAM-1 and in the recruitment of leukocytes to skeletal muscles. Furthermore, experimental jet lag eliminates leukocyte homing to the bone marrow during the active phase (Scheiermann et al. 2012). To date, the most thorough study demonstrating and describing circadian oscillations in immune cell trafficking was performed by He and colleagues (He et al. 2018). The authors first assessed oscillations in endothelial cell adhesion molecules in eight different tissues (thymus, spleen, lymph node, liver, skin, gut, lung, and Peyer's patches). They demonstrate that oscillations in endothelial cell adhesion molecules are tissue-specific. However, each tissue displays oscillations of at least one of the cell adhesion molecules examined (ICAM-1, ICAM-2, VCAM-1, P-selectin, E-selectin, CD44), and when assessed as a whole, endothelial cell adhesion molecule expression peaks during the active phase. This corresponds with previous studies demonstrating the nadir of blood leukocyte counts occurring during the active phase (Ackermann et al. 2012; Born et al. 1997; Dimitrov et al. 2009; He et al. 2018; Pelegrí et al. 2003; Prendergast et al. 2013). In addition to oscillations in endothelial cell adhesion molecules, there are also leukocyte-specific oscillations in pro-migratory factors (i.e., adhesion molecules and chemokine receptors (He et al. 2018)). Oscillations in leukocyte numbers were most sensitive to blockade of adhesion molecules, CD49d (α4-integrin) or L-selectin and chemokine receptor CXCR4. Administration of AMD3100, a CXCR4 antagonist, prevented oscillations in all assessed leukocytes (He et al. 2018). When examining tissue-specific homing following adoptive transfer, the authors demonstrate significant homing to the bone marrow, lymph nodes, spleen, liver, and lungs. However, there was little homing to the gut, skin, or thymus (He et al. 2018). Furthermore, each subset of leukocytes demonstrated a distinctive capacity in homing to tissues (detailed in He et al. 2018). Notably, knockout of BMAL-1 in leukocytes or endothelial cells abolished daily oscillations in immune cell trafficking, demonstrating that rhythmic recruitment is clock gene dependent (He et al. 2018).

In sum, it is apparent that there are circadian oscillations in blood leukocyte counts and migration of leukocytes into tissues. This occurs via multiple mechanisms working in concert. For example, signaling of the autonomic nervous system to the bone marrow to regulate egress and sympathetic nerves acting on β-adrenoreceptors expressed by endothelial cells, which leads to tissue-specific oscillations in endothelial cell adhesion molecules and chemokine expression, allowing the emigration of leukocytes. Additionally, daily cortisol rhythms can alter leukocyte expression of CXCR4 and redistribute these cells to the bone marrow. Both regulation of egress and emigration of leukocytes are clock-gene-dependent and sensitive to disrupted circadian rhythms.

11.6 Effects of Disrupted Circadian Rhythms on Disease

The immune system is heavily regulated by circadian rhythmicity, both directly, through cellular and molecular cues that either promote or inhibit the production and activity of immune cells, or indirectly, by environmental changes that modulate those cellular pathways. These findings are based on epidemiological studies, clinical observations (e.g., in those engaged in night shift work), and through translational research using nonhuman animals. In animal studies, disruption to the circadian clock via light at night, sleep deprivation, mistimed eating and jet-lag or selective disruption via core clock gene knockouts or SCN ablation indicates the important role the circadian system plays in the regulation of immune function.

Furthermore, this work has provided insights to the pathophysiological immune responses evoked by the disrupted clock. We primarily focus on cancer as a well-studied immunopathological disease with distinct circadian controls, but this is not the only disease affected by disrupted circadian rhythms (Hou et al. 2020). A number of studies also implicate direct effects of disrupted circadian rhythms on functional immune responses (e.g, Walker et al. 2019). Below we discuss data from epidemiological studies tracking health outcomes in nurses and other night shift workers, as these populations often experience multiple forms of circadian disruption, such as exposure to artificial light during the nighttime, sleep deprivation, mistimed eating, and social jet-lag.

Shift work is characterized by working during hours other than the commonplace "9 to 5." Night shift work impairs many physiological processes, including the incidence of various cancers, such as endometrial (Viswanathan et al. 2007), prostate (Arafa et al. 2021; Behrens et al. 2017; Conlon et al. 2007; Du et al. 2017; Gan et al. 2018; Kubo et al. 2006), colon (Kloog et al. 2009; Schwartzbaum et al. 2007; Wichert et al. 2020) and breast (Hansen 2017; Kloog et al. 2008, 2010) cancer. Further, the risk of development of these is elevated with the frequency of rotating night shifts worked (Schernhammer et al. 2001, 2006; Wegrzyn et al. 2017), as well as by the total of years working night shifts (S. Davis et al. 2001; Schernhammer et al. 2001, 2003; Viswanathan et al. 2007); women who worked night shifts for more than 6 months of a year had increased risk of developing breast cancer (Hansen 2001; Schernhammer et al. 2001). Marked circadian rhythms in cortisol (Abercrombie et al. 2004; Sephton et al. 2000) and distinct 24 h rest-activity phases (Mormont et al. 2000) are significant predictors of survival and quality of life. One caveat when interpreting these results is that night shift workers experience more than one simultaneous source of disrupted circadian rhythms, thus complicating identification of specific molecular pathways and mechanisms that underlie disease development and progression. Nonetheless, impaired biological rhythms contribute to the development of diseases, at least partially, through alterations to homeostatic immune pathways. However, controlled experimental studies, mostly in rodents, have begun to uncover some of the cellular mechanisms linking disrupted circadian rhythms and cancer.

In both human and rodent studies, sleep deprivation (Graves et al. 2003; Irwin et al. 2010; Mullington et al. 2010; Richardson and Churilla 2017), forced desynchrony (Cuesta et al. 2016) and simulated jet-lag (Chen et al. 2021) are common experimental tools used to elicit hindered rhythms, with output measures of immune responses, learning, memory and general activity, among others. Studies in humans of sleep deprivation between 24 and 69 h, and intermittently for up to 5 days, have reported increased neutrophils and monocytes (Born et al. 1997; Boudjeltia et al. 2008; Dinges et al. 1994; Heiser et al. 2000; Kuhn et al. 1969). Elevation of these primary immune response cells, along with elevated concentrations of IL-1β, IL-6, and TNF-α, reflected an exacerbated immune response that was associated with increased pain- sensitivity in healthy volunteers (Haack et al. 2007) and elevated C-reactive protein (CRP), an inflammatory marker associated with cardiovascular disease (Meier-Ewert et al. 2004). However, other studies assessing the same inflammatory markers in healthy volunteers present less consistent data. For instance, males allowed to sleep a total of 4 h per day, in 2 h intervals over 4 days, showed no increase in these markers (Shearer et al. 2001), however, another study that limited sleep to 4 h per night over 10 days reported elevated IL-6 concentrations with no effects on TNF-α (Haack et al. 2007). Another two studies subjecting participants to only 4 h of sleep per night reported increased IL-6 and CRP in both sexes (Meier-Ewert et al. 2004), with elevated TNF-α levels only in men (Vgontzas et al. 2004). Various factors may have contributed to the apparent discrepancies in these studies, despite all subjects being classified as *healthy* volunteers. For example, variations in activity and/or stress levels prior to initiation of the study, or other lifestyle factors such as smoker status or dietary habits (timing of meal intake and caloric composition) may have contributed to the disparate outcomes. Further, individual differences in endogenous neuroendocrine rhythms can modulate circulating cytokines and neutrophils. Finally, it is worth noting that most of these studies analyzed a very small number of participants (~20–30), which can accentuate individual differences.

Animal models have provided some perspective into the underlying association between disrupted circadian rhythms and cancer. At the molecular level, clock gene expression regulates various tumor suppressor genes (Davis et al. 2019; Fu and Lee 2003; Soták et al. 2013); thus, tumor progression successfully downregulates clock genes in the affected tissues. Similarly, disruption of the splenic molecular clock by hampered sympathetic innervation results in altered rhythms in cytokines and cytolytic factors in NK cells and splenocytes (Logan et al. 2011). Hindering of the molecular clock at the central (SCN) level, or by means of environmental disruption, similarly affect levels of circulating immune cells. SCN ablation or rhythm disruption by experimental chronic jet-lag alter daily rhythms in total circulating lymphocyte, which is associated with an exacerbated tumor growth rate (Aiello et al. 2020; Filipski et al. 2003, 2006; Guerrero-Vargas et al. 2017; Inokawa et al. 2020; Yasuniwa et al. 2010). On the other hand, inducing circadian rhythmicity in the tumor microenvironment by dexamethasone administration triggers renewed rhythmic clock gene expression, resulting in more cells in G1, rather than S phase (Kiessling et al. 2017). Moreover, modulating entrainment cues, such as exercise (Eschke et al. 2019) or timed-feeding (Walker et al. 2021), restrict growth (Eschke et al. 2019) and energetic supply (Walker et al. 2021), thus reducing the growth efficiency of tumors. Together, these data suggest that, although dysfunctional environmental cues and/or malfunctioning molecular circadian systems negatively affect immune cell response, this process is dynamic, and can be modulated at various levels.

11.7 Conclusions, Pitfalls, Future Directions

In summary, daily rhythms exist across all aspects of the immune system. The organization of these rhythms is determined by individuals' diurnal or nocturnal nature. For example, disrupting circadian rhythms by exposure to light at night impairs both innate and adaptive immune function in invertebrates, birds, and rodents with robust pineal melatonin rhythms. In contrast, in diurnal rodents and nocturnal species lacking robust pineal melatonin rhythms, light at night may enhance some features of innate and adaptive immune systems. Because nocturnal rodents are the most common species used in studies investigating the effects of nighttime light exposure on immune function, it is important to recognize their limitation in translating the results to humans. To the extent possible, future studies should pivot and focus on diurnal rodents with robust pineal melatonin rhythms to model interactions between circadian rhythms and immune function in humans. Also, future studies should ascertain whether diurnal animals display altered daily sensitivity to immune challenges and determine the phase relationship relative to nocturnal animals (Cermakian et al. 2013).

Another potential pitfall is that the vast majority of studies cited above were conducted on males. Future studies should be conducted on both sexes to provide a more realistic biological perspective on circadian rhythms and immune function. This is also critical because of the significant sex differences in both immune function and circadian rhythms, particularly given the role of sex hormones in regulating immune responses (Taneja 2018).

Finally, few studies examined report the time of day of tissue collection or immune challenge. Time-of-day is a crucial biological variable in biomedical research. Time-of-day should be considered in all analyses and reported to improve reproducibility of studies and to provide the appropriate context to the conclusions. Despite strong daily fluctuations in immune function, physiology, and behavior, most studies of neuroendocrinology and immune function conduct tissue and blood sampling during the light phase of nocturnal animal models. Given the circadian regulation of immune function and the renewed call for time-of-day reporting in all biological studies (e.g., Nelson et al. 2021), future studies must be required to report time of day. A long-term, but achievable, goal would involve clinical studies that elucidate the best time of day for patients to take anti-inflammatory or other drugs that affect immune function to optimize drug efficacy.

11.8 Key References

Partch et al. (2014). Molecular architecture of the mammalian circadian clock. *Trends in Cell Biology*: Key review detailing the mammalian circadian clock.

Pick et al. (2019). Time-of-Day-Dependent Trafficking and Function of Leukocyte Subsets. *Trends in Immunology*: Key review describing circadian dependent function and trafficking of immune cells.

References

Abercrombie HC, Giese-Davis J, Sephton S, Epel ES, Turner-Cobb JM, Spiegel D (2004) Flattened cortisol rhythms in metastatic breast cancer patients. Psychoneuroendocrinology 29(8): 1082–1092. https://doi.org/10.1016/j.psyneuen.2003.11.003

Abrahamson EE, Moore RY (2001) Suprachiasmatic nucleus in the mouse: retinal innervation, intrinsic organization and efferent projections. Brain Res 916(1–2):172–191. https://doi.org/10.1016/s0006-8993(01)02890-6

Ackermann K, Revell VL, Lao O, Rombouts EJ, Skene DJ, Kayser M (2012) Diurnal rhythms in blood cell populations and the effect of acute sleep deprivation in healthy young men. Sleep 35(7):933–940. https://doi.org/10.5665/sleep.1954

Aiello I, Fedele MLM, Román F, Marpegan L, Caldart C, Chiesa JJ, Golombek DA, Finkielstein CV, Paladino N (2020) Circadian disruption promotes tumor-immune microenvironment remodeling favoring tumor cell proliferation. Sci Adv 6(42). https://doi.org/10.1126/sciadv.aaz4530

Akashi M, Takumi T (2005) The orphan nuclear receptor RORalpha regulates circadian transcription of the mammalian core-clock Bmal1. Nat Struct Mol Biol 12(5):441–448. https://doi.org/10.1038/nsmb925

Albrecht U, Sun ZS, Eichele G, Lee CC (1997) A differential response of two putative mammalian circadian regulators, mper1 and mper2, to light. Cell 91(7):1055–1064. https://doi.org/10.1016/s0092-8674(00)80495-x

Allen NC, Philip NH, Hui L, Zhou X, Franklin RA, Kong Y, Medzhitov R (2019) Desynchronization of the molecular clock contributes to the heterogeneity of the inflammatory response. Sci Signal 12(571). https://doi.org/10.1126/scisignal.aau1851

Arafa A, Eshak ES, Iso H, Muraki I, Tamakoshi A (2021) Night work, rotating shift work, and the risk of cancer in Japanese men and women: the JACC study. J Epidemiol 31(12):585–592. https://doi.org/10.2188/jea.JE20200208

Arrigoni E, Chee MJS, Fuller PM (2019) To eat or to sleep: that is a lateral hypothalamic question. Neuropharmacology 154:34–49. https://doi.org/10.1016/j.neuropharm.2018.11.017

Asala SA, Okano Y, Honda K, Inoué S (1990) Effects of medial preoptic area lesions on sleep and wakefulness in unrestrained rats. Neurosci Lett 114(3):300–304. https://doi.org/10.1016/0304-3940(90)90580-3

Behrens T, Rabstein S, Wichert K, Erbel R, Eisele L, Arendt M, Dragano N, Brüning T, Jöckel K-H (2017) Shift work and the incidence of prostate cancer: a 10-year follow-up of a German population-based cohort study. Scand J Work Environ Health 43(6):560–568. https://doi.org/10.5271/sjweh.3666

Bellinger DL, Lorton D, Hamill RW, Felten SY, Felten DL (1993) Acetylcholinesterase staining and choline acetyltransferase activity in the young adult rat spleen: lack of evidence for cholinergic innervation. Brain Behav Immun 7(3):191–204. https://doi.org/10.1006/brbi.1993.1021

Bentivoglio M, Deng X, Nygård M, Sadki A, Kristensson K (2006) The aging suprachiasmatic nucleus and cytokines: functional, molecular, and cellular changes in rodents. Chronobiol Int 23(1–2):437–449. https://doi.org/10.1080/07420520500545797

Berson DM, Dunn FA, Takao M (2002) Phototransduction by retinal ganglion cells that set the circadian clock. Science 295(5557):1070–1073. https://doi.org/10.1126/science.1067262

Berthoud H-R, Münzberg H (2011) The lateral hypothalamus as integrator of metabolic and environmental needs: from electrical self-stimulation to opto-genetics. Physiol Behav 104(1): 29–39. https://doi.org/10.1016/j.physbeh.2011.04.051

Beynon AL, Coogan AN (2010) Diurnal, age, and immune regulation of interleukin-1β and interleukin-1 type 1 receptor in the mouse suprachiasmatic nucleus. Chronobiol Int 27(8): 1546–1563. https://doi.org/10.3109/07420528.2010.501927

Bhadra U, Thakkar N, Das P, Pal Bhadra M (2017) Evolution of circadian rhythms: from bacteria to human. Sleep Med 35:49–61. https://doi.org/10.1016/j.sleep.2017.04.008

Boggio VI, Castrillón PO, Perez Lloret S, Riccio P, Esquifino AI, Cardinali DP, Cutrera RA (2003) Cerebroventricular administration of interferon-gamma modifies locomotor activity in the golden hamster. Neurosignals 12(2):89–94. https://doi.org/10.1159/000071818

Bohnet SG, Traynor TR, Majde JA, Kacsoh B, Krueger JM (2004) Mice deficient in the interferon type I receptor have reduced REM sleep and altered hypothalamic hypocretin, prolactin and 2′,5′-oligoadenylate synthetase expression. Brain Res 1027(1-2):117–125. https://doi.org/10.1016/j.brainres.2004.08.041

Born J, Lange T, Hansen K, Mölle M, Fehm HL (1997) Effects of sleep and circadian rhythm on human circulating immune cells. J Immunol 158(9):4454–4464

Boudjeltia KZ, Faraut B, Stenuit P, Esposito MJ, Dyzma M, Brohée D, Ducobu J, Vanhaeverbeek M, Kerkhofs M (2008) Sleep restriction increases white blood cells, mainly neutrophil count, in young healthy men: a pilot study. Vasc Health Risk Manag 4(6):1467. https://doi.org/10.2147/vhrm.s3934

Bugge A, Feng D, Everett LJ, Briggs ER, Mullican SE, Wang F, Jager J, Lazar MA (2012) Rev-erbα and Rev-erbβ coordinately protect the circadian clock and normal metabolic function. Genes Dev 26(7):657–667. https://doi.org/10.1101/gad.186858.112

Buijs RM, la Fleur SE, Wortel J, Van Heyningen C, Zuiddam L, Mettenleiter TC, Kalsbeek A, Nagai K, Niijima A (2003a) The suprachiasmatic nucleus balances sympathetic and parasympathetic output to peripheral organs through separate preautonomic neurons. J Comp Neurol 464(1):36–48. https://doi.org/10.1002/cne.10765

Buijs RM, van Eden CG, Goncharuk VD, Kalsbeek A (2003b) The biological clock tunes the organs of the body: timing by hormones and the autonomic nervous system. J Endocrinol 177(1):17–26. https://doi.org/10.1677/joe.0.1770017

Cano G, Sved AF, Rinaman L, Rabin BS, Card JP (2001) Characterization of the central nervous system innervation of the rat spleen using viral transneuronal tracing. J Comp Neurol 439(1):1–18. https://doi.org/10.1002/cne.1331

Carrillo-Vico A, Lardone PJ, Alvarez-Sánchez N, Rodríguez-Rodríguez A, Guerrero JM (2013) Melatonin: buffering the immune system. Int J Mol Sci 14(4):8638–8683. https://doi.org/10.3390/ijms14048638

Casanova-Acebes M, Pitaval C, Weiss LA, Nombela-Arrieta C, Chèvre R, Noelia A, Kunisaki Y, Zhang D, Van Rooijen N, Silberstein LE, Weber C (2013) Rhythmic modulation of the hematopoietic niche through neutrophil clearance. Cell 153(5):1025–1035. https://doi.org/10.1016/j.cell.2013.04.040

Cavadini G, Petrzilka S, Kohler P, Jud C, Tobler I, Birchler T, Fontana A (2007) TNF-alpha suppresses the expression of clock genes by interfering with E-box-mediated transcription. Proc Natl Acad Sci U S A 104(31):12843–12848. https://doi.org/10.1073/pnas.0701466104

Cermakian N, Lange T, Golombek D, Sarkar D, Nakao A, Shibata S, Mazzoccoli G (2013) Crosstalk between the circadian clock circuitry and the immune system. Chronobiol Int 30(7):870–888. https://doi.org/10.3109/07420528.2013.782315

Cermakian N, Stegeman SK, Tekade K, Labrecque N (2021) Circadian rhythms in adaptive immunity and vaccination. Semin Immunopathol. 44(2):193–207. https://doi.org/10.1007/s00281-021-00903-7

Chaix A, Zarrinpar A, Panda S (2016) The circadian coordination of cell biology. J Cell Biol 215(1):15–25. https://doi.org/10.1083/jcb.201603076

Chen R, Weitzner AS, McKennon LA, Fonken LK (2021) Chronic circadian phase advance in male mice induces depressive-like responses and suppresses neuroimmune activation. Brain Behav Immun 17:100337. https://doi.org/10.1016/j.bbih.2021.100337

Cho H, Zhao X, Hatori M, Yu RT, Barish GD, Lam MT, Chong LW, DiTacchio L, Atkins AR, Glass CK, Liddle C (2012) Regulation of circadian behaviour and metabolism by REV-ERB-α and REV-ERB-β. Nature 485(7396):123–127. https://doi.org/10.1038/nature11048

Chong SZ, Evrard M, Devi S, Chen J, Lim JY, See P, Zhang Y, Adrover JM, Lee B, Tan L, Li JL (2016) CXCR4 identifies transitional bone marrow premonocytes that replenish the mature

monocyte pool for peripheral responses. J Exp Med 213(11):2293–2314. https://doi.org/10.1084/jem.20160800
Chou TC, Scammell TE, Gooley JJ, Gaus SE, Saper CB, Lu J (2003) Critical role of dorsomedial hypothalamic nucleus in a wide range of behavioral circadian rhythms. J Neurosci 23(33):10691–10702. https://www.ncbi.nlm.nih.gov/pubmed/14627654
Conlon M, Lightfoot N, Kreiger N (2007) Rotating shift work and risk of prostate cancer. Epidemiology 18(1):182–183. https://doi.org/10.1097/01.ede.0000249519.33978.31
Cuesta M, Boudreau P, Dubeau-Laramée G, Cermakian N, Boivin DB (2016) Simulated night shift disrupts circadian rhythms of immune functions in humans. J Immunol 196(6):2466–2475. https://doi.org/10.4049/jimmunol.1502422
Davis S, Mirick DK, Stevens RG (2001) Night shift work, light at night, and risk of breast cancer. J Natl Cancer Inst 93(20):1557–1562. https://doi.org/10.1093/jnci/93.20.1557
Davis K, Roden LC, Leaner VD, van der Watt PJ (2019) The tumour suppressing role of the circadian clock. IUBMB Life 71(7):771–780. https://doi.org/10.1002/iub.2005
Deng X-H, Bertini G, Palomba M, Xu Y-Z, Bonaconsa M, Nygård M, Bentivoglio M (2010) Glial transcripts and immune-challenged glia in the suprachiasmatic nucleus of young and aged mice. Chronobiol Int 27(4):742–767. https://doi.org/10.3109/07420521003681498
Dibner C, Schibler U, Albrecht U (2010) The mammalian circadian timing system: organization and coordination of central and peripheral clocks. Annu Rev Physiol 72:517–549. https://doi.org/10.1146/annurev-physiol-021909-135821
Dimitrov S, Benedict C, Heutling D, Westermann J, Born J, Lange T (2009) Cortisol and epinephrine control opposing circadian rhythms in T cell subsets. Blood 113(21):5134–5143. https://doi.org/10.1182/blood-2008-11-190769
Dinges DF, Douglas SD, Zaugg L, Campbell DE, McMann JM, Whitehouse WG, Orne EC, Kapoor SC, Icaza E, Orne MT (1994) Leukocytosis and natural killer cell function parallel neurobehavioral fatigue induced by 64 hours of sleep deprivation. J Clin Invest 93(5):1930–1939. https://doi.org/10.1172/JCI117184
Do MTH, Yau K-W (2010) Intrinsically photosensitive retinal ganglion cells. Physiol Rev 90(4):1547–1581. https://doi.org/10.1152/physrev.00013.2010
Dokur M, Boyadjieva N, Sarkar DK (2004) Catecholaminergic control of NK cell cytolytic activity regulatory factors in the spleen. J Neuroimmunol 151(1–2):148–157. https://doi.org/10.1016/j.jneuroim.2004.03.003
Druzd D, Matveeva O, Ince L, Harrison U, He W, Schmal C et al (2017) Lymphocyte circadian clocks control lymph node trafficking and adaptive immune responses. Immunity 46(1):120–132. https://doi.org/10.1016/j.immuni.2016.12.011
Du H-B, Bin K-Y, Liu W-H, Yang F-S (2017) Shift work, night work, and the risk of prostate cancer: a meta-analysis based on 9 cohort studies. Medicine 96(46):e8537. https://doi.org/10.1097/MD.0000000000008537
Duhart JM, Leone MJ, Paladino N, Evans JA, Castanon-Cervantes O, Davidson AJ, Golombek DA (2013) Suprachiasmatic astrocytes modulate the circadian clock in response to TNF-α. J Immunol 191(9):4656–4664. https://doi.org/10.4049/jimmunol.1300450
Dunlap JC (1999) Molecular bases for circadian clocks. Cell 96(2):271–290. https://doi.org/10.1016/s0092-8674(00)80566-8
Early JO, Menon D, Wyse CA, Cervantes-Silva MP, Zaslona Z, Carroll RG, Palsson-McDermott EM, Angiari S, Ryan DG, Corcoran SE, Timmons G (2018) Circadian clock protein BMAL1 regulates IL-1β in macrophages via NRF2. Proc Natl Acad Sci U S A 115(36):E8460–E8468. https://doi.org/10.1073/pnas.1800431115
Edgar RS, Green EW, Zhao Y, Van Ooijen G, Olmedo M, Qin X, Xu Y, Pan M, Valekunja UK, Feeney KA, Maywood ES (2012) Peroxiredoxins are conserved markers of circadian rhythms. Nature 485(7399):459–464. https://doi.org/10.1038/nature11088
Elenkov IJ, Wilder RL, Chrousos GP, Vizi ES (2000) The sympathetic nerve—an integrative interface between two supersystems: the brain and the immune system. Pharmacol Rev 52(4):595–638

Eschke R-CK-R, Lampit A, Schenk A, Javelle F, Steindorf K, Diel P, Bloch W, Zimmer P (2019) Impact of physical exercise on growth and progression of cancer in rodents—a systematic review and meta-analysis. Front Oncol 9:35. https://doi.org/10.3389/fonc.2019.00035

Filipski E, King VM, Li X, Granda TG, Mormont M-C, Claustrat B, Hastings MH, Lévi F (2003) Disruption of circadian coordination accelerates malignant growth in mice. Pathol Biol 51(4): 216–219. https://doi.org/10.1016/s0369-8114(03)00034-8

Filipski E, Li XM, Lévi F (2006) Disruption of circadian coordination and malignant growth. Cancer Causes Control 17(4):509–514. https://doi.org/10.1007/s10552-005-9007-4

Fortier EE, Rooney J, Dardente H, Hardy M-P, Labrecque N, Cermakian N (2011) Circadian variation of the response of T cells to antigen. J Immunol 187(12):6291–6300. https://doi.org/10.4049/jimmunol.1004030

Fu L, Lee CC (2003) The circadian clock: pacemaker and tumour suppressor. Nat Rev Cancer 3(5): 350–361. https://doi.org/10.1038/nrc1072

Furuzawa M, Kuwahara M, Ishii K, Iwakura Y, Tsubone H (2002) Diurnal variation of heart rate, locomotor activity, and body temperature in interleukin-1 alpha/beta doubly deficient mice. Exp Anim 51(1):49–56. https://doi.org/10.1538/expanim.51.49

Gallego M, Virshup DM (2007) Post-translational modifications regulate the ticking of the circadian clock. Nat Rev Mol Cell Biol 8(2):139–148. https://doi.org/10.1038/nrm2106

Gamble KL, Berry R, Frank SJ, Young ME (2014) Circadian clock control of endocrine factors. Nat Rev Endocrinol 10(8):466–475. https://doi.org/10.1038/nrendo.2014.78

Gan Y, Li L, Zhang L, Yan S, Gao C, Hu S, Qiao Y, Tang S, Wang C, Lu Z (2018) Association between shift work and risk of prostate cancer: a systematic review and meta-analysis of observational studies. Carcinogenesis 39(2):87–97. https://doi.org/10.1093/carcin/bgx129

Gao S, Sanchez C, Deviche PJ (2017) Corticosterone rapidly suppresses innate immune activity in the house sparrow (Passer domesticus). J Exp Biol 220(Pt 2):322–327. https://doi.org/10.1242/jeb.144378

García-García A, Korn C, García-Fernández M, Domingues O, Villadiego J, Martín-Pérez D, Isern J, Bejarano-García JA, Zimmer J, Pérez-Simón JA, Toledo-Aral JJ (2019) Dual cholinergic signals regulate daily migration of hematopoietic stem cells and leukocytes. Blood 133(3): 224–236. https://doi.org/10.1182/blood-2018-08-867648

Gau D, Lemberger T, Von Gall C, Kretz O, Le Minh N, Gass P, Schmid W, Schibler U, Korf HW, Schütz G (2002) Phosphorylation of CREB Ser142 regulates light-induced phase shifts of the circadian clock. Neuron 34(2):245–253. https://doi.org/10.1016/s0896-6273(02)00656-6

Gegout-Pottie P, Philippe L, Simonin MA, Guingamp C, Gillet P, Netter P, Terlain B (1999) Biotelemetry: an original approach to experimental models of inflammation. Inflamm Res 48(8): 417–424. https://doi.org/10.1007/s000110050481

Gibbs JE, Blaikley J, Beesley S, Matthews L, Simpson KD, Boyce SH, Farrow SN, Else KJ, Singh D, Ray DW, Loudon AS (2012) The nuclear receptor REV-ERBα mediates circadian regulation of innate immunity through selective regulation of inflammatory cytokines. Proc Natl Acad Sci U S A 109(2):582–587. https://doi.org/10.1073/pnas.1106750109

Ginty DD, Kornhauser JM, Thompson MA, Bading H, Mayo KE, Takahashi JS, Greenberg ME (1993) Regulation of CREB phosphorylation in the suprachiasmatic nucleus by light and a circadian clock. Science 260(5105):238–241. https://doi.org/10.1126/science.8097062

Graves LA, Heller EA, Pack AI, Abel T (2003) Sleep deprivation selectively impairs memory consolidation for contextual fear conditioning. Learn Mem 10(3):168–176. https://doi.org/10.1101/lm.48803

Griffin EA, Staknis D, Weitz CJ (1999) Light-independent role of CRY1 and CRY2 in the mammalian circadian clock. Science 286(5440):768–771. https://doi.org/10.1126/science.286.5440.768

Guerrero-Vargas NN, Salgado-Delgado R, del Carmen Basualdo M, García J, Guzmán-Ruiz M, Carrero JC, Escobar C, Buijs RM (2014) Reciprocal interaction between the suprachiasmatic nucleus and the immune system tunes down the inflammatory response to lipopolysaccharide. J Neuroimmunol 273(1–2):22–30. https://doi.org/10.1016/j.jneuroim.2014.05.012

Guerrero-Vargas NN, Navarro-Espíndola R, Guzmán-Ruíz MA, Basualdo MDC, Espitia-Bautista-E, López-Bago A, Lascurain R, Córdoba-Manilla C, Buijs RM, Escobar C (2017) Circadian disruption promotes tumor growth by anabolic host metabolism; experimental evidence in a rat model. BMC Cancer 17(1):625. https://doi.org/10.1186/s12885-017-3636-3

Haack M, Sanchez E, Mullington JM (2007) Elevated inflammatory markers in response to prolonged sleep restriction are associated with increased pain experience in healthy volunteers. Sleep 30(9):1145–1152. https://doi.org/10.1093/sleep/30.9.1145

Haimovich B, Calvano J, Haimovich AD, Calvano SE, Coyle SM, Lowry SF (2010) In vivo endotoxin synchronizes and suppresses clock gene expression in human peripheral blood leukocytes. Crit Care Med 38(3):751–758. https://doi.org/10.1097/CCM.0b013e3181cd131c

Halberg F, Johnson EA, Brown BW, Bittner JJ (1960) Susceptibility rhythm to E. coli endotoxin and bioassay. Proc Soc Exp Biol Med 103:142–144. https://doi.org/10.3181/00379727-103-25439

Hansen J (2001) Increased breast cancer risk among women who work predominantly at night. Epidemiology 12(1):74–77. https://doi.org/10.1097/00001648-200101000-00013

Hansen J (2017) Night shift work and risk of breast cancer. Curr Environ Health Rep 4:325–339

Hastings M, O'Neill JS, Maywood ES (2007) Circadian clocks: regulators of endocrine and metabolic rhythms. J Endocrinol 195(2):187–198. https://doi.org/10.1677/JOE-07-0378

Hastings MH, Maywood ES, Brancaccio M (2018) Generation of circadian rhythms in the suprachiasmatic nucleus. Nat Rev Neurosci 19(8):453–469. https://doi.org/10.1038/s41583-018-0026-z

He W, Holtkamp S, Hergenhan SM, Kraus K, de Juan A, Weber J et al (2018) Circadian expression of migratory factors establishes lineage-specific signatures that guide the homing of leukocyte subsets to tissues. Immunity 49(6):1175–1190.e7. https://doi.org/10.1016/j.immuni.2018.10.007

Heiser P, Dickhaus B, Schreiber W, Clement HW, Hasse C, Hennig J, Remschmidt H, Krieg JC, Wesemann W, Opper C (2000) White blood cells and cortisol after sleep deprivation and recovery sleep in humans. Eur Arch Psychiatry Clin Neurosci 250(1):16–23. https://doi.org/10.1007/pl00007534

Hogenesch JB, Chan WK, Jackiw VH, Brown RC, Gu YZ, Pray-Grant M, Perdew GH, Bradfield CA (1997) Characterization of a subset of the basic-helix-loop-helix-PAS superfamily that interacts with components of the dioxin signaling pathway. J Biol Chem 272(13):8581–8593. https://doi.org/10.1074/jbc.272.13.8581

Hoover DB, Brown TC, Miller MK, Schweitzer JB, Williams DL (2017) Loss of sympathetic nerves in spleens from patients with end stage sepsis. Front Immunol 8:1712. https://doi.org/10.3389/fimmu.2017.01712

Hopwood TW, Hall S, Begley N, Forman R, Brown S, Vonslow R et al (2018) The circadian regulator BMAL1 programmes responses to parasitic worm infection via a dendritic cell clock. Sci Rep 8(1):3782. https://doi.org/10.1038/s41598-018-22021-5

Hou Y, Liu L, Chen X, Li Q, Li J (2020) Association between circadian disruption and diseases: A narrative review. Life Sci 262:118512

Hrushesky WJ, Lester B, Lannin D (1999) Circadian coordination of cancer growth and metastatic spread. Int J Cancer 83(3):365–373

Ikeda M, Nomura M (1997) cDNA cloning and tissue-specific expression of a novel basic helix-loop-helix/PAS protein (BMAL1) and identification of alternatively spliced variants with alternative translation initiation site usage. Biochem Biophys Res Commun 233(1):258–264. https://doi.org/10.1006/bbrc.1997.6371

Inokawa H, Umemura Y, Shimba A, Kawakami E, Koike N, Tsuchiya Y, Ohashi M, Minami Y, Cui G, Asahi T, Ono R, Sasawaki Y, Konishi E, Yoo S-H, Chen Z, Teramukai S, Ikuta K, Yagita K (2020) Chronic circadian misalignment accelerates immune senescence and abbreviates lifespan in mice. Sci Rep 10(1):2569. https://doi.org/10.1038/s41598-020-59541-y

Inouye S-LT, Kawamura H (1979) Persistence of circadian rhythmicity in a mammalian hypothalamic "island" containing the suprachiasmatic nucleus. Proc Natl Acad Sci 76(11):5962–5966. https://www.pnas.org/content/76/11/5962.short

Irwin MR, Carrillo C, Olmstead R (2010) Sleep loss activates cellular markers of inflammation: sex differences. Brain Behav Immun 24(1):54–57. https://doi.org/10.1016/j.bbi.2009.06.001

Jones JR, Chaturvedi S, Granados-Fuentes D, Herzog ED (2021) Circadian neurons in the paraventricular nucleus entrain and sustain daily rhythms in glucocorticoids. Nat Commun 12(1):5763. https://doi.org/10.1038/s41467-021-25959-9

Kalsbeek A, Buijs RM (2002) Output pathways of the mammalian suprachiasmatic nucleus: coding circadian time by transmitter selection and specific targeting. Cell Tissue Res 309(1):109–118. https://doi.org/10.1007/s00441-002-0577-0

Kalsbeek A, Garidou ML, Palm IF, Van Der Vliet J, Simonneaux V, Pévet P, Buijs RM (2000) Melatonin sees the light: blocking GABA-ergic transmission in the paraventricular nucleus induces daytime secretion of melatonin. Eur J Neurosci 12(9):3146–3154. https://doi.org/10.1046/j.1460-9568.2000.00202.x

Kalsbeek A, Palm IF, La Fleur SE, Scheer FAJL, Perreau-Lenz S, Ruiter M, Kreier F, Cailotto C, Buijs RM (2006) SCN outputs and the hypothalamic balance of life. J Biol Rhythm 21(6): 458–469. https://doi.org/10.1177/0748730406293854

Keller M, Mazuch J, Abraham U, Eom GD, Herzog ED, Volk H-D et al (2009) A circadian clock in macrophages controls inflammatory immune responses. Proc Natl Acad Sci U S A 106(50): 21407–21412. https://doi.org/10.1073/pnas.0906361106

Kiessling S, Beaulieu-Laroche L, Blum ID, Landgraf D, Welsh DK, Storch K-F, Labrecque N, Cermakian N (2017) Enhancing circadian clock function in cancer cells inhibits tumor growth. BMC Biol 15(1):13. https://doi.org/10.1186/s12915-017-0349-7

King DP, Zhao Y, Sangoram AM, Wilsbacher LD, Tanaka M, Antoch MP et al (1997) Positional cloning of the mouse circadian clock gene. Cell 89(4):641–653. https://doi.org/10.1016/s0092-8674(00)80245-7

Kloog I, Haim A, Stevens RG, Barchana M, Portnov BA (2008) Light at night co-distributes with incident breast but not lung cancer in the female population of Israel. Chronobiol Int 25(1): 65–81. https://doi.org/10.1080/07420520801921572

Kloog I, Haim A, Stevens RG, Portnov BA (2009) Global Co-distribution of light at night (LAN) and cancers of prostate, colon, and lung in men. Chronobiol Int 26(1):108–125. https://doi.org/10.1080/07420520802694020

Kloog I, Stevens RG, Haim A, Portnov BA (2010) Nighttime light level co-distributes with breast cancer incidence worldwide. Cancer Causes Control 21(12):2059–2068. https://doi.org/10.1007/s10552-010-9624-4

Kondratov RV, Kondratova AA, Gorbacheva VY, Vykhovanets OV, Antoch MP (2006) Early aging and age-related pathologies in mice deficient in BMAL1, the core component of the circadian clock. Genes Dev 20(14):1868–1873. https://doi.org/10.1101/gad.1432206

Kriegsfeld LJ, Leak RK, Yackulic CB, LeSauter J, Silver R (2004) Organization of suprachiasmatic nucleus projections in Syrian hamsters (Mesocricetus auratus): an anterograde and retrograde analysis. J Comp Neurol 468(3):361–379. https://doi.org/10.1002/cne.10995

Kubo T, Ozasa K, Mikami K, Wakai K, Fujino Y, Watanabe Y, Miki T, Nakao M, Hayashi K, Suzuki K, Mori M, Washio M, Sakauchi F, Ito Y, Yoshimura T, Tamakoshi A (2006) Prospective cohort study of the risk of prostate cancer among rotating-shift workers: findings from the Japan collaborative cohort study. Am J Epidemiol 164(6):549–555. https://doi.org/10.1093/aje/kwj232

Kuhn E, Brodan V, Brodanová M, Rysánek K (1969) Metabolic reflection of sleep deprivation. Act Nerv Super 11(3):165–174. https://www.ncbi.nlm.nih.gov/pubmed/5798779

Kwak Y, Lundkvist GB, Brask J, Davidson A, Menaker M, Kristensson K, Block GD (2008) Interferon-gamma alters electrical activity and clock gene expression in suprachiasmatic nucleus neurons. J Biol Rhythm 23(2):150–159. https://doi.org/10.1177/0748730407313355

Labrecque N, Cermakian N (2015) Circadian clocks in the immune system. J Biol Rhythm 30(4): 277–290. https://doi.org/10.1177/0748730415577723
Laste G, Vidor L, de Macedo IC, Rozisky JR, Medeiros L, de Souza A et al (2013) Melatonin treatment entrains the rest-activity circadian rhythm in rats with chronic inflammation. Chronobiol Int 30(9):1077–1088. https://doi.org/10.3109/07420528.2013.800088
Lee B, Li A, Hansen KF, Cao R, Yoon JH, Obrietan K (2010) CREB influences timing and entrainment of the SCN circadian clock. J Biol Rhythm 25(6):410–420. https://doi.org/10.1177/0748730410381229
Leibowitz SF (1970) Reciprocal hunger-regulating circuits involving alpha-and beta-adrenergic receptors located, respectively, in the ventromedial and lateral hypothalamus. Proc Natl Acad Sci 67(2):1063–1070. https://doi.org/10.1073/pnas.67.2.1063
Leone MJ, Marpegan L, Bekinschtein TA, Costas MA, Golombek DA (2006) Suprachiasmatic astrocytes as an interface for immune-circadian signalling. J Neurosci Res 84(7):1521–1527. https://doi.org/10.1002/jnr.21042
Leone MJ, Marpegan L, Duhart JM, Golombek DA (2012) Role of proinflammatory cytokines on lipopolysaccharide-induced phase shifts in locomotor activity circadian rhythm. Chronobiol Int 29(6):715–723. https://doi.org/10.3109/07420528.2012.682681
Logan RW, Arjona A, Sarkar DK (2011) Role of sympathetic nervous system in the entrainment of circadian natural-killer cell function. Brain Behav Immun 25(1):101–109. https://doi.org/10.1016/j.bbi.2010.08.007
Long JE, Drayson MT, Taylor AE, Toellner KM, Lord JM, Phillips AC (2016) Morning vaccination enhances antibody response over afternoon vaccination: a cluster-randomised trial. Vaccine 34(24):2679–2685. https://doi.org/10.1016/j.vaccine.2016.04.032
Lu J, Zhang Y-H, Chou TC, Gaus SE, Elmquist JK, Shiromani P, Saper CB (2001) Contrasting effects of ibotenate lesions of the paraventricular nucleus and subparaventricular zone on sleep–wake cycle and temperature regulation. J Neurosci Off J Soc Neurosci 21(13):4864–4874. https://doi.org/10.1523/JNEUROSCI.21-13-04864.2001
Lucas D, Battista M, Shi PA, Isola L, Frenette PS (2008) Mobilized hematopoietic stem cell yield depends on species-specific circadian timing. Cell Stem Cell 3(4):364–366. https://doi.org/10.1016/j.stem.2008.09.004
Lundkvist GB, Hill RH, Kristensson K (2002) Disruption of circadian rhythms in synaptic activity of the suprachiasmatic nuclei by African trypanosomes and cytokines. Neurobiol Dis 11(1): 20–27
Man K, Loudon A, Chawla A (2016) Immunity around the clock. Science 354(6315):999–1003. https://doi.org/10.1126/science.aah4966
Marpegán L, Bekinschtein TA, Costas MA, Golombek DA (2005) Circadian responses to endotoxin treatment in mice. J Neuroimmunol 160(1–2):102–109. https://doi.org/10.1016/j.jneuroim.2004.11.003
McDearmon EL, Patel KN, Ko CH, Walisser JA, Schook AC, Chong JL et al (2006) Dissecting the functions of the mammalian clock protein BMAL1 by tissue-specific rescue in mice. Science 314(5803):1304–1308. https://doi.org/10.1126/science.1132430
Meier-Ewert HK, Ridker PM, Rifai N, Regan MM, Price NJ, Dinges DF, Mullington JM (2004) Effect of sleep loss on C-reactive protein, an inflammatory marker of cardiovascular risk. J Am Coll Cardiol 43(4):678–683. https://doi.org/10.1016/j.jacc.2003.07.050
Méndez-Ferrer S, Lucas D, Battista M, Frenette PS (2008) Haematopoietic stem cell release is regulated by circadian oscillations. Nature 452(7186):442–447. https://doi.org/10.1038/nature06685
Meyer-Bernstein EL, Jetton AE, Matsumoto SI, Markuns JF, Lehman MN, Bittman EL (1999) Effects of suprachiasmatic transplants on circadian rhythms of neuroendocrine function in golden hamsters. Endocrinology 140(1):207–218. https://doi.org/10.1210/endo.140.1.6428
Mohawk JA, Green CB, Takahashi JS (2012) Central and peripheral circadian clocks in mammals. Annu Rev Neurosci 35:445–462. https://doi.org/10.1146/annurev-neuro-060909-153128

Mondino A, Hambrecht-Wiedbusch VS, Li D, York AK, Pal D, González J, Torterolo P, Mashour GA, Vanini G (2021) Glutamatergic neurons in the preoptic hypothalamus promote wakefulness, destabilize NREM sleep, suppress REM sleep, and regulate cortical dynamics. J Neurosci 41(15):3462–3478. https://doi.org/10.1523/JNEUROSCI.2718-20.2021

Mormont M-C, Waterhouse J, Bleuzen P, Giacchetti S, Jami A, Bogdan A, Lellouch J, Misset J-L, Touitou Y, Lévi F (2000) Marked 24-h rest/activity rhythms are associated with better quality of life, better response, and longer survival in patients with metastatic colorectal cancer and good performance status. Clin Cancer Res 6(8):3038–3045. https://clincancerres.aacrjournals.org/content/6/8/3038.short

Mul Fedele ML, Aiello I, Caldart CS, Golombek DA, Marpegan L, Paladino N (2020) Differential thermoregulatory and inflammatory patterns in the circadian response to LPS-induced septic shock. Front Cell Infect Microbiol 10:100

Mullington JM, Simpson NS, Meier-Ewert HK, Haack M (2010) Sleep loss and inflammation. Best Pract Res Clin Endocrinol Metab 24(5):775–784. https://doi.org/10.1016/j.beem.2010.08.014

Mure LS, Le HD, Benegiamo G, Chang MW, Rios L, Jillani N et al (2018) Diurnal transcriptome atlas of a primate across major neural and peripheral tissues. Science 359(6381). https://doi.org/10.1126/science.aao0318

Murphy BA, Vick MM, Sessions DR, Cook RF, Fitzgerald BP (2007) Acute systemic inflammation transiently synchronizes clock gene expression in equine peripheral blood. Brain Behav Immun 21(4):467–476. https://doi.org/10.1016/j.bbi.2006.11.002

Nader N, Chrousos GP, Kino T (2010) Interactions of the circadian CLOCK system and the HPA axis. Trends Endocrinol Metab 21(5):277–286. https://doi.org/10.1016/j.tem.2009.12.011

Nakamura Y, Ishimaru K, Shibata S, Nakao A (2017) Regulation of plasma histamine levels by the mast cell clock and its modulation by stress. Sci Rep 7:39934. https://doi.org/10.1038/srep39934

Nelson RJ, Bumgarner JR, Walker WH II, DeVries AC (2021) Time-of-day as a critical biological variable. Neurosci Biobehav Rev 127:740–746. Chicago

Nobis CC, Dubeau Laramée G, Kervezee L, Maurice De Sousa D, Labrecque N, Cermakian N (2019) The circadian clock of CD8 T cells modulates their early response to vaccination and the rhythmicity of related signaling pathways. Proc Natl Acad Sci U S A 116(40):20077–20086. https://doi.org/10.1073/pnas.1905080116

Ohdo S, Koyanagi S, Suyama H, Higuchi S, Aramaki H (2001) Changing the dosing schedule minimizes the disruptive effects of interferon on clock function. Nat Med 7(3):356–360. https://doi.org/10.1038/85507

Okada K, Yano M, Doki Y, Azama T, Iwanaga H, Miki H et al (2008) Injection of LPS causes transient suppression of biological clock genes in rats. J Surg Res 145(1):5–12. https://doi.org/10.1016/j.jss.2007.01.010

Paladino N, Leone MJ, Plano SA, Golombek DA (2010) Paying the circadian toll: the circadian response to LPS injection is dependent on the Toll-like receptor 4. J Neuroimmunol 225(1–2):62–67. https://doi.org/10.1016/j.jneuroim.2010.04.015

Palomba M, Bentivoglio M (2008) Chronic inflammation affects the photic response of the suprachiasmatic nucleus. J Neuroimmunol 193(1–2):24–27. https://doi.org/10.1016/j.jneuroim.2007.09.002

Partch CL, Green CB, Takahashi JS (2014) Molecular architecture of the mammalian circadian clock. Trends Cell Biol 24(2):90–99. https://doi.org/10.1016/j.tcb.2013.07.002

Patke A, Young MW, Axelrod S (2020) Molecular mechanisms and physiological importance of circadian rhythms. Nat Rev Mol Cell Biol 21(2):67–84. https://doi.org/10.1038/s41580-019-0179-2

Pelegrí C, Vilaplana J, Castellote C, Rabanal M, Franch A, Castell M (2003) Circadian rhythms in surface molecules of rat blood lymphocytes. Am J Physiol Cell Physiol 284(1):C67–C76. https://doi.org/10.1152/ajpcell.00084.2002

Pett JP, Korenčič A, Wesener F, Kramer A, Herzel H (2016) Feedback loops of the mammalian circadian clock constitute repressilator. PLoS Comput Biol 12(12):e1005266. https://doi.org/10.1371/journal.pcbi.1005266

Pick R, He W, Chen C-S, Scheiermann C (2019) Time-of-day-dependent trafficking and function of leukocyte subsets. Trends Immunol 40(6):524–537. https://doi.org/10.1016/j.it.2019.03.010

Preitner N, Damiola F, Lopez-Molina L, Zakany J, Duboule D, Albrecht U, Schibler U (2002) The orphan nuclear receptor REV-ERBalpha controls circadian transcription within the positive limb of the mammalian circadian oscillator. Cell 110(2):251–260. https://doi.org/10.1016/s0092-8674(02)00825-5

Prendergast BJ, Cable EJ, Patel PN, Pyter LM, Onishi KG, Stevenson TJ et al (2013) Impaired leukocyte trafficking and skin inflammatory responses in hamsters lacking a functional circadian system. Brain Behav Immun 32:94–104. https://doi.org/10.1016/j.bbi.2013.02.007

Ramakrishnan R, DuBois DC, Almon RR, Pyszczynski NA, Jusko WJ (2002) Fifth-generation model for corticosteroid pharmacodynamics: application to steady-state receptor down-regulation and enzyme induction patterns during seven-day continuous infusion of methylprednisolone in rats. J Pharmacokinet Pharmacodyn 29(1):1–24. https://doi.org/10.1023/A:1015765201129

Reghunandanan V, Reghunandanan R (2006) Neurotransmitters of the suprachiasmatic nuclei. J Circadian Rhythms 4:2. https://doi.org/10.1186/1740-3391-4-2

Reiter RJ, Tan DX, Kim SJ, Cruz MHC (2014) Delivery of pineal melatonin to the brain and SCN: role of canaliculi, cerebrospinal fluid, tanycytes and Virchow–Robin perivascular spaces. Brain Struct Funct 219(6):1873–1887

Relógio A, Westermark PO, Wallach T, Schellenberg K, Kramer A, Herzel H (2011) Tuning the mammalian circadian clock: robust synergy of two loops. PLoS Comput Biol 7(12):e1002309. https://doi.org/10.1371/journal.pcbi.1002309

Ren W, Liu G, Chen S, Yin J, Wang J, Tan B, Wu G, Bazer FW, Peng Y, Li T, Reiter RJ, Yin Y (2017) Melatonin signaling in T cells: functions and applications. J Pineal Res 62(3). https://doi.org/10.1111/jpi.12394

Richardson MR, Churilla JR (2017) Sleep duration and C-reactive protein in US adults. South Med J 110(4):314–317. https://doi.org/10.14423/SMJ.0000000000000632

Rosas-Ballina M, Ochani M, Parrish WR, Ochani K, Harris YT, Huston JM, Chavan S, Tracey KJ (2008) Splenic nerve is required for cholinergic antiinflammatory pathway control of TNF in endotoxemia. Proc Natl Acad Sci U S A 105(31):11008–11013. https://doi.org/10.1073/pnas.0803237105

Rosato E, Tauber E, Kyriacou CP (2006) Molecular genetics of the fruit-fly circadian clock. Eur J Hum Genet 14(6):729–738. https://doi.org/10.1038/sj.ejhg.5201547

Sadki A, Bentivoglio M, Kristensson K, Nygård M (2007) Suppressors, receptors and effects of cytokines on the aging mouse biological clock. Neurobiol Aging 28(2):296–305. https://doi.org/10.1016/j.neurobiolaging.2005.12.007

Scheiermann C, Kunisaki Y, Lucas D, Chow A, Jang J-E, Zhang D et al (2012) Adrenergic nerves govern circadian leukocyte recruitment to tissues. Immunity 37(2):290–301. https://doi.org/10.1016/j.immuni.2012.05.021

Scheiermann C, Gibbs J, Ince L, Loudon A (2018) Clocking in to immunity. Nat Rev Immunol 18(7):423–437. https://doi.org/10.1038/s41577-018-0008-4

Schernhammer ES, Laden F, Speizer FE, Willett WC, Hunter DJ, Kawachi I, Colditz GA (2001) Rotating night shifts and risk of breast cancer in women participating in the nurses' health study. J Natl Cancer Inst 93(20):1563–1568. https://doi.org/10.1093/jnci/93.20.1563

Schernhammer ES, Laden F, Speizer FE, Willett WC, Hunter DJ, Kawachi I, Fuchs CS, Colditz GA (2003) Night-shift work and risk of colorectal cancer in the nurses' health study. J Natl Cancer Inst 95(11):825–828. https://doi.org/10.1093/jnci/95.11.825

Schernhammer ES, Kroenke CH, Laden F, Hankinson SE (2006) Night work and risk of breast cancer. Epidemiology 17(1):108–111. https://doi.org/10.1097/01.ede.0000190539.03500.c1

Schwartzbaum J, Ahlbom A, Feychting M (2007) Cohort study of cancer risk among male and female shift workers. Scand J Work Environ Health 33(5):336–343. https://doi.org/10.5271/sjweh.1150

Sehgal A (2017) Physiology flies with time. Cell 171(6):1232–1235. https://doi.org/10.1016/j.cell.2017.11.028

Sephton SE, Sapolsky RM, Kraemer HC, Spiegel D (2000) Diurnal cortisol rhythm as a predictor of breast cancer survival. J Natl Cancer Inst 92(12):994–1000. https://doi.org/10.1093/jnci/92.12.994

Shackelford PG, Feigin RD (1973) Periodicity of susceptibility to pneumococcal infection: influence of light and adrenocortical secretions. Science 182(4109):285–287. https://doi.org/10.1126/science.182.4109.285

Shearer WT, Reuben JM, Mullington JM, Price NJ, Lee B-N, Smith EO, Szuba MP, Van Dongen HPA, Dinges DF (2001) Soluble TNF-α receptor 1 and IL-6 plasma levels in humans subjected to the sleep deprivation model of spaceflight. J Allergy Clin Immunol 107(1):165–170. https://doi.org/10.1067/mai.2001.112270

Shearman LP, Zylka MJ, Weaver DR, Kolakowski LF, Reppert SM (1997) Two period homologs: circadian expression and photic regulation in the suprachiasmatic nuclei. Neuron 19(6):1261–1269. https://doi.org/10.1016/s0896-6273(00)80417-1

Shearman LP, Sriram S, Weaver DR, Maywood ES, Chaves I, Zheng B et al (2000) Interacting molecular loops in the mammalian circadian clock. Science 288(5468):1013–1019. https://doi.org/10.1126/science.288.5468.1013

Shimba A, Cui G, Tani-Ichi S, Ogawa M, Abe S, Okazaki F et al (2018) glucocorticoids drive diurnal oscillations in T cell distribution and responses by inducing interleukin-7 receptor and CXCR4. Immunity 48(2):286–298.e6. https://doi.org/10.1016/j.immuni.2018.01.004

Silver R, LeSauter J, Tresco PA, Lehman MN (1996) A diffusible coupling signal from the transplanted suprachiasmatic nucleus controlling circadian locomotor rhythms. Nature 382(6594):810–813. https://doi.org/10.1038/382810a0

Silver AC, Arjona A, Walker WE, Fikrig E (2012) The circadian clock controls toll-like receptor 9-mediated innate and adaptive immunity. Immunity 36(2):251–261. https://doi.org/10.1016/j.immuni.2011.12.017

Slominski RM, Reiter RJ, Schlabritz-Loutsevitch N, Ostrom RS, Slominski AT (2012) Melatonin membrane receptors in peripheral tissues: distribution and functions. Mol Cell Endocrinol 351(2):152–166

Soták M, Polidarová L, Ergang P, Sumová A, Pácha J (2013) An association between clock genes and clock-controlled cell cycle genes in murine colorectal tumors. Int J Cancer 132(5):1032–1041. https://doi.org/10.1002/ijc.27760

Stier KS, Almasi B, Gasparini J, Piault R, Roulin A, Jenni L (2009) Effects of corticosterone on innate and humoral immune functions and oxidative stress in barn owl nestlings. J Exp Biol 212(Pt 13):2085–2091. https://doi.org/10.1242/jeb.024406

Sun Y, Yang Z, Niu Z, Wang W, Peng J, Li Q et al (2006) The mortality of MOP3 deficient mice with a systemic functional failure. J Biomed Sci 13(6):845–851. https://doi.org/10.1007/s11373-006-9108-4

Suzuki K, Hayano Y, Nakai A, Furuta F, Noda M (2016) Adrenergic control of the adaptive immune response by diurnal lymphocyte recirculation through lymph nodes. J Exp Med 213(12):2567–2574. https://doi.org/10.1084/jem.20160723

Swanson LW, Cowan WM (1975) The efferent connections of the suprachiasmatic nucleus of the hypothalamus. J Comp Neurol 160(1):1–12. https://doi.org/10.1002/cne.901600102

Takahashi JS (2017) Transcriptional architecture of the mammalian circadian clock. Nat Rev Genet 18(3):164–179. https://doi.org/10.1038/nrg.2016.150

Taneja V (2018) Sex hormones determine immune response. Front Immunol 9:1931

Tonsfeldt KJ, Chappell PE (2012) Clocks on top: the role of the circadian clock in the hypothalamic and pituitary regulation of endocrine physiology. Mol Cell Endocrinol 349(1):3–12. https://doi.org/10.1016/j.mce.2011.07.003

Ukai-Tadenuma M, Yamada RG, Xu H, Ripperger JA, Liu AC, Ueda HR (2011) Delay in feedback repression by cryptochrome 1 is required for circadian clock function. Cell 144(2):268–281. https://doi.org/10.1016/j.cell.2010.12.019

Vgontzas AN, Zoumakis E, Bixler EO, Lin H-M, Follett H, Kales A, Chrousos GP (2004) Adverse effects of modest sleep restriction on sleepiness, performance, and inflammatory cytokines. J Clin Endocrinol Metab 89(5):2119–2126. https://doi.org/10.1210/jc.2003-031562

Viswanathan AN, Hankinson SE, Schernhammer ES (2007) Night shift work and the risk of endometrial cancer. Cancer Res 67(21):10618–10622. https://doi.org/10.1158/0008-5472.CAN-07-2485

Von Gall C, Duffield GE, Hastings MH, Kopp MD, Dehghani F, Korf HW, Stehle JH (1998) CREB in the mouse SCN: a molecular interface coding the phase-adjusting stimuli light, glutamate, PACAP, and melatonin for clockwork access. J Neurosci 18(24):10389–10397

Vrang N, Larsen PJ, Mikkelsen JD (1995) Direct projection from the suprachiasmatic nucleus to hypophysiotrophic corticotropin-releasing factor immunoreactive cells in the paraventricular nucleus of the hypothalamus demonstrated by means of Phaseolus vulgaris-leucoagglutinin tract tracing. Brain Res 684(1):61–69. https://www.sciencedirect.com/science/article/pii/000689939500425P

Vujovic N, Gooley JJ, Jhou TC, Saper CB (2015) Projections from the subparaventricular zone define four channels of output from the circadian timing system. J Comp Neurol 523(18): 2714–2737. https://doi.org/10.1002/cne.23812

Waggoner SN (2020) Circadian rhythms in immunity. Curr Allergy Asthma Rep 20(1):2. https://doi.org/10.1007/s11882-020-0896-9

Walker WH 2nd, Kaper AL, Meléndez-Fernández OH, Bumgarner JR, Liu JA, Walton JC, DeVries AC, Nelson RJ (2021) Time-restricted feeding alters the efficiency of mammary tumor growth. Chronobiol Int 39(4):535–546. https://doi.org/10.1080/07420528.2021.2011306

Walker WH, Meléndez-Fernández OH, Nelson RJ (2019) Prior exposure to dim light at night impairs dermal wound healing in female C57BL/6 mice. Arch Dermatol Res 311(7):573–576

Wang D, Cui L-N, Renaud LP (2003) Pre- and postsynaptic GABA(B) receptors modulate rapid neurotransmission from suprachiasmatic nucleus to parvocellular hypothalamic paraventricular nucleus neurons. Neuroscience 118(1):49–58. https://doi.org/10.1016/s0306-4522(02)00906-5

Wegrzyn LR, Tamimi RM, Rosner BA, Brown SB, Stevens RG, Eliassen AH, Laden F, Willett WC, Hankinson SE, Schernhammer ES (2017) Rotating night-shift work and the risk of breast cancer in the nurses' health studies. Am J Epidemiol 186(5):532–540. https://doi.org/10.1093/aje/kwx140

Welsh DK, Takahashi JS, Kay SA (2010) Suprachiasmatic nucleus: cell autonomy and network properties. Annu Rev Physiol 72:551–577. https://doi.org/10.1146/annurev-physiol-021909-135919

Wichert K, Rabstein S, Stang A, Erbel R, Eisele L, Arendt M, Keimer A, Dragano N, Hoffmann W, Lerch MM, Roskoden FC, Schmidt CO, Völzke H, Jöckel K-H, Brüning T, Behrens T (2020) Associations between shift work and risk of colorectal cancer in two German cohort studies. Chronobiol Int 37(8):1235–1243

Yamamura Y, Yano I, Kudo T, Shibata S (2010) Time-dependent inhibitory effect of lipopolysaccharide injection on Per1 and Per2 gene expression in the mouse heart and liver. Chronobiol Int 27(2):213–232. https://doi.org/10.3109/07420521003769111

Yang G, Chen L, Grant GR, Paschos G, Song W-L, Musiek ES et al (2016) Timing of expression of the core clock gene Bmal1 influences its effects on aging and survival. Sci Transl Med 8(324): 324ra16. https://doi.org/10.1126/scitranslmed.aad3305

Yasuniwa Y, Izumi H, Wang K-Y, Shimajiri S, Sasaguri Y, Kawai K, Kasai H, Shimada T, Miyake K, Kashiwagi E, Hirano G, Kidani A, Akiyama M, Han B, Wu Y, Ieiri I, Higuchi S, Kohno K (2010) Circadian disruption accelerates tumor growth and angio/stromagenesis through a Wnt signaling pathway. PLoS One 5(12):e15330. https://doi.org/10.1371/journal.pone.0015330

Zhang R, Lahens NF, Ballance HI, Hughes ME, Hogenesch JB (2014) A circadian gene expression atlas in mammals: implications for biology and medicine. Proc Natl Acad Sci U S A 111(45): 16219–16224. https://doi.org/10.1073/pnas.1408886111

Zhang D, Chen G, Manwani D, Mortha A, Xu C, Faith JJ et al (2015) Neutrophil ageing is regulated by the microbiome. Nature 525(7570):528–532. https://doi.org/10.1038/nature15367

Zhou YD, Barnard M, Tian H, Li X, Ring HZ, Francke U et al (1997) Molecular characterization of two mammalian bHLH-PAS domain proteins selectively expressed in the central nervous system. Proc Natl Acad Sci U S A 94(2):713–718. https://doi.org/10.1073/pnas.94.2.713

Zhuang X, Rambhatla SB, Lai AG, McKeating JA (2017) Interplay between circadian clock and viral infection. J Mol Med 95(12):1283–1289. https://doi.org/10.1007/s00109-017-1592-7

Influence of the Gut Microbiota on Neuroendocrine-Immune Interactions

12

Michael T. Bailey

Abstract

Body surfaces are colonized by an enormous array of microbes collectively referred to as the microbiota. In the past decade, there has been an increase in research activity directed toward understanding the importance of microbiota for human health and disease. This research activity has clearly established that the commensal microbiota play an essential role in the development and maintenance of physiological processes, including neuroendocrine and immune responses, throughout life. Although our understanding of exactly how these microbes exert their effects on the host is still in its infancy, multiple pathways have been identified in recent years. These pathways are described here, along with a description of some of the tools needed for their study.

Keywords

Brain-gut axis · Microbiota · Probiotic · Prebiotic · Neuroendocrine-immune interactions · Metabolites

M. T. Bailey (✉)
Center for Microbial Pathogenesis, Abigail Wexner Research Institute at Nationwide Children's Hospital, Columbus, OH, USA

Department of Pediatrics, Wexner Medical Center, The Ohio State University, Columbus, OH, USA

Oral and Gastrointestinal Microbiology Research Affinity Group, Abigail Wexner Research Institute at Nationwide Children's Hospital, Columbus, OH, USA

Institute for Behavioral Medicine Research, The Ohio state University, Columbus, OH, USA
e-mail: Michael.Bailey2@nationwidechildrens.org

J. P. Konsman, T. M. Reyes (eds.), *Neuroendocrine-Immune System Interactions*, Masterclass in Neuroendocrinology 13,
https://doi.org/10.1007/978-3-031-21358-8_12

12.1 Introduction

Body surfaces are colonized by a vast array of microbes that are collectively referred to as the microbiota. The relatively recent development of next-generation, high throughput "Omic" techniques has led to our understanding that these microbes contribute to nearly every aspect of host physiology, ranging from digestion and absorption of nutrients from the diet to regulation of immune responses and development of the brain and behavior. Despite our newfound appreciation of the importance of the microbiota for maintaining health, or contributing to disease, our understanding of the biological pathways through which the microbiota exert their effects is still burgeoning. This review will discuss the gastrointestinal (GI) microbiota and pathways through which they can impact neuroendocrine and immune responses.

12.2 The Microbiota

Every surface of the body is colonized by microbes, including the skin, oral cavity, respiratory tract, and reproductive tract. But, the vast majority of microbes in the body reside within the GI tract as part of the GI (or commonly referred to as the gut) microbiota. The upper regions of the GI tract, namely the stomach and duodenum, maintain relatively low bacterial numbers ranging from 10^2 bacteria in the stomach to 10^4 bacteria per gram in the duodenum. Low levels of bacteria are maintained in these regions through normal physiological processes, such as the low gastric pH, secretion of bile into the duodenum and the relatively rapid motility in these upper GI locations (Berg 1996). In more distal regions, however, such as the ileum and the colon, motility slows and bacterial numbers range from 10^6 in the ileum to over 10^{12} bacteria per gram in the colon (Berg 1996) (Fig. 12.1). When considered in total, there are approximately three times as many bacteria as there are human cells in the body (Jones et al. 2014).

In addition to having a large number of bacteria in the intestine, there is also a large diversity of bacteria. The human body can harbor between 500 and 1000 different species of bacteria. These bacteria primarily belong to 11 different phyla (i.e., Bacteroidetes, Firmicutes, Actinobacteria, Proteobacteria, Verrucomicrobia, Fusobacteria, Spirochaetes, Tenericutes, Chlamydiae and Synergistetes (Hugon et al. 2015)), but the majority of bacteria (~85%) belong to the Bacteroidetes and Firmicutes, with Proteobacteria, Actinobacteria, Verrocomicrobia and Fusobacteria also commonly found (Bilen et al. 2018; Eckburg et al. 2005; Hugon et al. 2015; Li et al. 2014). Despite belonging to a relatively small number of bacterial phyla, the bacterial species found in the body are genetically diverse. As a result, when all members of the microbiome are considered together, there are approximately 150 times as many bacterial genes that can be expressed in the body as there are genes in the human genome (Qin et al. 2010).

Although this chapter focuses on bacteria, and bacteria have been the most widely studied members of the microbiome in relation to host neuroendocrinology,

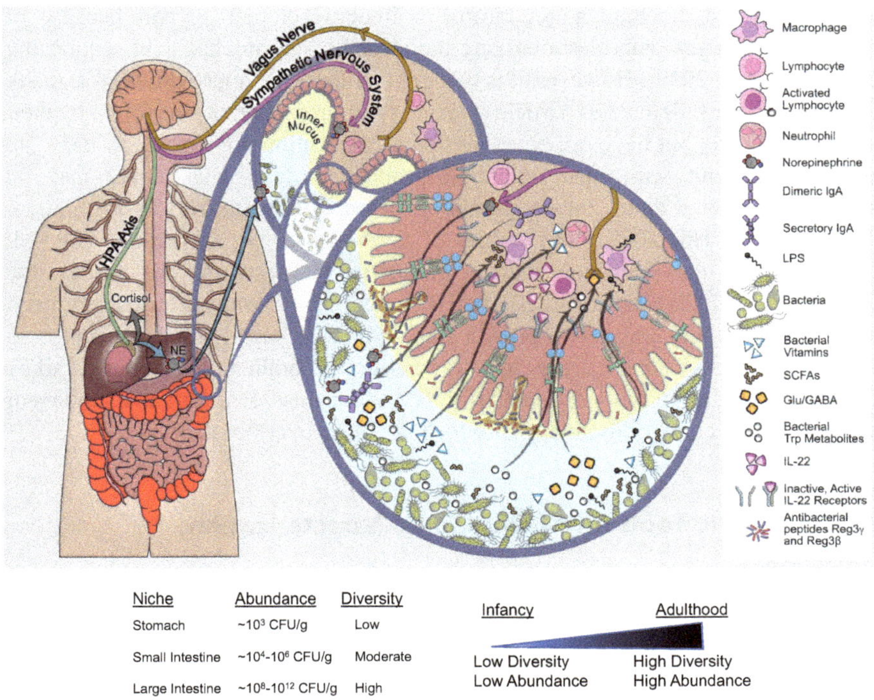

Fig. 12.1 Gut microbiota and neuroendocrine-immune interactions. Both the abundance and the diversity of microbes are low in proximal portions of the intestine (i.e., stomach and upper small intestine), but in more distal regions of the intestine (i.e., distal small intestine and large intestine) bacterial abundance and bacterial diversity significantly increase. In addition, intestinal microbial diversity and abundance are low in early infancy but increase into adulthood (bottom panel). There is a layer of mucus that lines intestinal epithelial cells and secretory immune components, such as secretory IgA and antimicrobial peptides, are embedded in the mucus to keep bacteria spatially separated from the intestinal epithelium. In addition, there are tight junctions between epithelial cells to prevent microbes from entering the body. Neuroendocrine activation can lead to reductions in mucus and secretory immunity and disrupt intestinal tight junctions allowing microbes or microbial products (like LPS) to translocate into the body and triggering a low-grade activation of macrophages, neutrophils, and lymphocytes. Neuroendocrine activation, or neuroendocrine-immune activation, can also disrupt microbial communities in the intestine leading to changes in the production of microbial metabolites, such as vitamins, tryptophan metabolites, glutamate/GABA, and short-chain fatty acids. These metabolites can significantly alter immune system activity (such as by altering the cytokine IL-22 which regulates antimicrobial peptide secretion) or the neuroendocrine system (such as by stimulating the vagus nerve)

neuroimmunology, and behavior, it is important to remember that prokaryotic microorganisms (that do not have a nucleus and includes Archaea in addition to bacteria), eukaryotic microorganisms (that do have a nucleus and includes the myco- (i.e., fungal) biome) and viruses (which are non-living entities and comprise the virome) are all part of the microbiome. The importance of the virome for health and disease has become apparent in recent years (Liang et al. 2022), but our understanding of the human gut virome is underdeveloped, in part because of technical challenges with characterizing the virome using nucleic acid sequencing (Garmaeva et al. 2019). However, it is thought that there are approximately as many viral-like particles (defined as particles that are presumed to be viruses (often by their genetic sequence) but have not been verified as replication competent) as there are bacteria in the gut, with many additional prophages (defined as genetic material predicted to be viral that is integrated into a bacterial genome) (Hoyles et al. 2014; Shkoporov and Hill 2019). The impact of these viruses on the host is not well understood, but they can have indirect effects by changing the host bacteriome (Gregory et al. 2020; Jahn et al. 2019). The fungal mycobiome is present in much lower numbers, but fungal overgrowth (e.g., *Candida* yeast) has been linked to various disease states and has the capacity to affect the brain and behavior (Markey et al. 2020). Future studies will benefit from a comprehensive view of all members of the microbiome community, many of which are often overlooked.

12.2.1 Genetic Tools for Microbiome Characterization

The human microbiome is highly individualized, and its structure is typically characterized in terms of diversity. Alpha diversity is an assessment of the richness (i.e., the number of types of bacteria in a community) and evenness (i.e., the distribution of individual bacterial types) in a specific sample or group of samples, whereas beta diversity assesses differences in microbial community composition between groups/individuals. In general, alpha diversity assesses how many different types of bacteria are found in a group of individuals, whereas beta diversity assesses how many different bacterial types are shared or distinct between groups of individuals. This diversity can be assessed using genetic tools, the most basic of which is 16S rRNA gene sequencing (Finotello et al. 2018) (Table 12.1). With this approach, a variable region of the bacterial 16S rRNA gene (often variable region 4, or V4) is amplified using PCR, and the PCR amplicons are then sequenced. This is done because the 16S rRNA gene is unique to bacteria (i.e., eukaryotes have an 18S rRNA that is genetically distinct from the bacterial 16S rRNA) and because the 16S rRNA gene consists of variable and non-variable regions for which designed primers are used to bind the non-variable region (found in all bacteria) and to amplify variable regions (that are unique to specific bacteria). The 16S rRNA sequences are then aligned to databases of known bacterial 16S rRNA gene sequences and the taxonomy (typically to the genus level) and the relative abundances of bacteria in the samples are determined. This allows one to characterize bacterial diversity and relative abundances within a single sample (Finotello et al. 2018).

Bacterial diversity in the intestine is very low early in the lifespan, but diversity significantly increases at later ages (Dominguez-Bello et al. 2011). The developing fetus is largely devoid of large numbers of bacteria. Although whether there is fetal bacterial colonization has been a subject of debate (Perez-Munoz et al. 2017; Zakis et al. 2022), what is clear is that during birth, the infant is exposed to maternal and environmental bacteria that are the first to colonize the newborn. These microbes

Table 12.1 Tools for microbiome research

"Omic" tools	What it captures	Strengths	Weaknesses
16S rRNA gene sequencing	• Sequence of short region of 16S rRNA gene	• Preferred for characterizing community composition	• Bacterial species are not able to be identified consistently
	• Overall microbial community structure		• Unable to determine what bacterial genes are present
	• Relative abundances of individual bacterial taxa		
Shotgun metagenomic sequencing	• Sequence of short regions of all bacterial DNA in a community	• Characterizes what bacteria are capable of doing based on gene abundances and MAGs	• Unable to determine functional state of bacteria (only their genetic potential)
	• Abundance of bacterial genes in a community		
	• Bacterial genomes inferred from metagenome-assembled genomes (MAGs)		
Metatranscriptomics	• Characterizes genes being expressed within an entire bacterial community	• Provides a snapshot of bacterial gene expression at the time of sampling	• Bacterial gene expression can be very transient and sensitive to multiple signals
	• Offers insight into what bacteria are doing		• Not all expressed genes are translated
Proteomics/ metabolomics	• Characterizes all of the proteins and/or small molecules within an environment	• Identifies compounds within a microenvironment that may have important biological outcomes	• Many proteins and metabolites are produced by both the host and its microbiota making it difficult to determine the nature of host–bacterial interactions.
Animal models	**Approach**	**Strengths**	**Weaknesses**
Germfree/ gnotobiotics	• Germfree (not colonized or previously exposed to any type of microbe) or gnotobiotic (colonized by a single microbe or a defined	• Germfree mice are relatively easy to colonize with a specific microbe or microbial consortium providing greater experimental control	• Host physiology develops abnormally in the absence of microbes

(continued)

Table 12.1 (continued)

	microbial consortium) animals compared to conventional animals		
			• Specialized vivarium and approaches required; labor-intensive
Antibiotic-treatment	• Antibiotic cocktails administered to conventional animals to disrupt the commensal microbiota	• Unlike germfree animals, conventional animals have normal physiological development prior to disrupting the microbiota	• Antibiotic cocktails do not completely eliminate the host microbiota
	• Used alone or prior to administering an experimental microbe/microbial consortium		• Newly administered microbes or microbial consortia may have difficulty engrafting and/or persisting
Pre/probiotics	• Prebiotics administered to alter bacterial abundances and/or bacterial activity	• Ability to modify beneficial bacteria is highly translational when successful	• Approaches are not always targeted and mechanisms of action are not always clear
	• Probiotic administered to study how specific microbes impact host immunity and physiology		• Probiotics do not easily colonize hosts and must be administered repeatedly

dominate in the infant, thus, alpha diversity in the neonate is very low. However, as the infant develops, the microbiome becomes more diverse and is shaped by both intrinsic factors (e.g., host genetics, digestive physiology, immune functioning) and extrinsic factors (e.g., mode of delivery, antibiotic use, and diet (formula fed vs. breastfed)). As the infant begins to eat a more consistent adult-like diet, diversity increases even more and a microbial structure develops that is relatively stable throughout later childhood and into adulthood (Dominguez-Bello et al. 2011).

When bacterial diversity and community composition are "altered" it is often referred to as "dysbiosis". Although widely found throughout both the lay and scientific literatures, the appropriateness and usefulness of this term has been debated (Brussow 2020; Hooks and O'Malley 2017; Tiffany and Baumler 2019). Olesen and Alm (2016) even argue that the term is based on prescientific thought and is not useful in more modern microbiome science (Olesen and Alm 2016). Many of the problems with the term dysbiosis revolve around the fact that the microbiome is highly individualized in humans. As a result, it is not yet clear what distinguishes a "normal" microbiome from a "dysbiotic" microbiome. We conceptualize dysbiosis as a difference or change in microbial community composition that has been experimentally verified to be causally linked to an effect on host physiology. This is the conceptualization that will be used here.

While determining microbial community structure can be insightful for determining whether a specific treatment or condition is related to changes in the microbiome (Fuhrman 2009), it is really differences in microbiome function that are important for understanding the microbiome's role in host physiology. Differences in microbial community structure imply differences in community function (Fuhrman 2009), with the function of a community referring to the biological effects of the microbiome on the microbial community itself (such as nutrient flow between bacteria) or the biological effects on the host (O'Hara and Shanahan 2006). It has been known for decades that the microbiome has positive (as well as negative) influences on the host, with perhaps the most well-studied influence being protection from invading pathogens (Bailey 2012). Common microbiota, including *Bifidobacterium* and *Lactobacillus* species are known to prevent pathogen colonization and can increase protective immune responses (Bailey 2012). More recently, however, it has become evident that the microbiota can impact multiple physiological processes that impinge on neuroendocrine-immune interactions.

12.2.2 Tools to Characterize Microbial Community Functions

Microbial community function can be inferred from microbial community composition using predictive software (for example Douglas et al. (2020)), but determining the genetic potential of the microbiome (i.e., the metagenome), the expression of genes (i.e., the metatranscriptome), or the production of proteins and small molecules (i.e., the proteome and metabolome) can provide greater insight into microbiome function (Franzosa et al. 2015, 2018) (Table 12.1). Shotgun metagenomic sequencing is a sequencing approach wherein all the bacterial genes within a sample can be sequenced and characterized (Sharpton 2014). This allows for the determination of gene abundances within a community, or can be used to assemble the genomes of specific bacteria to create metagenome-assembled genomes (MAGs) (Goussarov et al. 2022). If sampled deeply enough, these MAGs describe the genetic capability of all bacteria in a community. Thus, more information is gleaned from this approach, compared to traditional 16S rRNA gene sequencing, because it allows for the characterization of the taxonomy and the genomic functional potential from the bacterial population (Franzosa et al. 2015).

Metagenomics and the analysis of MAGs are powerful tools in microbiome research, but these approaches do not describe what bacteria are doing, only what bacteria are capable of doing. In contrast, metatranscriptomics, which entails directly sequencing the microbial community transcript pool, specifically messenger RNA (mRNA), can be used to identify which genes are being expressed within a microbial community (Shakya et al. 2019) (Table 12.1). This approach can provide an important snapshot of microbial community function, but this technology is still developing and does have some significant limitations. For example, bacterial mRNA transcripts exhibit varying levels of stability and genes may be expressed for varying lengths of time (ranging from several seconds to several minutes) based on bacterial replication and environmental conditions, such as temperature, nutrient availability, and oxygen levels (Vargas-Blanco and Shell 2020) that are commonly affected during disease and dysbiosis. Ultimately, this complicates metatranscriptome characterization and interpretation and makes sampling considerations crucial (Franzosa et al. 2015).

An alternative to metatranscriptomics is the use of proteomics and metabolomics, which avoid the issue of mRNA stability and can be used to characterize all of the proteins or metabolites in a specific microbial community (Misheva et al. 2021) (Table 12.1). The most commonly used metabolomics platforms involve characterizing all of the compounds in a sample based on mass and charge, using Mass Spectometry (MS). The mass and charge of the molecules are then compared to various databases to identify the compounds, classify metabolic functions, and reconstruct metabolic pathways within the sample. When used with metagenomic sequencing, it is possible to determine which genes are present (and possibly expressed), along with which gene products are produced (Franzosa et al. 2015). Through the use of inter-omic software (Liu et al. 2021), it is becoming increasingly possible to differentiate bacterial from host products and determine which bacterial products differ between groups and individuals. It is these bacterial products that have the capacity to affect the host.

12.2.3 Neuroendocrine Effects on Microbiome Stability

The adult intestinal microbiota form climax communities, wherein the members of the community have reached a steady/balanced state as a result of ecological successions involving the selection of microbes best adapted for their given niche (Huffnagle 2010). During homeostasis, climax communities are largely stable and the types of bacteria, and their relative proportions within the community, remain relatively constant and resistant to change (Allison and Martiny 2008). However, there are multiple factors that can disrupt microbial community composition, including disease, medications, antibiotics, and diet. Of interest to the topic of this chapter, neuroendocrine activation can impact gut microbiota composition. This has been studied in multiple contexts, including multiple studies in rodents, nonhuman primates, and humans, that demonstrate that exposure to stress can lead to changes in microbiome structure and function.

Multiple studies in rodents show that behavioral stressors, including prenatal restraint stress, maternal separation, social defeat, physical restraint, overcrowding, tail shock, and chronic subordinate housing, change gut microbiome structure and function (Antonson et al. 2020; Bharwani et al. 2016; Gur and Bailey 2016; Gur et al. 2016, 2018; Jasarevic et al. 2015, 2017; Maslanik et al. 2012; Reber et al. 2016). The mechanisms by which this occurs are not yet defined, but it is known that microbes can recognize and respond to neuroendocrine factors. Early studies focused on neuroendocrine factors related to the stress response, including those involved with the sympathetic nervous system (SNS) and the hypothalamic-pituitary-adrenal (HPA) axis. Activation of the SNS leads to the release of catecholamines, i.e., norepinephrine and epinephrine, from nerve terminals in innervated organs or from the medulla of the adrenal gland (respectively). The catecholamines are largely responsible for the classical fight or flight response to stress, including increasing vigilance, heart and breathing rate and blood flow to essential organs (e.g., brain, heart, and lungs) and peripheral muscles. This arm of the stress response is activated almost instantaneously, whereas activation of the HPA axis occurs over the course of several minutes and leads to the release of glucocorticoids (namely cortisol in humans or corticosterone in laboratory rodents) from the adrenal cortex. The glucocorticoids are necessary for maintaining high glucose levels in the body to help support the fight or flight response (Russell and Lightman 2019).

Bacteria are able to respond to stressor-induced neuroendocrine factors of the SNS and HPA axis. This was first described for pathogenic bacteria, such as *Yersinia enterocolitica* and pathogenic *Escherichia coli*, that have significant increases in growth and enhanced production of virulence factors when in vitro, serum-based growth media are supplemented with norepinephrine, epinephrine, or dopamine (Bansal et al. 2007; Freestone et al. 2002; Lyte et al. 1997; Vlisidou et al. 2004). While multiple studies have focused on catecholamines (Hughes et al. 2009; Moreira et al. 2016; Moreira and Sperandio 2012; Pullinger et al. 2010; Sandrini et al. 2010), it is also recognized that glucocorticoids can affect bacterial metabolism (Morris and Brem 2019; Morris and Ridlon 2017). Although it is recognized that these neuroendocrine factors can reach the lumen of the intestine, further work is needed to determine the extent to which hormones can impact gut microbiota *in vivo*. It is highly likely that luminal neurotransmitters and hormones directly affect bacteria *in vivo*, since early studies showed that chemical sympathectomy, which increased norepinephrine levels in the intestine acutely, also led to increases in commensal *E. coli* (Lyte and Bailey 1997). More recently, Moreira et al. (2016) showed that the murine pathogen, *Citrobacter rodentium*, which is very closely related to enteropathogenic *E. coli*, has a decreased ability to colonize mice lacking dopamine β-hydroxylase, which is needed to produce norepinephrine and epinephrine (Moreira et al. 2016). This study demonstrates that host catecholamines directly affect intestinal luminal bacteria.

12.2.4 Involvement of the Immune System in Neuroendocrine Effects on the Microbiome

The neuroendocrine system may also have a more indirect effect on the microbiome by acting on host immune and physiological factors known to shape the microbiome. A single layer of epithelial cells separates intestinal microbes from the interior of the body. Given the enormous number of bacteria in the intestine, it is not surprising that the epithelial barrier is reinforced. The intestine is lined with a protective layer of mucus comprised of glycosylated mucins that are constitutively produced and secreted by goblet cells. Exposing mice to a social defeat stressor significantly reduces mucus thickness (Allen et al. 2022; Jaggers et al. 2022). This reduction in mucus thickness was associated with significant changes in intestinal epithelium gene transcription, especially of genes related to immune activation and the production of antimicrobial peptides (Allen et al. 2022; Jaggers et al. 2022). Antimicrobial compounds, such as defensins, lysozym, and regenerating islet-derived protein 3 (Reg3) are secreted into the intestinal mucus layer to help prevent bacterial penetration (Bevins and Salzman 2011; Meyer-Hoffert et al. 2008; Vaishnava et al. 2011). As a result, intestinal bacteria are kept spatially separated from intestinal tissue, and antimicrobial peptides shape the microbiome (Hooper 2015). Although stress has been shown to alter the production of antimicrobial peptides (Allen et al. 2022), whether this is due to the effects of glucocorticoids and catecholamines has not been extensively studied. However, glucocorticoids have the potential to affect antimicrobial peptides. For example, glucocorticoid treatment increased beta-defensin production by respiratory epithelial cells in the absence of pathogen challenge but decreased beta-defensins in the presence of a pathogen (Marin-Luevano et al. 2021).

Antimicrobial peptide secretion is strongly influenced by cytokine signals, with the cytokine IL-22 playing a prominent role. IL-22 belongs to the IL-10 family of cytokines (Parks et al. 2015) and exerts its action by binding to the IL-22 receptor (Dumoutier et al. 2000; Kotenko et al. 2001; Xie et al. 2000). When IL-22 binds to its receptor on epithelial cells, it leads to the secretion of antimicrobial peptides (e.g., Reg3β and Reg3γ) by epithelial cells (Zheng et al. 2008), as well as expansion of intestinal stem cells (Lindemans et al. 2015). Thus, it is not surprising that IL-22 has been found to have multiple effects on microbiome composition. For example, administering IL-22 antibody to ex-germfree mice significantly microbiome composition (Nagao-Kitamoto et al. 2020). Likewise, mice lacking the IL-22 receptor have significant differences in microbiome composition (Gaudino et al. 2021) that increase susceptibility to intestinal pathogens (Pham et al. 2014). Although the effects of hormones and neurotransmitters on IL-22 secretion have not been well studied, it is known that calcitonin-gene related peptide upregulates IL-22 production by γ/δ T cells (Peng et al. 2022), whereas glucocorticoids can induce apoptosis in IL-22 producing CD45+CD90+ innate lymphoid cells (Shaler et al. 2021). In animals exposed to experimental stressors, IL-22 levels have been found to be increased (Elkhatib et al. 2022; Li et al. 2022) or decreased (Shaler et al. 2021). The discrepant effects of stress exposure on IL-22 levels may be dependent upon gut microbes, which can produce metabolites of the amino acid tryptophan to stimulate IL-22 release through activation of the aryl-hydrocarbon receptor (AhR) (Pernomian et al. 2020; Safe et al. 2020; Zelante et al. 2013). Undoubtedly, IL-22 and subsequent antimicrobial peptides are important components of the brain-gut axis, but further research is needed to fully understand how these factors impact health and disease.

In addition to antimicrobial peptides, immunoglobulin A (IgA) helps to shape microbial communities. In response to the cytokines IL-4, IL-5, IL-6, and IL-10, IgA + plasmablasts synthesize and release dimeric IgA (Cerutti and Rescigno 2008;

Pabst 2012). In the intestine, the IgA is passed into the lumen of the intestine as secretory IgA (sIgA) (Cerutti and Rescigno 2008; Kaetzel et al. 2017; Mowat and Agace 2014; Pabst 2012). The sIgA plays a role in neutralizing pathogens as well as commensal microbes. Although the majority of commensal microbes bind to sIgA with low affinity, during inflammatory conditions coating of gut bacteria with sIgA is increased to reduce bacterial invasion (Chen et al. 2020; Li et al. 2020; Macpherson et al. 2015; Pietrzak et al. 2020). Multiple studies demonstrate that stress can change sIgA in the intestine, with most (Eriksson et al. 2004; Jarillo-Luna et al. 2007; Liu et al. 2012; Martinez-Carrillo et al. 2011), but not all (Aguilera et al. 2013; Reyna-Garfias et al. 2010) studies showing that stress decreases intestinal sIgA. Both the HPA axis and the sympathetic nervous system are thought to contribute to stress-induced reductions in sIgA, since adrenalectomy completely prevented, and treating mice with a glucocorticoid receptor antagonist (i.e., RU-486) or with a sympathetic neurotoxin (i.e., 6-hydroxydopamine) partially prevented the stress-induced reduction in sIgA (Jarillo-Luna et al. 2007). In studies assessing the microbiome, concomitant changes in the microbiome suggest that the stress-induced effects on sIgA impact the microbiome (Aoki-Yoshida et al. 2016; Rengarajan et al. 2020).

12.2.5 Neuroendocrine-Immune Effects on the Microbiome in Humans in Humans

The effects of neuroendocrine hormones and immune system activity on the human bacteriome have not been studied, but there is evidence to support that stress affects microbiome composition in humans. For example, in a study of Japanese medical students, it was found that exam stress significantly reduced *Bifidobacterium* relative abundance (Nishida et al. 2019). Although few studies have examined the microbiome in response to specific stressors, there have been multiple studies that have tested whether perceived stress is related to microbiome composition, and the results have been mixed. In one study, individuals with different levels of perceived stress were not found to have any differences in microbiome composition (Kleiman et al. 2017). However, others have found perceived stress to be related to microbiome differences. Interestingly, these associations were dependent upon the ancestry of the host. In caucasian women, higher levels of perceived stress were associated with higher relative abundances of *Clostridium* and *Ruminococcus*, but lower relative abundances of *Fusobacterium*. However, this pattern was reversed in black women (Carson et al. 2018). It is important to note, however, that this study did not control for factors, such as diet, that may be related to race and ethnicity, making it difficult to discern whether there are differences in the relationship between perceived stress and the microbiome in these populations. However, other studies have also found a relationship between perceived stress and the microbiome. For example, perceived stress has been studied in individuals with inflammatory bowel disease, which is characterized by chronic intestinal inflammation. In both adult (Humbel et al. 2020) and pediatric patients (Mackner et al. 2020) with inflammatory bowel disease, perceived stress was associated with significant changes in microbiome composition. Although the specific bacterial genera that were associated with perceived stress were different in adults and children, the study still demonstrates significant associations between perceived stress and the microbiome in individuals with chronic intestinal inflammation.

Exposure to stressful stimuli may have stronger effects on microbiome composition early in the life span compared to later in life, since stable climax communities have not yet developed in the young infant. In support of this, early life stressors have significant effects on microbiome composition in animal models (Bailey and Coe 1999; Jasarevic et al. 2015; O'Mahoney et al. 2009; Park et al. 2021). Support for early life stress effects on the microbiome is evident in humans as well. For example, a study of neonates in a neonatal intensive care unit found that higher levels of neonatal infant stressors were associated with differences in the microbiome compared to infants with lower exposure to neonatal infant stressors (D'Agata et al. 2019). Studies of adverse childhood events have also found that individuals with higher childhood adversity had differences in the microbiome than individuals with lower childhood adversity. Interestingly, these microbiome differences were evident in adulthood even though the adversity occurred in childhood (Hantsoo et al. 2019), demonstrating that early life stress can have lasting effects on microbiome composition.

Anxiety and depression are common sequelae to chronic stress, and it is now evident that in addition to stress, both anxiety and depression have been linked to changes in the gut microbiome. For example, depression has been associated with increases in *Bacteroidales* and *Alistipes* (Jiang et al. 2015; Naseribafrouei et al. 2014), whereas *Prevotella* have been found to be decreased (Jiang et al. 2015; Kelly et al. 2016). Microbiome composition has also been found to be distinct in patients with irritable bowel syndrome and comorbid anxiety (De Palma et al. 2017), and in patients with obsessive compulsive disorder, the relative abundances of *Oscillospira, Odoribacter*, and *Anaerostipes* were found to be lower than in healthy controls (Turna et al. 2020). Although there is an increasing number of studies demonstrating differences in intestinal bacteria in individuals with high levels of stress, anxiety, depression, and other emotional and neurological disorders, the mechanisms by which this occurs are not yet clear. Because stress, anxiety, and depression are associated with changes in neuroendocrine axes and sympathetic nervous system activity, it is tempting to speculate that changes in intestinal bacteria occur through effects of hormones and neurotransmitters directly on bacteria, or indirectly through changes in immune system activity. However, these neuroendocrine factors can also change gastrointestinal physiology. For example, the secretion of gastric acid in the stomach (Yang et al. 2000), motility in the small and large intestine (Tache et al. 2001; Tache and Perdue 2004; O'Malley et al. 2010; Shigeshiro et al. 2012), are altered through HPA and SNS activation. Because gastrointestinal physiology strongly influences which types of bacteria can reside within the intestine, as well as the level to which those microbes can grow (Berg 1996), it is likely that changes in gastrointestinal physiology are additional factors linking the brain and behavior to changes in the gut microbiota. Importantly, these interactions are often bidirectional. Early studies regarding interactions between the brain and gut microbes were correlational in nature; thus, addressing causality of the interactions has been difficult (Cryan and Mazmanian 2022). As a result, researchers have developed additional tools to determine pathways by which neuroendocrine systems and gut microbiota interact.

12.2.6 Tools to Assess the Impact of the Microbiome on the Host

"Omic" studies provide a rationale for suggesting that gut bacteria are related to neuroendocrine functioning, but these studies are limited by their largely correlational nature. Thus, experimental methods have been developed to test causal relationships between gut microbes and host physiology. The most direct way to test whether a microbe affects host physiology is to simply administer the microbe and test the host. However, it is extremely difficult to introduce a new bacterium into an established microbial ecosystem due to colonization resistance, defined as the resistance to colonization by ingested bacteria or inhibition of overgrowth of resident bacteria normally present at low levels (Lawley and Walker 2013). In order for newly administered/ingested bacteria to colonize the intestine, they must survive the low pH of the stomach, locate a suitable niche for growth in the intestine, and outcompete the already established microbiota for access to nutrients (Lawley and Walker 2013). This is a formidable task for commensal microbes. An alternative approach would be to selectively remove bacteria to study their effects on the host. However, it is currently not possible to selectively remove specific bacterial types from a microbial community due to a lack of specificity of available tools (e.g., antibiotics). As a result, researchers have relied on the use of germfree mice, which can be colonized by a single microbe or by a defined microbial consortium (Table 12.1). Germfree mice have been born and raised in sterile environments and fed sterilized food and water. They have never come into contact with any microbes, and as a result are easily colonized by experimental bacteria. Mice colonized with defined microbes, which are then referred to as gnotobiotic mice, can be tested to determine the effects that a specific microbe or a specific microbial consortium has on host physiology (Table 12.1). Although these mice are incredibly useful and are increasingly used in microbiome research, it must be noted that microbes are necessary for normal animal development. As a result, germfree mice have underdeveloped immune systems and multiple abnormalities in brain and behavioral responses (Kennedy et al. 2018). Consequently, studies exclusively using germfree and/or gnotobiotic models must be interpreted with caution.

An alternative approach to the use of germfree mice is the use of antibiotics to disrupt the microbiome (Table 12.1). Antibiotics are not specific enough to eliminate a single species (or even genus) of microbes, and contrary to popular belief, antibiotic cocktails do not remove all bacteria from a host. However, through the use of broad-spectrum antibiotic cocktails over prolonged periods of time, it is possible to significantly reduce the number of bacteria in the intestine (Kennedy et al. 2018). This reduction in bacterial numbers leaves the host susceptible to colonization by new microbes. Thus, studies can use broad-spectrum antibiotic cocktails to either determine if bacterial load is related to neuroendocrine-immune outcomes, or to leave mice susceptible to colonization by newly administered bacteria (or bacterial consortia). After the new colonization, the host can be studied to determine if the newly administered bacteria have effects on host physiology. Although this approach has the advantage that normal, conventional mice are utilized, and thus the mice have fully developed immune and physiological systems, the newly administered microbes do not engraft in antibiotic-treated mice as well as they do in germfree mice (Kennedy et al. 2018). As a result, microbial community composition is not always easily transferred in antibiotic-treated mice, which can lead to results that are difficult to interpret.

The ability of gut microbes to impact neuroendocrine-immune interactions has also been studied through the use of probiotics and prebiotics (Table 12.1). Probiotics are traditionally defined as live microorganisms that when administered in adequate amounts confer a health benefit on the host (Hill et al. 2014). These microbes are often isolated from healthy humans and are commonly members of the genus *Lactobacillus* or *Bifidobacterium*. Prebiotics, on the other hand, are substrates that are selectively used by host microorganisms that confer a health benefit (Gibson et al. 2017). Prebiotics are often fibers that cannot be digested by the host but can be broken down by bacteria. Commensal probiotic bacteria commonly break down these fibers. Thus, prebiotics often have beneficial effects because they enhance the growth and/or activity of probiotic bacteria (Gibson et al. 2017). There are now multiple studies suggesting that prebiotics and probiotics can positively impact the neuroendocrine and immune systems to improve mental health and wellbeing through mechanisms involving the brain-gut-microbiota axis which are described below.

12.3 The Brain-Gut Microbiota Axis

12.3.1 The Brain-Gut Axis

There is extensive bidirectional communication between the brain and the gut, such that activity in one impacts activity in the other. Gastrointestinal functioning is highly influenced by the autonomic nervous system (both the sympathetic and parasympathetic branches) as well as the HPA axis. The autonomic nervous system interacts with the enteric nervous system (ENS), which is the intrinsic nervous system within the gastrointestinal tract. The ENS is comprised of two neuronal plexi in the intestine, i.e., the submucosal and the myenteric plexi, which coordinate intestinal motility and intestinal secretions (Furness et al. 2014). Interneurons that signal between the plexi allow the intestine to function independently of the brain through reflex circuits that detect physiological conditions of the gastrointestinal tract, integrate signals pertaining to physiological conditions, and trigger appropriate secretory and motility responses. Homeostasis in the intestine is maintained by the enteric nervous system, but when challenged, intestinal dysregulation induces autonomic nervous system activity that can lead to altered motility and intestinal secretions (Wehrwein et al. 2016). Because the microbiome is affected by gut motility and secretions, enteric nervous system activity can have significant effects on the gut microbiome. These effects, though, are bidirectional, and stimulation of the enteric nervous system, such as through mechanoreceptors or chemoreceptors, can lead to activation of the autonomic nervous system (Furness et al. 2014). Thus, the enteric nervous system is a key component of microbe-host interactions with specific mechanisms discussed below.

The CNS and the ENS are strongly influenced by gastrointestinal hormones that are mainly secreted by endocrine cells throughout the epithelium of the stomach and small intestine in response to specific nutrients in the intestinal lumen. These enteroendocrine cells are diverse, with each cell type secreting different molecules. For example, cholecystokinin (CCK) is released by duodenum and jejunum enteroendocrine cells called I cells in response to the presence of fatty acids in food being digested (called chyme). The CCK binds to receptors in the stomach (to reduce gastric emptying) and in the pancreas (to increase pancreatic enzyme secretion), which ultimately facilitates digestion and absorption (Borovicka et al. 1997; Schwizer et al. 1997). The CCK also binds to receptors in pancreatic ducts to relax the ducts while stimulating the contraction of the gallbladder to enhance the passage of the digestive enzymes from the pancreas into the duodenum where they will break down fats (Borovicka et al. 1997). The CCK can also stimulate vagal afferent pathways in the CNS to reduce food intake (Smith et al. 1981). Once fats are broken down into monoglycerides or free fatty acids that can be absorbed into the body, CCK is no longer secreted. Similarly, glucagon-like peptide 1 (GLP-1) and peptide YY (PYY) are released from jejunum enteroendocrine cells (specifically L cells) (Habib et al. 2013) in response to luminal glucose and protein (respectively) (Spreckley and Murphy 2015). The GLP-1 and PYY can bind to receptors in the hypothalamus to inhibit food intake (Abbott et al. 2005; Turton et al. 1996) as well as in the stomach to decrease gastric emptying and the pancreas to increase insulin secretion (Spreckley and Murphy 2015). While the importance of these hormones for digestive physiology and related diseases (e.g., obesity) has been well studied (Kuhne and Stengel 2019), interactions with gut microbes have been relatively unexplored. However, it has been shown that CCK antagonists can affect gut microbes in a murine model of nonalcoholic steatohepatitis (NASH) (Gay et al. 2022 and that a GLP-1 receptor agonist can change bacterial abundance in the murine cecum (Kato et al. 2021).

12.3.2 Influence of Microbes on Gut-Derived Peptide Hormones and Immune System Activity

There is increasing recognition that microbes and their products can influence the secretion of gut-derived hormones. For example, many bacteria can ferment polysaccharides leading to the production of short-chain fatty acids (e.g., acetate, propionate, and butyrate) (Tingirikari 2018). These SCFAs are known to have beneficial effects on the host, including anti-inflammatory effects, energy homeostasis effects and effects on lipid and carbohydrate metabolism (De Vadder et al. 2014; Tan et al. 2014). Intestinal L cells express G-protein coupled receptors for SCFA's (Holst 2007; Holzer et al. 2012; Samuel et al. 2008), and bacterial colonization in the gut leads to the release of GLP-1 and PYY through G-protein coupled receptor-dependent mechanisms (Samuel et al. 2008). In addition, administering prebiotics such as inulin-type fructans that result in increased bacterial SCFA production also increases GLP-1 and PYY in healthy individuals (Cani and Delzenne 2009; Cani et al. 2009). These data indicate that bacterial SCFAs can increase GLP-1 and PYY. Short-chain fatty acids are not the only way in which the microbiota may impact gut-derived peptide hormones. For example, bacterial-produced caseinolytic protease B stimulates the secretion of PYY in culture (Dominique et al. 2019a, b) and reduces food intake, body weight gain and fat mass in obese mice, which are physiological effects related to PYY signaling (Legrand et al. 2020). In addition, lipopolysaccharide (LPS) produced by Gram-negative bacteria can reduce CCK signaling (de La Serre et al. 2015), demonstrating multiple ways in which gut bacteria can influence gut hormone release and functions.

Not only are there interactions between gut-derived hormones and the microbiome, but also between these hormones and immune system activity. Although it is not heavily studied, mice lacking the T cell receptor alpha chain have a decrease in CCK, 5-HT, and PYY enteroendocrine cells in infancy, showing that there are immune to enteroendocrine cell interactions (Rubin et al. 2000). However, there is more evidence for enteroendocrine-to immune effects. For example, CCK has been shown to disrupt cytokine production in dendritic cells (Jia et al. 2014) and helper T cells (Oiry et al. 1997; Zhang et al. 2014). CCK can also affect B cells (Zhang et al. 2011), and interestingly the reduction in IgA that commonly occurs with parenteral nutrition can be prevented via the administration of CCK (Genton and Kudsk 2003; Hanna and Kudsk 2000). Agonists of the GLP-1 receptor have been studied for use as anti-inflammatory therapies (Yang et al. 2021). Activation of the GLP-1 receptor can significantly reduce inflammatory cytokine production by LPS-stimulated neutrophils (including IL-1β, IL-6, TNF-α, and IFN-γ (Yanay et al. 2015; Yusta et al. 2015) as well Th2-type cytokines such as IL-4, IL-8, and IL-13 produced by eosinophils (Mitchell et al. 2017). Fewer studies have focused on the effects of PYY on immune system activity, but in early studies, physiologically relevant concentrations of PYY were shown to significantly increase peritoneal macrophage phagocytosis (De la Fuente et al. 1993). It is likely that gut-derived peptide hormones can mediate some of the interactions between gut microbes and immune system activity, but the extent of these interactions and their importance for maintaining health or precipitating disease needs further study.

12.3.3 Gut Microbes and the Gut-Brain Axis

While there have been ample studies showing that exposure to stress, and the ensuing neuroendocrine response, can impact the gut microbiome, it was not until the pioneering study by Sudo et al. that it became evident that the gut microbes also impact the neuroendocrine system (Sudo et al. 2004). In this study, germfree mice had higher adrenocorticotrophic hormone (ACTH) and corticosterone responses to a restraint stressor than did conventional mice. Interestingly, colonizing the mice with commensal bacteria, but not with bacteria that were potentially pathogenic, normalized ACTH and corticosterone responses to stress (Sudo et al. 2004). This colonization, however, had to occur early in life; colonization later in life did not normalize the HPA response to stress (Sudo et al. 2004). There are now multiple studies demonstrating that gut microbes can impact HPA axis activity (Clarke et al. 2013; Crumeyrolle-Arias et al. 2014; Neufeld et al. 2011). These studies spurred research into the role of gut microbes in the gut-brain axis.

12.3.3.1 The Vagus Nerve and Gut Microbe to Brain Signaling

Early studies of microbial influences on gut-brain signaling suggested that bacterial stimulation of the vagus nerve may be a key component (Figure 12.1). Colonizing specific pathogen-free mice with a bacterium not typically found in the murine gut, i.e., *Campylobacter jejuni,* led to the development of anxiety-like behavior. In addition, c-Fos expression in the cell bodies of vagal afferents suggested that bacterial stimulation of the vagus nerve could lead to behavioral changes (Goehler et al. 2005, 2007, 2008). In support of this contention, studies on the mechanisms of action of prebiotics and probiotics have provided evidence of bacterial activation of the vagus nerve. For example, healthy rats given the prebiotic 2'-fucosyllactose have improvements in cognitive behavior (as assessed by operant conditioning), but these effects could be abolished by subdiaphragmatic vagotomy (Vazquez et al. 2016). Similarly, Shank3B-/- mice, which are commonly used to model behavioral deficits consistent with autism spectrum disorder, have improvements in social behavior when given *L. reuteri*; this effect is prevented by subdiaphragmatic vagotomy (Sgritta et al. 2019). From a cellular level, it is evident that probiotic bacteria can stimulate neurons in the myenteric plexus, which contains vagal afferents (Perez-Burgos et al. 2013). It is important to note, however, that not all bacterial effects on the brain and behavior are due to signaling through the vagus nerve. This was evident in early studies when Bercik et al showed that colonizing germfree Balb/C mice with microbiota from NIH Swiss mice increased exploratory behavior and hippocampal brain-derived neurotrophic factor (BDNF). Interestingly, subdiaphragmatic vagotomy did not prevent this effect (Bercik et al. 2011). While more research is needed to understand how gut microbes can impact the brain and behavior through vagal-dependent pathways, this study illustrated that there are additional mechanisms by which gut microbes can impact the brain and behavior.

12.3.4 Interkingdom Communication Through the Use of Common Ligands and Receptors

Molecules used by mammals as neurotransmitters and hormones are found throughout nature. For example, catecholamines have been identified in plants (Kulma and

Szopa 2007), insects (Pitman 1971), and fish (Guerrero et al. 1990). As a result, these molecules in higher-level organisms are thought to have evolved from lower-level organisms, even single-cell organisms like bacteria, that have the genetic capacity to produce them (Roth et al. 1982). For example, acetylcholine (Stephenson and Rowatt 1947), histamine (Devalia et al. 1989; Masson et al. 1996; Thomas et al. 2012), serotonin (Hurley et al. 1971; Kuley and Ozogul 2011; Kuley et al. 2011), and the catecholamines (Asano et al. 2012; Kuley and Ozogul 2011; Kuley et al. 2011; Tsavkelova et al. 2000) have been identified in a wide diversity of bacteria. Germfree mice have significantly lower levels of biologically active catecholamines in the intestinal lumen compared to conventional mice, but colonizing the germfree mice normalized luminal catecholamine levels (Asano et al. 2012). Although these studies suggest that the secretion of these molecules may directly contribute to host behavioral and immune responses, the extent to which this occurs is influenced by multiple factors, as outlined below.

12.3.4.1 Tryptophan and Its Metabolites

The neurotransmitter serotonin (5-HT) is produced from tryptophan and is a prominent signaling molecule throughout the body. Intestinal enterochromaffin cells are the major source of 5-HT in the body, which is released in response to chemical and mechanical stimulation in the intestine itself. Serotonin is used to control intestinal physiology, such as peristaltic reflexes. Although intestinal 5-HT plays multiple roles in the intestine, it is not likely that intestinal 5-HT influences behavioral states, since it cannot pass the blood-brain barrier. However, gut microbes may be able to influence behavioral states by indirectly influencing 5-HT through the metabolism of tryptophan (Trp). The amount of 5-HT produced in the brain is largely affected by the amount of Trp that is transported from the intestine into the blood and eventually into the brain through L-type amino acid transporters (Fotiadis et al. 2013; Ruddick et al. 2006). Bacterial metabolism significantly decreases Trp levels in the host, as exemplified by findings that germfree mice have higher Trp levels than do conventional mice (Clarke et al. 2013). This is because many bacteria can metabolize Trp. For example, bacteria can express tryptophan decarboxylases that decarboxylate Trp to form tryptamine (Agus et al. 2018). This enzyme is commonly found in bacteria such as *Clostridium, Ruminococcus, Blautia,* and *Lactobacillus* spp., all of which are known Trp metabolizers (Williams et al. 2014). Tryptophan can also be metabolized by gut bacteria to form indoles, such as indole-3-aldehyde, indole-3-acetic acid, and indole-3-propionic acid (Agus et al. 2018). In some bacteria, this is regulated by bacterial expression of the enzyme tryptophanase, which is found in a wide diversity of bacteria (Lee et al. 2012; Smith and Macfarlane 1996). Multiple bacteria can also express an aromatic amino acid transferase, which converts Trp to indole-3-pyruvate, which can be further converted into indole-3-lactate, indole-3-acetaldehyde, or indole-3-acetate (Kumavath et al. 2010).

Bacterial metabolism of Trp may have important implications for the brain, behavior, and immunity beyond just reducing the amount of Trp available for the synthesis of 5-HT in the brain. For example, bacterial-derived Trp metabolites have been shown to affect astrocyte activation. Rothhammer et al. (2016) showed that indoxyl-3-sulfate (produced when bacterial-derived indole is metabolized in the liver), as well as bacterial metabolites indole-3-aldehyde and indole-3-pyruvate, reduce astrocyte activation (Rothhammer et al. 2016). Bacterial-derived tryptophan derivatives have also been shown to suppress immune system activity throughout the body, and some of the anti-inflammatory effects of probiotic bacteria such as *Lactobacillus reuteri* have been shown to occur through the production of Trp metabolites (Zelante et al. 2013). For example, under low sugar conditions and high Trp conditions, *L. reuteri* utilizes Trp as an energy source, leading to the production of indole derivatives. The indole derivatives bind to the aryl hydrocarbon receptor (AhR), which has a myriad of effects on the immune system. Bacterial indoles are known to stimulate the production of IL-22 through activation of the AhR on innate lymphoid cells (Zelante et al. 2013). As described earlier, IL-22 leads to the production of antimicrobial peptides that can alter microbiome composition, showing how gut bacteria can influence neuroendocrine-immune interactions (Zheng et al. 2008) (Fig. 12.1). In addition to these interactions, bacterial indole acid-induced activation of the AhR can have potent anti-inflammatory effects, which has been well illustrated in studies of necrotizing enterocolitis (Lu et al. 2021; Meng et al. 2020; Nolan et al. 2021) involving hyperinflammatory responses in the intestines of newborn infants, particularly low birthweight/premature infants (Mara et al. 2018). In animal models, activation of the AhR prevents necrotizing enterocolitis-associated inflammation (Nolan et al. 2021). This does not occur in mice lacking AhR expression on dendritic cells. In contrast, these mice have significant increases in necrotizing enterocolitis (Nolan et al. 2021). Interestingly, probiotic microbes, such as *L. rueteri* and *B. longum*, which are known to metabolize tryptophan and produce AhR ligands, can also prevent necrotizing enterocolitis in animal models (Meng et al. 2020; Olson et al. 2016, 2018). Thus, probiotic metabolism of tryptophan is an exciting area of research that is likely to lead to treatment strategies to affect neuroendocrine and immune responses.

12.3.4.2 Glutamate and GABA: Bacterial Influences on Excitatory-Inhibitory Balance

The balance of the main excitatory (i.e., glutamate (Glu)) and the main inhibitory (i.e., gamma-amino butyric acid (GABA)) neurotransmitters in the body is thought to contribute to a wide range of diseases and conditions, such as autism spectrum disorder, epilepsy, and anxiety/depression (Braat et al. 2015; Braat and Kooy 2015; Luscher et al. 2011; Wallace et al. 2001; Wong et al. 2003). Both of these are produced from dietary glutamine. In the host, Glu is produced through transamination of α-ketoglutarate and by hydrolytic deamination of glutamine by glutaminase (McKenna 2013). Bacteria can also produce Glu. In fact, many probiotic, lactic acid-producing bacteria, such as *L. paracasei* and *L. lactis*, are capable of synthesizing Glu through deamination of glutamine (Sanchez et al. 2017). Although Glu does not typically pass across the intestinal barrier to reach circulation, Glu receptors, including both ionotropic and metabotropic Glu receptors, are found through the length of the intestine in neurons that control the regulation of intestinal sensory, secretory and motor functions (Filpa et al. 2016; Julio-Pieper et al. 2011). The effects of bacterial-derived Glu on host behavior and neuroendocrine responses have yet to be fully understood, but it is known that intragastric administration of monosodium glutamate activates the hippocampus and amygdala, two brain regions known to be involved with memory, learning, and emotion. Interestingly, vagotomy reduced this response (Tsurugizawa et al. 2010, 2014; Tsurugizawa and Uneyama 2014) providing a plausible pathway through which bacterial Glu could signal through the vagus nerve to impact the neuroendocrine system (Fig. 12.1).

In the brain, Glu and GABA are maintained in a homeostatic balance, referred to as the excitatory-inhibitory balance (Fee et al. 2017). Although this has not been studied in the intestine, GABA is produced throughout the mammalian nervous system, including the enteric nervous system, and it is also widely produced by bacteria. In mammals, as well as bacteria, GABA can be produced from Glu by the enzyme glutamate decarboxylase (GAD) (Feehily and Karatzas 2013; Feehily et al. 2013; Strausbauch and Fischer 1970). GABA can also be derived from putrescine, arginine, and ornithine in bacteria (Strandwitz et al. 2019). In many bacteria, resistance to low pH involves the activation of a GABA shunt and the subsequent production of GABA, which helps to buffer the bacteria from harsh acidic environments (Feehily and Karatzas 2013; Feehily et al. 2013). The importance of bacterial-derived GABA for the host is not completely understood, but it is known that the administration of GABA-producing bacteria, such as probiotic *Bifidobacterium dentium*, but not of genetically similar *Bifidobacterium* spp. that lack the GAD gene (e.g., *B. breve*), inhibits visceral pain in rats (Pokusaeva et al. 2017). It is currently not clear whether bacterial-produced GABA can reach the brain (Boonstra et al. 2015), but there are GABA transporters that are expressed in the BBB as well as the intestinal epithelial barrier (Takanaga et al. 2001). Bacterial influences on brain GABA need not occur only through passage of GABA into the brain. Administration of L. *rhamnosus* was shown to increase the expression of GABA receptors in the brain, which was associated with reductions in anxiety-like behavior (Bravo et al. 2011). This effect was blocked by vagotomy, again suggesting that bacterial stimulation of the vagus nerve can directly impact the brain and subsequent behavioral responses.

Both Glu and GABA can significantly impact immune system functioning. Feeding Glu to laboratory animals significantly increased the expression of intestinal tight junctions and mucins, which was associated with changes in the microbiome and lower levels of IL-6 and TNF-α in the blood (Kyoung et al. 2021). GABA can also affect immune system activity, with studies showing that GABA signaling can have either detrimental or protective effects, depending upon multiple factors. In murine models of ulcerative colitis induced by administering dextran sulfate sodium salt (DSS), activation of the GABA A receptors in colonic epithelial cells significantly increased colonic inflammation (Ma et al. 2018). In human patients with ulcerative colitis, however, GABA levels (and the abundance of GABA-producing bacteria) were lower, suggesting that GABA-producing bacteria may be protective (Aggarwal et al. 2017). Protective effects of GABA-producing bacteria were also evident in mice challenged with the intestinal pathogen enterotoxigenic *E. coli* (ETEC). In this model, resistance to the pathogen is dependent upon the intestinal

IL-17 immune response. Interestingly, the IL-17 response was enhanced by bacterial-produced GABA, demonstrating important bacterial-neuroendocrine-immune interactions (Ren et al. 2016). Bacterial-produced GABA may also have important effects on the intestinal barrier by stimulating mucus production and autophagy-mediated calcium signaling in intestinal epithelial cells (Engevik et al. 2019). The extent to which bacterial-produced Glu or GABA affect immune responses outside of the intestine has not yet been well studied, but it is evident that mucosal immune responses are widely affected (Fig. 12.1).

12.3.4.3 Bacterial Stimulation of Oxytocin

Studies showing that orally administering antibiotic cocktails that are not easily absorbed from the intestine in mice reduces oxytocin levels in the hypothalamus were among the first to suggest that gut microbes can impact brain oxytocin (Desbonnet et al. 2015). Administering antibiotics also disrupted social behavior, which is not surprising when considering that oxytocin plays an essential role in social behavior (Desbonnet et al. 2015). This finding was consistent with studies in germfree mice that have abnormal social behavior that can be normalized by colonization with normal microbiota (Desbonnet et al. 2014). Environmental influences on the microbiome can have downstream effects on host oxytocin. For example, the offspring of pregnant mice with diet-induced obesity were found to have lower oxytocin, and lower social behavior, than offspring from normal-weight mice. The dams with diet-induced obesity had lower levels of *L. reuteri*, and when *L. reuteri* were administered to their offspring, oxytocin and social behavior were normalized (Buffington et al. 2016). Interestingly, the probiotic *L. reuteri* has now been shown to increase host oxytocin in a variety of different studies (Buffington et al. 2016, 2021; Poutahidis et al. 2013; Sgritta et al. 2019; Varian et al. 2017), including in studies assessing animal models of autism spectrum disorder (Sgritta et al. 2019). The mechanisms by which *L. reuteri* can increase oxytocin are not yet clear, but findings that *L. reuteri* can increase oxytocin and social behavior, even in genetic murine models of autism spectrum disorder, have led to clinical trials involving *L. reuteri* (for example Clinical Trial: NCT04944901) to determine whether *L. reuteri* may be useful in the treatment of autism spectrum disorders.

Beyond its role in social behaviors, oxytocin has been gaining recognition for its ability to alter the immune response. The oxytocin receptor is found on both innate and adaptive leukocytes, and stimulation of the oxytocin receptor is largely thought to be anti-inflammatory (Mehdi et al. 2022). For example, oxytocin can reduce Toll-like receptor expression on neutrophils and can inhibit IL-6 production (Iseri et al. 2005a, b; Szeto et al. 2017). Interestingly, activation of the transcription factor NF-κβ, which leads to the expression of multiple inflammatory cytokines and chemokines, also increases the expression of the oxytocin receptor (Szeto et al. 2017). Stimulation of the oxytocin receptor, in turn, inhibits the mitogen-activated protein (MAP) kinases and downstream NADPH oxidases (Rashed et al. 2011; Szeto et al. 2008). The extent to which gut microbes influence immune system activity by altering oxytocin levels is not yet clearly known, but this possibility seems highly likely given the ability of probiotic microbes to both alter immunity and stimulate oxytocin.

12.3.5 Other Microbial Metabolites that Can Affect the Brain, Behavior, and Immunity

12.3.5.1 Short-Chain Fatty Acids (SCFA)

Alterations in molecules that can be used as hormones and neurotransmitters are not the only ways in which bacteria can communicate with the host. Wikoff's study from 2009 showed that there are hundreds of metabolites in the plasma of conventional mice that are not found in germfree mice (Wikoff et al. 2009), and many of these bacterial metabolites can impact the immune system, brain, and behavior. For example, the short-chain fatty acids, primarily butyrate, propionate, and acetate, are produced by bacterial fermentation of carbohydrates in the diet (Champ 2004; den Besten et al. 2013; Louis and Flint 2009, 2017; Louis et al. 2010). These SCFAs are metabolized by the liver, but compounds that escape this metabolism are able to affect the brain and behavior through multiple routes. One of these routes involves SCFA-induced increases in tyrosine hydroxylase, which is needed for the synthesis of dopamine, norepinephrine, and epinephrine (DeCastro et al. 2005; Stilling et al. 2016). In addition, treating rat pheochromocytoma (PC12) cell lines with SCFA (primarily butyrate and propionate) leads to significant changes in dopamine and serotonin signaling pathways (Nankova et al. 2014). The SCFAs can also affect depressive-like behavior through effects on histone acetylation (Schroeder et al. 2007) and can affect microglial activity to impact behavior (Erny et al. 2015; Huuskonen et al. 2004). The SCFAs are also well known for their effects on the immune system. For example, butyrate can increase the production of antimicrobial peptides by intestinal epithelial cells (Hase et al. 2002; Zhao et al. 2018) and macrophages (Schulthess et al. 2019). Butyrate also increases neutrophil chemotaxis and microbicidal activity (Sina et al. 2009; Vinolo et al. 2011a, b). Interestingly, butyrate can have different effects on monocytes/macrophage inflammatory profiles, with a general inhibitory/anti-inflammatory effect on colonic lamina propria macrophages (Chang et al. 2014), but pro-inflammatory effects on peripheral blood monocytes (Mirmonsef et al. 2012). Butyrate can also have anti-inflammatory effects on dendritic cells (Liu et al. 2012; Millard et al. 2002 and facilitates the differentiation of naïve T cells into regulatory T cells (Kespohl et al. 2017). Thus, there are multiple routes through which bacterial-derived SCFA's can impact immune system activity, the brain, and behavior.

12.3.5.2 Bacterial Vitamins

There is a broad range of gut bacteria that have the capacity for synthesis and metabolism of essential vitamins, including multiple B vitamins (i.e., B1, B2, B3, B5, B6, B7, B9, and B12), vitamin K and vitamin A (Das et al. 2019; Iyer and Vaishnava 2019; Magnusdottir et al. 2015). The importance of bacteria for the maintenance of host vitamins was first realized when it was observed that germfree mice quickly develop vitamin K deficiency if their diets are not supplemented with vitamin K or if they are not colonized by conventional microbiota (Gustafsson 1959; Gustafsson et al. 1962). Diseases and conditions that involve alterations in the gut microbiome, such as type 2 diabetes or inflammatory bowel disease, have been associated with abnormalities in the abundance of bacterial genes necessary for the synthesis and metabolism of B vitamins (Das et al. 2019; Magnusdottir et al. 2015). In addition, exposure to social stress in mice (Allen et al. 2019) or high levels of perceived stress in pediatric patients with inflammatory bowel disease (Mackner et al. 2020) has been shown to lead to lower levels of B vitamins, as well as the abundance and diversity of bacteria capable of vitamin B biosynthesis (Allen et al. 2019). This suggests that activation of the endocrine system can impact bacteria capable of vitamin biosynthesis.

Vitamin deficiencies can result in severe symptoms and conditions, including conditions that affect the brain, behavior, and immune responses. For example, deficiency of vitamin B3 (niacin) leads to a condition called pellagra that can involve the development of dementia, delirium, and psychosis (Badawy 2014; Rudzki et al. 2021). Vitamin B1 (thiamine) deficiency can also lead to confusion, psychosis, delirium, and memory loss (Rudzki et al. 2021), whereas vitamin B6 (pyridoxal) deficiency may contribute to major depressive disorder, bipolar disorder, schizophrenia, autism, dementia, or Parkinson's disease (Douaud et al. 2013; Rudzki et al. 2021; Sato 2018; Skarupski et al. 2010) The B vitamins are particularly important for brain health and proper endocrine functioning, because they are co-factors for hundreds of enzymic reactions in the host, including neurotransmitter and hormone metabolism, such as of serotonin, which is dependent upon vitamin B6 and involves vitamin B3 (Kennedy 2016). Moreover, B vitamins can also strongly affect immune system activity (Lee et al. 2015; Lipszyc et al. 2013). For example, vitamin B3 can attenuate inflammation, including colonic inflammation (Singh et al. 2014). This is in part due to the fact that niacin shares a receptor with butyrate (namely Niacr1 or GPR109a) which promotes anti-inflammatory responses and attenuates experimental, chemical-induced colitis (Li et al. 2017; Singh et al. 2014). In fact, studies in humans suggest that daily enema with niacin can improve colonic inflammation (Li et al. 2017). Given that essential vitamins can affect both the brain and the immune system, bacterial-produced vitamins may be an important mechanism by which the gut microbiota impact neuroendocrine-immune interactions.

Prebiotics have the potential to increase the abundance of bacteria capable of vitamin B metabolism (Allen et al. 2019), and studies have suggested that probiotics can be used to increase host vitamin B (Das et al. 2019; Hill 1997; LeBlanc et al. 2011, 2013; Magnusdottir et al. 2015). The extent to which prebiotics or probiotics can have beneficial effects on neuroendocrine-immune interactions by sustaining essential vitamin levels is not yet known, but bacterial production of essential vitamins for the maintenance of brain health is a new horizon for brain-gut-microbiota axis research (Rudzki et al. 2021).

12.4 Importance of the Intestinal Epithelium in the Brain-Gut-Microbiota Axis

The intestinal epithelium forms an effective barrier that prevents the luminal contents of the intestine from diffusing across the barrier into the blood. As a result, bacterial-produced compounds, including amino acids and their derivatives, vitamins, biogenic amines, and polyamines have to bind to epithelial receptors or be passed through transporters from the lumen of the intestine to the interior of the body to have biological effects on the host. These interactions are complex and multi-stepped, as elegantly demonstrated by Ye et al. (2021) who used zebrafish to show that bacterial tryptophan metabolites bind to the receptor Trpa1 on intestinal epithelial enteroendocrine cells to stimulate the release of serotonin. Intestinal serotonin has been shown to significantly alter mucosal immune responses (Gross Margolis et al. 2017; Margolis and Gershon 2016), but in this study the authors also found that the serotonin stimulates enteric and vagal neurons, demonstrating the multistep process for bacterial signals to be propagated from the lumen of the intestines to the interior of the body.

The epithelial barrier in the context of neuroendocrine-immune interactions has been understudied but is likely a crucial player in this axis (Kelly et al. 2015; Rudzki and Maes 2020, 2021). When the epithelial barrier becomes leaky, bacteria and/or bacterial products can migrate from the lumen of the intestine to the interior of the body to affect host physiology and immune responses. While intestinal mucus keeps bacteria spatially separated from the lumen of the intestine, if bacteria are able to penetrate the mucus layer, tight junctional proteins between epithelial cells are essential for preventing intercellular passage of bacteria/bacterial products across the epithelium (Hooper 2015), but can be affected by the endocrine and immune systems (Zong et al. 2019). For example, in rats exposed to heat stress, stress-related increases in norepinephrine were related to decreased expression of tight junction proteins (e.g., zona occludin 1 and occludin). Interestingly, in epithelial cell cultures, exposure to norepinephrine acutely (6 h) increased these tight junction proteins, but prolonged exposure (24 h) led to decreases (Luo et al. 2021). Cortisol induces similar effects on occludin in epithelial cell cultures (Zong et al. 2019), indicating that the catecholamines and glucocorticoids can downregulate tight junction protein expression.

It has long been recognized that stressor exposure can increase gut permeability in animal models, leading to inflammatory responses. Low-grade inflammation is well known to affect host physiology, including the brain and behavior. In fact, it is now appreciated that low-grade inflammation can lead to depression in susceptible individuals. In our studies, treating stressor-exposed mice with a broad-spectrum antibiotic could prevent stressor-induced disruption in cognitive behavior (Jaggers et al. 2022). This disruption in cognitive behavior was correlated with the expression of LPS-binding protein (LBP), which is released by intestinal epithelial cells (as well as by cells in the liver) as part of the acute phase response to bacterial challenge (Tobias et al. 1995; Zweigner et al. 2006). When mice lacking the receptor to LBP (i.e., CD14-/- mice) were exposed to stress, they were resistant to stressor-induced changes in cognitive behavior (Jaggers et al. 2022). When taken together, these data show that stressor-induced disruption of the protective mucosal barrier leads to immune activation and the expression of LBP which are part of the natural response to bacterial infection. This low level of immune activation originating in the gut epithelium leads to behavioral changes in stressor-exposed mice, thus illustrating the importance of the intestinal epithelium in the stress response.

Stressor-induced alterations in epithelial integrity are not limited to murine studies. There are now a number of studies that have found increases in markers of epithelial disruption (such as LBP, as well as tight junction proteins such as zonulin) in individuals with higher levels of stress (Kiecolt-Glaser et al. 2018, 2021; Shrout et al. 2022) or in persons with psychiatric disease, such as major depressive disorder (Maes et al. 2013; Rudzki and Maes 2020; Stevens et al. 2018). Markers of gut permeability have also been identified in bipolar disorder and autism spectrum disorder (Doney et al. 2021; Piras et al. 2022).

The importance of the epithelium extends beyond just the regulation of protective and potentially deleterious immune responses. Bacterial metabolites and food components are not able to diffuse directly across the intestinal epithelium. If the gut is not leaky, as described above, then microbial metabolites and food components must either bind to an active receptor or be transported across the epithelium. Receptors and transporters for some bacterial metabolites have been well studied, such as the 5-HT transporters (e.g., SLC6A4, a.k.a. SERT), which are found in the intestinal epithelium, as well as on enteric neurons. And although this transporter is known to be important for neuronal development, its importance for bacterial signaling to the endocrine system is just beginning to be understood (Margolis 2017). As we identify additional, bacterial-produced metabolites that have the capacity to impact neuroendocrine-immune interactions, it will become increasingly important to understand how the bacterial metabolites propagate a signal across the formidable intestinal barrier.

12.5 Conclusion

There are extensive bidirectional interactions between the host and its microbiota through the shared production of ligands and their receptors. Activation of the host neuroendocrine system can affect the gut microbiome through direct neuroendocrine-bacterial interactions, through effects on gastrointestinal physiology, or through effects on the host immune system. The gut microbiota are metabolically active, and alterations in microbiome structure and function in turn influence the endocrine and immune systems. The field of psychoneuroimmunology has studied these neuroendocrine-immune interactions for approximately 50 years, with the first edition of the book "Psychoneuroimmunology" published in 1981 (Ader 1981). However, it has been only recently that the contributing role of the gut microbiota in neuroendocrine-immune interactions has been considered. We now know that gut bacteria are essential for the development and maintenance of the nervous, endocrine, and immune systems. This has been suggested through genomic and metabolomic studies that have demonstrated associations between gut bacteria, their products and host responses to a variety of behavioral, endocrine, and immune outcomes. However, the mechanisms responsible for these associations have been harder to define. This is, in part, due to the fact that bacteria are metabolically very active and produce hundreds of molecules that can impact host physiology. While some of these molecules have been identified, such as the short-chain fatty acids, Trp derivatives, Glu and GABA, as well as essential vitamins, we are simply at the tip of the iceberg when it comes to understanding how microbial metabolites support host homeostasis or contribute to disease. As we continue to identify pathways by which microbes can impact the host, understanding the primary gatekeeper separating microbes and their products from the interior of the body (i.e., the intestinal epithelium) should be a high priority. Understanding how microbes and their metabolites interact with these barriers to impact the neuroendocrine and immune systems will provide for new approaches, including holistic/dietary approaches, to improve health and prevent disease.

12.6 Key References

Bravo et al. (2011)—Bacteria had been suggested to impact the brain and endocrine responses through stimulation of the vagus nerve. However, this was one of the first studies to demonstrate that some of the beneficial effects of probiotic bacteria on emotional behavior are dependent upon signaling through the vagus nerve.

Jaggers et al. (2022)—Stressor-induced changes in immune system activity have been linked to behavioral responses to stress. However, whether neuroendocrine-immune-behavioral interactions involve the gut microbiota has not been as well studied. This study showed that intestinal bacteria are needed for stressor-induced changes in cognitive behavior and that this effect was dependent upon bacterial LPS signaling.

Rothheammer et al. (2016)—This study demonstrated that microbial tryptophan metabolites not only affect mucosal immunity, but can also affect brain glia. Together with the Zalente et al. (2013) study, this study demonstrates how microbial metabolites can alter neuroimmune responses.

Sgritta et al. (2019)—There is increasing interest in the potential of probiotic bacteria to treat brain and gut disorders. This mechanistic study shows that the probiotic bacterium *L. reuteri* improves social behavior in a murine model of autism spectrum disorder by stimulating oxytocin release.

Zelante et al. (2013)—This mechanistic study was among the first to demonstrate how a microbial metabolite of an amino acid shapes mucosal immunity. This study was among the first to demonstrate that bacterial metabolites of tryptophan have strong immunomodulatory effects.

References

Abbott CR, Monteiro M, Small CJ, Sajedi A, Smith KL, Parkinson JR, Ghatei MA, Bloom SR (2005) The inhibitory effects of peripheral administration of peptide YY(3-36) and glucagon-like peptide-1 on food intake are attenuated by ablation of the vagal-brainstem-hypothalamic pathway. Brain Res 1044:127–131

Ader R (1981) Psychoneuroimmunology, 1st edn. Academic Press, New York

Aggarwal S, Ahuja V, Paul J (2017) Attenuated GABAergic signaling in intestinal epithelium contributes to pathogenesis of ulcerative colitis. Dig Dis Sci 62:2768–2779

Aguilera M, Vergara P, Martinez V (2013) Stress and antibiotics alter luminal and wall-adhered microbiota and enhance the local expression of visceral sensory-related systems in mice. Neurogastroenterol Motil 25:e515–e529

Agus A, Planchais J, Sokol H (2018) Gut microbiota regulation of tryptophan metabolism in health and disease. Cell Host Microbe 23:716–724

Allen JM, Jaggers RM, Solden LM, Loman BR, Davies RH, Mackos AR, Ladaika CA, Berg BM, Chichlowski M, Bailey MT (2019) Dietary oligosaccharides attenuate stress-induced disruptions in immune reactivity and microbial B-vitamin metabolism. Front Immunol 10:1774

Allen JM, Mackos AR, Jaggers RM, Brewster PC, Webb M, Lin CH, Ladaika C, Davies R, White P, Loman BR, Bailey MT (2022) Psychological stress disrupts intestinal epithelial cell function and mucosal integrity through microbe and host-directed processes. Gut Microbes 14: 2035661

Allison SD, Martiny JB (2008) Colloquium paper: resistance, resilience, and redundancy in microbial communities. Proc Natl Acad Sci USA 105:11512–11519

Antonson AM, Evans MV, Galley JD, Chen HJ, Rajasekera TA, Lammers SM, Hale VL, Bailey MT, Gur TL (2020) Unique maternal immune and functional microbial profiles during prenatal stress. Sci Rep 10:20288

Aoki-Yoshida A, Aoki R, Moriya N, Goto T, Kubota Y, Toyoda A, Takayama Y, Suzuki C (2016) Omics studies of the murine intestinal ecosystem exposed to subchronic and mild social defeat stress. J Proteome Res 15(9):3126–3138

Asano Y, Hiramoto T, Nishino R, Aiba Y, Kimura T, Yoshihara K, Koga Y, Sudo N (2012) Critical role of gut microbiota in the production of biologically active, free catecholamines in the gut lumen of mice. Am J Physiol Gastrointest Liver Physiol 303:G1288–G1295

Badawy AA (2014) Pellagra and alcoholism: a biochemical perspective. Alcohol Alcohol 49:238–250

Bailey MT (2012) The contributing role of the intestinal microbiota in stressor-induced increases in susceptibility to enteric infection and systemic immunomodulation. Horm Behav 62:286–294

Bailey MT, Coe CL (1999) Maternal separation disrupts the integrity of the intestinal microflora in infant rhesus monkeys. Dev Psychobiol 35:146–155

Bansal T, Englert D, Lee J, Hegde M, Wood TK, Jayaraman A (2007) Differential effects of epinephrine, norepinephrine, and indole on Escherichia coli O157:H7 chemotaxis, colonization, and gene expression. Infect Immun 75:4597–4607

Bercik P, Park AJ, Sinclair D, Khoshdel A, Lu J, Huang X, Deng Y, Blennerhassett PA, Fahnestock M, Moine D, Berger B, Huizinga JD, Kunze W, McLean PG, Bergonzelli GE, Collins SM, Verdu EF (2011) The anxiolytic effect of Bifidobacterium longum NCC3001 involves vagal pathways for gut-brain communication. Neurogastroenterol Motil 23:1132–1139

Berg RD (1996) The indeginous gastrointestinal microflora. Trends Microbiol 4:430–435

Bevins CL, Salzman NH (2011) Paneth cells, antimicrobial peptides and maintenance of intestinal homeostasis. Nat Rev Microbiol 9:356–368

Bharwani A, Mian MF, Foster JA, Surette MG, Bienenstock J, Forsythe P (2016) Structural & functional consequences of chronic psychosocial stress on the microbiome & host. Psychoneuroendocrinology 63:217–227

Bilen M, Dufour JC, Lagier JC, Cadoret F, Daoud Z, Dubourg G, Raoult D (2018) The contribution of culturomics to the repertoire of isolated human bacterial and archaeal species. Microbiome 6: 94

Boonstra E, de Kleijn R, Colzato LS, Alkemade A, Forstmann BU, Nieuwenhuis S (2015) Neurotransmitters as food supplements: the effects of GABA on brain and behavior. Front Psychol 6:1520

Borovicka J, Schwizer W, Mettraux C, Kreiss C, Remy B, Asal K, Jansen JB, Douchet I, Verger R, Fried M (1997) Regulation of gastric and pancreatic lipase secretion by CCK and cholinergic mechanisms in humans. Am J Physiol 273:G374–G380

Braat S, Kooy RF (2015) The GABAA receptor as a therapeutic target for neurodevelopmental disorders. Neuron 86:1119–1130

Braat S, D'Hulst C, Heulens I, De Rubeis S, Mientjes E, Nelson DL, Willemsen R, Bagni C, Van Dam D, De Deyn PP, Kooy RF (2015) The GABAA receptor is an FMRP target with therapeutic potential in fragile X syndrome. Cell Cycle 14:2985–2995

Bravo JA, Forsythe P, Chew MV, Escaravage E, Savignac HM, Dinan TG, Bienenstock J, Cryan JF (2011) Ingestion of Lactobacillus strain regulates emotional behavior and central GABA receptor expression in a mouse via the vagus nerve. Proc Natl Acad Sci U S A 108:16050–16055

Brussow H (2020) Problems with the concept of gut microbiota dysbiosis. Microb Biotechnol 13: 423–434

Buffington SA, Di Prisco GV, Auchtung TA, Ajami NJ, Petrosino JF, Costa-Mattioli M (2016) Microbial reconstitution reverses maternal diet-induced social and synaptic deficits in offspring. Cell 165:1762–1775

Buffington SA, Dooling SW, Sgritta M, Noecker C, Murillo OD, Felice DF, Turnbaugh PJ, Costa-Mattioli M (2021) Dissecting the contribution of host genetics and the microbiome in complex behaviors. Cell 184:1740–1756

Cani PD, Delzenne NM (2009) The role of the gut microbiota in energy metabolism and metabolic disease. Curr Pharm Des 15:1546–1558

Cani PD, Lecourt E, Dewulf EM, Sohet FM, Pachikian BD, Naslain D, De Backer F, Neyrinck AM, Delzenne NM (2009) Gut microbiota fermentation of prebiotics increases satietogenic and incretin gut peptide production with consequences for appetite sensation and glucose response after a meal. Am J Clin Nutr 90:1236–1243

Carson TL, Wang F, Cui X, Jackson BE, Van Der Pol WJ, Lefkowitz EJ, Morrow C, Baskin ML (2018) Associations between race, perceived psychological stress, and the gut microbiota in a sample of generally healthy black and white women: a pilot study on the role of race and perceived psychological stress. Psychosom Med 80:640–648

Cerutti A, Rescigno M (2008) The biology of intestinal immunoglobulin A responses. Immunity 28:740–750

Champ MM (2004) Physiological aspects of resistant starch and in vivo measurements. J AOAC Int 87:749–755

Chang PV, Hao L, Offermanns S, Medzhitov R (2014) The microbial metabolite butyrate regulates intestinal macrophage function via histone deacetylase inhibition. Proc Natl Acad Sci U S A 111:2247–2252

Chen K, Magri G, Grasset EK, Cerutti A (2020) Rethinking mucosal antibody responses: IgM, IgG and IgD join IgA. Nat Rev Immunol 20:427–441

Clarke G, Grenham S, Scully P, Fitzgerald P, Moloney RD, Shanahan F, Dinan TG, Cryan JF (2013) The microbiome-gut-brain axis during early life regulates the hippocampal serotonergic system in a sex-dependent manner. Mol Psychiatry 18:666–673

Crumeyrolle-Arias M, Jaglin M, Bruneau A, Vancassel S, Cardona A, Dauge V, Naudon L, Rabot S (2014) Absence of the gut microbiota enhances anxiety-like behavior and neuroendocrine response to acute stress in rats. Psychoneuroendocrinology 42:207–217

Cryan JF, Mazmanian SK (2022) Microbiota-brain axis: context and causality. Science 376:938–939

D'Agata AL, Wu J, Welandawe MKV, Dutra SVO, Kane B, Groer MW (2019) Effects of early life NICU stress on the developing gut microbiome. Dev Psychobiol 61:650–660
Das P, Babaei P, Nielsen J (2019) Metagenomic analysis of microbe-mediated vitamin metabolism in the human gut microbiome. BMC Genomics 20:208
De la Fuente M, Bernaez I, Del Rio M, Hernanz A (1993) Stimulation of murine peritoneal macrophage functions by neuropeptide Y and peptide YY. Involvement of protein kinase C. Immunology 80:259–265
de La Serre CB, de Lartigue G, Raybould HE (2015) Chronic exposure to low dose bacterial lipopolysaccharide inhibits leptin signaling in vagal afferent neurons. Physiol Behav 139:188–194
De Palma G, Lynch MD, Lu J, Dang VT, Deng Y, Jury J, Umeh G, Miranda PM, Pigrau Pastor M, Sidani S, Pinto-Sanchez MI, Philip V, McLean PG, Hagelsieb MG, Surette MG, Bergonzelli GE, Verdu EF, Britz-McKibbin P, Neufeld JD, Collins SM, Bercik P (2017) Transplantation of fecal microbiota from patients with irritable bowel syndrome alters gut function and behavior in recipient mice. Sci Transl Med 9:eaaf6397
De Vadder F, Kovatcheva-Datchary P, Goncalves D, Vinera J, Zitoun C, Duchampt A, Backhed F, Mithieux G (2014) Microbiota-generated metabolites promote metabolic benefits via gut-brain neural circuits. Cell 156:84–96
DeCastro M, Nankova BB, Shah P, Patel P, Mally PV, Mishra R, La Gamma EF (2005) Short chain fatty acids regulate tyrosine hydroxylase gene expression through a cAMP-dependent signaling pathway. Brain Res Mol Brain Res 142:28–38
den Besten G, van Eunen K, Groen AK, Venema K, Reijngoud DJ, Bakker BM (2013) The role of short-chain fatty acids in the interplay between diet, gut microbiota, and host energy metabolism. J Lipid Res 54:2325–2340
Desbonnet L, Clarke G, Shanahan F, Dinan TG, Cryan JF (2014) Microbiota is essential for social development in the mouse. Mol Psychiatry 19:146–148
Desbonnet L, Clarke G, Traplin A, O'Sullivan O, Crispie F, Moloney RD, Cotter PD, Dinan TG, Cryan JF (2015) Gut microbiota depletion from early adolescence in mice: implications for brain and behaviour. Brain Behav Immun 48:165–173
Devalia JL, Grady D, Harmanyeri Y, Tabaqchali S, Davies RJ (1989) Histamine synthesis by respiratory tract micro-organisms: possible role in pathogenicity. J Clin Pathol 42:516–522
Dominguez-Bello MG, Blaser MJ, Ley RE, Knight R (2011) Development of the human gastrointestinal microbiota and insights from high-throughput sequencing. Gastroenterology 140:1713–1719
Dominique M, Breton J, Guerin C, Bole-Feysot C, Lambert G, Dechelotte P, Fetissov S (2019a) Effects of macronutrients on the in vitro production of ClpB, a bacterial mimetic protein of alpha-MSH and its possible role in satiety signaling. Nutrients 11:2115
Dominique M, Legrand R, Galmiche M, Azhar S, Deroissart C, Guerin C, do Rego JL, Leon F, Nobis S, Lambert G, Lucas N, Dechelotte P (2019b) Changes in microbiota and bacterial protein caseinolytic peptidase B during food restriction in mice: relevance for the onset and perpetuation of anorexia nervosa. Nutrients 11:2514
Doney E, Cadoret A, Dion-Albert L, Lebel M, Menard C (2021) Inflammation-driven brain and gut barrier dysfunction in stress and mood disorders. Eur J Neurosci 55(9–10):2851–2894
Douaud G, Refsum H, de Jager CA, Jacoby R, Nichols TE, Smith SM, Smith AD (2013) Preventing Alzheimer's disease-related gray matter atrophy by B-vitamin treatment. Proc Natl Acad Sci U S A 110:9523–9528
Douglas GM, Maffei VJ, Zaneveld JR, Yurgel SN, Brown JR, Taylor CM, Huttenhower C, Langille MGI (2020) PICRUSt2 for prediction of metagenome functions. Nat Biotechnol 38:685–688
Dumoutier L, Van Roost E, Ameye G, Michaux L, Renauld JC (2000) IL-TIF/IL-22: genomic organization and mapping of the human and mouse genes. Genes Immun 1:488–494
Eckburg PB, Bik EM, Bernstein CN, Purdom E, Dethlefsen L, Sargent M, Gill SR, Nelson KE, Relman DA (2005) Diversity of the human intestinal microbial flora. Science 308:1635–1638

Elkhatib SK, Moshfegh CM, Watson GF, Case AJ (2022) T-lymphocyte tyrosine hydroxylase regulates TH17 T-lymphocytes during repeated social defeat stress. Brain Behav Immun 104: 18–28

Engevik MA, Luk B, Chang-Graham AL, Hall A, Herrmann B, Ruan W, Endres BT, Shi Z, Garey KW, Hyser JM, Versalovic J (2019) Bifidobacterium dentium fortifies the intestinal mucus layer via autophagy and calcium signaling pathways. mBio 10(3):e01087

Eriksson E, Royo F, Lyberg K et al (2004) Effect of metabolic cage housing on immunoglobulin A and corticosterone excretion in faeces and urine of young male rats. Exp Physiol 89:427–433

Erny D, Hrabe de Angelis AL, Jaitin D, Wieghofer P, Staszewski O, David E, Keren-Shaul H, Mahlakoiv T, Jakobshagen K, Buch T, Schwierzeck V, Utermohlen O, Chun E, Garrett WS, McCoy KD, Diefenbach A, Staeheli P, Stecher B, Amit I, Prinz M (2015) Host microbiota constantly control maturation and function of microglia in the CNS. Nat Neurosci 18:965–977

Fee C, Banasr M, Sibille E (2017) Somatostatin-positive gamma-aminobutyric acid interneuron deficits in depression: cortical microcircuit and therapeutic perspectives. Biol Psychiatry 82: 549–559

Feehily C, Karatzas KA (2013) Role of glutamate metabolism in bacterial responses towards acid and other stresses. J Appl Microbiol 114:11–24

Feehily C, O'Byrne CP, Karatzas KA (2013) Functional gamma-aminobutyrate shunt in listeria monocytogenes: role in acid tolerance and succinate biosynthesis. Appl Environ Microbiol 79: 74–80

Filpa V, Moro E, Protasoni M, Crema F, Frigo G, Giaroni C (2016) Role of glutamatergic neurotransmission in the enteric nervous system and brain-gut axis in health and disease. Neuropharmacology 111:14–33

Finotello F, Mastrorilli E, Di Camillo B (2018) Measuring the diversity of the human microbiota with targeted next-generation sequencing. Brief Bioinform 19:679–692

Fotiadis D, Kanai Y, Palacin M (2013) The SLC3 and SLC7 families of amino acid transporters. Mol Aspects Med 34:139–158

Franzosa EA, Hsu T, Sirota-Madi A, Shafquat A, Abu-Ali G, Morgan XC, Huttenhower C (2015) Sequencing and beyond: integrating molecular 'omics' for microbial community profiling. Nat Rev Microbiol 13:360–372

Franzosa EA, McIver LJ, Rahnavard G, Thompson LR, Schirmer M, Weingart G, Lipson KS, Knight R, Caporaso JG, Segata N, Huttenhower C (2018) Species-level functional profiling of metagenomes and metatranscriptomes. Nat Methods 15:962–968

Freestone PP, Williams PH, Haigh RD, Maggs AF, Neal CP, Lyte M (2002) Growth stimulation of intestinal commensal Escherichia coli by catecholamines: a possible contributory factor in trauma-induced sepsis. Shock 18:465–470

Fuhrman JA (2009) Microbial community structure and its functional implications. Nature 459: 193–199

Furness JB, Callaghan BP, Rivera LR, Cho HJ (2014) The enteric nervous system and gastrointestinal innervation: integrated local and central control. Adv Exp Med Biol 817:39–71

Garmaeva S, Sinha T, Kurilshikov A, Fu J, Wijmenga C, Zhernakova A (2019) Studying the gut virome in the metagenomic era: challenges and perspectives. BMC Biol 17:84

Gaudino SJ, Beaupre M, Lin X, Joshi P, Rathi S, McLaughlin PA, Kempen C, Mehta N, Eskiocak O, Yueh B, Blumberg RS, van der Velden AWM, Shroyer KR, Bialkowska AB, Beyaz S, Kumar P (2021) IL-22 receptor signaling in Paneth cells is critical for their maturation, microbiota colonization, Th17-related immune responses, and anti-Salmonella immunity. Mucosal Immunol 14:389–401

Gay MD, Cao H, Shivapurkar N, Dakshanamurthy S, Kallakury B, Tucker RD, Kwagyan J, Smith JP (2022) Proglumide reverses nonalcoholic steatohepatitis by interaction with the Farnesoid X receptor and altering the microbiome. Int J Mol Sci 23:1899

Genton L, Kudsk KA (2003) Interactions between the enteric nervous system and the immune system: role of neuropeptides and nutrition. Am J Surg 186:253–258

Gibson GR, Hutkins R, Sanders ME, Prescott SL, Reimer RA, Salminen SJ, Scott K, Stanton C, Swanson KS, Cani PD, Verbeke K, Reid G (2017) Expert consensus document: the International Scientific Association for Probiotics and Prebiotics (ISAPP) consensus statement on the definition and scope of prebiotics. Nat Rev Gastroenterol Hepatol 14:491–502

Goehler LE, Gaykema RP, Opitz N, Reddaway R, Badr N, Lyte M (2005) Activation in vagal afferents and central autonomic pathways: early responses to intestinal infection with Campylobacter jejuni. Brain Behav Immun 19:334–344

Goehler LE, Lyte M, Gaykema RP (2007) Infection-induced viscerosensory signals from the gut enhance anxiety: implications for psychoneuroimmunology. Brain Behav Immun 21:721–726

Goehler LE, Park SM, Opitz N, Lyte M, Gaykema RP (2008) Campylobacter jejuni infection increases anxiety-like behavior in the holeboard: possible anatomical substrates for viscerosensory modulation of exploratory behavior. Brain Behav Immun 22:354–366

Goussarov G, Claesen J, Mysara M, Cleenwerck I, Leys N, Vandamme P, Van Houdt R (2022) Accurate prediction of metagenome-assembled genome completeness by MAGISTA, a random forest model built on alignment-free intra-bin statistics. Environ Microbiome 17:9

Gregory AC, Zablocki O, Zayed AA, Howell A, Bolduc B, Sullivan MB (2020) The gut virome database reveals age-dependent patterns of virome diversity in the human gut. Cell Host Microbe 28(724-740):e728

Gross Margolis K, Vittorio J, Talavera M, Gluck K, Li Z, Iuga A, Stevanovic K, Saurman V, Israelyan N, Welch MG, Gershon MD (2017) Enteric serotonin and oxytocin: endogenous regulation of severity in a murine model of necrotizing enterocolitis. Am J Physiol Gastrointest Liver Physiol 313:G386–G398

Guerrero HY, Caceres G, Paiva CL, Marcano D (1990) Hypothalamic and telencephalic catecholamine content in the brain of the teleost fish, Pygocentrus notatus, during the annual reproductive cycle. Gen Comp Endocrinol 80:257–263

Gur TL, Bailey MT (2016) Effects of stress on commensal microbes and immune system activity. Adv Exp Med Biol 874:289–300

Gur TL, Shay L, Palkar AV, Fisher S, Varaljay VA, Dowd S, Bailey MT (2016) Prenatal stress affects placental cytokines and neurotrophins, commensal microbes, and anxiety-like behavior in adult female offspring. Brain Behav Immun 64:50–58

Gur TL, Palkar AV, Rajasekera T, Allen J, Niraula A, Godbout J, Bailey MT (2018) Prenatal stress disrupts social behavior, cortical neurobiology and commensal microbes in adult male offspring. Behav Brain Res 359:886–894

Gustafsson BE (1959) Vitamin K deficiency in germfree rats. Ann N Y Acad Sci 78:166–174

Gustafsson BE, Daft FS, McDaniel EG, Smith JC, Fitzgerald RJ (1962) Effects of vitamin K-active compounds and intestinal microorganisms in vitamin K-deficient germfree rats. J Nutr 78:461–468

Habib AM, Richards P, Rogers GJ, Reimann F, Gribble FM (2013) Co-localisation and secretion of glucagon-like peptide 1 and peptide YY from primary cultured human L cells. Diabetologia 56:1413–1416

Hanna MK, Kudsk KA (2000) Nutritional and pharmacological enhancement of gut-associated lymphoid tissue. Can J Gastroenterol 14:145D–151D

Hantsoo L, Jasarevic E, Criniti S, McGeehan B, Tanes C, Sammel MD, Elovitz MA, Compher C, Wu G, Epperson CN (2019) Childhood adversity impact on gut microbiota and inflammatory response to stress during pregnancy. Brain Behav Immun 75:240–250

Hase K, Eckmann L, Leopard JD, Varki N, Kagnoff MF (2002) Cell differentiation is a key determinant of cathelicidin LL-37/human cationic antimicrobial protein 18 expression by human colon epithelium. Infect Immun 70:953–963

Hill MJ (1997) Intestinal flora and endogenous vitamin synthesis. Eur J Cancer Prev 6(Suppl 1):S43–S45

Hill C, Guarner F, Reid G, Gibson GR, Merenstein DJ, Pot B, Morelli L, Canani RB, Flint HJ, Salminen S, Calder PC, Sanders ME (2014) Expert consensus document. The International

Scientific Association for Probiotics and Prebiotics consensus statement on the scope and appropriate use of the term probiotic. Nat Rev Gastroenterol Hepatol 11:506–514
Holst JJ (2007) The physiology of glucagon-like peptide 1. Physiol Rev 87:1409–1439
Holzer P, Reichmann F, Farzi A (2012) Neuropeptide Y, peptide YY and pancreatic polypeptide in the gut-brain axis. Neuropeptides 46:261–274
Hooks KB, O'Malley MA (2017) Dysbiosis and its discontents. mBio 8:e01492–e01417
Hooper LV (2015) Epithelial cell contributions to intestinal immunity. Adv Immunol 126:129–172
Hoyles L, McCartney AL, Neve H, Gibson GR, Sanderson JD, Heller KJ, van Sinderen D (2014) Characterization of virus-like particles associated with the human faecal and caecal microbiota. Res Microbiol 165:803–812
Huffnagle GB (2010) The microbiota and allergies/asthma. PLoS Pathog 6:e1000549
Hughes DT, Clarke MB, Yamamoto K, Rasko DA, Sperandio V (2009) The QseC adrenergic signaling cascade in Enterohemorrhagic E. coli (EHEC). PLoS Pathog 5:e1000553
Hugon P, Dufour JC, Colson P, Fournier PE, Sallah K, Raoult D (2015) A comprehensive repertoire of prokaryotic species identified in human beings. Lancet Infect Dis 15:1211–1219
Humbel F, Rieder JH, Franc Y, Juillerat P, Scharl M, Misselwitz B, Schreiner P, Begre S, Rogler G, von Kanel R, Yilmaz B, Biedermann L, Swiss IBDCSG (2020) Association of alterations in intestinal microbiota with impaired psychological function in patients with inflammatory bowel diseases in remission. Clin Gastroenterol Hepatol 18:2019–2029.e2011
Hurley R, Leask BG, Ruthven CR, Sandler M, Southgate J (1971) Investigation of 5-hydroxytryptamine production by Candida albicans in vitro and in vivo. Microbios 4:133–143
Huuskonen J, Suuronen T, Nuutinen T, Kyrylenko S, Salminen A (2004) Regulation of microglial inflammatory response by sodium butyrate and short-chain fatty acids. Br J Pharmacol 141:874–880
Iseri SO, Sener G, Saglam B, Gedik N, Ercan F, Yegen BC (2005a) Oxytocin ameliorates oxidative colonic inflammation by a neutrophil-dependent mechanism. Peptides 26:483–491
Iseri SO, Sener G, Saglam B, Gedik N, Ercan F, Yegen BC (2005b) Oxytocin protects against sepsis-induced multiple organ damage: role of neutrophils. J Surg Res 126:73–81
Iyer N, Vaishnava S (2019) Vitamin A at the interface of host-commensal-pathogen interactions. PLoS Pathog 15:e1007750
Jaggers RM, Disabato DJ, Loman BR, Kontic D, Spencer KD, Allen JM, Godbout JP, Quan N, Gur TL, Bailey MT (2022) Stressor-induced reduction in cognitive behavior is associated with impaired colonic mucus layer integrity and is dependent upon the LPS-binding protein receptor CD14. J Inflamm Res 15:18
Jahn MT, Arkhipova K, Markert SM, Stigloher C, Lachnit T, Pita L, Kupczok A, Ribes M, Stengel ST, Rosenstiel P, Dutilh BE, Hentschel U (2019) A phage protein aids bacterial symbionts in eukaryote immune evasion. Cell Host Microbe 26:542–550.e545
Jarillo-Luna A, Rivera-Aguilar V, Garfias HR, Lara-Padilla E, Kormanovsky A, Campos-Rodriguez R (2007) Effect of repeated restraint stress on the levels of intestinal IgA in mice. Psychoneuroendocrinology 32:681–692
Jasarevic E, Howerton CL, Howard CD, Bale TL (2015) Alterations in the vaginal microbiome by maternal stress are associated with metabolic reprogramming of the offspring gut and brain. Endocrinology 156:3265–3276
Jasarevic E, Howard CD, Misic AM, Beiting DP, Bale TL (2017) Stress during pregnancy alters temporal and spatial dynamics of the maternal and offspring microbiome in a sex-specific manner. Sci Rep 7:44182
Jia X, Cong B, Zhang J, Li H, Liu W, Chang H, Dong M, Ma C (2014) CCK8 negatively regulates the TLR9-induced activation of human peripheral blood pDCs by targeting TRAF6 signaling. Eur J Immunol 44:489–499
Jiang H et al (2015) Altered fecal microbiota composition in patients with major depressive Disorder. Brain Behav Immun 48:186–194
Jones ML, Ganopolsky JG, Martoni CJ, Labbe A, Prakash S (2014) Emerging science of the human microbiome. Gut Microbes 5:446–457

Julio-Pieper M, Flor PJ, Dinan TG, Cryan JF (2011) Exciting times beyond the brain: metabotropic glutamate receptors in peripheral and non-neural tissues. Pharmacol Rev 63:35–58
Kaetzel CS, Mestecky J, Johansen FE (2017) Two cells, one antibody: the discovery of the cellular origins and transport of secretory IgA. J Immunol 198:1765–1767
Kato S, Sato T, Fujita H, Kawatani M, Yamada Y (2021) Effects of GLP-1 receptor agonist on changes in the gut bacterium and the underlying mechanisms. Sci Rep 11:9167
Kelly JR, Kennedy PJ, Cryan JF, Dinan TG, Clarke G, Hyland NP (2015) Breaking down the barriers: the gut microbiome, intestinal permeability and stress-related psychiatric disorders. Front Cell Neurosci 9:392
Kelly JR, Borre Y, Patterson E, El Aidy S, Deane J, Kennedy PJ, Beers S, Scott K, Moloney G, Hoban AE, Scott L, Fitzgerald P, Ross P, Stanton C, Clarke G, Cryan JF, Dinan TG (2016) Transferring the blues: depression-associated gut microbiota induces neurobehavioural changes in the rat. J Psychiatr Res 82:109–118
Kennedy DO (2016) B vitamins and the brain: mechanisms, dose and efficacy: a review. Nutrients 8:68
Kennedy EA, King KY, Baldridge MT (2018) Mouse microbiota models: comparing germ-free mice and antibiotics treatment as tools for modifying gut bacteria. Front Physiol 9:1534
Kespohl M, Vachharajani N, Luu M, Harb H, Pautz S, Wolff S, Sillner N, Walker A, Schmitt-Kopplin P, Boettger T, Renz H, Offermanns S, Steinhoff U, Visekruna A (2017) The microbial metabolite butyrate induces expression of Th1-associated factors in CD4(+) T cells. Front Immunol 8:1036
Kiecolt-Glaser JK, Wilson SJ, Bailey ML, Andridge R, Peng J, Jaremka LM, Fagundes CP, Malarkey WB, Laskowski B, Belury MA (2018) Marital distress, depression, and a leaky gut: translocation of bacterial endotoxin as a pathway to inflammation. Psychoneuroendocrinology 98:52–60
Kiecolt-Glaser JK, Wilson SJ, Shrout MR, Madison AA, Andridge R, Peng J, Malarkey WB, Bailey MT (2021) The gut reaction to couples' relationship troubles: a route to gut dysbiosis through changes in depressive symptoms. Psychoneuroendocrinology 125:105132
Kleiman SC, Bulik-Sullivan EC, Glenny EM, Zerwas SC, Huh EY, Tsilimigras MC, Fodor AA, Bulik CM, Carroll IM (2017) The gut-brain axis in healthy females: lack of significant association between microbial composition and diversity with psychiatric measures. PLoS One 12: e0170208
Kotenko SV, Izotova LS, Mirochnitchenko OV, Esterova E, Dickensheets H, Donnelly RP, Pestka S (2001) Identification of the functional interleukin-22 (IL-22) receptor complex: the IL-10R2 chain (IL-10Rbeta) is a common chain of both the IL-10 and IL-22 (IL-10-related T cell-derived inducible factor, IL-TIF) receptor complexes. J Biol Chem 276:2725–2732
Kuhne SG, Stengel A (2019) Alteration of peptidergic gut-brain signaling under conditions of obesity. J Physiol Pharmacol 70. https://doi.org/10.26402/jpp.2019.5.01
Kuley E, Ozogul F (2011) Synergistic and antagonistic effect of lactic acid bacteria on tyramine production by food-borne pathogenic bacteria in tyrosine decarboxylase broth. Food Chem 127: 1163–1168
Kuley E, Ozogul F, Ozogul Y, Akyol I (2011) The function of lactic acid bacteria and brine solutions on biogenic amine formation by foodborne pathogens in trout fillets. Food Chem 129: 1211–1216
Kulma A, Szopa J (2007) Catecholamines are active compounds in plants. Plant Sci 172:8
Kumavath RN, Ramana C, V., Sasikala, C. (2010) L-Tryptophan catabolism by Rubrivivax benzoatilyticus JA2 occurs through indole 3-pyruvic acid pathway. Biodegradation 21:825–832
Kyoung H, Lee JJ, Cho JH, Choe J, Kang J, Lee H, Liu Y, Kim Y, Kim HB, Song M (2021) Dietary glutamic acid modulates immune responses and gut health of weaned pigs. Animals (Basel) 11: 504
Lawley TD, Walker AW (2013) Intestinal colonization resistance. Immunology 138:1–11

LeBlanc JG, Laino JE, del Valle MJ, Vannini V, van Sinderen D, Taranto MP, de Valdez GF, de Giori GS, Sesma F (2011) B-group vitamin production by lactic acid bacteria--current knowledge and potential applications. J Appl Microbiol 111:1297–1309

LeBlanc JG, Milani C, de Giori GS, Sesma F, van Sinderen D, Ventura M (2013) Bacteria as vitamin suppliers to their host: a gut microbiota perspective. Curr Opin Biotechnol 24:160–168

Lee JH, Kim YG, Kim CJ, Lee JC, Cho MH, Lee J (2012) Indole-3-acetaldehyde from Rhodococcus sp. BFI 332 inhibits Escherichia coli O157:H7 biofilm formation. Appl Microbiol Biotechnol 96:1071–1078

Lee HJ, Hong YS, Jun W, Yang SJ (2015) Nicotinamide riboside ameliorates hepatic metaflammation by modulating NLRP3 inflammasome in a rodent model of type 2 diabetes. J Med Food 18:1207–1213

Legrand R, Lucas N, Dominique M, Azhar S, Deroissart C, Le Solliec MA, Rondeaux J, Nobis S, Guerin C, Leon F, do Rego JC, Pons N, Le Chatelier E, Ehrlich SD, Lambert G, Dechelotte P, Fetissov SO (2020) Commensal Hafnia alvei strain reduces food intake and fat mass in obese mice-a new potential probiotic for appetite and body weight management. Int J Obes (Lond) 44: 1041–1051

Li J, Jia H, Cai X, Zhong H, Feng Q, Sunagawa S, Arumugam M, Kultima JR, Prifti E, Nielsen T, Juncker AS, Manichanh C, Chen B, Zhang W, Levenez F, Wang J, Xu X, Xiao L, Liang S, Zhang D, Zhang Z, Chen W, Zhao H, Al-Aama JY, Edris S, Yang H, Wang J, Hansen T, Nielsen HB, Brunak S, Kristiansen K, Guarner F, Pedersen O, Dore J, Ehrlich SD, Meta HITC, Bork P, Wang J, Meta HITC (2014) An integrated catalog of reference genes in the human gut microbiome. Nat Biotechnol 32:834–841

Li J, Kong D, Wang Q, Wu W, Tang Y, Bai T, Guo L, Wei L, Zhang Q, Yu Y, Qian Y, Zuo S, Liu G, Liu Q, Wu S, Zang Y, Zhu Q, Jia D, Wang Y, Yao W, Ji Y, Yin H, Nakamura M, Lazarus M, Breyer RM, Wang L, Yu Y (2017) Niacin ameliorates ulcerative colitis via prostaglandin D2-mediated D prostanoid receptor 1 activation. EMBO Mol Med 9:571–588

Li Y, Jin L, Chen T (2020) The effects of secretory IgA in the mucosal immune system. Biomed Res Int 2020:2032057

Li Q, Zhou H, Ouyang J, Guo S, Zheng J, Li G (2022) Effects of dietary tryptophan supplementation on body temperature, hormone, and cytokine levels in broilers exposed to acute heat stress. Trop Anim Health Prod 54:164

Liang G, Gao H, Bushman FD (2022) The pediatric virome in health and disease. Cell Host Microbe 30:639–649

Lindemans CA, Calafiore M, Mertelsmann AM, O'Connor MH, Dudakov JA, Jenq RR, Velardi E, Young LF, Smith OM, Lawrence G, Ivanov JA, Fu YY, Takashima S, Hua G, Martin ML, O'Rourke KP, Lo YH, Mokry M, Romera-Hernandez M, Cupedo T, Dow L, Nieuwenhuis EE, Shroyer NF, Liu C, Kolesnick R, van den Brink MRM, Hanash AM (2015) Interleukin-22 promotes intestinal-stem-cell-mediated epithelial regeneration. Nature 528:560–564

Lipszyc PS, Cremaschi GA, Zorrilla-Zubilete M, Bertolino ML, Capani F, Genaro AM, Wald MR (2013) Niacin modulates pro-inflammatory cytokine secretion. A potential mechanism involved in its anti-atherosclerotic effect. Open Cardiovasc Med J 7:90–98

Liu L, Li L, Min J, Wang J, Wu H, Zeng Y, Chen S, Chu Z (2012) Butyrate interferes with the differentiation and function of human monocyte-derived dendritic cells. Cell Immunol 277:66–73

Liu Z, Ma A, Mathe E, Merling M, Ma Q, Liu B (2021) Network analyses in microbiome based on high-throughput multi-omics data. Brief Bioinform 22:1639–1655

Louis P, Flint HJ (2009) Diversity, metabolism and microbial ecology of butyrate-producing bacteria from the human large intestine. FEMS Microbiol Lett 294:1–8

Louis P, Flint HJ (2017) Formation of propionate and butyrate by the human colonic microbiota. Environ Microbiol 19:29–41

Louis P, Young P, Holtrop G, Flint HJ (2010) Diversity of human colonic butyrate-producing bacteria revealed by analysis of the butyryl-CoA:acetate CoA-transferase gene. Environ Microbiol 12:304–314

Lu P, Yamaguchi Y, Fulton WB, Wang S, Zhou Q, Jia H, Kovler ML, Salazar AG, Sampah M, Prindle T Jr, Wipf P, Sodhi CP, Hackam DJ (2021) Maternal aryl hydrocarbon receptor activation protects newborns against necrotizing enterocolitis. Nat Commun 12:1042
Luo Y, Ma H, Niu S, Li X, Nie L, Li G (2021) Effects of norepinephrine on colonic tight junction protein expression during heat stress. Exp Ther Med 21:421
Luscher B, Shen Q, Sahir N (2011) The GABAergic deficit hypothesis of major depressive disorder. Mol Psychiatry 16:383–406
Lyte M, Bailey MT (1997) Neuroendocrine-bacterial interactions in a neurotoxin-induced model of trauma. J Surg Res 70:195–201
Lyte M, Arulanandam B, Nguyen K, Frank C, Erickson A, Francis D (1997) Norepinephrine induced growth and expression of virulence associated factors in enterotoxigenic and enterohemorrhagic strains of Escherichia coli. Adv Exp Med Biol 412:331–339
Ma X, Sun Q, Sun X, Chen D, Wei C, Yu X, Liu C, Li Y, Li J (2018) Activation of GABAA receptors in colon epithelium exacerbates acute colitis. Front Immunol 9:987
Mackner LM, Hatzakis E, Allen JM, Davies RH, Kim SC, Maltz RM, Bailey MT (2020) Fecal microbiota and metabolites are distinct in a pilot study of pediatric Crohn's disease patients with higher levels of perceived stress. Psychoneuroendocrinology 111:104469
Macpherson AJ, Koller Y, McCoy KD (2015) The bilateral responsiveness between intestinal microbes and IgA. Trends Immunol 36:460–470
Maes M, Kubera M, Leunis JC, Berk M, Geffard M, Bosmans E (2013) In depression, bacterial translocation may drive inflammatory responses, oxidative and nitrosative stress (O&NS), and autoimmune responses directed against O&NS-damaged neoepitopes. Acta Psychiatr Scand 127:344–354
Magnusdottir S, Ravcheev D, de Crecy-Lagard V, Thiele I (2015) Systematic genome assessment of B-vitamin biosynthesis suggests co-operation among gut microbes. Front Genet 6:148
Mara MA, Good M, Weitkamp JH (2018) Innate and adaptive immunity in necrotizing enterocolitis. Semin Fetal Neonatal Med 23:394–399
Margolis KG (2017) A role for the serotonin reuptake transporter in the brain and intestinal features of autism spectrum disorders and developmental antidepressant exposure. J Chem Neuroanat 83–84:36–40
Margolis KG, Gershon MD (2016) Enteric neuronal regulation of intestinal inflammation. Trends Neurosci 39:614–624
Marin-Luevano SP, Rodriguez-Carlos A, Jacobo-Delgado Y, Valdez-Miramontes C, Enciso-Moreno JA, Rivas-Santiago B (2021) Steroid hormone modulates the production of cathelicidin and human beta-defensins in lung epithelial cells and macrophages promoting Mycobacterium tuberculosis killing. Tuberculosis (Edinb) 128:102080
Markey L, Hooper A, Melon LC, Baglot S, Hill MN, Maguire J, Kumamoto CA (2020) Colonization with the commensal fungus Candida albicans perturbs the gut-brain axis through dysregulation of endocannabinoid signaling. Psychoneuroendocrinology 121:104808
Martinez-Carillo BE, Godinez-Victoria M, Jarillo-Luna A, Oros-Pantoja R, Abarca-Rojano E, Rivera-Aguilar V, Yepez JP, Sanchez-Torres LE, Campos-Rodriguez R (2011) Repeated restraint stress reduces the number of IgA-producing cells in Peyer's paches. Neuroimmunomodulation 18(3):131–141
Maslanik T, Tannura K, Mahaffey L, Loughridge AB, Beninson L, Ursell L, Greenwood BN, Knight R, Fleshner M (2012) Commensal bacteria and MAMPs are necessary for stress-induced increases in IL-1β and IL-18 but not IL-6, IL-10 or MCP-1. PLoS One 7:e50636
Masson F, Talon R, Montel MC (1996) Histamine and tyramine production by bacteria from meat products. Int J Food Microbiol 32:199–207
McKenna MC (2013) Glutamate pays its own way in astrocytes. Front Endocrinol (Lausanne) 4: 191
Mehdi SF, Pusapati S, Khenhrani RR, Farooqi MS, Sarwar S, Alnasarat A, Mathur N, Metz CN, LeRoith D, Tracey KJ, Yang H, Brownstein MJ, Roth J (2022) Oxytocin and related peptide

hormones: candidate anti-inflammatory therapy in early stages of sepsis. Front Immunol 13: 864007

Meng D, Sommella E, Salviati E, Campiglia P, Ganguli K, Djebali K, Zhu W, Walker WA (2020) Indole-3-lactic acid, a metabolite of tryptophan, secreted by Bifidobacterium longum subspecies infantis is anti-inflammatory in the immature intestine. Pediatr Res 88:209–217

Meyer-Hoffert U, Hornef MW, Henriques-Normark B, Axelsson LG, Midtvedt T, Putsep K, Andersson M (2008) Secreted enteric antimicrobial activity localises to the mucus surface layer. Gut 57:764–771

Millard AL, Mertes PM, Ittelet D, Villard F, Jeannesson P, Bernard J (2002) Butyrate affects differentiation, maturation and function of human monocyte-derived dendritic cells and macrophages. Clin Exp Immunol 130:245–255

Mirmonsef P, Zariffard MR, Gilbert D, Makinde H, Landay AL, Spear GT (2012) Short-chain fatty acids induce pro-inflammatory cytokine production alone and in combination with toll-like receptor ligands. Am J Reprod Immunol 67:391–400

Misheva M, Llott NE, McCullagh JSO (2021) Recent advances and future directions in microbiome metabolomics. Curr Opin Endocr Metab Res 20:100283

Mitchell PD, Salter BM, Oliveria JP, El-Gammal A, Tworek D, Smith SG, Sehmi R, Gauvreau GM, Butler M, O'Byrne PM (2017) Glucagon-like peptide-1 receptor expression on human eosinophils and its regulation of eosinophil activation. Clin Exp Allergy 47:331–338

Moreira CG, Sperandio V (2012) Interplay between the QseC and QseE bacterial adrenergic sensor kinases in Salmonella enterica serovar Typhimurium pathogenesis. Infect Immun 80:4344–4353

Moreira CG, Russell R, Mishra AA, Narayanan S, Ritchie JM, Waldor MK, Curtis MM, Winter SE, Weinshenker D, Sperandio V (2016) Bacterial adrenergic sensors regulate virulence of enteric pathogens in the gut. MBio 7:e00826–e00816

Morris DJ, Brem AS (2019) Role of gut metabolism of adrenal corticosteroids and hypertension: clues gut-cleansing antibiotics give us. Physiol Genomics 51:83–89

Morris DJ, Ridlon JM (2017) Glucocorticoids and gut bacteria: “the GALF hypothesis” in the metagenomic era. Steroids 125:1–13

Mowat AM, Agace WW (2014) Regional specialization within the intestinal immune system. Nat Rev Immunol 14:667–685

Nagao-Kitamoto H, Leslie JL, Kitamoto S, Jin C, Thomsson KA, Gillilland MG 3rd, Kuffa P, Goto Y, Jenq RR, Ishii C, Hirayama A, Seekatz AM, Martens EC, Eaton KA, Kao JY, Fukuda S, Higgins PDR, Karlsson NG, Young VB, Kamada N (2020) Interleukin-22-mediated host glycosylation prevents Clostridioides difficile infection by modulating the metabolic activity of the gut microbiota. Nat Med 26:608–617

Nankova BB, Agarwal R, MacFabe DF, La Gamma EF (2014) Enteric bacterial metabolites propionic and butyric acid modulate gene expression, including CREB-dependent catecholaminergic neurotransmission, in PC12 cells - possible relevance to autism spectrum disorders. PLoS One 9:e103740

Naseribafrouei A, Hestad K, Avershina E, Sekelja M, Linlokken A, Wilson R, Rudi K (2014) Correlation between the human fecal microbiota and depression. Neurogastroenterol Motil 26: 1155–1162

Neufeld KM, Kang N, Bienenstock J, Foster JA (2011) Reduced anxiety-like behavior and central neurochemical change in germ-free mice. Neurogastroenterol Motil 23(255-264):e119

Nishida K, Sawada D, Kuwano Y, Tanaka H, Rokutan K (2019) Health benefits of Lactobacillus gasseri CP2305 tablets in young adults exposed to chronic stress: a randomized, double-blind, placebo-controlled study. Nutrients 11:1859

Nolan LS, Mihi B, Agrawal P, Gong Q, Rimer JM, Bidani SS, Gale SE, Goree M, Hu E, Lanik WE, Huang E, Bando JK, Liu V, Lewis AN, Bustos A, Hodzic Z, Laury ML, Good M (2021) Indole-3-carbinol-dependent aryl hydrocarbon receptor signaling attenuates the inflammatory response in experimental necrotizing enterocolitis. Immunohorizons 5:193–209

O’Hara AM, Shanahan F (2006) The gut flora as a forgotten organ. EMBO Rep 7:688–693

O'Mahoney SM, Marchesi JR, Scully P et al (2009) Early life stress alters behavior, immunity, and microbiota in rats: implications for irritable bowel syndrome and psychiatric illnesses. Biol Psychiatry 65:263–267

Oiry C, Gagne D, Cottin E, Bernad N, Galleyrand JC, Berge G, Lignon MF, Eldin P, Le Cunff M, Leger J, Clerc P, Fourmy D, Martinez J (1997) CholecystokininB receptor from human Jurkat lymphoblastic T cells is involved in activator protein-1-responsive gene activation. Mol Pharmacol 52:292–299

Olesen SW, Alm EJ (2016) Dysbiosis is not an answer. Nat Microbiol 1:16228

Olson JK, Rager TM, Navarro JB, Mashburn-Warren L, Goodman SD, Besner GE (2016) Harvesting the benefits of biofilms: A novel probiotic delivery system for the prevention of necrotizing enterocolitis. J Pediatr Surg 51:936–941

Olson JK, Navarro JB, Allen JM, McCulloh CJ, Mashburn-Warren L, Wang Y, Varaljay VA, Bailey MT, Goodman SD, Besner GE (2018) An enhanced Lactobacillus reuteri biofilm formulation that increases protection against experimental necrotizing enterocolitis. Am J Physiol Gastrointest Liver Physiol 315:G408–G419

O'Malley D, Julio-Piper M, Gibney SM, Dinan TG, Cryan JF (2010) Distinct alterations in colonic morphology and physiology in two rat models of enhanced stress-induced anxiety and depression-like behaviour. Stress 13:114–122

Pabst O (2012) New concepts in the generation and functions of IgA. Nat Rev Immunol 12:821–832

Park HJ, Kim SA, Kang WS, Kim JW (2021) Early-life stress modulates gut microbiota and peripheral and central inflammation in a sex-dependent manner. Int J Mol Sci 22(4):1899

Parks OB, Pociask DA, Hodzic Z, Kolls JK, Good M (2015) Interleukin-22 signaling in the regulation of intestinal health and disease. Front Cell Dev Biol 3:85

Peng F, Zhao S, Zhang X, Long S, He Y (2022) Calcitonin gene-related peptide upregulates IL-17A and IL-22 in gammadelta-T cells through the paracrine effect of langerhans cells on LC/gammadelta-T co-culture model. J Neuroimmunol 364:577792

Perez-Burgos A, Wang B, Mao YK, Mistry B, McVey Neufeld KA, Bienenstock J, Kunze W (2013) Psychoactive bacteria Lactobacillus rhamnosus (JB-1) elicits rapid frequency facilitation in vagal afferents. Am J Physiol Gastrointest Liver Physiol 304:G211–G220

Perez-Munoz ME, Arrieta MC, Ramer-Tait AE, Walter J (2017) A critical assessment of the "sterile womb" and "in utero colonization" hypotheses: implications for research on the pioneer infant microbiome. Microbiome 5:48

Pernomian L, Duarte-Silva M, de Barros Cardoso CR (2020) The aryl hydrocarbon receptor (AHR) as a potential target for the control of intestinal inflammation: insights from an immune and bacteria sensor receptor. Clin Rev Allergy Immunol 59:382–390

Pham TA, Clare S, Goulding D, Arasteh JM, Stares MD, Browne HP, Keane JA, Page AJ, Kumasaka N, Kane L, Mottram L, Harcourt K, Hale C, Arends MJ, Gaffney DJ, The Sanger Mouse Genetics Project, Dougan G, Lawley TD (2014) Epithelial IL-22RA1-mediated fucosylation promotes intestinal colonization resistance to an opportunistic pathogen. Cell Host Microbe 16:504–516

Pietrzak B, Tomela K, Olejnik-Schmidt A, Mackiewicz A, Schmidt M (2020) Secretory IgA in intestinal mucosal secretions as an adaptive barrier against microbial cells. Int J Mol Sci 21:9254

Piras C, Mussap M, Noto A, De Giacomo A, Cristofori F, Spada M, Fanos V, Atzori L, Francavilla R (2022) Alterations of the intestinal permeability are reflected by changes in the urine metabolome of young autistic children: preliminary results. Metabolites 12:104

Pitman RM (1971) Transmitter substances in insects: a review. Comp Gen Pharmacol 2:347–371

Pokusaeva K, Johnson C, Luk B, Uribe G, Fu Y, Oezguen N, Matsunami RK, Lugo M, Major A, Mori-Akiyama Y, Hollister EB, Dann SM, Shi XZ, Engler DA, Savidge T, Versalovic J (2017) GABA-producing Bifidobacterium dentium modulates visceral sensitivity in the intestine. Neurogastroenterol Motil 29:e12904

Poutahidis T, Kearney SM, Levkovich T, Qi P, Varian BJ, Lakritz JR, Ibrahim YM, Chatzigiagkos A, Alm EJ, Erdman SE (2013) Microbial symbionts accelerate wound healing via the neuropeptide hormone oxytocin. PLoS One 8:e78898
Pullinger GD, Carnell SC, Sharaff FF et al (2010) Norepinephrine augments Salmonella enterica-induced enteritis in a manner associated with increased net replication but independent of the putative adrenergic sensor kinasa QseC and QseE. Infect Immun 78:372–380
Qin J, Li R, Raes J, Arumugam M, Burgdorf KS, Manichanh C, Nielsen T, Pons N, Levenez F, Yamada T, Mende DR, Li J, Xu J, Li S, Li D, Cao J, Wang B, Liang H, Zheng H, Xie Y, Tap J, Lepage P, Bertalan M, Batto JM, Hansen T, Le Paslier D, Linneberg A, Nielsen HB, Pelletier E, Renault P, Sicheritz-Ponten T, Turner K, Zhu H, Yu C, Li S, Jian M, Zhou Y, Li Y, Zhang X, Li S, Qin N, Yang H, Wang J, Brunak S, Dore J, Guarner F, Kristiansen K, Pedersen O, Parkhill J, Weissenbach J, Meta HITC, Bork P, Ehrlich SD, Wang J (2010) A human gut microbial gene catalogue established by metagenomic sequencing. Nature 464:59–65
Rashed LA, Hashem RM, Soliman HM (2011) Oxytocin inhibits NADPH oxidase and P38 MAPK in cisplatin-induced nephrotoxicity. Biomed Pharmacother 65:474–480
Reber SO, Siebler PH, Donner NC, Morton JT, Smith DG, Kopelman JM, Lowe KR, Wheeler KJ, Fox JH, Hassell JE Jr, Greenwood BN, Jansch C, Lechner A, Schmidt D, Uschold-Schmidt N, Fuchsl AM, Langgartner D, Walker FR, Hale MW, Lopez Perez G, Van Treuren W, Gonzalez A, Halweg-Edwards AL, Fleshner M, Raison CL, Rook GA, Peddada SD, Knight R, Lowry CA (2016) Immunization with a heat-killed preparation of the environmental bacterium Mycobacterium vaccae promotes stress resilience in mice. Proc Natl Acad Sci U S A 113:E3130–E3139
Ren W, Yin J, Xiao H, Chen S, Liu G, Tan B, Li N, Peng Y, Li T, Zeng B, Li W, Wei H, Yin Z, Wu G, Hardwidge PR, Yin Y (2016) Intestinal microbiota-derived GABA mediates interleukin-17 expression during enterotoxigenic Escherichia coli infection. Front Immunol 7:685
Rengarajan S, Knoop KA, Rengarajan A, Chai JN, Grajales-Reyes JG, Samineni VK, Russler-Germain EV, Ranganathan P, Fasano A, Sayuk GS, Gereau RWT, Kau AL, Knights D, Kashyap PC, Ciorba MA, Newberry RD, Hsieh CS (2020) A potential role for stress-induced microbial alterations in IgA-associated irritable Bowel syndrome with diarrhea. Cell Rep Med 1(7):100124
Reyna-Garfias H, Miliar A, Jarillo-Luna A, Rivera-Aguilar V, Pacheco-Yepez J, Baeza I, Campos-Rodriguez R (2010) Repeated restraint stress increases IgA concentration in rat small intestine. Brain Behav Immun 24:110–118
Roth J, LeRoith D, Shiloach J, Rosenzweig JL, Lesniak MA, Havrankova J (1982) The evolutionary origins of hormones, neurotransmitters, and other extracellular chemical messengers: implications for mammalian biology. N Engl J Med 306:523–527
Rothhammer V, Mascanfroni ID, Bunse L, Takenaka MC, Kenison JE, Mayo L, Chao CC, Patel B, Yan R, Blain M, Alvarez JI, Kebir H, Anandasabapathy N, Izquierdo G, Jung S, Obholzer N, Pochet N, Clish CB, Prinz M, Prat A, Antel J, Quintana FJ (2016) Type I interferons and microbial metabolites of tryptophan modulate astrocyte activity and central nervous system inflammation via the aryl hydrocarbon receptor. Nat Med 22:586–597
Rubin DC, Zhang H, Qian P, Lorenz RG, Hutton K, Peters MG (2000) Altered enteroendocrine cell expression in T cell receptor alpha chain knock-out mice. Microsc Res Tech 51:112–120
Ruddick JP, Evans AK, Nutt DJ, Lightman SL, Rook GA, Lowry CA (2006) Tryptophan metabolism in the central nervous system: medical implications. Expert Rev Mol Med 8:1–27
Rudzki L, Maes M (2020) The microbiota-gut-immune-glia (MGIG) axis in major depression. Mol Neurobiol 57:4269–4295
Rudzki L, Maes M (2021) From "leaky gut" to impaired glia-neuron communication in depression. Adv Exp Med Biol 1305:129–155
Rudzki L, Stone TW, Maes M, Misiak B, Samochowiec J, Szulc A (2021) Gut microbiota-derived vitamins - underrated powers of a multipotent ally in psychiatric health and disease. Prog Neuropsychopharmacol Biol Psychiatry 107:110240
Russell G, Lightman S (2019) The human stress response. Nat Rev Endocrinol 15:525–534

Safe S, Jayaraman A, Chapkin RS (2020) Ah receptor ligands and their impacts on gut resilience: structure-activity effects. Crit Rev Toxicol 50:463–473

Samuel BS, Shaito A, Motoike T, Rey FE, Backhed F, Manchester JK, Hammer RE, Williams SC, Crowley J, Yanagisawa M, Gordon JI (2008) Effects of the gut microbiota on host adiposity are modulated by the short-chain fatty-acid binding G protein-coupled receptor, Gpr41. Proc Natl Acad Sci U S A 105:16767–16772

Sanchez S, Rodriguez-Sanoja R, Ramos A, Demain AL (2017) Our microbes not only produce antibiotics, they also overproduce amino acids. J Antibiot (Tokyo). https://doi.org/10.1038/ja.2017.142

Sandrini SM, Shergill R, Woodward J, Muralikuttan R, Haigh RD, Lyte M, Freestone PP (2010) Elucidation of the mechanism by which catecholamine stress hormones liberate iron from the innate immune defense proteins transferrin and lactoferrin. J Bacteriol 192:587–594

Sato K (2018) Why is vitamin B6 effective in alleviating the symptoms of autism? Med Hypotheses 115:103–106

Schroeder FA, Lin CL, Crusio WE, Akbarian S (2007) Antidepressant-like effects of the histone deacetylase inhibitor, sodium butyrate, in the mouse. Biol Psychiatry 62:55–64

Schulthess J, Pandey S, Capitani M, Rue-Albrecht KC, Arnold I, Franchini F, Chomka A, Ilott NE, Johnston DGW, Pires E, McCullagh J, Sansom SN, Arancibia-Carcamo CV, Uhlig HH, Powrie F (2019) The short chain fatty acid butyrate imprints an antimicrobial program in macrophages. Immunity 50:432–445.e437

Schwizer W, Borovicka J, Kunz P, Fraser R, Kreiss C, D'Amato M, Crelier G, Boesiger P, Fried M (1997) Role of cholecystokinin in the regulation of liquid gastric emptying and gastric motility in humans: studies with the CCK antagonist loxiglumide. Gut 41:500–504

Sgritta M, Dooling SW, Buffington SA, Momin EN, Francis MB, Britton RA, Costa-Mattioli M (2019) Mechanisms underlying microbial-mediated changes in social behavior in mouse models of autism spectrum disorder. Neuron 101:246–259.e246

Shakya M, Lo CC, Chain PSG (2019) Advances and challenges in metatranscriptomic analysis. Front Genet 10:904

Shaler CR, Parco AA, Elhenawy W, Dourka J, Jury J, Verdu EF, Coombes BK (2021) Psychological stress impairs IL22-driven protective gut mucosal immunity against colonising pathobionts. Nat Commun 12:6664

Sharpton TJ (2014) An introduction to the analysis of shotgun metagenomic data. Front Plant Sci 5: 209

Shigeshiro M, Tanabe S, Suzuki T (2012) Repeated exposure to water immersion stress reduced the Muc2 gene level in the rat colon via two distinct mechanisms. Brain Behav Immun 26:1061–1065

Shkoporov AN, Hill C (2019) Bacteriophages of the human gut: the "known unknown" of the microbiome. Cell Host Microbe 25:195–209

Shrout MR, Madison AA, Renna ME, Alfano CM, Povoski SP, Lipari AM, Agnese DM, Carson WE 3rd, Malarkey WB, Bailey MT, Kiecolt-Glaser JK (2022) The gut connection: intestinal permeability as a pathway from breast cancer survivors' relationship satisfaction to inflammation across treatment. Brain Behav Immun 100:145–154

Sina C, Gavrilova O, Forster M, Till A, Derer S, Hildebrand F, Raabe B, Chalaris A, Scheller J, Rehmann A, Franke A, Ott S, Hasler R, Nikolaus S, Folsch UR, Rose-John S, Jiang HP, Li J, Schreiber S, Rosenstiel P (2009) G protein-coupled receptor 43 is essential for neutrophil recruitment during intestinal inflammation. J Immunol 183:7514–7522

Singh N, Gurav A, Sivaprakasam S, Brady E, Padia R, Shi H, Thangaraju M, Prasad PD, Manicassamy S, Munn DH, Lee JR, Offermanns S, Ganapathy V (2014) Activation of Gpr109a, receptor for niacin and the commensal metabolite butyrate, suppresses colonic inflammation and carcinogenesis. Immunity 40:128–139

Skarupski KA, Tangney C, Li H, Ouyang B, Evans DA, Morris MC (2010) Longitudinal association of vitamin B-6, folate, and vitamin B-12 with depressive symptoms among older adults over time. Am J Clin Nutr 92:330–335

Smith EA, Macfarlane GT (1996) Enumeration of human colonic bacteria producing phenolic and indolic compounds: effects of pH, carbohydrate availability and retention time on dissimilatory aromatic amino acid metabolism. J Appl Bacteriol 81:288–302
Smith GP, Jerome C, Cushin BJ, Eterno R, Simansky KJ (1981) Abdominal vagotomy blocks the satiety effect of cholecystokinin in the rat. Science 213:1036–1037
Spreckley E, Murphy KG (2015) The L-cell in nutritional sensing and the regulation of appetite. Front Nutr 2:23
Stephenson M, Rowatt E (1947) The production of acetylcholine by a strain of Lactobacillus plantarum. J Gen Microbiol 1:279–298
Stevens BR, Goel R, Seungbum K, Richards EM, Holbert RC, Pepine CJ, Raizada MK (2018) Increased human intestinal barrier permeability plasma biomarkers zonulin and FABP2 correlated with plasma LPS and altered gut microbiome in anxiety or depression. Gut 67: 1555–1557
Stilling RM, van de Wouw M, Clarke G, Stanton C, Dinan TG, Cryan JF (2016) The neuropharmacology of butyrate: The bread and butter of the microbiota-gut-brain axis? Neurochem Int 99: 110–132
Strandwitz P, Kim KH, Terekhova D, Liu JK, Sharma A, Levering J, McDonald D, Dietrich D, Ramadhar TR, Lekbua A, Mroue N, Liston C, Stewart EJ, Dubin MJ, Zengler K, Knight R, Gilbert JA, Clardy J, Lewis K (2019) GABA-modulating bacteria of the human gut microbiota. Nat Microbiol 4:396–403
Strausbauch PH, Fischer EH (1970) Structure of the binding site of pyridoxal 5′-phosphate to Escherichia coli glutamate decarboxylase. Biochemistry 9:233–238
Sudo N, Chida Y, Aiba Y, Sonoda J, Oyama N, Yu XN, Kubo C, Koga Y (2004) Postnatal microbial colonization programs the hypothalamic-pituitary-adrenal system for stress response in mice. J Physiol 558:263–275
Szeto A, Nation DA, Mendez AJ, Dominguez-Bendala J, Brooks LG, Schneiderman N, McCabe PM (2008) Oxytocin attenuates NADPH-dependent superoxide activity and IL-6 secretion in macrophages and vascular cells. Am J Physiol Endocrinol Metab 295:E1495–E1501
Szeto A, Sun-Suslow N, Mendez AJ, Hernandez RI, Wagner KV, McCabe PM (2017) Regulation of the macrophage oxytocin receptor in response to inflammation. Am J Physiol Endocrinol Metab 312:E183–E189
Tache Y, Perdue MH (2004) Role of peripheral CRF signalling pathways in stress-related alterations of gut motibility and mucosal function. Neurogastroenterol Motil 16:137–142
Tache Y, Martinez V, Million M, Wang L (2001) Stress and the gastrointestinal tract III. Stress-related alterations of gut motor function: role of brain corticotropin-releasing factor receptors. Am J Physiol Gastrointest Liver Physiol 280:G173–G177
Takanaga H, Ohtsuki S, Hosoya K, Terasaki T (2001) GAT2/BGT-1 as a system responsible for the transport of gamma-aminobutyric acid at the mouse blood-brain barrier. J Cereb Blood Flow Metab 21:1232–1239
Tan J, McKenzie C, Potamitis M, Thorburn AN, Mackay CR, Macia L (2014) The role of short-chain fatty acids in health and disease. Adv Immunol 121:91–119
Thomas CM, Hong T, van Pijkeren JP, Hemarajata P, Trinh DV, Hu W, Britton RA, Kalkum M, Versalovic J (2012) Histamine derived from probiotic Lactobacillus reuteri suppresses TNF via modulation of PKA and ERK signaling. PLoS One 7:e31951
Tiffany CR, Baumler AJ (2019) Dysbiosis: from fiction to function. Am J Physiol Gastrointest Liver Physiol 317:G602–G608
Tingirikari JMR (2018) Microbiota-accessible pectic poly- and oligosaccharides in gut health. Food Funct 9:5059–5073
Tobias PS, Soldau K, Gegner JA, Mintz D, Ulevitch RJ (1995) Lipopolysaccharide binding protein-mediated complexation of lipopolysaccharide with soluble CD14. J Biol Chem 270:10482–10488

Tsavkelova EA, Botvinko IV, Kudrin VS, Oleskin AV (2000) Detection of neurotransmitter amines in microorganisms with the use of high-performance liquid chromatography. Dokl Biochem 372:115–117
Tsurugizawa T, Uneyama H (2014) Differences in BOLD responses to intragastrically infused glucose and saccharin in rats. Chem Senses 39:683–691
Tsurugizawa T, Uematsu A, Uneyama H, Torii K (2010) Effects of isoflurane and alpha-chloralose anesthesia on BOLD fMRI responses to ingested L-glutamate in rats. Neuroscience 165:244–251
Tsurugizawa T, Uneyama H, Torii K (2014) Brain amino acid sensing. Diabetes Obes Metab 16 (Suppl 1):41–48
Turna J, Grosman Kaplan K, Anglin R, Patterson B, Soreni N, Bercik P, Surette MG, Van Ameringen M (2020) The gut microbiome and inflammation in obsessive-compulsive disorder patients compared to age- and sex-matched controls: a pilot study. Acta Psychiatr Scand 142: 337–347
Turton MD, O'Shea D, Gunn I, Beak SA, Edwards CM, Meeran K, Choi SJ, Taylor GM, Heath MM, Lambert PD, Wilding JP, Smith DM, Ghatei MA, Herbert J, Bloom SR (1996) A role for glucagon-like peptide-1 in the central regulation of feeding. Nature 379:69–72
Vaishnava S, Yamamoto M, Severson KM, Ruhn KA, Yu X, Koren O, Ley R, Wakeland EK, Hooper LV (2011) The antibacterial lectin RegIIIgamma promotes the spatial segregation of microbiota and host in the intestine. Science 334:255–258
Vargas-Blanco DA, Shell SS (2020) Regulation of mRNA stability during bacterial stress responses. Front Microbiol 11:2111
Varian BJ, Poutahidis T, DiBenedictis BT, Levkovich T, Ibrahim Y, Didyk E, Shikhman L, Cheung HK, Hardas A, Ricciardi CE, Kolandaivelu K, Veenema AH, Alm EJ, Erdman SE (2017) Microbial lysate upregulates host oxytocin. Brain Behav Immun 61:36–49
Vazquez E, Barranco A, Ramirez M, Gruart A, Delgado-Garcia JM, Jimenez ML, Buck R, Rueda R (2016) Dietary 2'-fucosyllactose enhances operant conditioning and long-term potentiation via gut-brain communication through the vagus nerve in rodents. PLoS One 11:e0166070
Vinolo MA, Rodrigues HG, Hatanaka E, Sato FT, Sampaio SC, Curi R (2011a) Suppressive effect of short-chain fatty acids on production of proinflammatory mediators by neutrophils. J Nutr Biochem 22:849–855
Vinolo MA, Rodrigues HG, Nachbar RT, Curi R (2011b) Regulation of inflammation by short chain fatty acids. Nutrients 3:858–876
Vlisidou I, Lyte M, van Diemen PM, Hawes P, Monaghan P, Wallis TS, Stevens MP (2004) The neuroendocrine stress hormone norepinephrine augments Escherichia coli O157:H7-induced enteritis and adherence in a bovine ligated ileal loop model of infection. Infect Immun 72:5446–5451
Wallace RH, Marini C, Petrou S, Harkin LA, Bowser DN, Panchal RG, Williams DA, Sutherland GR, Mulley JC, Scheffer IE, Berkovic SF (2001) Mutant GABA(A) receptor gamma2-subunit in childhood absence epilepsy and febrile seizures. Nat Genet 28:49–52
Wehrwein EA, Orer HS, Barman SM (2016) Overview of the anatomy, physiology, and pharmacology of the autonomic nervous system. Compr Physiol 6:1239–1278
Wikoff WR, Anfora AT, Liu J, Schultz PG, Lesley SA, Peters EC, Siuzdak G (2009) Metabolomics analysis reveals large effects of gut microflora on mammalian blood metabolites. Proc Natl Acad Sci U S A 106:3698–3703
Williams BB, Van Benschoten AH, Cimermancic P, Donia MS, Zimmermann M, Taketani M, Ishihara A, Kashyap PC, Fraser JS, Fischbach MA (2014) Discovery and characterization of gut microbiota decarboxylases that can produce the neurotransmitter tryptamine. Cell Host Microbe 16:495–503
Wong CG, Bottiglieri T, Snead OC 3rd. (2003) GABA, gamma-hydroxybutyric acid, and neurological disease. Ann Neurol 54(Suppl 6):S3–S12

Xie MH, Aggarwal S, Ho WH, Foster J, Zhang Z, Stinson J, Wood WI, Goddard AD, Gurney AL (2000) Interleukin (IL)-22, a novel human cytokine that signals through the interferon receptor-related proteins CRF2-4 and IL-22R. J Biol Chem 275:31335–31339

Yanay O, Bailey AL, Kernan K, Zimmerman JJ, Osborne WR (2015) Effects of exendin-4, a glucagon like peptide-1 receptor agonist, on neutrophil count and inflammatory cytokines in a rat model of endotoxemia. J Inflamm Res 8:129–135

Yang H, Yuan PQ, Wang L, Tache Y (2000) Activation of the parapyramidal region in the vetral medulla stimulates gastric acid secretion through vagal pathways in rats. Neuroscience 95:773–779

Yang F, Zeng F, Luo X, Lei Y, Li J, Lu S, Huang X, Lan Y, Liu R (2021) GLP-1 receptor: a new target for sepsis. Front Pharmacol 12:706908

Ye L, Bae M, Cassilly CD, Jabba SV, Thorpe DW, Martin AM, Lu HY, Wang J, Thompson JD, Lickwar CR, Poss KD, Keating DJ, Jordt SE, Clardy J, Liddle RA, Rawls JF (2021) Enteroendocrine cells sense bacterial tryptophan catabolites to activate enteric and vagal neuronal pathways. Cell Host Microbe 29:179–196.e179

Yusta B, Baggio LL, Koehler J, Holland D, Cao X, Pinnell LJ, Johnson-Henry KC, Yeung W, Surette MG, Bang KW, Sherman PM, Drucker DJ (2015) GLP-1R agonists modulate enteric immune responses through the intestinal intraepithelial lymphocyte GLP-1R. Diabetes 64: 2537–2549

Zakis DR, Paulissen E, Kornete L, Kaan AMM, Nicu EA, Zaura E (2022) The evidence for placental microbiome and its composition in healthy pregnancies: a systematic review. J Reprod Immunol 149:103455

Zelante T, Iannitti RG, Cunha C, De Luca A, Giovannini G, Pieraccini G, Zecchi R, D'Angelo C, Massi-Benedetti C, Fallarino F, Carvalho A, Puccetti P, Romani L (2013) Tryptophan catabolites from microbiota engage aryl hydrocarbon receptor and balance mucosal reactivity via interleukin-22. Immunity 39:372–385

Zhang JG, Cong B, Li QX, Chen HY, Qin J, Fu LH (2011) Cholecystokinin octapeptide regulates lipopolysaccharide-activated B cells co-stimulatory molecule expression and cytokines production in vitro. Immunopharmacol Immunotoxicol 33:157–163

Zhang JG, Liu JX, Jia XX, Geng J, Yu F, Cong B (2014) Cholecystokinin octapeptide regulates the differentiation and effector cytokine production of CD4(+) T cells in vitro. Int Immunopharmacol 20:307–315

Zhao Y, Chen F, Wu W, Sun M, Bilotta AJ, Yao S, Xiao Y, Huang X, Eaves-Pyles TD, Golovko G, Fofanov Y, D'Souza W, Zhao Q, Liu Z, Cong Y (2018) GPR43 mediates microbiota metabolite SCFA regulation of antimicrobial peptide expression in intestinal epithelial cells via activation of mTOR and STAT3. Mucosal Immunol 11:752–762

Zheng Y, Valdez PA, Danilenko DM, Hu Y, Sa SM, Gong Q, Abbas AR, Modrusan Z, Ghilardi N, de Sauvage FJ, Ouyang W (2008) Interleukin-22 mediates early host defense against attaching and effacing bacterial pathogens. Nat Med 14:282–289

Zong Y, Zhu S, Zhang S, Zheng G, Wiley JW, Hong S (2019) Chronic stress and intestinal permeability: lubiprostone regulates glucocorticoid receptor-mediated changes in colon epithelial tight junction proteins, barrier function, and visceral pain in the rodent and human. Neurogastroenterol Motil 31:e13477

Zweigner J, Schumann RR, Weber JR (2006) The role of lipopolysaccharide-binding protein in modulating the innate immune response. Microbes Infect 8:946–952

Part IV

Disease

Energy Balance and Neuroendocrine-Immune Regulation in Chronic Inflammatory and Neoplastic Diseases: An Evolutionary Perspective

13

Jan Pieter Konsman and Rainer H. Straub

Abstract

The central nervous and immune systems are often in competition for energy substrates. After infection or injury, energy expenditure typically increases while exploratory behavior and food intake decrease (sickness behavior). However, in the absence of food intake, animal bodies can only sustain increased expenditure for 3–7 weeks before energy substrates run out. Nevertheless, symptoms reminiscent of wasting and sickness behavior can be observed in autoimmunity, and in chronic inflammatory or brain-related disorders. The hypothesis defended here is that chronic diseases can exist because they mostly occur after reproductive age and involve responses that were selected during evolution in response to acute infection and injury. Indeed, fever and reduced food intake can increase survival in response to acute bacterial infection and are brought about by actions of pro-inflammatory cytokines on the brain, resulting in autonomic, behavioral, and neuroendocrine responses in a context of energy trade-offs. While these responses may be adaptive and actively brought about when an organism responds to acute infection, they seem maladaptive when lasting too long, such as in chronic diseases, as they contribute to a negative energy balance.

Keywords

Chronic disease energy balance · Evolution · Inflammation

J. P. Konsman (✉)
Immunology from Concepts and Experiments to Translation, University of Bordeaux, Bordeaux, France
e-mail: jan-pieter.konsman@u-bordeaux.fr

R. H. Straub
Laboratory of Experimental Rheumatology and Neuroendocrine Immunology, Department of Internal Medicine I, University Hospital Regensburg, Regensburg, Germany

J. P. Konsman, T. M. Reyes (eds.), *Neuroendocrine-Immune System Interactions*, Masterclass in Neuroendocrinology 13,
https://doi.org/10.1007/978-3-031-21358-8_13

13.1 Introduction

The often cited title: "Nothing makes sense in biology except in the light of evolution" was intended to emphasize the importance of "biological integration [at] levels above the molecular one" (Dobzhansky 1964, p. 445). The light of evolution hit medicine much later, presumably because medicine was considered to deal only with the maladaptive pathological processes. However, when in the 1970s and 1980s evidence indicating that non-specific disease symptoms were, under certain conditions, positively linked to survival of infected animals, these symptoms started to be considered as potentially evolutionarily selected inflammatory responses. While these responses may be adaptive and actively brought about when an organism responds to acute infection, they seem maladaptive when long-lasting, as they can then contribute to a negative energy balance. One of the main challenges of evolutionary medicine is therefore to explain why symptoms and pathophysiological processes that occur during chronic inflammatory and neoplastic diseases can be so prevalent. Here, the argument will be developed that these symptoms and pathophysiological processes are adaptive in response to acute infection and injury in young reproducing individuals and have been retained during evolution, even though they may be detrimental later in life during chronic conditions. Although there are many ways in which elements of the neuroendocrine and immune systems interact during disease, the focus here will be on inflammation and neuroendocrine signaling involved in corticosteroid secretion and energy balance. Indeed, these interactions have been the most widely studied and can thus be compared to some extent to acute responses to infection or injury and chronic disease conditions.

13.2 Evolutionary Medicine and Energy Regulation

13.2.1 Evolutionary Medicine Fundamentals

So-called Darwinian medicine was proposed in the early 1990s with the promise to "provide . . . new insights into the causes of medical disorders" by considering the signs and symptoms of infection or responses to injuries and toxins as adaptive (Williams and Nesse 1991, p. 1). For example, susceptibility to breast and ovarian cancer has been suggested to be related to "genotypes selected for high fertility" (Greaves 2007, p. 219) and predisposition to chronic inflammatory disease to "immunodysregulation resulting from lack of exposure to microorganisms" in childhood (Rook 2010, p. 70). Although the terms "Darwinian medicine" and "evolutionary medicine" have been used in parallel for some time, the latter has been adopted progressively because it "is more general and acknowledges other important aspects of the theory of evolution [other than natural selection and adaptation] such as symbiosis [and] the role of epigenetic processes" (Methot 2011, p. 76).

"Evolutionary medicine thus consists of all areas in which evolutionary thought productively informs medical and epidemiological issues," including physiology and metabolism (Stearns 2012, p. 4305). In this context, the principle of antagonistic pleiotropy, according to which "genes [can] have opposite effects on fitness in the young compared to later ages" (Williams 1957, p. 400), has been put forward to render explicit the idea that "genes being adaptive at an early age, [can be] maladaptive at older age" and lead, for example, to chronic inflammatory disorders (Straub and Schradin 2016, p. 42). Most recently, proponents of evolutionary medicine have reached a consensus, according to which, (1) in addition to natural selection, constraints and trade-offs are among the core evolutionary processes, (2) "disease risks can be altered for organisms living in environments that differ from those in which their ancestors evolved," and (3) disease symptoms and signs "are useful defenses, which can be pathological if dysregulated" and long-lasting (Grunspan et al. 2018, Table 2).

13.2.2 Energy Balance in Health and Acute Disease

Animal life comes with challenges such as being a potential prey and the necessity of obtaining food, which, in turn, imply risks of injury, infection, and other stressful events. It has been argued that animals weigh their "energy landscape" that includes food sources against their "fear landscape" regarding predators when they explore and move about in their environment (Gallagher et al. 2017). It is therefore not surprising that animals seem to be equipped with food and threat detection systems (Illius et al. 2002; Douglas et al. 2005; Forbey et al. 2018; Fendt et al. 2020). In addition, animals generate and respond to their own bodily signals like hormones and cytokines after having consumed food or being injured or infected (Medzhitov 2001; Kono and Rock 2008; Keestra-Gounder et al. 2015; Baral et al. 2019; Donnelly et al. 2020; Olson et al. 2020).

Infections, either foodborne or injury-related, can develop into life-threatening situations, including for humans (Lederberg 1997; High 2004; Smith and Lewin 2009; Francisco et al. 2018; Mcnamara et al. 2018). Responses to injury or infection are energetically costly because inflammation and tissue repair increase energy expenditure and, in parallel, energy intake is often reduced due to event-associated anorexia (Hart 1988; Goldstein and Elwyn 1989; Archie 2013). Indeed, "[m]ultiple bone fractures increase heat production by 15–30% ... [,] sepsis [systemic inflammatory response to infection]... by 50% ... [and] [e]xtensive burns ... well in excess of 100%" (Blaxter 1989, p. 218). Even though considerations of energy balance have been part of a long tradition in endocrinology and have recently been incorporated into eco-immunology (Martin et al. 2006), this seems to have been less the case for medicine.

The fact that several of the responses to injury and infection, such as reduced activity and fever, are present in many different animals considered to represent various stages of evolution, such as lizards, mice, and humans (Hart 1988; Kluger 1991; Walters 1994; Sylvia and Demas 2017; Hite et al. 2020; Lopes et al. 2021), suggests that these are evolutionarily conserved responses. In this context, it is important to keep in mind that such responses can be transferred to offspring only if they did not put at risk the parents' survival prior to or during reproduction. Given the negative effects of the responses to injury or infection on energy balance, fever, reduced food intake, and low activity can therefore be hypothesized to increase survival of young animals. Since the administration of antipyretics, force-feeding or sleep deprivation increases mortality in experimental models of bacterial infection in young adult rodents (Murray and Murray 1979; Vaughn et al. 1980; Friese et al. 2009; Wang et al. 2016), fever, reduced food intake and increased sleep seem indeed to enhance survival when animals capable of reproduction are subject to bacterial infection.

Among biological systems and organs, the "threat-detecting" nervous and immune systems are capable of shunting energy flows to assure their own functions, to the potential detriment of other bodily organs and systems, and have therefore been termed "selfish" (Straub 2017). If one considers the brain as an "information-processing" system it is not surprising that, in particular in primates and human beings, it requires a lot of energy relative to its weight and that techniques sensitive to glucose and oxygen consumption have been proposed to image brain "information-processing" activities (Shulman et al. 2004; Herculano-Houzel 2012; Magistretti and Allaman 2015; Bordone et al. 2019). For example, during acute visual activity or a card-sorting task, brain glucose uptake increases between 10 and 50%, and often more than oxygen consumption (Fox et al. 1988; Madsen et al. 1995). Interestingly, restricting caloric intake lowers plasma concentrations of glucose and creatine, a nitrogenous organic acid that can serve as a high-energy substrate, but increases cerebral concentrations of creatine, while reducing brain glucose uptake (Wijeyesekera et al. 2012; Guo et al. 2015). Moreover, caloric restriction has repeatedly been shown to induce only minor changes in brain mass, as opposed to important reductions in body mass (Sprengell et al. 2021). These findings can be interpreted as suggesting that the brain can use different energy substrates and obtain these from peripheral tissues, for example, from muscle in the case of creatine.

Although the immune system is less easy to circumscribe than the nervous system, it can be estimated to be composed of 5.8×10^{11} leukocytes, which, when activated, require a 25–30% increase of basal metabolic rate (Straub et al. 2010). While the initial inflammatory response is typically fueled by glucose and glycolysis, later phases are often characterized by the mobilization and utilization of a variety of energy-providing substrates and pathways, including proteolysis and fatty acid oxidation (Straub et al. 2010; Liu et al. 2012). Chronic caloric restriction attenuates fever and reduces circulating interleukin-6 concentrations after systemic administration of bacterial lipopolysaccharide (Macdonald et al. 2011; Macdonald et al. 2012; Macdonald et al. 2014). However, although a 20 h fast or chronic intermittent fasting in mice reduces the number of circulating monocytes and induces a more "quiescent" metabolism, monocyte mobilization after inoculation with *Listeria monocytogenes* or a skin injury is not altered (Jordan et al. 2019). Thus the immune system, like the central nervous system, seems to assure different energy substrate flows, and both are capable of maintaining functional reactivity under conditions of limited energy intake.

13.2.3 Why Chronic Disease?

The central nervous and immune systems can be considered to be often in competition for energy provided by the rest of the body. This competition, in the case of a response to acute infection or injury, may be expressed as sickness behavior, characterized by decreased energy expenditure in the form of locomotor activity. However, even though reduced locomotor behavior can mitigate the negative energy balance of responding to infection or injury with fever and tissue repair, animal bodies, in the absence of food intake, can typically only sustain such responses for 3–7 weeks before energy substrates run out (Straub 2012). Nevertheless it is well known that many disease symptoms and processes occur beyond this time frame. Indeed, chronic symptoms, reminiscent of sickness behavior or chronic inflammation and tissue growth, can be observed in chronic brain-related disorders, autoimmunity, and cancer. Before addressing the question of how this can be brought about (proximate causation in medicine), it is interesting to consider why this may be the case (ultimate causation). Indeed, given the suffering, handicaps and costs associated with autoimmune and chronic inflammatory diseases, cancer, and depression, one may well wonder why these conditions seem to be so frequent and have not been selected against more during evolution.

One way to start addressing the "why question" involves (1) postulating that these chronic disease symptoms and processes are related to positively selected responses to acute tissue infection, acute injury or acute stressful events, (2) appealing to the antagonistic pleiotropy principle (Williams 1957) to propose that "genes being adaptive at an early age, [can be] maladaptive at older age" (Straub and Schradin 2016, p. 42), and (3) keeping in mind that many of the chronic disease symptoms and processes emerge after reproductive age and may therefore have little influence on the transmission of genes to offspring. So the same kind of responses may occur both in acute and chronic conditions and may be adaptive in the former and maladaptive in the latter. But often the triggering events seem to be different between acute and chronic diseases. In the case of an acute infection or injury, the host may detect so-called pathogen- or danger-associated molecular patterns (PAMPs or DAMPs), while in the case of chronic infections, microorganisms often seem to have developed strategies to escape such detection. In chronic autoimmune disease, the host seems to be mounting immune responses against so-called self-components, whereas in cancer the immune system does not seem to detect the supposedly transformed self. Indeed, in cancer, reactivity of the immune system often fails because of cancer-driven immunosuppression.

Notwithstanding the capacity of the immune system to efficiently detect tumor cells, many cancers have been related to chronic infection, inflammation, and tissue repair (Furman et al. 2019; Fishbein et al. 2021; Iriana et al. 2021; Okada et al. 2021). Interestingly, tumors can shunt energy streams towards cancer tissue in ways similar to those of the immune system in chronic inflammatory conditions and autoimmune diseases (Cheng et al. 2014; Goretzki et al. 2021; Schuster et al. 2021; Suchard and Savulescu 2021; Vaupel and Multhoff 2021). Moreover, all of these conditions can lead to states of cachexia or lean tissue wasting, even though this is sometimes hidden by obesity (Tisdale 2002; Delano and Moldawer 2006; Straub et al. 2010; Baracos et al. 2018; Santo et al. 2018; Biswas and Acharyya 2020; Olson et al. 2020; Berardi et al. 2021). Importantly, caloric restriction can improve anti-tumor immunity (Farazi et al. 2014; Kishton et al. 2017). Overall, this suggests that evolution may have selected the "selfish" bodily systems of the brain and the immune system, provided that their activation does not last beyond 3–7 weeks. However longer activation of such systems may occur due to erroneous continuous stimulation and result in systems that seem to be on an "ego-trip," for example, in chronic inflammatory disease (immunity against an autoantigen) and cancer (erroneous growth pathways and missing cell death).

In the case of chronic activation of these systems, when they are on an 'ego-trip,' reducing behavioral activity is not the only way to save energy and mitigate tissue wasting. Indeed, different forms of storage and memory may allow the saving of energy, and thus more flexible responses of the organism, and may therefore have been selected for during evolution. In particular, in a terrestrial environment, where food sources may be more scattered than in a marine environment, the possibility of an organism being able to store nutrients in some form would be an obvious advantage for survival. Interestingly, while many organisms, from yeast to *C. elegans,* can synthesize triacylglycerol and form lipid droplets and insects contain a fat body, only vertebrates have adipose tissue (Ottaviani et al. 2011). Although storage in fat tissues has been most widely studied, it is important to keep in mind that nutrient storage also occurs in the liver and skeletal muscle of mammals (Efeyan et al. 2015). These forms of storage correspond to what Walter Cannon called "storage by segregation" and that he proposed "to be subject to nervous or neuro-endocrine government" (Cannon 1929, p. 407).

In addition to nutrient storage, it has also been argued that "[e]volutionary pressure to optimize decision-making has led to the inevitable exploitation of past history" or memory, which "can be defined as experience-dependent modification of internal structure, in a stimulus-specific manner that alters the way the system will respond to stimuli in the future as a function of its past" (Baluska and Levin 2016, pp. 1–2). Neural memory enables an organism to rapidly access already encountered food sources while "immune memory ... leads to shorter, more effective and, finally, less energy-consuming reactions towards microbes" (Straub and Schradin 2016, p. 44). In this context, it is also interesting to note that dietary restriction results in an accumulation of memory T-cells in the bone marrow and enhanced protection against infections and tumors (Collins et al. 2019). Similarly, "immunological tolerance versus harmless foreign antigens (e.g. of microbes on the skin) or harmless autoantigens" can be considered "a memory function that spares energy reserves" (Straub and Schradin 2016, p. 44). Besides these neural and immune memories, other "environmental and/or physiological events" may "leave[] traces in a labile intracellular or extracellular medium which can be read as memories in the future by cells making decisions" (Baluska and Levin 2016, p. 4), for example in the form of epigenetic modifications of chromatin (Ginsburg and Jablonka 2009). The possibility that fatty acids, in addition to constituting a nutrient storage system in adipose tissues, could represent some memory system has recently been put forward as a stimulating perspective (Straub 2020).

Taken together, the evidence discussed above suggests that chronic diseases ultimately exist because they mostly occur after reproductive age, when natural selection no longer has an effect on the genes that are transmitted to offspring, and involve responses that were selected during evolution in response to acute infection and injury. These responses are typically driven by the immune system and/or the brain, which take control of the organism's physiology and behavior. While the immune system and the brain can thus be considered to act 'selfishly' when the responses are adaptive to overcome acutely dangerous events, such as infection, it seems that they are on an 'ego-trip' during chronic inflammatory diseases or chronic brain activation and thus put the organism's energy balance and tissue integrity at stake. Even though forms of memory and reduced behavioral activity can mitigate the consequences of chronic immune or central nervous activation or cancer somewhat, survival is nevertheless often threatened in these chronic conditions.

13.3 Neuroendocrine-Immune System Interactions: Focus on the Hypothalamo-Pituitary-Adrenal Axis and Energy Balance Signals

13.3.1 Possible Origins of Neuroendocrine-Immune System Interactions

After considering why (ultimate causation) chronic symptoms reminiscent of sickness behavior or chronic inflammation and tissue repair can be observed in cancer, autoimmune and chronic inflammatory diseases, it is now time to address the question of how (proximate causation) this can be brought about by interactions between the neuroendocrine and immune systems. In this context, it is worth considering that neuroendocrine and immune systems may have a common cellular ancestor (see also Chap. 1). This hypothesis was first put forward based on the observation that a common set of molecules involved in inflammation, phagocytosis, and stress response, such as proopiomelanocortin-derived peptides and cytokines, exist in invertebrates and vertebrates, and it has subsequently been corroborated by protein-folding recognition algorithms (Ottaviani and Franceschi 1997; Ottaviani et al. 2007). Another interesting and related hypothesis, indicating evolutionary ancient interactions between the neuroendocrine and immune system, emerged after the finding that neurons of the primitive nervous system of the freshwater polyp *Hydra* can not only alter the production of antimicrobial peptides in other cell types (Kasahara and Bosch 2003), but can themselves also secrete antimicrobial peptides in some cases (Augustin et al. 2017; Klimovich et al. 2020).

Although evolutionarily ancient, less complex organisms are useful in indicating a possible common origin or the first types of interaction between neuroendocrine and immune cells, it is important to keep in mind that these labels may not correspond neatly to the functions of cells and tissues of these organisms. Indeed, the labels “neuroendocrine” and “immune” have emerged mainly in the context of the study of mammals, for which it makes sense to talk about neuroendocrine and immune systems. However, in mammals the wealth of potential interactions between neuroendocrine and immune system elements may be perceived as close to overwhelming, making it challenging to determine where to start. Here, some of these interactions will be considered, first for healthy organisms and then for animals (including humans) in response to an acute infection or injury and finally for chronic disease conditions.

Under physiological conditions, animals show rhythms of neuroendocrine and immune responses that may be linked. Indeed, ectotherm vertebrates show seasonal variations in immune system components that seem to be mediated by corticosteroid and sex hormones as a result of neuroendocrine modulation (Zapata et al. 1992). In endotherm mammals like rodents, circadian patterns of peripheral immune cells, mediators and responses, such as the number of peripheral leukocytes, cytokine concentrations, and natural killer cell function, are regulated by the superchiasmatic nucleus of the hypothalamus, indicating communication through the autonomic nervous system or neuroendocrine systems (Arjona and Sarkar 2008; Mavroudis et al. 2013; Prendergast et al. 2013; Jacquelot et al. 2021). In healthy human beings, correlations have been found between circadian variations in neuroendocrine hormones and immune cell populations, indicating causal relationships or a common cause (Mazzoccoli et al. 2010a; Straub et al. 2010; Mazzoccoli et al. 2011).

13.3.2 Interactions between the Hypothalamo-Pituitary-Adrenal Axis and pro-Inflammatory Cytokines in Acute Disease Models

Work in animal models of acute infection and injury has established some causal interactions between the neuroendocrine and immune system. Several decades ago it was shown that exposure of rats to new antigens increased the firing rate of hypothalamic neurons and that administration of the pro-inflammatory cytokine interleukin-1 resulted in activation of corticotropin-releasing hormone (CRH)-containing hypothalamic neurons, which, in turn, stimulated adrenocorticotropic hormone (ACTH) release from the pituitary (Besedovsky et al. 1977; Berkenbosch et al. 1987). These findings gave rise to a set of studies showing that prostaglandin synthesis at the interface between the nervous system and peripheral tissues and brainstem to hypothalamus catecholaminergic projection are involved in activation of hypothalamic CRH neurons after administration of interleukin-1 (Ericsson et al. 1994; Ericsson et al. 1997; Lacroix and Rivest 1997; Ek et al. 1998; Matsuwaki et al. 2014). The resulting release of glucocorticoids from the adrenal glands (activation of the Hypothalamo-Pituitary Adrenal (HPA-)axis), in turn, mobilizes energy from liver, adipose tissue and muscle to sustain the inflammatory response, and, at higher concentrations, inhibits the synthesis and action of pro-inflammatory mediators (Straub 2014). Although other neuroendocrine axes are also altered in response to acute infection or injury, the most detailed knowledge of interactions between the immune and neuroendocrine systems has been obtained regarding activation of the HPA-axis in animal models of infection.

13.3.3 Interactions Between Neuroendocrine Signaling Underlying Energy Balance and Pro-Inflammatory Cytokines in Acute Disease Models

Another set of neuroendocrine pathways that are modified during host immune activation in response to detection of microbial fragments are those involved in the regulation of food intake. As outlined above, reduced food intake in response to bacterial infection can be considered an adaptive response. The pathways underlying different aspects of food intake can therefore be expected to be altered by immune mediators. Gastrointestinal tract-derived ghrelin and adipose tissue-produced leptin are, respectively, orexigenic and anorexigenic hormones through their actions on the nervous system, but also have anabolic and catabolic effects on metabolism, respectively (Shan and Yeo 2011; Frago and Chowen 2015; Klockars et al. 2019). Interestingly, bacterial lipopolysaccharide and interleukin-1 increase the synthesis of leptin and decrease that of ghrelin (Grunfeld and Feingold 1996; Sarraf et al. 1997; Faggioni et al. 1998; Asakawa et al. 2001; Basa et al. 2003; Wang et al. 2006)). In addition, it has been shown recently that interleukin-1 can also act directly on orexigenic and anabolic hypothalamic neurons that constitute targets for ghrelin and leptin (Chaskiel et al. 2019). Thus, pro-inflammatory immune mediators interact with neuroendocrine systems regulating energy intake and expenditure, but the converse also occurs, as part of the reactions to limit inflammatory responses (see above). Finally, and in regard to the previous section, it is important to point out that neuroendocrine and immune responses and interactions during acute infection-induced inflammation and injury-related wound healing seem to occur in a context of energy trade-offs in a wide variety of animals (French et al. 2011; Ashley and Demas 2017; Sylvia and Demas 2017).

13.3.4 Interactions Between Neuroendocrine and Immune Systems in Chronic Disease

In chronic diseases, such as arthritis and cancer, the circadian rhythms of neuroendocrine hormones, cytokines, and leukocyte populations as well as their phasic or antiphasic relationship are often altered (Lissoni et al. 2007; Cutolo and Straub 2008; Meyer-Hermann et al. 2009; Mazzoccoli et al. 2010b; Sierakowski and Cutolo 2011). These observations, along with the energy costs of long-term activation of immune responses, raise the question of how this is brought about. However, and in contrast to acute diseases, much less is known about the etiology of chronic diseases and about the possible transitions from acute to chronic conditions.

While evidence in favor of a link between a gastrointestinal infectious episode and chronic gut-related symptoms certainly exists, it is not yet clear whether and how it can cause a first flare of chronic inflammatory bowel disease (Chervy et al. 2020; Axelrad et al. 2021). Similarly, epidemiology has linked *Helicobacter Pylori* infection and autoimmune diseases such as rheumatoid arthritis, but biomedical research has yet to elucidate the mechanisms that could explain such a connection (Youssefi et al. 2021). As outlined above, many cancers have been related to chronic infection, inflammation, and tissue repair/growth, with some of the mediating mechanisms being progressively unraveled (Furman et al. 2019; Fishbein et al. 2021; Iriana et al. 2021; Okada et al. 2021). So beyond some similarity in terms of symptoms and pathophysiological processes, like inflammation, between the acute response to tissue infection or injury and chronic conditions, such as arthritis, cancer, and inflammatory bowel disease, discussed above, there are also epidemiological associations between infection and these chronic conditions.

13.3.5 Interactions Between the HPA-Axis and pro-Inflammatory Cytokines in Chronic Disease and their Animal Models

Although bacterial sepsis has long been considered a subacute condition, characterized by systemic inflammatory responses and high mortality, successful reduction of short-term mortality (< 30 days) has recently revealed long-term mortality and morbidity as emerging clinical challenges (Delano and Ward 2016). Interestingly, plasma cortisol concentrations during the first 24 h after diagnosis correlate with both short- and long-term mortality in sepsis (De Castro et al. 2019) and septic shock-related long-term mortality is associated with less frequent administration of corticosteroids (Nesseler et al. 2013). But circulating corticosteroid concentrations should not be taken to reflect the activity of the HPA-axis in sepsis, as lower ACTH concentrations than expected and loss of diurnal cortisol and ACTH rhythms are often observed (Kanczkowski et al. 2015; Peeters et al. 2017). Instead, increased corticosteroid concentrations may be due to direct adrenal or peripheral immune cell action of bacterial fragments and pro-inflammatory cytokines (Engstrom et al. 2008; Kanczkowski et al. 2013) and reduced breakdown of corticosteroids (Peeters et al. 2017).

Contrary to sepsis, the clinical benefits of corticosteroid administration on symptoms and disease processes in arthritis have been well established (Da Silva and Bijlsma 2000; Spies et al. 2014). In chronic rheumatoid arthritis "cortisol secretion appears to be inadequate in relation to inflammation" with cortisol/ACTH and cortisol/pro-inflammatory cytokine ratios being lower than expected (Spies et al. 2014, p. 2). Interestingly, the diurnal rhythm of cortisol secretion in rheumatoid arthritis patients with low to moderate disease activity does not differ from that of healthy individuals with pro-inflammatory cytokine levels, peaking when cortisol is low (Straub and Cutolo 2007). These observations thus provide a good rationale for attempts to 'time' the therapeutic effects of corticosteroids during the day-night cycle (Spies et al. 2014). But regardless of the issue of the therapeutic use of corticosteroids, it is clear that both in sepsis and arthritis, the lower cortisol/ACTH ratio indicates endocrine dysfunction. However, it does not yet seem clear that the whole neuroendocrine axis would be involved. Although patient-specific psychological stress-induced activation of the HPA-axis can be hypothesized to influence rheumatoid arthritis disease and symptoms, daily stressor- and worrying-associated exacerbated disease activity and symptoms have been found to be independent of plasma cortisol concentrations (Evers et al. 2014). Similarly, psychological stress has been proposed to predispose to tumor growth in part via HPA-axis activation (Shin et al. 2016; Colon-Echevarria et al. 2019).

While research on patients suffering from chronic conditions is certainly most relevant, animal models of such conditions can also provide insight into early events that typically occur before a patient consults a physician. Thus, it has been shown in animal models that neurons of the paraventricular nucleus of the hypothalamus, many of which control HPA-axis activity, express Fos immediate-early gene cellular activation markers at the onset of sepsis, arthritis-related hyperalgesia and cancer-associated anorexia, but also during chronic disease phases (Harbuz and Jessop 1999; Konsman and Blomqvist 2005; Carlson et al. 2007; Nishimura et al. 2020). So transcriptional activation at the hypothalamic stage of the HPA-axis may occur both during the initial and chronic phases of animal models of arthritis, cancer, and sepsis.

13.3.6 Interactions Between Neuroendocrine Signaling Underlying Energy Balance and Cytokines in Chronic Disease

In terms of energy balance, cancer, chronic inflammatory bowel disease and rheumatoid arthritis are all conditions in which food intake often does not match energy expenditure and this can lead to wasting of lean tissue or cachexia. In fact, reduced food intake is not adaptive when energy expenditure is chronically increased. Interestingly, rodents fully recover their food intake after a couple of days in standardized experimental models of infection (Valles et al. 2000; Crowell et al. 2017), but display a progressive loss of recovery of food consumption in cancer models (Konsman and Blomqvist 2005; Pourtau et al. 2011). Indeed, the lack of compensatory food intake in response to weight loss seems to characterize the cancer-associated anorexia cachexia syndrome (Olson et al. 2020) and the same seems to be the case in autoimmune diseases, such as arthritis. In terms of interactions between neuroendocrine and immune systems, it is important to point out that many tumors seem to be capable of "hijacking" cytokine and leptin signaling to promote angiogenesis and growth (Le Bitoux and Stamenkovic 2008; Ray and Cleary 2017; Shalapour and Karin 2019; Angelucci et al. 2020). In this respect, the hypothalamic action of a higher leptin concentration than expected for body weight and of relatively lower plasma ghrelin, along with brainstem action of the macrophage inhibitory cytokine-1/growth differentiation factor 15 (MIC-1/GDF15), may explain the lack of compensatory food intake in response to weight loss in slowly-growing tumors (Pourtau et al. 2011; Borner et al. 2017). However, muscle protein degradation does not seem to be fully related to overall nutrient intake in cancer cachexia (Kawamura et al. 1982; Del Fabbro 2019; Yang et al. 2020), indicating that food intake is not the sole factor determining the outcome when chronically activated bodily systems seem to be on an "ego-trip."

Similarly, both clinical and experimental arthritis are characterized by changes in body weight and cell mass that cannot be solely explained by reduced or lack of compensatory food intake (Roubenoff et al. 1994; Roubenoff et al. 1997; Skurlova et al. 2010; Olson et al. 2020). However, loss of body weight in humans may often not be apparent in arthritis, despite increased protein catabolism, because reduced activity may lead to fat gain (Roubenoff et al. 1994; Straub et al. 2010). Elevated pro-inflammatory cytokine production, along with normal to slightly raised circulating leptin levels and lower than expected increased plasma ghrelin concentrations (Roubenoff et al. 1997; Skurlova et al. 2010), constitute neuroendocrine signals that will most likely not increase food intake as to mitigate protein catabolism. Moreover, the inflammation-driven activation of the neuroendocrine HPA-axis and sympathetic nervous system will not only result in gluconeogenesis in the liver and lipolysis in adipose tissue, but also in breakdown of muscle protein, and provide further energy substrates for sustained inflammation (Straub et al. 2010). The activation of these neuroendocrine systems can thus be considered to underlie an energy appeal reaction providing "fuels for the inflammatory processes" (Straub et al. 2010, p. 6) that serves the immune system, which seems to be on an 'ego-trip,' but at the expense of a progressive loss of function, for example when it comes to energy balance and muscle function. Therefore it seems reasonable to postulate that sustained or repeated activation of the immune system, in interaction with the neuroendocrine system, plays an important role in the symptoms and disease processes of chronic inflammatory diseases (Straub 2014, 2017).

13.4 Conclusion

In this chapter, energy balance was considered for acute and chronic diseases in the context of evolutionary medicine and neuroendocrine-immune regulations. Fever and reduced food intake seem to increase the survival of bacterially infected animals and are brought about by direct and indirect actions of pro-inflammatory cytokines on the brain. This, in turn, also gives rise to activation of the HPA-axis, allowing for the mobilization of energy reserves and ultimately for mitigating inflammatory responses. While these responses may be adaptive and actively brought about when an organism responds to acute infection, they seem maladaptive when lasting too long, as they contribute to a negative energy balance and cachexia. One of the main challenges is therefore to explain why symptoms and pathophysiological processes, which occur during chronic inflammatory and neoplastic diseases and put the organism's energy balance at stake, can be so prevalent. The argument was developed that these symptoms and pathophysiological processes are adaptive in response to acute infection and injury in young reproducing individuals, and may therefore have been retained during evolution, but can be detrimental later in life during chronic conditions. In terms of potential immune-neuroendocrine interactions, it seems that components of the HPA-axis and energy balance-related signals are less or differently activated in arthritis and cancer, compared to acute infectious disease models. For example, the HPA-axis may be readily activated by acute inflammatory signals, but much less so when these same signals occur chronically.

13.5 Key References

Grunspan et al., *Evol Med Public Health,* 2018. This review provides an insightful overview of the core principles of evolutionary medicine.

Olson et al., *J Cachexia Sarcopenia Muscle,* 2020. This review discusses metabolic and behavioral responses during starvation, protein malnutrition and cachexia.

Straub, *Nat Rev Rheumatol,* 2017. This review lays out how the central nervous and immune systems can induce energy shortage during chronic inflammation and aging.

References

Angelucci A, Clementi L, Alesse E (2020) Leptin in tumor microenvironment. Adv Exp Med Biol 1259:89–112

Archie EA (2013) Wound healing in the wild: stress, sociality and energetic costs affect wound healing in natural populations. Parasite Immunol 35:374–385

Arjona A, Sarkar DK (2008) Are circadian rhythms the code of hypothalamic-immune communication? Insights from natural killer cells. Neurochem Res 33:708–718

Asakawa A, Inui A, Kaga T, Yuzuriha H, Nagata T, Ueno N, Makino S, Fujimiya M, Niijima A, Fujino MA, Kasuga M (2001) Ghrelin is an appetite-stimulatory signal from stomach with structural resemblance to motilin. Gastroenterology 120:337–345

Ashley NT, Demas GE (2017) Neuroendocrine-immune circuits, phenotypes, and interactions. Horm Behav 87:25–34

Augustin R, Schroder K, Murillo Rincon AP, Fraune S, Anton-Erxleben F, Herbst EM, Wittlieb J, Schwentner M, Grotzinger J, Wassenaar TM, Bosch TCG (2017) A secreted antibacterial neuropeptide shapes the microbiome of hydra. Nat Commun 8:698

Axelrad JE, Cadwell KH, Colombel JF, Shah SC (2021) The role of gastrointestinal pathogens in inflammatory bowel disease: a systematic review. Ther Adv Gastroenterol 14: 17562848211004493

Baluska F, Levin M (2016) On having no head: cognition throughout biological systems. Front Psychol 7:902

Baracos VE, Martin L, Korc M, Guttridge DC, Fearon KCH (2018) Cancer-associated cachexia. Nat Rev Dis Primers 4:17105

Baral P, Udit S, Chiu IM (2019) Pain and immunity: implications for host defence. Nat Rev Immunol 19:433–447

Basa NR, Wang L, Arteaga JR, Heber D, Livingston EH, Tache Y (2003) Bacterial lipopolysaccharide shifts fasted plasma ghrelin to postprandial levels in rats. Neurosci Lett 343:25–28

Berardi E, Madaro L, Lozanoska-Ochser B, Adamo S, Thorrez L, Bouche M, Coletti D (2021) A pound of flesh: what cachexia is and what it is not. Diagnostics (Basel) 11(1):116
Berkenbosch F, Van Oers J, Del Rey A, Tilders F, Besedovsky H (1987) Corticotropin-releasing factor-producing neurons in the rat activated by interleukin-1. Science 238:524–526
Besedovsky H, Sorkin E, Felix D, Haas H (1977) Hypothalamic changes during the immune response. Eur J Immunol 7:323–325
Biswas AK, Acharyya S (2020) Understanding cachexia in the context of metastatic progression. Nat Rev Cancer 20:274–284
Blaxter K (1989) Energy metabolism in animals and man. Cambridge University Press, Cambrdige
Bordone MP, Salman MM, Titus HE, Amini E, Andersen JV, Chakraborti B, Diuba AV, Dubouskaya TG, Ehrke E, Espindola De Freitas A, Braga De Freitas G, Goncalves RA, Gupta D, Gupta R, Ha SR, Hemming IA, Jaggar M, Jakobsen E, Kumari P, Lakkappa N, Marsh APL, Mitlohner J, Ogawa Y, Paidi RK, Ribeiro FC, Salamian A, Saleem S, Sharma S, Silva JM, Singh S, Sulakhiya K, Tefera TW, Vafadari B, Yadav A, Yamazaki R, Seidenbecher CI (2019) The energetic brain - a review from students to students. J Neurochem 151:139–165
Borner T, Arnold M, Ruud J, Breit SN, Langhans W, Lutz TA, Blomqvist A, Riediger T (2017) Anorexia-cachexia syndrome in hepatoma tumour-bearing rats requires the area postrema but not vagal afferents and is paralleled by increased MIC-1/GDF15. J Cachexia Sarcopenia Muscle 8:417–427
Cannon WB (1929) Organization for Physiological Homeostasis. Physiol Rev 9:399–431
Carlson DE, Chiu WC, Fiedler SM, Hoffman GE (2007) Central neural distribution of immunoreactive Fos and CRH in relation to plasma ACTH and corticosterone during sepsis in the rat. Exp Neurol 205:485–500
Chaskiel L, Bristow AD, Bluthe RM, Dantzer R, Blomqvist A, Konsman JP (2019) Interleukin-1 reduces food intake and body weight in rat by acting in the arcuate hypothalamus. Brain Behav Immun 81:560–573
Cheng SC, Joosten LA, Netea MG (2014) The interplay between central metabolism and innate immune responses. Cytokine Growth Factor Rev 25:707–713
Chervy M, Barnich N, Denizot J (2020) Adherent-Invasive E. coli: update on the lifestyle of a troublemaker in Crohn's disease. Int J Mol Sci 21(10):3734
Collins N, Han SJ, Enamorado M, Link VM, Huang B, Moseman EA, Kishton RJ, Shannon JP, Dixit D, Schwab SR, Hickman HD, Restifo NP, Mcgavern DB, Schwartzberg PL, Belkaid Y (2019) The bone marrow protects and optimizes immunological memory during dietary restriction. Cell 178(1088–1101):e15
Colon-Echevarria CB, Lamboy-Caraballo R, Aquino-Acevedo AN, Armaiz-Pena GN (2019) Neuroendocrine regulation of tumor-associated immune cells. Front Oncol 9:1077
Crowell KT, Soybel DI, Lang CH (2017) Restorative mechanisms regulating protein balance in skeletal muscle during recovery from sepsis. Shock 47:463–473
Cutolo M, Straub RH (2008) Circadian rhythms in arthritis: hormonal effects on the immune/inflammatory reaction. Autoimmun Rev 7:223–228
Da Silva JA, Bijlsma JW (2000) Optimizing glucocorticoid therapy in rheumatoid arthritis. Rheum Dis Clin N Am 26:859–880
De Castro R, Ruiz D, Lavin BA, Lamsfus JA, Vazquez L, Montalban C, Marcano G, Sarabia R, Paz-Zulueta M, Blanco C, Santibanez M (2019) Cortisol and adrenal androgens as independent predictors of mortality in septic patients. PLoS One 14:e0214312
Del Fabbro E (2019) Combination therapy in cachexia. Ann Palliat Med 8:59–66
Delano MJ, Moldawer LL (2006) The origins of cachexia in acute and chronic inflammatory diseases. Nutr Clin Pract 21:68–81
Delano MJ, Ward PA (2016) The immune system's role in sepsis progression, resolution, and long-term outcome. Immunol Rev 274:330–353
Dobzhansky T (1964) Biology, molecular and organismic. Am Zool 4:443–452
Donnelly CR, Chen O, Ji RR (2020) How do sensory neurons sense danger signals? Trends Neurosci 43:822–838

Douglas SJ, Dawson-Scully K, Sokolowski MB (2005) The neurogenetics and evolution of food-related behaviour. Trends Neurosci 28:644–652

Efeyan A, Comb WC, Sabatini DM (2015) Nutrient-sensing mechanisms and pathways. Nature 517:302–310

Ek M, Kurosawa M, Lundeberg T, Ericsson A (1998) Activation of vagal afferents after intravenous injection of interleukin-1beta: role of endogenous prostaglandins. J Neurosci 18:9471–9479

Engstrom L, Rosen K, Angel A, Fyrberg A, Mackerlova L, Konsman JP, Engblom D, Blomqvist A (2008) Systemic immune challenge activates an intrinsically regulated local inflammatory circuit in the adrenal gland. Endocrinology 149:1436–1450

Ericsson A, Kovács KJ, Sawchenko PE (1994) A functional anatomical analysis of central pathways subserving the effects of interleukin-1 on stress-related neuroendocrine neurons. J Neurosci 14:897–913

Ericsson A, Arias C, Sawchenko PE (1997) Evidence for an intramedullary prostaglandin-dependent mechanism in the activation of stress-related neuroendocrine circuitry by intravenous interleukin-1. J Neurosci 17:7166–7179

Evers AW, Verhoeven EW, Van Middendorp H, Sweep FC, Kraaimaat FW, Donders AR, Eijsbouts AE, Van Laarhoven AI, De Brouwer SJ, Wirken L, Radstake TR, Van Riel PL (2014) Does stress affect the joints? Daily stressors, stress vulnerability, immune and HPA axis activity, and short-term disease and symptom fluctuations in rheumatoid arthritis. Ann Rheum Dis 73:1683–1688

Faggioni R, Fantuzzi G, Fuller J, Dinarello CA, Feingold KR, Grunfeld C (1998) IL-1 beta mediates leptin induction during inflammation. Am J Physiol 274:R204–R208

Farazi M, Nguyen J, Goldufsky J, Linnane S, Lukaesko L, Weinberg AD, Ruby CE (2014) Caloric restriction maintains OX40 agonist-mediated tumor immunity and CD4 T cell priming during aging. Cancer Immunol Immunother 63:615–626

Fendt M, Parsons MH, Apfelbach R, Carthey AJR, Dickman CR, Endres T, Frank ASK, Heinz DE, Jones ME, Kiyokawa Y, Kreutzmann JC, Roelofs K, Schneider M, Sulger J, Wotjak CT, Blumstein DT (2020) Context and trade-offs characterize real-world threat detection systems: a review and comprehensive framework to improve research practice and resolve the translational crisis. Neurosci Biobehav Rev 115:25–33

Fishbein A, Hammock BD, Serhan CN, Panigrahy D (2021) Carcinogenesis: failure of resolution of inflammation? Pharmacol Ther 218:107670

Forbey JS, Liu R, Caughlin TT, Matocq MD, Vucetich JA, Kohl KD, Dearing MD, Felton AM (2018) Review: using physiologically based models to predict population responses to phytochemicals by wild vertebrate herbivores. Animal 12:s383–s398

Fox PT, Raichle ME, Mintun MA, Dence C (1988) Nonoxidative glucose consumption during focal physiologic neural activity. Science 241:462–464

Frago LM, Chowen JA (2015) Hypothalamic leptin and ghrelin signaling as targets for improvement in metabolic control. Curr Pharm Des 21:3596–3605

Francisco J, Aragao I, Cardoso T (2018) Risk factors for long-term mortality in patients admitted with severe infection. BMC Infect Dis 18:161

French SS, Dearing MD, Demas GE (2011) Leptin as a physiological mediator of energetic trade-offs in ecoimmunology: implications for disease. Integr Comp Biol 51:505–513

Friese RS, Bruns B, Sinton CM (2009) Sleep deprivation after septic insult increases mortality independent of age. J Trauma 66:50–54

Furman D, Campisi J, Verdin E, Carrera-Bastos P, Targ S, Franceschi C, Ferrucci L, Gilroy DW, Fasano A, Miller GW, Miller AH, Mantovani A, Weyand CM, Barzilai N, Goronzy JJ, Rando TA, Effros RB, Lucia A, Kleinstreuer N, Slavich GM (2019) Chronic inflammation in the etiology of disease across the life span. Nat Med 25:1822–1832

Gallagher AJ, Creel S, Wilson RP, Cooke SJ (2017) Energy landscapes and the landscape of fear. Trends Ecol Evol 32:88–96

Ginsburg S, Jablonka E (2009) Epigenetic learning in non-neural organisms. J Biosci 34:633–646

Goldstein SA, Elwyn DH (1989) The effects of injury and sepsis on fuel utilization. Annu Rev Nutr 9:445–473
Goretzki A, Lin YJ, Schulke S (2021) Immune metabolism in allergies, does it matter?-a review of immune metabolic basics and adaptations associated with the activation of innate immune cells in allergy. Allergy 76:3314
Greaves M (2007) Darwinian medicine: a case for cancer. Nat Rev Cancer 7:213–221
Grunfeld C, Feingold KR (1996) Regulation of lipid metabolism by cytokines during host defense. Nutrition 12:S24–S26
Grunspan DZ, Nesse RM, Barnes ME, Brownell SE (2018) Core principles of evolutionary medicine: a Delphi study. Evol Med Public Health 2018:13–23
Guo J, Bakshi V, Lin AL (2015) Early shifts of brain metabolism by caloric restriction preserve white matter integrity and long-term memory in aging mice. Front Aging Neurosci 7:213
Harbuz MS, Jessop DS (1999) Dissociation between c-fos mRNA in the paraventricular nucleus and corticosterone secretion in rats with adjuvant-induced arthritis. J Endocrinol 163:107–113
Hart BL (1988) Biological basis of the behavior of sick animals. Neurosci Biobehav Rev 12:123–137
Herculano-Houzel S (2012) The remarkable, yet not extraordinary, human brain as a scaled-up primate brain and its associated cost. Proc Natl Acad Sci U S A 109(Suppl 1):10661–10668
High KP (2004) Infection as a cause of age-related morbidity and mortality. Ageing Res Rev 3:1–14
Hite JL, Pfenning AC, Cressler CE (2020) Starving the enemy? Feeding behavior shapes host-parasite interactions. Trends Ecol Evol 35:68–80
Illius AW, Tolkamp BJ, Yearsley J (2002) The evolution of the control of food intake. Proc Nutr Soc 61:465–472
Iriana S, Asha K, Repak M, Sharma-Walia N (2021) Hedgehog signaling: implications in cancers and viral infections. Int J Mol Sci 22(3):1042
Jacquelot N, Belz GT, Seillet C (2021) Neuroimmune interactions and rhythmic regulation of innate lymphoid cells. Front Neurosci 15:657081
Jordan S, Tung N, Casanova-Acebes M, Chang C, Cantoni C, Zhang D, Wirtz TH, Naik S, Rose SA, Brocker CN, Gainullina A, Hornburg D, Horng S, Maier BB, Cravedi P, Leroith D, Gonzalez FJ, Meissner F, Ochando J, Rahman A, Chipuk JE, Artyomov MN, Frenette PS, Piccio L, Berres ML, Gallagher EJ, Merad M (2019) Dietary intake regulates the circulating inflammatory monocyte Pool. Cell 178(1102–1114):e17
Kanczkowski W, Alexaki VI, Tran N, Grossklaus S, Zacharowski K, Martinez A, Popovics P, Block NL, Chavakis T, Schally AV, Bornstein SR (2013) Hypothalamo-pituitary and immune-dependent adrenal regulation during systemic inflammation. Proc Natl Acad Sci U S A 110: 14801–14806
Kanczkowski W, Sue M, Zacharowski K, Reincke M, Bornstein SR (2015) The role of adrenal gland microenvironment in the HPA axis function and dysfunction during sepsis. Mol Cell Endocrinol 408:241–248
Kasahara S, Bosch TC (2003) Enhanced antibacterial activity in hydra polyps lacking nerve cells. Dev Comp Immunol 27:79–85
Kawamura I, Moldawer LL, Keenan RA, Batist G, Bothe A Jr, Bistrian BR, Blackburn GL (1982) Altered amino acid kinetics in rats with progressive tumor growth. Cancer Res 42:824–829
Keestra-Gounder AM, Tsolis RM, Baumler AJ (2015) Now you see me, now you don't: the interaction of salmonella with innate immune receptors. Nat Rev Microbiol 13:206–216
Kishton RJ, Sukumar M, Restifo NP (2017) Metabolic regulation of T cell longevity and function in tumor immunotherapy. Cell Metab 26:94–109
Klimovich A, Giacomello S, Bjorklund A, Faure L, Kaucka M, Giez C, Murillo-Rincon AP, Matt AS, Willoweit-Ohl D, Crupi G, De Anda J, Wong GCL, D'amato M, Adameyko I, Bosch TCG (2020) Prototypical pacemaker neurons interact with the resident microbiota. Proc Natl Acad Sci U S A 117:17854–17863
Klockars A, Levine AS, Olszewski PK (2019) Hypothalamic integration of the endocrine signaling related to food intake. Curr Top Behav Neurosci 43:239–269

Kluger MJ (1991) Fever: role of pyrogens and cryogens. Physiol Rev 71:93–127
Kono H, Rock KL (2008) How dying cells alert the immune system to danger. Nat Rev Immunol 8: 279–289
Konsman JP, Blomqvist A (2005) Forebrain patterns of c-Fos and FosB induction during cancer-associated anorexia-cachexia in rat. Eur J Neurosci 21:2752–2766
Lacroix S, Rivest S (1997) Functional circuitry in the brain of immune-challenged rats: partial involvement of prostaglandins. J Comp Neurol 387:307–324
Le Bitoux MA, Stamenkovic I (2008) Tumor-host interactions: the role of inflammation. Histochem Cell Biol 130:1079–1090
Lederberg J (1997) Infectious disease as an evolutionary paradigm. Emerg Infect Dis 3:417–423
Lissoni P, Brivio F, Fumagalli L, Messina G, Secreto G, Romelli B, Fumagalli G, Rovelli F, Colciago M, Brera G (2007) Immune and endocrine mechanisms of advanced cancer-related hypercortisolemia. In Vivo 21:647–650
Liu TF, Brown CM, El Gazzar M, Mcphail L, Millet P, Rao A, Vachharajani VT, Yoza BK, Mccall CE (2012) Fueling the flame: bioenergy couples metabolism and inflammation. J Leukoc Biol 92:499–507
Lopes PC, French SS, Woodhams DC, Binning SA (2021) Sickness behaviors across vertebrate taxa: proximate and ultimate mechanisms. J Exp Biol 224(9):jeb225847
Macdonald L, Radler M, Paolini AG, Kent S (2011) Calorie restriction attenuates LPS-induced sickness behavior and shifts hypothalamic signaling pathways to an anti-inflammatory bias. Am J Physiol Regul Integr Comp Physiol 301:R172–R184
Macdonald L, Begg D, Weisinger RS, Kent S (2012) Calorie restricted rats do not increase metabolic rate post-LPS, but do seek out warmer ambient temperatures to behaviourally induce a fever. Physiol Behav 107:762–772
Macdonald L, Hazi A, Paolini AG, Kent S (2014) Calorie restriction dose-dependently abates lipopolysaccharide-induced fever, sickness behavior, and circulating interleukin-6 while increasing corticosterone. Brain Behav Immun 40:18–26
Madsen PL, Hasselbalch SG, Hagemann LP, Olsen KS, Bulow J, Holm S, Wildschiodtz G, Paulson OB, Lassen NA (1995) Persistent resetting of the cerebral oxygen/glucose uptake ratio by brain activation: evidence obtained with the Kety-Schmidt technique. J Cereb Blood Flow Metab 15: 485–491
Magistretti PJ, Allaman I (2015) A cellular perspective on brain energy metabolism and functional imaging. Neuron 86:883–901
Martin LB 2nd, Weil ZM, Nelson RJ (2006) Refining approaches and diversifying directions in ecoimmunology. Integr Comp Biol 46:1030–1039
Matsuwaki T, Eskilsson A, Kugelberg U, Jonsson JI, Blomqvist A (2014) Interleukin-1beta induced activation of the hypothalamus-pituitary-adrenal axis is dependent on interleukin-1 receptors on non-hematopoietic cells. Brain Behav Immun 40:166–173
Mavroudis PD, Scheff JD, Calvano SE, Androulakis IP (2013) Systems biology of circadian-immune interactions. J Innate Immun 5:153–162
Mazzoccoli G, De Cata A, Greco A, Carughi S, Giuliani F, Tarquini R (2010a) Circadian rhythmicity of lymphocyte subpopulations and relationship with neuro-endocrine system. J Biol Regul Homeost Agents 24:341–350
Mazzoccoli G, Vendemiale G, De Cata A, Carughi S, Tarquini R (2010b) Altered time structure of neuro-endocrine-immune system function in lung cancer patients. BMC Cancer 10:314
Mazzoccoli G, Muscarella LA, Fazio VM, Piepoli A, Pazienza V, Dagostino MP, Giuliani F, Polyakova VO, Kvetnoy I (2011) Antiphase signalling in the neuroendocrine-immune system in healthy humans. Biomed Pharmacother 65:275–279
Mcnamara JF, Righi E, Wright H, Hartel GF, Harris PNA, Paterson DL (2018) Long-term morbidity and mortality following bloodstream infection: a systematic literature review. J Infect 77:1–8
Medzhitov R (2001) Toll-like receptors and innate immunity. Nat Rev Immunol 1:135–145

Methot PO (2011) Research traditions and evolutionary explanations in medicine. Theor Med Bioeth 32:75–90

Meyer-Hermann M, Figge MT, Straub RH (2009) Mathematical modeling of the circadian rhythm of key neuroendocrine-immune system players in rheumatoid arthritis: a systems biology approach. Arthritis Rheum 60:2585–2594

Murray MJ, Murray AB (1979) Anorexia of infection as a mechanism of host defense. Am J Clin Nutr 32:593–596

Nesseler N, Defontaine A, Launey Y, Morcet J, Malledant Y, Seguin P (2013) Long-term mortality and quality of life after septic shock: a follow-up observational study. Intensive Care Med 39: 881–888

Nishimura H, Kawasaki M, Matsuura T, Suzuki H, Motojima Y, Baba K, Ohnishi H, Yamanaka Y, Fujitani T, Yoshimura M, Maruyama T, Ueno H, Sonoda S, Nishimura K, Tanaka K, Sanada K, Onaka T, Ueta Y, Sakai A (2020) Acute mono-arthritis activates the Neurohypophysial system and Hypothalamo-pituitary adrenal Axis in rats. Front Endocrinol (Lausanne) 11:43

Okada F, Izutsu R, Goto K, Osaki M (2021) Inflammation-related carcinogenesis: lessons from animal models to clinical aspects. Cancers (Basel) 13(4):921

Olson B, Marks DL, Grossberg AJ (2020) Diverging metabolic programmes and behaviours during states of starvation, protein malnutrition, and cachexia. J Cachexia Sarcopenia Muscle 11:1429–1446

Ottaviani E, Franceschi C (1997) The invertebrate phagocytic immunocyte: clues to a common evolution of immune and neuroendocrine systems. Immunol Today 18:169–174

Ottaviani E, Malagoli D, Franceschi C (2007) Common evolutionary origin of the immune and neuroendocrine systems: from morphological and functional evidence to in silico approaches. Trends Immunol 28:497–502

Ottaviani E, Malagoli D, Franceschi C (2011) The evolution of the adipose tissue: a neglected enigma. Gen Comp Endocrinol 174:1–4

Peeters B, Langouche L, Van Den Berghe G (2017) Adrenocortical stress response during the course of critical illness. Compr Physiol 8:283–298

Pourtau L, Leemburg S, Roux P, Leste-Lasserre T, Costaglioli P, Garbay B, Drutel G, Konsman JP (2011) Hormonal, hypothalamic and striatal responses to reduced body weight gain are attenuated in anorectic rats bearing small tumors. Brain Behav Immun 25:777–786

Prendergast BJ, Cable EJ, Patel PN, Pyter LM, Onishi KG, Stevenson TJ, Ruby NF, Bradley SP (2013) Impaired leukocyte trafficking and skin inflammatory responses in hamsters lacking a functional circadian system. Brain Behav Immun 32:94–104

Ray A, Cleary MP (2017) The potential role of leptin in tumor invasion and metastasis. Cytokine Growth Factor Rev 38:80–97

Rook GA (2010) 99th Dahlem conference on infection, inflammation and chronic inflammatory disorders: darwinian medicine and the 'hygiene' or 'old friends' hypothesis. Clin Exp Immunol 160:70–79

Roubenoff R, Roubenoff RA, Cannon JG, Kehayias JJ, Zhuang H, Dawson-Hughes B, Dinarello CA, Rosenberg IH (1994) Rheumatoid cachexia: cytokine-driven hypermetabolism accompanying reduced body cell mass in chronic inflammation. J Clin Investig 93:2379–2386

Roubenoff R, Freeman LM, Smith DE, Abad LW, Dinarello CA, Kehayias JJ (1997) Adjuvant arthritis as a model of inflammatory cachexia. Arthritis Rheum 40:534–539

Santo RCE, Fernandes KZ, Lora PS, Filippin LI, Xavier RM (2018) Prevalence of rheumatoid cachexia in rheumatoid arthritis: a systematic review and meta-analysis. J Cachexia Sarcopenia Muscle 9:816–825

Sarraf P, Frederich RC, Turner EM, Ma G, Jaskowiak NT, Rivet DJR, Flier JS, Lowell BB, Fraker DL, Alexander HR (1997) Multiple cytokines and acute inflammation raise mouse leptin levels: potential role in inflammatory anorexia. J Exp Med 185:171–175

Schuster S, Ewald J, Kaleta C (2021) Modeling the energy metabolism in immune cells. Curr Opin Biotechnol 68:282–291

Shalapour S, Karin M (2019) Pas de Deux: control of anti-tumor immunity by cancer-associated inflammation. Immunity 51:15–26
Shan X, Yeo GS (2011) Central leptin and ghrelin signalling: comparing and contrasting their mechanisms of action in the brain. Rev Endocr Metab Disord 12:197–209
Shin KJ, Lee YJ, Yang YR, Park S, Suh PG, Follo MY, Cocco L, Ryu SH (2016) Molecular mechanisms underlying psychological stress and cancer. Curr Pharm Des 22:2389–2402
Shulman RG, Rothman DL, Behar KL, Hyder F (2004) Energetic basis of brain activity: implications for neuroimaging. Trends Neurosci 27:489–495
Sierakowski S, Cutolo M (2011) Morning symptoms in rheumatoid arthritis: a defining characteristic and marker of active disease. Scand J Rheumatol Suppl 125:1–5
Skurlova M, Stofkova A, Kiss A, Belacek J, Pecha O, Deykun K, Jurcovicova J (2010) Transient anorexia, hyper-nociception and cognitive impairment in early adjuvant arthritis in rats. Endocr Regul 44:165–173
Smith ES, Lewin GR (2009) Nociceptors: a phylogenetic view. J Comp Physiol A Neuroethol Sens Neural Behav Physiol 195:1089–1106
Spies CM, Straub RH, Cutolo M, Buttgereit F (2014) Circadian rhythms in rheumatology--a glucocorticoid perspective. Arthritis Res Ther 16(Suppl 2):S3
Sprengell M, Kubera B, Peters A (2021) Brain more resistant to energy restriction than body: a systematic review. Front Neurosci 15:639617
Stearns SC (2012) Evolutionary medicine: its scope, interest and potential. Proc Biol Sci 279:4305–4321
Straub RH (2012) Evolutionary medicine and chronic inflammatory state--known and new concepts in pathophysiology. J Mol Med (Berl) 90:523–534
Straub RH (2014) Insulin resistance, selfish brain, and selfish immune system: an evolutionarily positively selected program used in chronic inflammatory diseases. Arthritis Res Ther 16(Suppl 2):S4
Straub RH (2017) The brain and immune system prompt energy shortage in chronic inflammation and ageing. Nat Rev Rheumatol 13:743–751
Straub RH (2020) The memory of the fatty acid system. Prog Lipid Res 79:101049
Straub RH, Cutolo M (2007) Circadian rhythms in rheumatoid arthritis: implications for pathophysiology and therapeutic management. Arthritis Rheum 56:399–408
Straub RH, Schradin C (2016) Chronic inflammatory systemic diseases: an evolutionary trade-off between acutely beneficial but chronically harmful programs. Evol Med Public Health 2016(1): 37–51
Straub RH, Cutolo M, Buttgereit F, Pongratz G (2010) Energy regulation and neuroendocrine-immune control in chronic inflammatory diseases. J Intern Med 267:543–560
Suchard MS, Savulescu DM (2021) Nicotinamide pathways as the root cause of sepsis - an evolutionary perspective on macrophage energetic shifts. FEBS J 289:955
Sylvia KE, Demas GE (2017) A return to wisdom: using sickness behaviors to integrate ecological and translational research. Integr Comp Biol 57:1204–1213
Tisdale MJ (2002) Cachexia in cancer patients. Nat Rev Cancer 2:862–871
Valles A, Marti O, Garcia A, Armario A (2000) Single exposure to stressors causes long-lasting, stress-dependent reduction of food intake in rats. Am J Physiol Regul Integr Comp Physiol 279: R1138–R1144
Vaughn LK, Veale WL, Cooper KE (1980) Antipyresis: its effect on mortality rate of bacterially infected rabbits. Brain Res Bull 5:69–73
Vaupel P, Multhoff G (2021) Revisiting the Warburg effect: historical dogma versus current understanding. J Physiol 599:1745–1757
Walters ET (1994) Injury-related behavior and neuronal plasticity: an evolutionary perspective on sensitization, hyperalgesia, and analgesia. Int Rev Neurobiol 36:325–427
Wang W, Andersson M, Iresjo BM, Lonnroth C, Lundholm K (2006) Effects of ghrelin on anorexia in tumor-bearing mice with eicosanoid-related cachexia. Int J Oncol 28:1393–1400

Wang A, Huen SC, Luan HH, Yu S, Zhang C, Gallezot JD, Booth CJ, Medzhitov R (2016) Opposing effects of fasting metabolism on tissue tolerance in bacterial and viral inflammation. Cell 166(1512–1525):e12

Wijeyesekera A, Selman C, Barton RH, Holmes E, Nicholson JK, Withers DJ (2012) Metabotyping of long-lived mice using 1H NMR spectroscopy. J Proteome Res 11:2224–2235

Williams GC (1957) Pleiotropy, natural selection and the evolution of senescence. Evolution 11: 398–411

Williams GC, Nesse RM (1991) The dawn of Darwinian medicine. Q Rev Biol 66:1–22

Yang W, Huang J, Wu H, Wang Y, Du Z, Ling Y, Wang W, Wu Q, Gao W (2020) Molecular mechanisms of cancer cachexia-induced muscle atrophy (review). Mol Med Rep 22:4967–4980

Youssefi M, Tafaghodi M, Farsiani H, Ghazvini K, Keikha M (2021) Helicobacter pylori infection and autoimmune diseases; is there an association with systemic lupus erythematosus, rheumatoid arthritis, autoimmune atrophy gastritis and autoimmune pancreatitis? A systematic review and meta-analysis study. J Microbiol Immunol Infect 54:359–369

Zapata AG, Varas A, Torroba M (1992) Seasonal variations in the immune system of lower vertebrates. Immunol Today 13:142–147

Autonomic, Immune, Metabolic, and Neuroendocrine Dimensions of Anorexia Nervosa: An Integrative View

14

Lucas De Zorzi, Stéphane Ranfaing, Henrique Sequeira, and Odile Viltart

Abstract

Anorexia nervosa (AN) is the psychiatric disorder with the highest mortality rate, whose etiology remains largely unknown. It mainly concerns women and is characterized notably by a voluntary food restriction leading to a state of undernutrition often associated with excessive physical activity. Despite specialized care, relapse is common and affects approximately 40% of patients. AN is comorbid with several other psychiatric diseases such as depression, anxiety, or compulsivity. Patients suffering from AN also present autonomic, immune, metabolic, and neuroendocrine alterations that participate in the worsening of the disease. We describe here how the autonomic and hormonal systems, main contributors to the brain-body homeostasis, participate directly or indirectly in the modulation of the immune system in AN. The complexity of the interactions between these processes reflects the complexity of AN. The understanding of such complexity could help adjust procedures to personalize medical approaches, integrating the diversity of factors leading to AN manifestations.

Keywords

Anorexia nervosa · Autonomic nervous system · Heart rate variability · Skin conductance · Cytokines · Ghrelin · Leptin · Cortisol

L. De Zorzi · S. Ranfaing · H. Sequeira
Université de Lille, Laboratoire SCALab, CNRS UMR, Lille, France

O. Viltart (✉)
Université de Lille, Laboratoire SCALab, CNRS UMR, Lille, France

Université Paris Cité, IPNP, INSERM UMR-S, Paris, France
e-mail: odile.viltart@univ-lille.fr

J. P. Konsman, T. M. Reyes (eds.), *Neuroendocrine-Immune System Interactions*, Masterclass in Neuroendocrinology 13,
https://doi.org/10.1007/978-3-031-21358-8_14

Abbreviations

ACTH	adrenocorticotropic hormone
AN	anorexia nervosa
AN-BP	anorexia nervosa binge-purging type
AN-R	anorexia nervosa restrictive type
ANS	autonomic nervous system
BMI	body mass index
BP	blood pressure
CRH	corticotropin-releasing hormone
ED	eating disorders
EDA	electrodermal activity
ENS	enteric nervous system
GC	glucocorticoid hormones
GHS-R	growth hormone secretagogue receptor
GR	glucocorticoid receptor
GWAS	genome-wide association study
HPA	hypothalamo-pituitary-adrenal axis
HF-HRV	high-frequency heart rate variability
HR	heart rate
HRV	heart rate variability
Ig	immunoglobulin
IFNg	interferon gamma
IL	interleukin
LF-HRV	low-frequency heart rate variability
PBMC	peripheral blood mononuclear cells
PNS	parasympathetic nervous system
qPCR	quantitative polymerase chain reaction
RIA	radioimmunoassay
SC	skin conductance
SCRs	skin conductance responses
SNS	sympathetic nervous system
SNP	single nucleotide polymorphism
TGFβ	transforming growth factor beta
TNFα	tumor necrosis factor alpha
Treg	regulatory T lymphocytes

14.1 Introduction

Anorexia nervosa (AN) is a complex and multifactorial psychiatric disorder belonging to the category of eating disorders (Box 14.1). It mainly affects women, with sex ratios of approximately 10/1 to 15/1 (Treasure et al. 2015), and is characterized by a voluntary restriction of caloric intake leading to a state of undernutrition with a high morbidity and mortality, with a mortality rate of 5 per 1000 person/year (Arcelus et al. 2011). After specialized care, it takes several months for the patient to return to normal eating habits. There is significant variability across patients in this normalization process and relapse is common, rendering AN among the most serious and potentially lethal disorders in psychiatry and clinical psychology (Arcelus et al. 2011; Fichter and Quadflieg 2016; Fichter et al. 2017). The etiology of this disorder is complex, due to the contribution of genetic factors, as revealed by twin studies, where 58–70% of variance is due to additive genetic factors (Bulik et al. 2006), or recent genome-wide association studies (GWAS) (Duncan et al. 2017; Watson et al. 2019), but also environmental, metabolic, immune, and neurobiological factors (Schaumberg et al. 2017; Viltart et al. 2018; Duriez et al. 2019). As the pathophysiology underlying AN is still poorly understood, there are no current medications to target the core biology of the disorder. In these conditions, even though various psychotherapeutic approaches are currently offered to patients, it becomes urgent to better understand how biological factors could influence the gravity of the illness and the duration of the recovery. In this context, it becomes necessary to consider an integrative view of neurobiological, immune and metabolic factors associated with the AN.

Box 14.1 Criteria for and Characteristics of Anorexia Nervosa

Anorexia nervosa (AN) is one category of eating disorders that also encompass: anorexia avoidant restrictive food-intake disorder, bulimia nervosa, binge-eating disorder, pica, rumination disorder, other specified feeding or eating disorder, and unspecified feeding or eating disorder.

The criteria for AN established by the DSM-5 are the following:

1. Restriction of energy intake relative to requirements, leading to a significantly low body weight in the context of age, sex, developmental trajectory, and physical health. Significantly low weight is defined as a body weight that is less than minimally normal or, for children and adolescents, less than that minimally expected.
2. Intense fear of gaining weight or of becoming fat, or persistent behavior that interferes with weight gain, even though at a significantly low weight.
3. Disturbance in the way in which one's body weight or shape is experienced, undue influence of body weight or shape on self-evaluation, or persistent lack of recognition of the seriousness of the current low body weight.

Two subtypes are classically described: restricting type and binge eating/purging type. The restricting type concerns individuals whose weight loss is accomplished primarily through dieting, fasting, and/or excessive exercise and who have no recurrent episodes of binge eating or purging behavior (i.e., self-induced vomiting or the misuse of laxatives, diuretics, or enemas) during the last 3 months. In the binge eating/purging type, individuals have recurrent episodes of binge eating (i.e., eating large amounts of food in a discrete period

(continued)

Box 14.1 (continued)
of time when not feeling physically hungry, accompanied by a sense of lack of control) or purging behavior (as mentioned above) lasting longer than 3 months.

Additionally, some features related to personality or temperament traits, not mentioned in the DSM-5, also characterize AN: cognitive rigidity, compulsivity, low self-directedness, perfectionism, and anxiety that can amend coping strategies to a changing environment. Compared with restrictive type AN, patients with binge eating/purging type have higher rates of impulsivity and may be more likely to develop substance abuse (Treasure et al. 2015).

Anorexia nervosa is mainly characterized by a dramatic weight loss, self-induced by a chronic alteration of feeding behavior from a severe food restriction (restrictive subtype) to a binge eating consumption of food followed by various purging strategies (Box 14.1). One of the difficulties encountered in this disease is to decipher, among the psychiatric and somatic dysfunctions, what would be the causes and the consequences. Indeed, the refusal to maintain a normal weight can be related to an intense fear of gaining weight, a dysregulation in the emotional and nutritional valence (taste, texture...) given to food, a strategy for coping with negative emotions through weight loss, or a need to exercise in excess. Another difficulty is highlighted by neuroimaging studies performed with inpatients after several months of recovery. Imaging studies, using, for example, functional magnetic resonance, positron emission tomography, indicate differences of activation or in the microstructural environment revealed by magnetic resonance diffusion tensor imaging in a wide range of brain areas such as the hypothalamus (food intake), prefrontal cortex (executive function, cognitive flexibility), dorsal and ventral striatum (habit versus goal-directed behavior, reward), amygdala (emotion, fear), insula (taste, interoceptive emotional evaluation), or ventral tegmental area (motivation, reward), to cite only the most relevant (Frank 2012; Monteleone et al. 2018; Florent et al. 2020; Iorio-Morin et al. 2022). This reveals the heterogeneity of implicated cerebral regions and the potential complexity of subsequent physiological and behavioral manifestations.

Besides differences in activation or inhibition of specific brain areas that can sustain symptoms classically described in clinic, such as cognitive rigidity, body distortion, habit-driven behavior, anxiety, or depression, patients suffering from AN display autonomic, immune, metabolic, and neuroendocrine alterations that need to be considered as key pathophysiological determinants of the disorder as emphasized by recent GWAS (Duncan et al. 2017; Watson et al. 2019). Indeed, patients suffering from AN are also characterized by autonomic nervous system dysregulations (see point 14.2). In particular, cardiovascular complications, especially bradycardia, occur in up to 80% of these patients and account for up to 30% of mortality (Spaulding-Barclay et al. 2016). In addition, and as observed in several other psychiatric disorders, AN is accompanied by a dysregulated immune system (Gibson and Mehler 2019; Nilsson et al. 2020, see point 14.3). Moreover, restrictive-type AN is associated with numerous endocrine dysregulations, such as growth hormone resistance, hypercortisolemia, low levels of T3 (non-thyroidal illness syndrome), or reduced hypothalamic-pituitary-gonadal axis function (hypothalamic amenorrhea) (Schorr and Miller 2017). Furthermore, peripheral metabolic sensors, also challenged in this pathology, are considered as partly adaptive to the state of chronic caloric restriction, as weight restoration normalizes some of these dysregulations (see point 14.4). The endocrine and metabolic changes contribute to low bone mineral density and increased risk of fractures, which are common complications of AN (Steinman and Shibli-Rahhal 2019).

All these dysregulations lead clinicians and researchers to consider AN as a metabo-psychiatric disorder, making treatment options difficult (Fig. 14.1). Patients with AN are followed in specialized centers that use different therapeutic approaches, such as family or nutritional interventions (Watson and Bulik 2013), phone application devices (Esfandiari et al. 2018) and cognitive, behavioral or interpersonal therapies (Brockmeyer et al. 2018).

In brief, and as already evoked, despite the quality of the psychotherapeutic approach, a significant relapse rate, between 72.2% to 67.5%, is noted (Eddy et al. 2008). Therefore, there is an urgent need for a better understanding of the nature of interactions between several interconnected systems, such as the central and the autonomic nervous systems, the digestive and the immune systems, all of which could maintain the vicious circle of undernutrition.

14.2 Autonomic Alterations in AN

Autonomic nervous system activity corresponds to involuntary physiological processes that ensure body regulation. Such activity facilitates adaptive responses of neuroendocrine, immune, sensorimotor, and cognitive systems (Porges 1995). In addition, autonomic nervous system activity provides support to complex behaviours by reacting to specific stimuli, as in emotional reactions, or by anticipatory adjustments, as in preparation to the action.

14.2.1 Autonomic Organization and Control

The autonomic nervous system is a component of the peripheral nervous system that regulates the sensorimotor traffic to cardiac and smooth muscles (most visceral targets), glands (endocrine and exocrine), and sensory systems (eyes, skin) (Table 14.1).

14.2.1.1 Autonomic Components

The ANS is anatomically divided into sympathetic (SNS), parasympathetic (PNS), and enteric nervous systems (ENS). The SNS and the PNS contain both afferent and efferent fibers that interact with the ENS and provide sensory inputs to the central nervous system and motor outputs to the target organs (Fig. 14.2).

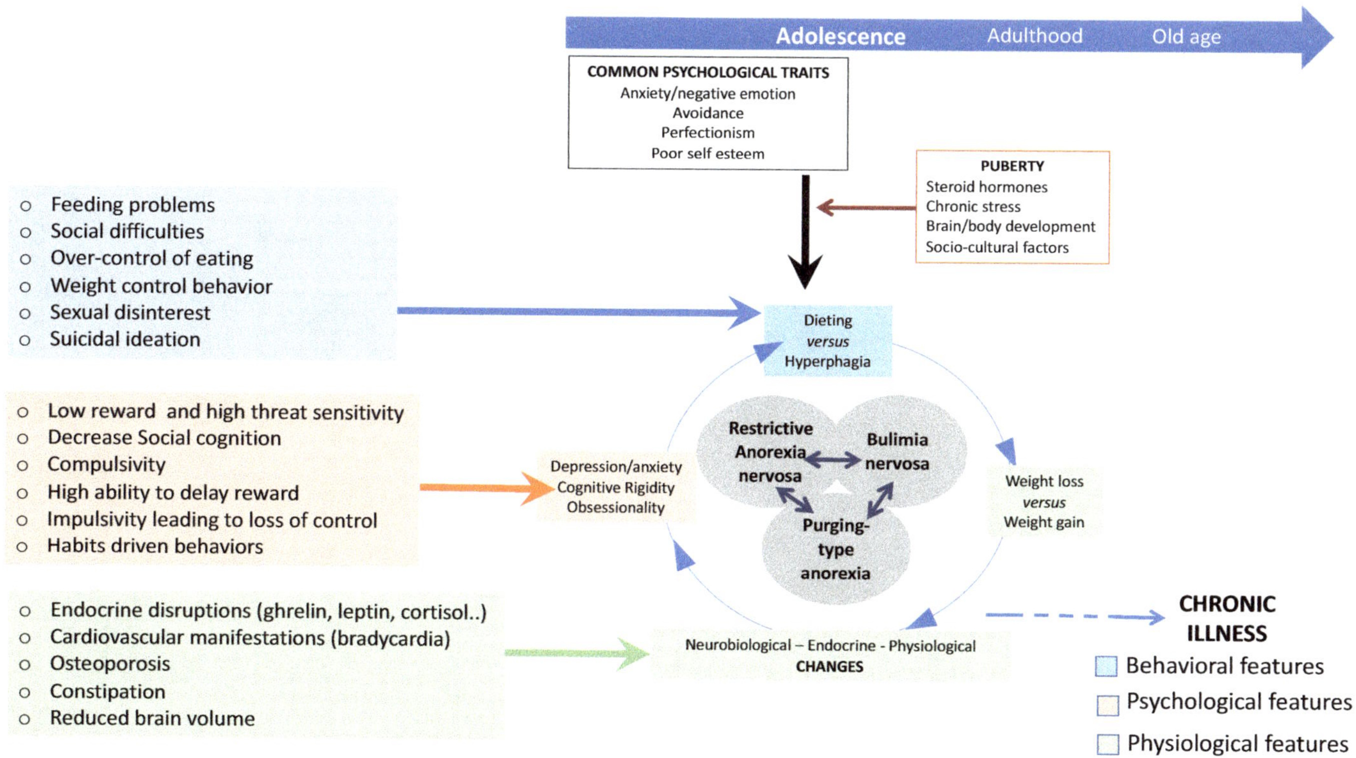

Fig. 14.1 Main behavioral, psychological, and physiological characteristics of anorexia nervosa (AN). Patients have common psychological traits that contribute to the onset of the disease, which occurs mainly at puberty. It is also of note that, as an alternative to the traditionally different eating disorder diagnoses, Fairburn et al. (2003) proposed a transdiagnostic approach whereby patients with AN can go through different phases, from restrictive AN, binge eating and purging behaviors to bulimia nervosa. Significant changes in body weight occur, thus causing physiological, psychological and behavioral consequences that contribute to make the disorder chronic

Table 14.1 Major Functions of the Autonomic Nervous System

Organs/Effectors	Sympathetic division		Parasympathetic division *(muscarinic receptors)*	
	Receptor	Effect	Nerve	Effect
Eye				
Ciliary muscle	β	Relaxation	III	Constriction
Radial iris muscle *Iris sphincter muscles*	α	Constriction (mydriasis)	III	Constriction (myosis)
Cephalic glands				
Lacrimal glands	α	Inhibition	VII	↑ secretion
Salivary glands	β	↓ secretion	VII, IX	↑ secretion
Heart				
Sinoatrial node	β_1	↑ rate	X	↓ rate
Bronchi				
Smooth muscle	β_2	Dilation	X	Constriction
Blood vessels				
Viscera	α	Constriction	X	Dilation
Skin	α	Constriction	X	Dilation
Skeletal muscle	β_2 α	Dilation Constriction		*No effect*
Veins	β_2	Dilation		*No effect*
Gastrointestinal tract				
Smooth muscle	α_2, β_2	↓ mobility	X	↑ mobility
Sphincters	α_2, β_2	Constriction	X	Relaxation
Colon, rectum	β	Relaxation	Sacral	Contraction
Pancreas				
Glucagon	α	Secretion		
Insulin	α	Inhibition	X	↑ secretion
Exocrine	β	Secretion	X	↑ secretion

(continued)

Table 14.1 (continued)

Organs/Effectors	Sympathetic division		Parasympathetic division (*muscarinic receptors*)	
	Receptor	Effect	Nerve	Effect
Uterus				
Non-pregnant	β_2	Relaxation		
Pregnant	α	Contraction		
Male sex organs	α	Ejaculation	Sacral	↑ secretion
Erectile organs	α	Vasoconstriction	Sacral	Vasodilation, erection
Liver, muscles	α, β_2	↑ gluconeogenesis ↑ glycogenolysis		↑ glycogenesis
Kidney	β_2	Renin secretion		*No effect*
Bladder				
Detrusor	β	Inhibition	Sacral	Contraction
Sphincter	α	Contraction		Relaxation
Skin				
Sweat glands	M	Secretion		*No effect*
Piloerection	α	Contraction		*No effect*
Adipocytes	α, β_1	↑ lipolysis		↑ lipogenesis
Lymphoid organs				
Immune responses		Stimulates		Local defenses
Epiphysis	β	Melatonin synthesis		

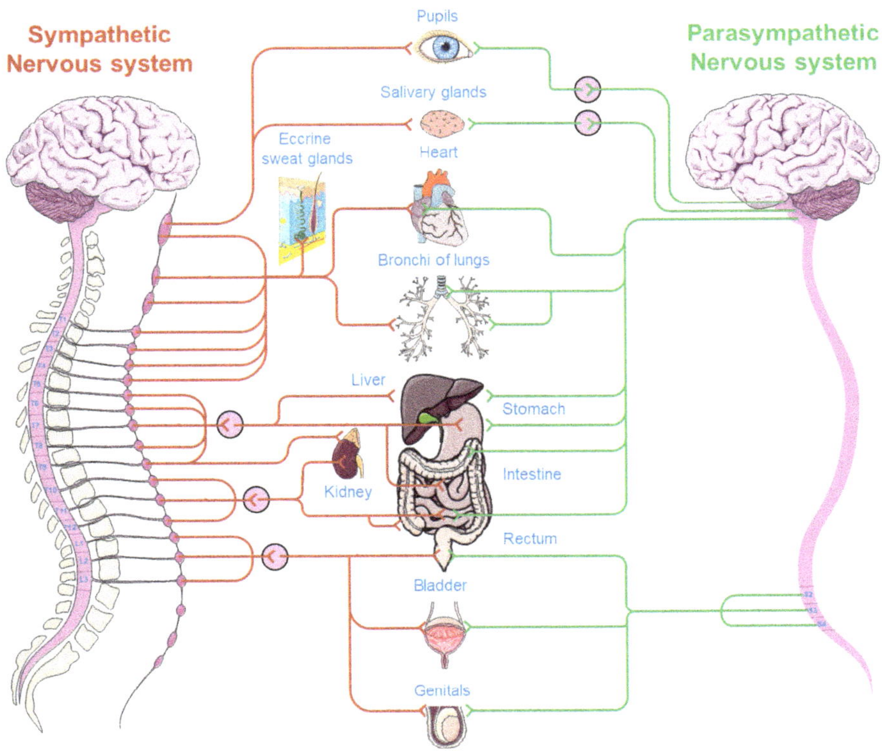

Fig. 14.2 The sympathetic (left, red) and parasympathetic (right, green) division of the autonomic nervous system innervating target tissues. The sympathetic nervous system (SNS) is a thoracolumbar (T1–L3) system arising from the intermediolateral cell column of the lateral horn of the spinal cord, acting through chain ganglia and collateral ganglia. It is a system enabling fight-or-flight reactions in case of an emergency. The parasympathetic nervous system (PNS) is a craniosacral system arising from brain stem nuclei associated with cranial nerves (CNs) III, VII, IX, and X and from the intermediate gray in the S2–S4 spinal cord. Connections from CNs III, VII, and IX act through cranial nerve ganglia; connections from the vagal system and sacral system act through intramural ganglia in or near the target tissue. The PNS is a homeostatic reparative system

The ENS is a complex, extensive, and quite autonomous network of neurons located in the wall of the gut and in accessory visceral organs such as the pancreas and gallbladder. This network interacts with SNS and PNS, ensuring the control of motility (contraction/absorption), exocrine and endocrine secretions and the microcirculation of the gastrointestinal tract, and is implicated in the regulation of immune and inflammatory processes (Goyal and Hirano 1996; Lake and Heuckeroth 2013). The ENS neurons also produce a large range of neurotransmitters with the possibility to modify digestive functions (Furness 2012).

The SNS and PNS exchange motor and sensory information with the central autonomic network which includes cortical, subcortical, and neural circuits of the brainstem and the spinal cord. These circuits ensure neural processes constituting the basis for the general body regulation, expressed by homeostatic and allostatic adaptations.

Motor pathways of SNS and PNS consist of a preganglionic neuron, for which the cell body is located at spinal or brain stem levels, and a postganglionic neuron, for which the cell body, located at the peripheral nervous system, directly innervates target tissues. The efferent components of SNS and PNS generally have an opposite excitatory and inhibitory control on most target organs, except on those receiving only SNS innervation (e.g., sweat and adrenal glands). Neurochemically, acetylcholine is the preganglionic neurotransmitter for both SNS and PNS and norepinephrine/noradrenalin and acetylcholine are neurotransmitters at postganglionic level, respectively, for SNS and PNS. In addition to these classical neurotransmitters, preganglionic and postganglionic transmissions include others as well, such as substance P, neuropeptide Y, and vasoactive intestinal peptide (Cameron 2009).

Concerning sensory pathways, SNS and PNS continuously transmit spontaneous or induced activity from autonomic targets to the cardiovascular brainstem circuits, the amygdala and the basal forebrain networks (Berntson et al. 2003). Overall, there are fewer sympathetic fibers than parasympathetic ones (Hardy and Naftel 1997); in particular, the vagus nerve is the main parasympathetic pathway sending afferents from a wide range of viscera (heart, lungs, esophagus, abdominal organs, etc.) to several key brain stem nuclei (tractus solitarius, dorsal vagal, ambiguus), which are linked to limbic and cortical areas. Considering its extended sensorimotor impact, the vagus nerve has been implicated in several psychosomatic disorders involving cardiovascular, immunological, and endocrine dysfunctions (Cameron 2009).

14.2.1.2 Autonomic Control

The ANS plays a crucial role in the maintenance of homeostasis through the induction of adaptive physiological responses, particularly in stressful or emotional contexts. The capacity of an organism to maintain homeostasis is based on its ability to initiate adaptive physiological responses to cope, for example, with the stressor, and to inhibit these responses when the stressor is no longer present (Sterling and Eyer 1988; McEwen 1998). In this context, the modulation of autonomic nervous activities is involved in the alarm phase of the stress response, allowing a quick adaptation. More particularly, the SNS activity mobilizes energy resources, preparing individuals to act in a so-called "fight or flight" response, while the PNS is associated with "rest-and-digest" or "feed-and-breed" activities. Sympathetic fibers innervate many effectors, including adrenal glands. Chromaffin cells of the adrenal medulla synthesize epinephrine/adrenalin and norepinephrine/noradrenalin, which provide a burst of energy. Consequently, SNS increases heart rate (HR), eccrine sweating, classically measured as skin conductance (SC) variations, the redness or pallor (subcutaneous vasodilation or vasoconstriction), measured by temperature variations in skin regions, and pupil diameter. This alarm phase is followed by a resistance phase, underpinned by hormonal responses depending on the neuroendocrine Hypothalamic-Pituitary-Adrenal (HPA) axis (Selye 1956). Since the burst of energy mediated by epinephrine/adrenalin and norepinephrine/noradrenalin cannot be sustained, other hormones come into play in a delayed response to stressors.

Thus, the hypothalamus releases corticotropin-releasing hormone (CRH), which travels to the pituitary gland, triggering the release of adrenocorticotropic hormone (ACTH). The blood diffusion of this hormone then reaches the adrenal glands, prompting them to release cortisol (in humans) and corticosterone (in rodents). The body thus stays aroused and on high alert (see point 14.4.3). Interestingly, physiological activities at rest, during a baseline period, can predict autonomic responses to a stressor or to emotional stimuli. Therefore, having a high autonomic arousal at rest would lead to a smaller stress-induced arousal, and probably to a maladaptive autonomic response.

14.2.1.3 Autonomic Responses

The ANS ensures the control of a large range of physiological activities providing monitoring, response and regulation of all systems in the body. Here, we will focus on responses that have been associated with clinical expressions of AN: heart rate (HR) and heart rate variability (HRV), skin conductance, and skin temperature.

The HR corresponds to the number of contractions the heart makes per minute. For adults, a normal resting HR normally ranges from 60 to 100; thus, HR over 100 is considered as tachycardia and below 60 as bradycardia, both corresponding to the risk of various cardiovascular dysfunctions (Fox et al. 2007). Cardiac activity is significantly modulated by physical efforts, preparation for action, environmental detection of or for certain stimuli and by emotion. Both branches of the ANS ensure such modulation: HR increases or the inter-beat interval decreases with the SNS activity and the HR decreases or the inter-beat interval increases with the PNS activity (Berntson et al. 1997). The parasympathetic-mediated initial deceleration occurs in the context of environmental detection, leading to a sensory intake and an enhanced orienting response (Bradley 2009), while the subsequent sympathetic-mediated acceleration is a response to the preparation for action, indicating sensory rejection and defensive arousal (Bradley et al. 2001; Bradley 2009; Jennings and van der Molen 2005). The HR slows further in response to unpleasant cues while it increases with emotional arousal and pleasant stimuli (Lang et al. 1993; Bradley 2009).

The HRV measures fluctuations in the time intervals between successive heartbeats. HRV seems critical for physiological functioning, is sensitive to pathological conditions, and extreme values have been associated with psychopathology (Heiss et al. 2021). HRV is used to investigate the sympatho-vagal balance and therefore the physiological flexibility (Thayer et al. 2012). More precisely, the HRV is often assessed by temporal and frequency measurements: in the time domain, i.e., how much HR varies in the time intervals between consecutive heartbeats, higher values meaning higher or increased HRV; in the frequency domain, i.e., how much signal power is associated with a specific frequency band, observed low (LF-HRV) and high (HF-HRV) frequencies indicating respectively a sympathetic and parasympathetic (vagal) modulation of the HR. In short-term resting recordings, the primary source of HR variation is parasympathetically-mediated respiratory sinus arrhythmia, which refers to the respiration-driven speeding and slowing of HR via the vagus nerve.

Electrodermal activity, usually recorded as skin conductance responses (SCRs), is under the exclusive control of the SNS and constitutes a robust index of central levels of activation. SCRs are directly related to eccrine sweat gland activity, are preferentially recorded at the skin level of hands, and reflect increases of sympathetic discharges related to emotional induction, initial attentional capture and preparation for action (Bradley 2009; Critchley 2002; De Zorzi et al. 2021; Sequeira et al. 2009).

Another index of emotional arousal, also under sympathetic control, is the thermal variation of the skin, which depends on cutaneous blood perfusion controlled by the ANS innervating the vessels that irrigate the skin (Kosonogov et al. 2017). Though the PNS has an influence through the endothelial cells, in glabrous skin (palmar and plantar surfaces, tip of the nose), the vasomotion appeared regulated principally by sympathetic noradrenergic fibers, whose activation leads to vasoconstriction and, therefore, to a decrease in local temperature (Donadio et al. 2006; Westcott and Segal 2013). The functional salience of peripheral skin temperature had been shown by Kosonogov et al. (2017): these authors observed that the temperature of the tip of the nose decreases with emotional arousal, regardless of the valence, and is correlated with the subjective evaluation of emotional stimulation. Some authors (Chudecka and Lubkowska 2016) also have proposed that the skin temperature, as an indicator of sympathetic influences on blood flow to the skin, could also be used to explore further eating disorders. Thus, some studies have attempted to link other autonomic indices, such as blood pressure (Sachs et al. 2016) and pupil response (Couton et al. 2022) to AN.

14.2.2 Autonomic Activity in AN

Both SNS and PNS are tonically active and constitute a dynamic sympatho-vagal balance important for organism flexibility and adaptability. In contrast, autonomic imbalances, characterized by maladaptive physiological responses to stressors or emotional induction, lead to pathophysiological consequences (McEwen and Stellar 1993). Autonomic imbalance, in particular, is associated with a higher risk of cardiovascular and other physical health complications (Malliani et al. 1991; Tonhajzerova et al. 2016), and contributes to the risk of development of psychiatric disorders (Tonhajzerova and Mestanik 2017).

In this context, disturbances in cardiac autonomic regulation may contribute to the increased cardiovascular complications and mortality observed in AN (Mazurak et al. 2011). Indeed, the energy deprivation and malnutrition associated with AN place immense pressure on the cardiovascular system, with up to 80% of patients suffering from cardiovascular complications (Spaulding-Barclay et al. 2016). There are also specific cardiac complications of AN that arise during the process of refeeding, such as arrhythmia, tachycardia, and congestive heart failure (Casiero and Frishman 2006; Vignaud et al. 2010). Based on this observation, a recent systematic review (Jenkins et al. 2021) aimed to synthesize the evidence of basal ANS function in individuals with AN.

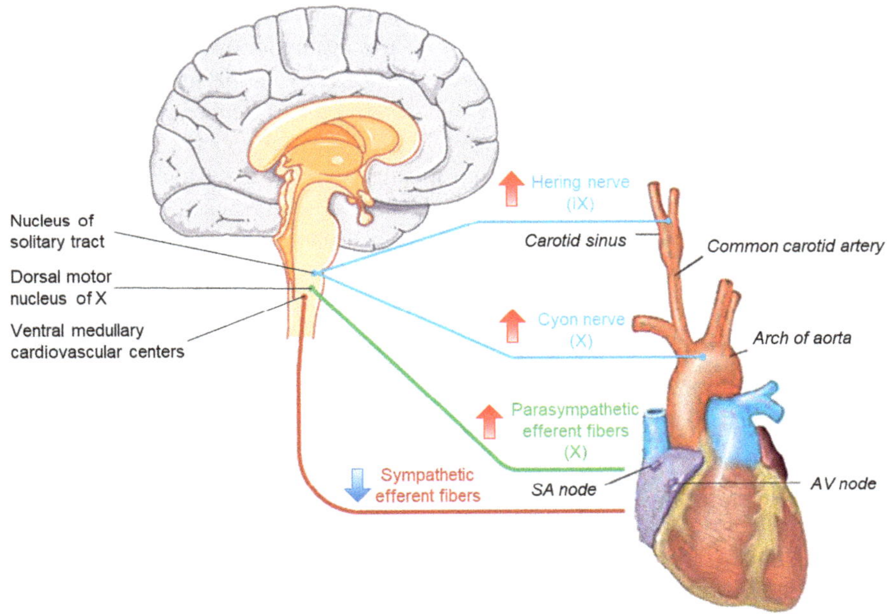

Fig. 14.3 Autonomic control of the heart rate. Heart rate (HR) is controlled by the two branches of the autonomic nervous system: the sympathetic nervous system (SNS) and the parasympathetic nervous system (PNS). Sympathetic efferent fibers increase HR via cardiac sympathetic nerves, while parasympathetic efferent fibers decrease HR via the vagal nerve (X). Baroreceptors are mechanoreceptors located in both the carotid sinus and the aortic arch. They detect changes in arterial pressure and communicate this to the nucleus of solitary tract of the medulla oblongata through Hering (IX) and Cyon (X) nerves. The nucleus solitarius is a major integrative center for ascending (limbic and hypothalamic), local brain stem and descending regulation of autonomic preganglionic responses (dorsal motor nucleus of X for parasympathetic, intermediolateral cell column for sympathetic). In anorexia nervosa (AN), sympathetic influence on HR seems decreased (blue arrow), while parasympathetic influence on HR seems increased (red arrow). Besides AN seems associated with increased blood pressure sensitivity (red arrows)

Concerning measures of baseline HR, patients with AN generally show a significantly lower HR compared to healthy individuals (Palomba et al. 2017). Otherwise, some results showed that chronic AN patients have a significantly higher HR when compared to acute AN patients (Platisa et al. 2006). These results could suggest that the duration of AN potentially influences cardiac regulation and also contribute to explaining contradictory results reporting baseline rates lower or higher than normal (e.g., Farasat et al. 2020; Krantz and Mehler 2004).

Regarding the HRV, which is used to investigate the sympatho-vagal balance and therefore the physiological flexibility (Thayer et al. 2012), Jenkins et al. (2021) reported increased heart rate variance, increased parasympathetic activity and decreased sympathetic activity in AN. More precisely, assessment of HRV in the time domain revealed increased beat-to-beat variability in HR in the acute state of AN, which does not continue following weight restoration. In short-term resting recordings, the primary source of such variation is parasympathetically-mediated respiratory sinus arrhythmia, which refers to the respiration-driven speeding and slowing of HR via the vagus nerve. Assessment of HRV in the frequency domain showed increased high-frequency HRV and decreased low-frequency HRV, which was reflected in a trend toward decreased LF/HF ratios in patients with a current diagnosis of AN. Interestingly, data from Jenkins et al. (2021) suggest that weight restoration normalizes HRV parameters, with either no difference or levels trending toward controls. In brief, most findings indicate that, in patients with AN, the HRV seems characterized by an increased parasympathetic modulation and a decreased sympathetic modulation.

Some studies recorded blood pressure (BP) or baroreflex sensitivity to assess autonomic repercussions of AN. Decreased mean BP (hypotension) is a typical finding for AN (Sachs et al. 2016). More precisely, results indicate decreased systolic BP (Gross et al. 1979), decreased diastolic BP (Casu et al. 2002) or both decreased systolic and diastolic BP (Lesem et al. 1989; Sánchez-Muniz et al. 1991; Awazu et al. 2000; Murialdo et al. 2007). Based on BP variability analysis, studies revealed sympathetic hypofunction in both the resting and the standing positions in patients with AN (Ishizawa et al. 2008; Takimoto et al. 2014). For baroreflex sensitivity, individuals with AN show increased indices of sensitivity (Kollai et al. 1994; Ishizawa et al. 2008; Takimoto et al. 2014) even if one study did not show any differences between AN and control participants (Tonhajzerova et al. 2020). Associated with higher HRV, increased baroreflex sensitivity argues for an enhanced parasympathetic reflex of HR control in AN (Fig. 14.3).

Regarding other indices of ANS, individuals with AN do not show the expected increased adrenergic outflow in response to orthostasis (Gross et al. 1979; Lechin et al. 2010), yet are comparable to controls after weight restoration (Gross et al. 1979; Lesem et al. 1989). Besides, studies indicate decreased plasma and urinary levels of noradrenaline or MHPG (3-methyl-4-hydroxyphenylglycol: the major catecholamine metabolite) in individuals with AN (Jenkins et al. 2021). Furthermore, there is some evidence of altered SC level between AN subtypes (Calloway et al. 1983) and SC level alterations in AN are observed to be associated with psychological factors, including anxiety (Léonard et al. 1998; Palomba et al. 2017). Especially, patients with AN show a greater frequency of SC responses than controls when categorizing underweight body shapes, and greater frequency in responses to underweight than normal body shapes (Clarke et al. 2016). In general, patients with AN manifest lower overall SC than healthy individuals (Knejzlíková et al. 2021). Studies that explored peripheral skin temperature show lower values for AN than for healthy controls (Bär et al. 2006) or no significant differences (Papezová et al. 2005). Finally, concerning pupil response, highly modulated by both branches of the ANS, one study showed decreased sympathetic and increased parasympathetic pupil response in a light reflex paradigm (Bär et al. 2006). In response to different types of body shapes, another study has shown differences in pupillary response that correlate with the subjective rating of emotional arousal that was observed in controls but not in AN (Couton et al. 2022).

Overall, autonomic dysregulation in AN seems characterized by an increased parasympathetic and a decreased sympathetic influence.

14.3 Immune Alterations in Anorexia Nervosa

Drastic metabolic changes as observed in obesity, cachexia associated with somatic disorders (Tan and Fearon 2008) or malnutrition (Rodríguez et al. 2007) strongly alter the immune system, rendering affected individuals more vulnerable to infections. The links between the immune system and psychiatric disorders have intrigued researchers for decades. Indeed, GWAS analyses demonstrated genomic regions common in both psychiatric and immune system disorders (Wang et al. 2015). More specifically, in the case of AN, a GWAS performed on 3495 cases with this eating disorder and 10,982 healthy controls identified for the first time a robust locus for anorexia, the single nucleotide polymorphism (SNP) rs4622308 located on chromosome 12. This SNP is in high linkage disequilibrium with rs11171739, which has been found to be associated in GWASs of type 1 diabetes and rheumatoid arthritis (autoimmune disease) (Duncan et al. 2017). As mentioned by the authors (p. 858, discussion): "Multiple reports of shared effects between anorexia nervosa and immune phenotypes fit into a broader pattern of above-chance comorbidity across psychiatric and immune phenotypes."

The immune system must be considered to be a central and pivotal element in AN since it is involved in physiological and psychological adaptation to stimuli. Many studies have highlighted the link between the abnormal production of certain cytokines, an inappropriate immune response and the response to stress, whether metabolic or psychosocial. For example, in obesity, the metabolic changes inherent in weight gain can lead to low-grade inflammation (Hotamisligil 2006). Similarly, acute or repeated psychosocial stress, anxiety or depression are often associated with multiple impairments of the immune function in humans and other animal species (Brambilla 2001; Dantzer et al. 1993; Viltart and Vanbesien 2013).

The literature on the immune system of patients with AN contains many discrepancies. These differences may be due to patient selection (small sample, age, body mass index or BMI, subtypes of AN, length of illness, pharmacological treatment, nutritional feeding habits), method of assay (RIA, ELISA, or qPCR), protocol of investigation (patients at the beginning of hospitalization or in advanced nutritional rehabilitation) or to the type of compartment examined (changes in lymphocyte subsets, in vivo circulatory concentrations of cytokines or in vitro stimulated (production of cytokines from peripheral blood mononuclear cells or PBMC). Furthermore, as will be discussed below, a range of processes that are frequently comorbid with AN are themselves known to adversely impact immune function, e.g., nutritional deficiencies, neuroendocrine changes, anxiety, depression, chronic stressful life events and excessive exercise. Considering these limitations and the complexity of the immune system, evaluation of the immune modifications in AN remains a difficult exercise. In the following sections, we try to indicate the main changes that can impact directly or indirectly the metabolic and psychological health of AN patients.

14.3.1 Immune Disturbances in AN

As reported in the review of Brown et al. (2008), mild changes occur in a range of immune measures in AN. Indeed, patients display several abnormalities in the proportion of immune cells, including in particular decreased or normal peripheral blood leukocyte counts, decreased or normal natural killer cells and total T-lymphocyte counts. The CD4/CD8 ratio (Box 14.2) was found to be either increased or decreased. In addition, granulocyte-macrophage colony-forming cell numbers in peripheral blood are decreased in patients with AN, relative to controls. The number of naïve T cells is decreased relative to memory T cell numbers, although selective depletion of memory T cells has also been reported. B lymphocyte numbers are reported to be normal.

Box 14.2 Cytokines and Inflammation

Cytokines are small secreted proteins released by specific cells and responsible for most of the biological effects in the immune system. They participate in interactions and communications between cells. Cytokine is a general name, within which can be distinguished lymphokines, or cytokines produced by lymphocytes, monokines, or cytokines made by monocytes, chemokines, or cytokines with chemotactic activities, adipokines, or cytokines released by adipocytes and interleukins (ILs) or cytokines made by one leukocyte and acting on other leukocytes.

Cytokines are produced by many cell populations, but the predominant producers are macrophages and helper T cells (Th). T lymphocytes have been divided into two main subsets, according to the presence of cell surface molecules known as CD4 and CD8. T lymphocytes expressing CD4 or helper T cells are regarded as being the most prolific cytokine producers (Mosmann et al., 1987). This subset of T lymphocytes can be further subdivided into Th1, Th2, Th9, Th17, and more recently in Th22. Briefly, the *Th1 subtype* type induces pro-inflammatory responses to kill intracellular pathogens, including protozoa, bacteria, and viruses. Interferon-gamma, TNFα, and IL-2 are the main Th1 cytokines. *Th2 subtypes* are primarily important in helping to mount a defense against extracellular pathogens such as helminth infections. They also participate in different types of allergic diseases (asthma, atopic dermatitis, allergic rhinitis, or food allergy). They release principally IL-4, IL-5, and IL-13. More recently other subtypes have been discovered, such as *Th17 subtype*, which is involved in inflammatory autoimmune pathologies (Crohn's disease, for example), carcinogenesis (pro- or anti-tumorigenic effects,

(continued)

Box 14.2 (continued)
depending on the tumor) and produces mainly IL-17, IL-21, and IL-26; *Th9 subtype* is implicated in tissue inflammation and immunity against parasites and cancer, and able to release IL-9 and IL-21 or the *Th22 subtype, which is* able to secrete IL-22 and whose main function is to protect epithelial barrier organs (skin, lung), and to modulate inflamed and injured tissue.

The activation of these cells requires the release of several cytokines that are classically subdivided into two categories: pro-inflammatory and anti-inflammatory cytokines. *The pro-inflammatory* cytokines are involved in the acute phase response. They concern mostly TNFα (Tumor Necrosis Factor α), IL-1β, and IL-6. *The anti-inflammatory cytokines* are a series of immunoregulatory molecules that control the pro-inflammatory cytokine response to regulate the immune response. Their physiological role in inflammation and pathological role in systemic inflammatory states are increasingly recognized. Major anti-inflammatory cytokines include the IL-1 receptor antagonist, IL-4, IL-6, IL-10, IL-11, and IL-13.

Brown and colleagues also reported abnormalities in immune function with decreased natural killer cell activity, deficits in T-lymphocyte cytotoxicity, decreased or normal granulocyte adherence and neutrophil chemotaxis and increased platelet aggregation. Lymphoproliferative responses to mitogenic stimulation did not reveal significant differences between AN patients and controls. Finally, data pertaining to delayed-type hypersensitivity skin responses showed slightly decreased, normal, or increased activity in AN, while anergy, which is a mechanism of immunologic self-tolerance in which T cells become functionally inactivated after previous stimulation, is elevated in patients with AN relative to controls.

Most of these immune disturbances do appear to normalize with refeeding and clinical recovery (Allende et al. 1998; Nagata et al. 1999). Besides these differences observed in production of immune cells, patients suffering from AN also present alterations in the release of several inflammatory cytokines (Box 14.2).

14.3.2 Cytokine Disturbances in AN

Anorexia nervosa is associated with various abnormalities in the spontaneous (in plasma) and stimulated (from PBMC) concentration of different cytokines (Box 14.2).

14.3.2.1 Comparison of AN Patients and Healthy Control Subjects

Data obtained on the plasma levels of IL-1β, IL-6, tumor necrosis factor (TNFα) and transforming growth factor β are not wholly consistent and seem to depend on BMI (Brown et al. 2008). Nevertheless, two meta-analyses have reported increased plasma concentrations of IL-6 and TNFα and decreased TGF β (Solmi et al. 2015;

Dalton et al. 2018) in patients with AN, compared to controls. When only patients suffering from restrictive AN were considered, IL-1β was found to be elevated compared to healthy controls (Solmi et al. 2015). However, a recent study using an "Olink Proteomics inflammatory panel" on 113 plasma samples from women with active and recovered AN and 114 control subjects did not support these results (Nilsson et al. 2020), as previously indicated by Brambilla (2001).

In addition, different results for cytokine production by PBMC have been reported. Indeed, in the study of Nova et al. (2002), in which cytokine production by phytohaemagglutinin-stimulated PBMC was assessed in two AN groups (AN-R and AN-BP) upon admission to the hospital, the production of TNF-α and IL-6 was lower, whereas production of IL-1β, was higher in patients with AN than in the control group. In another study, in which isolated PBMC were stimulated with concanavalin A, Raymond et al. (2000) reported that only IL-6 production tended to be higher in the AN group than in the controls. These differences could be attributed to (1) the population of patients considered (AN-R vs. AN-BP; age), (2) the medication taken by the patients, since it is known that, for example, benzodiazepines would be more likely to inhibit pro-inflammatory cytokine production and (3) the agent used to stimulate the PBMC (concanavalin A or phytohemagglutinin, two lectins that stimulate cellular mitogen capacity of immune cells).

The evaluation of plasma cytokine levels can be assessed by another method, highly sensitive, reliable, and easy to perform, using whole blood mRNA analysis by qPCR (quantitative real-time polymerase chain reaction). The mRNA expression patterns for TNF-α and IL-6 in whole blood have been found to be higher in patients with AN than in controls (Kahl et al. 2004). These data are thus in a good correlation with protein production.

One can note that very few studies reported changes in other cytokines (interferon-γ, IL-2, Il-4, IL-5, IL-7, IL-10, macrophage inhibitory cytokine-1...). However, as an example, IL-17, another pro-inflammatory cytokine, which induces activation and mobilization of neutrophils to sites of inflammation, is also considered to be a potent mediator of inflammatory responses in multiple sclerosis and other autoimmune diseases (Steinman 2007; Waisman et al. 2015). IL-17 is a main cytokine in central nervous system inflammatory disorders (Tzartos et al. 2008). Its plasma concentrations are increased and decreased in depression and schizophrenia, respectively, without the mechanistic consequences being elucidated at present, though its pro-inflammatory action is described in the brain through the activation of glial cells (Borovcanin et al. 2012; Davami et al. 2016; Waisman et al. 2015). Similarly, IL-7, a constitutive pleiotropic cytokine, is also involved in energy metabolism. In rodents, it protects from obesity through regulation of adipose tissue via a lymphocyte-independent mechanism, whereas administration of exogenous IL-7 decreased food intake in a situation of refeeding after fasting, by modulating the activity of hypothalamic neurons expressing the anorexigenic neuropeptide pro-opiomelanocortin (Macia et al. 2010; Lucas et al. 2012). Furthermore, the 24 h mean IL-7 plasma concentrations are decreased in patients with AN-R compared to controls (Germain et al. 2016). Finally, using a Proteomics inflammatory panel, Nilsson et al. (2020) showed a dysregulated inflammatory profile in acute AN, with

18 proteins having lower plasma concentrations compared to controls and 6 proteins displaying higher plasma concentrations. The concentrations of several plasma markers were also found to be correlated with variations in BMI. These data encourage us to extend future studies to a broader range of cytokines to better understand their role in the evolution of the illness. Interestingly, most of the studies showed that in women who had recovered from AN (after a period of weight gain), the plasma concentrations of most of the inflammatory markers had normalized (Nilsson et al. 2020; Solmi et al. 2015; Brown et al. 2008; Kahl et al. 2004; Nagata et al. 1999), but that TNF-α mRNA expression remained high even after refeeding. This suggests that this cytokine could contribute to later metabolic abnormalities in AN (see point 14.4).

To conclude, data from the literature confirm that alterations in the inflammatory profile in AN, particularly the restrictive subtype, are associated with low BMI and that some of these alterations may maintain or even reinforce metabolic abnormalities.

14.3.2.2 Are the Cytokine Disturbances in AN Caused by Inflammatory Conditions or by Nutritional Deprivation?

Despite the changes observed in various plasma proteins involved in inflammatory processes, several studies revealed that underweight AN patients are surprisingly free from common viral infections (Armstrong-Esther et al. 1978; Golla et al. 1981; Wade et al. 1985). A bidirectional relationship between inflammation and food intake is now commonly accepted: (1) reduced food intake is one symptom of the sickness behavior (Dantzer et al. 2008); (2) several cytokines, such as IL1β, IL6, and TNFα, are known to profoundly suppress food intake in animals and humans by acting at different brain sites involved in the homeostatic and hedonic control of food intake (Brown et al. 2008; Gautron and Layé 2010); (3) obesity and being overweight are now well accepted to be a state of chronic low-grade systemic inflammation, characterized by increased pro-inflammatory cytokine secretion from adipose tissue and infiltration of leukocytes, including macrophages, into this tissue (Hotamisligil 2006); (4) people suffering from undernutrition or malnutrition are at increased risk for infections and they also show a different cytokine profile compared to AN patients (Rytter et al. 2014; Gibson and Mehler 2019). Altogether, these data are intriguing. Patients suffering from AN experience metabolic and psychological conditions that should have an impact on their immune dynamic capacities. Severe malnutrition, anxiety, depression, and socially related tension that are present in AN-R lead to striking neurotransmitter and neuroendocrine changes, all of which can result in modulation of immune activity. As an example, the HPA axis, known to regulate stress response, metabolic homeostasis and to inhibit inflammatory activity, is upregulated in AN. Indeed, the continuous stress of nutritional deprivation leads to chronic HPA axis stimulation resulting in hypercortisolism in patients with AN and a blunted HPA axis reactivity in response to acute stress (Estour et al. 2010; Het et al. 2015; Sekaninova et al. 2020). In view of this data, one can hypothesize that in AN the immune function has become unresponsive to internal stimuli. The relative paucity of viral infections in AN in spite of these changes could be explained by the lack of nutrients which, in association with a low body temperature, might provide an environment that does not support pathogen survival. The interrelations between the function of the immune and central nervous system are complex and the differences in the ability of patients with AN to establish a compensatory mechanism through the autonomic nervous or the neuroendocrine systems could explain the variability of the results obtained.

14.4 Metabolic Alterations in Anorexia Nervosa

Patients with AN-R present several metabolic and endocrine abnormalities that need to be considered as key pathophysiological determinants of AN (Viltart et al. 2018; Duncan et al. 2017; Duriez et al. 2019; Estour et al. 2010). These alterations are partly adaptive to the state of chronic starvation since weight restoration usually normalizes many of those dysfunctions. Nevertheless, during the active phase of AN, these abnormalities worsen the patient's condition through their direct or indirect action on the immune system and ANS.

We focus here on three hormones implicated in the regulation of food intake, energy homeostasis, immune system and emotional/motivational aspects, dysregulated in AN: leptin, ghrelin, and glucocorticoids.

14.4.1 Ghrelin and Anorexia Nervosa

Ghrelin mRNA codes for the 117-amino acid preproghrelin, which is then enzymatically cleaved to produce ghrelin and obestatin. Ghrelin (based on "*ghre*" a word root in Proto-Indo-European languages meaning "grow") is thus a 28-amino acid hormone with a unique fatty acid modification on Ser-3, mainly produced by the gastric mucosa. This hormone was first identified as the endogenous ligand of the growth hormone secretagogue receptor (GHS-R) and for its ability to stimulate growth hormone secretion (Kojima et al. 1999). It has also been detected in other organs such as pancreas and gastrointestinal tract (Méquinion et al. 2015a, 2015b). This peptide is post-translationally modified with an eight-carbon fatty acid moiety by the enzyme ghrelin-O-acyl-transferase, giving rise to acyl-ghrelin. This acylation is necessary to activate the GHS-R (Wren et al. 2000). Its receptors are expressed in several brain structures, in many organs such as kidney, heart, intestine, liver, and adipose tissue, and in immune cells such as monocytes, macrophages, monocyte-derived dendritic cells, and T cells (Müller et al. 2015; Mathur et al. 2021). Ghrelin is a pleiotropic hormone whose primary role is to trigger growth hormone secretion, to modulate energy balance and to ensure that an organism seeks out and consumes food. Its main targets are hypothalamic neurons involved in the homeostatic regulation of food intake, and dopaminergic neurons located in the ventral tegmental area (brainstem) which is implicated in the motivational and rewarding aspects of feeding (Stoltenborg et al. 2022; Müller et al. 2015; Méquinion et al. 2015a, 2015b). Ghrelin is also pivotal in metabolic processes such as glucose homeostasis, gastric emptying or fat oxidation (Müller et al. 2015; Gorwood et al. 2016), and in immune regulation, since a number of reports have described ghrelin as a potent anti-inflammatory agent exerting inhibitory effects in colonic inflammation, arthritis, pancreatitis, and sepsis (Baatar et al. 2011; Pintér et al. 2014). A recent review suggests ghrelin may be considered an "anti-sepsis peptide" (Mathur et al. 2021). This gastric hormone downregulates the expression of pro-inflammatory cytokines such as TNF-α, IL-6, and IL-1β, and upregulates the expression of anti-inflammatory cytokines, e.g., IL-10 and TGFβ via a GHS-R-specific mechanism (Taub 2008), thus limiting the cytokine firing induced by the septic shock. Indeed, pre-clinical studies demonstrate that treatment with either ghrelin or ghrelin agonists decreases inflammation, by downregulating pro-inflammatory cytokines through activation of the vagus nerve or inhibition of the SNS activity (Wu et al. 2007a; Wu et al. 2007b) thus reducing the severity of inflammatory processes in animal models of sepsis, inflammatory bowel disease, arthritis, pancreatitis, or obesity, which is a state of low-grade inflammation (Mathur et al. 2021). Furthermore, its anti-inflammatory properties also lead to neuroprotective effects, making ghrelin a potentially effective therapeutic agent in disorders where chronic inflammation and neuroinflammation are present.

Patients suffering from AN-R display increased 24 h plasma ghrelin concentrations (Tolle et al. 2003; Germain et al. 2009). Elevated levels of this orexigenic hormone in undernourished patients appear paradoxical, because patients with AN do not respond to this meal initiation signal and maintain a self-restriction for food; the mechanisms responsible for such an increase are unknown. Indeed, in pathophysiological conditions, decrease in food intake associated or not with anorexia-cachexia is one of the most common symptoms of illness, injury, or inflammation. The cachectic state is usually associated with high plasma levels of ghrelin, as in the case of chronic heart failure (Nagaya et al. 2001). Because increased ghrelin failed to compensate for increased appetite and regulation of metabolic state, numerous authors suggest a ghrelin-resistant state. Similarly, the simultaneous presence of elevated plasma ghrelin concentration and a rather pro-inflammatory state in AN is counter-intuitive. Given that, one may wonder about potential central and/or peripheral resistance to ghrelin or antagonism of ghrelin's effects by other molecules involved in the regulation of appetite and energy balance. Indeed, on the one hand, immune cells are very sensitive to changes in energy balance in a tissue microenvironment and require an appropriate signal to alert or adjust immune responses and, on the other hand, inflammatory cytokines released from immune cells act on the central nervous system to modulate food intake and energy homeostasis. Ghrelin was also found to significantly inhibit leptin-induced increase in pro-inflammatory cytokines and Th1 responses in human mononuclear and T cells (Dixit et al. 2004; Taub 2008).

14.4.2 Leptin and Anorexia Nervosa

In humans, leptin (from the Greek word *leptos*, meaning "thin") is produced from the *lep* gene, located on chromosome 7, which transcribes a 167-amino acid peptide (see review, Münzberg and Morrison 2015). From its discovery in 1994 by Friedman and colleagues (Halaas et al. 1995), this adiposity hormone, produced mainly in white adipose tissues, is now well recognized as a valuable marker of long-term energy stores and is highly sensitive to changes in metabolic status. Circulating leptin positively reflects adipose tissue size and thus belongs to the family of adipokines, with increased and decreased levels in obesity and fasting or anorexia, respectively (Frederich et al. 1995; Ahima et al. 1996; Schwartz et al. 1996). This hormone plays an important role in body weight homeostasis through communication of the energy storage status to the brain, through leptin receptors that are located in several brain regions, peripheral tissue and immune cells (Münzberg and Morrison 2015; Gainsford et al. 1996; Dixit et al. 2004). One of the most important functions is to reduce food intake by stimulating the anorexigenic hypothalamic pathway and inhibiting the orexigenic pathway. Furthermore, leptin has been shown to exert pleiotropic effects and influences a wide spectrum of biological functions, such as reproduction, since it is required for the normal onset of puberty (Salem 2021), regulation of adaptive thermogenesis (Genchi et al. 2021), muscle and bone metabolism (Kirk et al. 2020), or immunity (Maurya et al. 2021). Indeed, this adipokine is an essential link between energy metabolism and optimal immune function, since its excessive release in obesity contributes to chronic low-grade inflammation, by inducing the production of inflammatory cytokines, including IL-1β, IL-6, and MCP-1 in eosinophils (Wong et al. 2007). Conversely, during fasting, starvation and AN, the decreased plasma concentrations of leptin are associated with impairments in cell-mediated immunity, proliferation of monocytes, and decreased release of IL1β, IL6, and TNFα (Wong et al. 2007, see Taylor 2021). Many immune cells, such as immature granulocytes, mature monocytes, macrophages, and lymphocytes, express leptin receptors (Gainsford et al. 1996). Leptin affects both innate and adaptive immunity through modulation of immune cell survival and proliferation, as well as activity (Francisco et al. 2018). Briefly, regarding innate immunity, leptin increases the cytotoxicity of natural killer cells and promotes the activation of granulocytes, macrophages and dendritic cells. As for adaptive immunity, leptin increases the proliferation of naïve T cells and B cells while reducing that of regulatory T cells (Treg). Leptin also promotes pro-inflammatory Th1 profile (Box 14.2) and facilitates Th17 responses. Furthermore, most immune cells express leptin receptors on their surface, which supports a direct effect of leptin in the modulation of the immune response (Procaccini et al. 2017). In mouse models deficient in leptin or in leptin-receptor, an augmented number and activity of Treg cells together with a resistance to autoimmune diseases were noted. This can be rescued by leptin treatment of leptin-deficient mice (Matarese et al. 2010).

These data, once again, indicate the complexity of immune system regulation in patients with AN, where the severe weight loss induces shrinking of adipose tissue and thus altered leptin release. The decreased circulating leptin levels can either directly impact immune cell functioning (innate and adaptive) or affect their functioning indirectly through a central action on the hypothalamus mediated, in part, by inhibition of HPA axis (see part 14.4.3) and activation of the sympatho-adrenal axis (see part 14.2.2) (Pérez-Pérez et al. 2017).

14.4.3 Glucocorticoids and Anorexia Nervosa: Involvement of the Hypothalamo-Pituitary-Adrenal Axis

Glucocorticoid hormones (GC) belong to a class of steroids that are synthesized by the adrenal cortex, primarily under the control of the HPA axis. Briefly, in a stressful situation, neurons located in the hypothalamic paraventricular nucleus release corticotropin-releasing hormone and vasopressin. They synergistically stimulate the secretion of stored adrenocorticotropic hormone (ACTH) from anterior pituitary gland cells. Then, ACTH is released into the blood to target the adrenal cortex, where it rapidly stimulates biosynthesis of corticosteroids such as cortisol from cholesterol (Del Rey and Besedovsky 2008). The function of GC is mediated by the GC receptor (GR), which is a member of the nuclear receptor superfamily of ligand-dependent transcription factors, widely distributed into the brain and numerous organs and by mineralocorticoid receptors (MRs) which are mainly found in the hippocampus (Kalinyak et al. 1989; Reul and de Kloet 1985; Reul et al. 2000). They both participate in the regulation of the HPA axis. Briefly, GRs mediate negative feedback signals of elevated GC levels, whereas MRs control the inhibitory tone of the hippocampus on HPA axis activity (Reul et al. 2000). Adrenal secretions are modulated by various hormones such as leptin, which directly regulates adrenal secretions via its receptors on adrenocortical cells by exerting an inhibitory effect on cortisol release (Bornstein et al. 1997).

In AN, evidence supports the idea that an overactive HPA axis may contribute to maintaining the neuroendocrine, emotional, and behavioral alterations observed in this disorder (Viltart et al. 2018). In the Minnesota Starvation Experiment, Keys (1950) was the first to link stress, anxious and depressive symptoms with starvation, in healthy male volunteers. The co-morbidity between AN and anxiety disorders, obsessive-compulsive disorders, and depression are often reported (Mattar et al. 2012), also because the latter can be characterized by altered HPA axis activity. When an individual is confronted with a stressful event or stressor, activation of the HPA axis and release of GCs enable coping strategies to face the stressful situation. This phenomenon becomes deleterious in case of chronic psychological and/or metabolic stress and regulation of the HPA axis is altered, as is the case in AN (Lo Sauro et al. 2008). Indeed, patients suffering from AN show elevated concentrations of cortisol as reported in various clinical studies (Estour et al. 2010; Bou Khalil et al. 2017). Several factors may account in this deregulation, such as (1) altered stress tolerance that is proposed to participate in the onset and course of AN and thus in the occurrence of anxious or depressive behavior; (2) pre/perinatal vulnerabilities that may program anxious/depressive diseases later in life; (3) metabolic stress generated by chronic starvation that prompts activation of the HPA axis to restore metabolic needs and (4) inappropriately elevated physical activity currently described in AN patients that might be considered as a strategy to lose weight or to be a reinforcer to alleviate anxious feelings.

Like the hormones previously detailed, GC has pleiotropic effects, being involved in metabolic, inflammatory, cardiovascular, and behavioral processes. The metabolic effects of GC are associated with hepatic and peripheral insulin resistance, hyperglycemia and dyslipidemia. As an example, after a fast GC stimulates lipolysis in adipocytes, resulting in generation of glycerol that is then used in gluconeogenesis, and free fatty acids that are oxidized. In chronic caloric restriction, such as in AN, GC actively participates in several biochemical processes that collectively serve to increase and maintain normal concentrations of glucose in the blood (Akalestou et al. 2020). The immune effects of GC are also well documented. They play a role in the development and homeostasis of T lymphocytes, up-regulating the expression of anti-inflammatory proteins and downregulating the expression of pro-inflammatory proteins. Indeed, chronic stress is known to suppress immune function, increase the susceptibility to infections and cancer, but also to exacerbate asthma, allergy, autoimmune, and inflammatory diseases (Dhabhar 2009). In this view, the consequences of stress might be beneficial or harmful depending on the type of immune reactions elicited and for the duration of the stress. In the case of acute stress, a biphasic effect on blood leukocytes is noted. Interestingly, Dhabhar and McEwen (1997) have described an initial increase in the number of blood leukocytes, occurring within minutes of the beginning of the stress response and correlated with high levels of catecholamines released through the activation of the SNS. Then, activation of the HPA axis results in a decreased number of blood leukocytes, paralleled by their extravasation from blood vessels to reach peripheral targets (skin, lungs, gastrointestinal, and urinary-genital tracts) to prepare the organism for potential lasting immune challenges as a consequence of the stressor (Dhabhar and McEwen 1997). Such a biphasic immune response is considered adaptive, with the release of GC producing anti-inflammatory effects through the production of IL-10, for example, which is both immunosuppressive and immunomodulatory. When individuals are subject to chronic stress, a downregulation of various components of the immune system occurs. The sustained release of GC occurring during such long-lasting stress might exert opposite effects depending on their targets. In the periphery, GC inhibit lymphocyte proliferation and dendritic cell maturation, induce the apoptosis of basophils, eosinophils, and T cells, whereas in the central nervous system, GC are not systematically anti-inflammatory and can even have pro-inflammatory effects, promoting extravasation and migration of immune cells and inducing the release and accumulation of nitric oxide and cytokines (TNFα, IL-1β, IL-6, or IFNγ) into the brain (Sorrells and Sapolsky 2007).

In AN, the metabolic stress induced by chronic caloric restriction is associated with altered HPA axis activity, as mentioned above, which can result in a dysregulation of the immune system. The anti-inflammatory role of GC might fluctuate according to the gravity of the disorder, established with the BMI, rendering the patients more or less vulnerable to the effect of pro-inflammatory cytokines and thus aggravating potentially their mood state. The plasma concentrations of these hormones vary according to the extent of weight loss, and as suggested in the previous parts, the immune system is not regulated only by one process. The divergent results concerning the functioning of the immune system in AN can be explained by taking into account the severity of weight loss, which generates a certain variability in the plasma concentrations of those hormones primarily involved in metabolic adaptations. Also, the variations of the immune system in

AN could reflect the adaptive flexibility of each patient according to its metabolic state. Finally, parameters such as age and duration of the disorder should also be taken into consideration. Maturation of the immune system continues during adolescence, where it is modulated by the surge of sex hormones and consequently has a direct impact on the maturation of several brain areas through neuroimmune interactions (Brenhouse and Schwarz 2016). We can thus hypothesize that the longer the duration of exposure to caloric restriction, the more chronic stress sets in, the more metabolic dysregulations worsens and the more immune alterations appear, leading the individual into a vicious circle that is difficult to break.

14.5 Conclusion and Perspectives

The difficulty encountered by clinicians in finding effective therapeutic solutions to the behavioral, psychological, and physiological problems of patients suffering from AN may be a consequence of the multi-faceted aspects of this disorder. The etiology of this disease is still poorly understood, although recent evidence points to a gene-environment interaction; indeed, GWAS and other genetic approaches support the existence of strong heritability (Duncan et al. 2017; Watson et al. 2019). Furthermore, the involvement of personal history or socio-cultural factors should not be neglected, as they may be key elements in the onset and the evolution of the disease, particularly in terms of relapse. Patients with AN often enter a vicious circle in which drastic variations in body weight induce significant physiological and behavioral consequences, including ANS manifestations and variations in various hormones involved in the regulation of energy expenditure and eating behavior. As a consequence, alterations of the immune system occur, which seem to be directly linked to the severity of weight loss. Indeed, the implication of PNS and SNS in immunity is well documented with a "yin-yang" participation in the inflammatory process, the PNS inhibiting the production of pro-inflammatory cytokines and SNS favoring it. In addition to this modulation, there are effects induced by metabolic hormones whose plasma concentrations fluctuate according to the patient's nutritional state. Finally, it is also essential to consider the patient's emotional state, measured in particular by variations in the autonomic nervous system and the HPA. The latter is likely to be activated both by metabolic stress as a consequence of food restriction and by the anxiety linked to food intake. Consequently, the chronic increases in plasma cortisol levels contribute to the disorganization of the immune system. The coping strategies implemented to reduce this anxiety by purging or intense physical activity will be factors modulating the activity of the ANS, the HPA and the metabolic hormones. Finally, intense physical activity, commonly described in patients with AN, should be considered carefully, since high physical activity level is associated with high HRV, as observed in high-level athletes, rendering the PNS more efficient (Fig. 14.4).

Understanding the involvement of these different systems is complicated by the individual variations of each patient, over time and according to the coping strategies they have adopted. Finding an appropriate therapeutic strategy therefore requires a

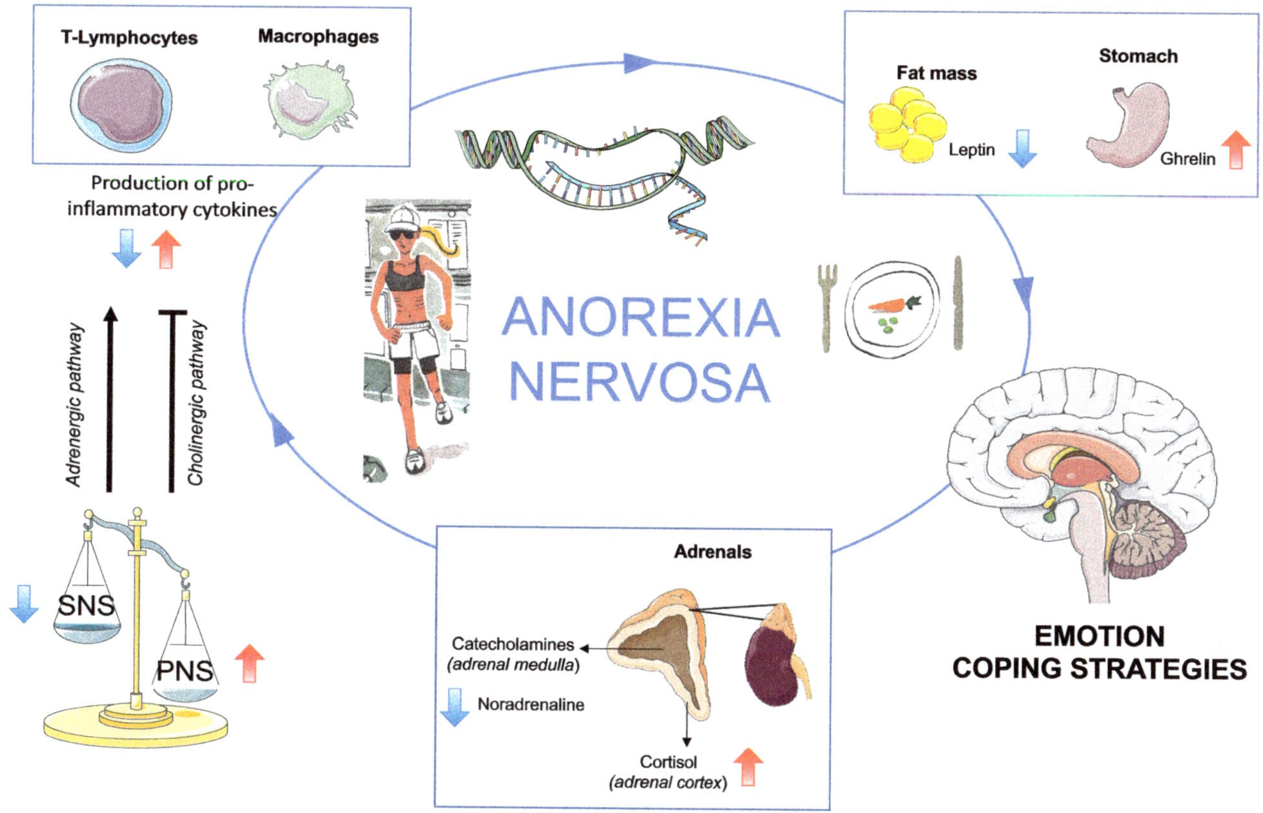

Fig. 14.4 Crosstalk between autonomic, immune, metabolic and neuroendocrine systems in AN. The complexity of anorexia nervosa (AN) is characterized by many interrelationships between different systems. The course of the disease thus corresponds to a vicious circle in which various circulating factors related to

greater understanding of such systems and holistic and personalized management of the disorder.

14.6 Key References: See Main List for Reference Details

Jenkins et al. (2021)—A systematic review synthesizing the evidence for basal ANS function in individuals with a current diagnosis of anorexia nervosa and those with a previous diagnosis who had achieved weight restoration.

De Zorzi et al. (2021)—Article that explores autonomic measures associated with emotions, in particular variations in electrodermal activity, temperature, heart rate and pupillary diameter.

Nilsson et al. (2020)—The most recent and complete analysis of the inflammation alterations in patients with active anorexia nervosa and those recovered from anorexia nervosa, using a specific proteomics inflammatory panel allowing the investigation of 92 preselected inflammation-related proteins.

Viltart et al. (2018)—Review of metabolic and neuroendocrine modifications in anorexia nervosa. They describe how variations in key hormones involved in the energy metabolism might support the various symptoms observed in patients with anorexia nervosa.

Treasure et al. (2015)—This review describes all that you need to know about anorexia nervosa, from epidemiologic studies to existing therapeutic treatments. The authors detail the mechanisms and the pathophysiology of this disorder using genetics, neuroimaging, and metabolic data obtained from clinical research and also present the therapeutic options (psychotherapy, pharmacology, transcranial stimulation) and preventative approaches.

References

Ahima RS, Prabakaran D, Mantzoros C et al (1996) Role of leptin in the neuroendocrine response to fasting. Nature 382:250–252. https://doi.org/10.1038/382250a0

Akalestou E, Genser L, Rutter GA (2020) Glucocorticoid metabolism in obesity and following weight loss. Front Endocrinol 11:59

Allende LM, Corell A, Manzanares J et al (1998) Immunodeficiency associated with anorexia nervosa is secondary and improves after refeeding. Immunology 94:543–551

Fig. 14.4 (continued) ANS (catecholamines), immune system (cytokines), metabolic and neuroendocrine systems (glucocorticoids, ghrelin, leptin) participate and interact in a subtle way. Their synthesis and release depend not only on genetic polymorphisms but also on variations in body weight associated or not with physical activity. All of these disturbances can contribute directly or indirectly to the inappropriate management of emotions and the subsequent need to develop coping strategies. Red arrow: increase observed in patients with AN. Blue arrow: decrease observed in patients with AN. ANS: autonomic nervous system. Created with the use of Servier Smart Medical Art

Arcelus J, Mitchell AJ, Wales J, Nielsen S (2011) Mortality rates in patients with anorexia nervosa and other eating disorders. A meta-analysis of 36 studies. Arch Gen Psychiatry 68:724–731. https://doi.org/10.1001/archgenpsychiatry.2011.74

Armstrong-Esther CA, Lacey JH, Crisp AH, Bryant TN (1978) An investigation of the immune response of patients suffering from anorexia nervosa. Postgrad Med J 54:395–399. https://doi.org/10.1136/pgmj.54.632.395

Awazu M, Matsuoka S, Kamimaki T et al (2000) Absent circadian variation of blood pressure in patients with anorexia nervosa. J Pediatr 136:524–527. https://doi.org/10.1016/S0022-3476(00)90017-9

Baatar D, Patel K, Taub DD (2011) The effects of ghrelin on inflammation and the immune system. Mol Cell Endocrinol 340:44–58. https://doi.org/10.1016/j.mce.2011.04.019

Bär K-J, Boettger S, Wagner G et al (2006) Changes of pain perception, autonomic function, and endocrine parameters during treatment of anorectic adolescents. J Am Acad Child Adolesc Psychiatry 45:1068–1076. https://doi.org/10.1097/01.chi.0000227876.19909.48

Berntson GG, Bigger JT, Eckberg DL et al (1997) Heart rate variability: origins, methods, and interpretive caveats. Psychophysiology 34:623–648. https://doi.org/10.1111/j.1469-8986.1997.tb02140.x

Berntson G, Sarter M, Cacioppo J (2003) Autonomic nervous system. In: Encyclopedia of cognitive science. Nature Publishing Group, pp 301–308

Bornstein SR, Uhlmann K, Haidan A et al (1997) Evidence for a novel peripheral action of leptin as a metabolic signal to the adrenal gland: leptin inhibits cortisol release directly. Diabetes 46:1235–1238. https://doi.org/10.2337/diab.46.7.1235

Borovcanin M, Jovanovic I, Radosavljevic G et al (2012) Elevated serum level of type-2 cytokine and low IL-17 in first episode psychosis and schizophrenia in relapse. J Psychiatr Res 46:1421–1426. https://doi.org/10.1016/j.jpsychires.2012.08.016

Bradley MM (2009) Natural selective attention: orienting and emotion. Psychophysiology 46:1–11. https://doi.org/10.1111/j.1469-8986.2008.00702.x

Bradley MM, Codispoti M, Cuthbert BN, Lang PJ (2001) Emotion and motivation I: defensive and appetitive reactions in picture processing. Emotion 1:276

Brambilla F (2001) Social stress in anorexia nervosa: a review of immuno-endocrine relationships. Physiol Behav 73:365–369. https://doi.org/10.1016/s0031-9384(01)00457-7

Brenhouse HC, Schwarz JM (2016) Immunoadolescence: Neuroimmune development and adolescent behavior. Neurosci Biobehav Rev 70:288–299. https://doi.org/10.1016/j.neubiorev.2016.05.035

Brockmeyer T, Friederich H-C, Schmidt U (2018) Advances in the treatment of anorexia nervosa: a review of established and emerging interventions. Psychol Med 48:1228–1256. https://doi.org/10.1017/S0033291717002604

Brown RF, Bartrop R, Birmingham CL (2008) Immunological disturbance and infectious disease in anorexia nervosa: a review. Acta Neuropsychiatr 20:117–128. https://doi.org/10.1111/j.1601-5215.2008.00286.x

Calloway P, Fonagy P, Wakeling A (1983) Autonomic arousal in eating disorders: further evidence for the clinical subdivision of anorexia nervosa. Br J Psychiatry J Ment Sci 142:38–42. https://doi.org/10.1192/bjp.142.1.38

Cameron OG (2009) Visceral brain-body information transfer. NeuroImage 47:787–794. https://doi.org/10.1016/j.neuroimage.2009.05.010

Casiero D, Frishman WH (2006) Cardiovascular complications of eating disorders. Cardiol Rev 14:227–231. https://doi.org/10.1097/01.crd.0000216745.96062.7c

Casu M, Patrone V, Gianelli MV et al (2002) Spectral analysis of R-R interval variability by short-term recording in anorexia nervosa. Eat Weight Disord EWD 7:239–243. https://doi.org/10.1007/BF03327462

Chudecka M, Lubkowska A (2016) Thermal imaging of body surface temperature distribution in women with anorexia nervosa. Eur Eat Disord Rev J Eat Disord Assoc 24:57–61. https://doi.org/10.1002/erv.2388

Critchley HD (2002) Electrodermal responses: what happens in the brain. Neuroscientist 8:11. https://doi.org/10.1177/107385840200800209
Dalton B, Bartholdy S, Robinson L et al (2018) A meta-analysis of cytokine concentrations in eating disorders. J Psychiatr Res 103:252–264. https://doi.org/10.1016/j.jpsychires.2018.06.002
Dantzer R, Bluthe R-M, Kent S, Goodall G (1993) 8 - behavioral effects of cytokines: an insight into mechanisms of sickness behavior. In: De Souza EB (ed) Methods in neurosciences. Academic Press, pp 130–150
Dantzer R, O'Connor JC, Freund GG et al (2008) From inflammation to sickness and depression: when the immune system subjugates the brain. Nat Rev Neurosci 9:46–56. https://doi.org/10.1038/nrn2297
Davami MH, Baharlou R, Ahmadi Vasmehjani A et al (2016) Elevated IL-17 and TGF-β serum levels: a positive correlation between T-helper 17 cell-related pro-inflammatory responses with major depressive disorder. Basic Clin Neurosci 7:137–142. https://doi.org/10.15412/J.BCN.03070207
De Zorzi L, Ranfaing S, Honoré J, Sequeira H (2021) Autonomic reactivity to emotion: a marker of sub-clinical anxiety and depression symptoms? Psychophysiology 58:e13774
Del Rey A, Besedovsky H (2008) The hypothalamus-pituitary-adrenal axis, volume 7 - 1st edition. https://www.elsevier.com/books/the-hypothalamus-pituitary-adrenal-axis/del-rey/978-0-444-53040-0. Accessed 5 Apr 2022
Dhabhar FS (2009) Enhancing versus suppressive effects of stress on immune function: implications for Immunoprotection and immunopathology. Neuroimmunomodulation 16:300–317. https://doi.org/10.1159/000216188
Dhabhar FS, McEwen BS (1997) Acute stress enhances while chronic stress suppresses cell-mediated immunity in vivo: a potential role for leukocyte trafficking. Brain Behav Immun 11(4):286–306
Diagnostic and statistical manual of mental disorders: DSM-5 (2013). Arlington, VA: American Psychiatric Association
Dixit VD, Schaffer EM, Pyle RS et al (2004) Ghrelin inhibits leptin- and activation-induced proinflammatory cytokine expression by human monocytes and T cells. J Clin Invest 114:57–66. https://doi.org/10.1172/JCI21134
Donadio V, Nolano M, Provitera V et al (2006) Skin sympathetic adrenergic innervation: an immunofluorescence confocal study. Ann Neurol 59:376–381. https://doi.org/10.1002/ana.20769
Duncan L, Yilmaz Z, Gaspar H et al (2017) Significant locus and metabolic genetic correlations revealed in genome-wide association study of anorexia nervosa. Am J Psychiatry 174:850–858. https://doi.org/10.1176/appi.ajp.2017.16121402
Duriez P, Ramoz N, Gorwood P et al (2019) A metabolic perspective on reward abnormalities in anorexia nervosa. Trends Endocrinol Metab TEM 30:915–928. https://doi.org/10.1016/j.tem.2019.08.004
Eddy KT, Dorer DJ, Franko DL et al (2008) Diagnostic crossover in anorexia nervosa and bulimia nervosa: implications for DSM-V. Am J Psychiatry 165:245–250. https://doi.org/10.1176/appi.ajp.2007.07060951
Esfandiari M, Papapanagiotou V, Diou C, et al (2018) Control of eating behavior using a novel feedback system. J Vis Exp JoVE (135): 57432. https://doi.org/10.3791/57432
Estour B, Germain N, Diconne E et al (2010) Hormonal profile heterogeneity and short-term physical risk in restrictive anorexia nervosa. J Clin Endocrinol Metab 95:2203–2210. https://doi.org/10.1210/jc.2009-2608
Farasat M, Watters A, Bendelow T et al (2020) Long-term cardiac arrhythmia and chronotropic evaluation in patients with severe anorexia nervosa (LACE-AN): a pilot study. J Cardiovasc Electrophysiol 31:432–439. https://doi.org/10.1111/jce.14338
Fichter MM, Quadflieg N (2016) Mortality in eating disorders - results of a large prospective clinical longitudinal study. Int J Eat Disord 49:391–401. https://doi.org/10.1002/eat.22501

Florent V, Baroncini M, Jissendi-Tchofo P et al (2020) Hypothalamic structural and functional imbalances in anorexia nervosa. Neuroendocrinology 110:552–562. https://doi.org/10.1159/000503147
Fox K, Borer JS, Camm AJ et al (2007) Resting heart rate in cardiovascular disease. J Am Coll Cardiol 50:823–830. https://doi.org/10.1016/j.jacc.2007.04.079
Frank GKW (2012) Advances in the diagnosis of anorexia nervosa and bulimia nervosa using brain imaging. Expert Opin Med Diagn 6:235–244. https://doi.org/10.1517/17530059.2012.673583
Frederich RC, Hamann A, Anderson S et al (1995) Leptin levels reflect body lipid content in mice: evidence for diet-induced resistance to leptin action. Nat Med 1:1311–1314. https://doi.org/10.1038/nm1295-1311
Furness JB (2012) The enteric nervous system and neurogastroenterology. Nat Rev Gastroenterol Hepatol 9:286–294. https://doi.org/10.1038/nrgastro.2012.32
Gainsford T, Willson TA, Metcalf D et al (1996) Leptin can induce proliferation, differentiation, and functional activation of hemopoietic cells. Proc Natl Acad Sci U S A 93:14564–14568. https://doi.org/10.1073/pnas.93.25.14564
Genchi VA, D'Oria R, Palma G et al (2021) Impaired leptin Signalling in obesity: is leptin a new Thermolipokine? Int J Mol Sci 22:6445. https://doi.org/10.3390/ijms22126445
Gibson D, Mehler PS (2019) Anorexia nervosa and the immune system—a narrative review. J Clin Med 8:1915. https://doi.org/10.3390/jcm8111915
Golla JA, Larson LA, Anderson CF et al (1981) An immunological assessment of patients with anorexia nervosa. Am J Clin Nutr 34:2756–2762. https://doi.org/10.1093/ajcn/34.12.2756
Goyal RK, Hirano I (1996) The enteric nervous system. N Engl J Med 334:1106–1115. https://doi.org/10.1056/NEJM199604253341707
Gross HA, Lake CR, Ebert MH et al (1979) Catecholamine metabolism in primary anorexia nervosa. J Clin Endocrinol Metab 49:805–809. https://doi.org/10.1210/jcem-49-6-805
Halaas JL, Gajiwala KS, Maffei M et al (1995) Weight-reducing effects of the plasma protein encoded by the obese gene. Science 269:543–546. https://doi.org/10.1126/science.7624777
Heiss S, Vaschillo B, Vaschillo EG et al (2021) Heart rate variability as a biobehavioral marker of diverse psychopathologies: a review and argument for an "ideal range.". Neurosci Biobehav Rev 121:144–155. https://doi.org/10.1016/j.neubiorev.2020.12.004
Het S, Vocks S, Wolf JM, Hammelstein P, Herpertz S, Wolf OT (2015) Blunted neuroendocrine stress reactivity in young women with eating disorders. J Psychosom Res 78(3):260–267. https://doi.org/10.1016/j.jpsychores.2014.11.001
Hotamisligil GS (2006) Inflammation and metabolic disorders. Nature 444:860–867. https://doi.org/10.1038/nature05485
Iorio-Morin C, Sarica C, Elias GJB et al (2022) Neuroimaging of psychiatric disorders. Prog Brain Res 270:149–169. https://doi.org/10.1016/bs.pbr.2021.12.001
Ishizawa T, Yoshiuchi K, Takimoto Y et al (2008) Heart rate and blood pressure variability and baroreflex sensitivity in patients with anorexia nervosa. Psychosom Med 70:695–700. https://doi.org/10.1097/PSY.0b013e31817bb090
Jenkins ZM, Eikelis N, Phillipou A et al (2021) Autonomic nervous system function in anorexia nervosa: a systematic review. Front Neurosci 15:682208. https://doi.org/10.3389/fnins.2021.682208
Jennings JR, van der Molen MW (2005) Preparation for speeded action as a psychophysiological concept. Psychol Bull 131:434
Kahl KG, Kruse N, Rieckmann P, Schmidt MH (2004) Cytokine mRNA expression patterns in the disease course of female adolescents with anorexia nervosa. Psychoneuroendocrinology 29:13–20. https://doi.org/10.1016/s0306-4530(02)00131-2
Kalinyak JE, Griffin CA, Hamilton RW et al (1989) Developmental and hormonal regulation of glucocorticoid receptor messenger RNA in the rat. J Clin Invest 84:1843–1848. https://doi.org/10.1172/JCI114370
Keys A (1950) The residues of malnutrition and starvation. Science 112:371–373. https://doi.org/10.1126/science.112.2909.371

Kirk B, Feehan J, Lombardi G, Duque G (2020) Muscle, bone, and fat crosstalk: the biological role of Myokines, Osteokines, and Adipokines. Curr Osteoporos Rep 18:388–400. https://doi.org/10.1007/s11914-020-00599-y

Knejzlíková T, Světlák M, Malatincová T et al (2021) Electrodermal response to Mirror exposure in relation to subjective emotional responses, emotional competences and affectivity in adolescent girls with restrictive anorexia and healthy controls. Front Psychol 12:673597

Kollai M, Bonyhay I, Jokkel G, Szonyi L (1994) Cardiac vagal hyperactivity in adolescent anorexia nervosa. Eur Heart J 15:1113–1118. https://doi.org/10.1093/oxfordjournals.eurheartj.a060636

Kosonogov V, De Zorzi L, Honoré J et al (2017) Facial thermal variations: a new marker of emotional arousal. PLoS One 12:e0183592

Krantz MJ, Mehler PS (2004) Resting tachycardia, a warning sign in anorexia nervosa: case report. BMC Cardiovasc Disord 4:10. https://doi.org/10.1186/1471-2261-4-10

Lake JI, Heuckeroth RO (2013) Enteric nervous system development: migration, differentiation, and disease. Am J Physiol Gastrointest Liver Physiol 305:G1–G24. https://doi.org/10.1152/ajpgi.00452.2012

Lang PJ, Greenwald MK, Bradley MM, Hamm AO (1993) Looking at pictures: affective, facial, visceral, and behavioral reactions. Psychophysiology 30:261–273. https://doi.org/10.1111/j.1469-8986.1993.tb03352.x

Léonard T, Pepinà C, Bond A, Treasure J (1998) Assessment of test-meal induced autonomic arousal in anorexic, bulimic and control females. Eur Eat Disord Rev 6:188–200. https://doi.org/10.1002/(SICI)1099-0968(199809)6:3<188::AID-ERV227>3.0.CO;2-G

Lesem MD, George DT, Kaye WH et al (1989) State-related changes in norepinephrine regulation in anorexia nervosa. Biol Psychiatry 25:509–512. https://doi.org/10.1016/0006-3223(89)90208-4

Lo Sauro C, Ravaldi C, Cabras PL et al (2008) Stress, hypothalamic-pituitary-adrenal axis and eating disorders. Neuropsychobiology 57:95–115. https://doi.org/10.1159/000138912

Malliani A, Pagani M, Lombardi F et al (1991) Spectral analysis to assess increased sympathetic tone in arterial hypertension. Hypertens Dallas Tex 1979 17:III36–III42. https://doi.org/10.1161/01.hyp.17.4_suppl.iii36

Matarese G, Procaccini C, De Rosa V et al (2010) Regulatory T cells in obesity: the leptin connection. Trends Mol Med 16:247–256. https://doi.org/10.1016/j.molmed.2010.04.002

Mathur N, Mehdi SF, Anipindi M et al (2021) Ghrelin as an anti-sepsis peptide: review. Front Immunol 11:610363

Mattar L, Thiébaud M-R, Huas C et al (2012) Depression, anxiety and obsessive-compulsive symptoms in relation to nutritional status and outcome in severe anorexia nervosa. Psychiatry Res 200:513–517. https://doi.org/10.1016/j.psychres.2012.04.032

Maurya R, Sebastian P, Namdeo M et al (2021) COVID-19 severity in obesity: leptin and inflammatory cytokine interplay in the link between high morbidity and mortality. Front Immunol 12:649359

Mazurak N, Enck P, Muth E et al (2011) Heart rate variability as a measure of cardiac autonomic function in anorexia nervosa: a review of the literature. Eur Eat Disord Rev J Eat Disord Assoc 19:87–99. https://doi.org/10.1002/erv.1081

McEwen BS (1998) Stress, adaptation, and disease. Allostasis and allostatic load. Ann N Y Acad Sci 840:33–44. https://doi.org/10.1111/j.1749-6632.1998.tb09546.x

McEwen BS, Stellar E (1993) Stress and the individual: mechanisms leading to disease. Arch Intern Med 153:2093–2101

Monteleone AM, Castellini G, Volpe U et al (2018) Neuroendocrinology and brain imaging of reward in eating disorders: a possible key to the treatment of anorexia nervosa and bulimia nervosa. Prog Neuro-Psychopharmacol Biol Psychiatry 80:132–142. https://doi.org/10.1016/j.pnpbp.2017.02.020

Müller TD, Nogueiras R, Andermann ML et al (2015) Ghrelin Mol Metab 4:437–460. https://doi.org/10.1016/j.molmet.2015.03.005

Münzberg H, Morrison CD (2015) Structure, production and signaling of leptin. Metabolism 64: 13–23. https://doi.org/10.1016/j.metabol.2014.09.010

Murialdo G, Casu M, Falchero M et al (2007) Alterations in the autonomic control of heart rate variability in patients with anorexia or bulimia nervosa: correlations between sympathovagal activity, clinical features, and leptin levels. J Endocrinol Investig 30:356–362. https://doi.org/10.1007/BF03346310

Nagata T, Kiriike N, Tobitani W et al (1999) Lymphocyte subset, lymphocyte proliferative response, and soluble interleukin-2 receptor in anorexic patients. Biol Psychiatry 45:471–474. https://doi.org/10.1016/s0006-3223(98)00082-1

Nagaya N, Uematsu M, Kojima M et al (2001) Elevated circulating level of ghrelin in cachexia associated with chronic heart failure: relationships between ghrelin and anabolic/catabolic factors. Circulation 104:2034–2038. https://doi.org/10.1161/hc4201.097836

Nilsson IAK, Millischer V, Göteson A et al (2020) Aberrant inflammatory profile in acute but not recovered anorexia nervosa. Brain Behav Immun 88:718–724. https://doi.org/10.1016/j.bbi.2020.05.024

Nova E, Gómez-Martínez S, Morandé G, Marcos A (2002) Cytokine production by blood mononuclear cells from in-patients with anorexia nervosa. Br J Nutr 88:183–188. https://doi.org/10.1079/BJNBJN2002608

Palomba D, Venturini M, Rausa M et al (2017) Reduced sympathetic activity and dysfunctional metacognition in patients with anorexia nervosa: a preliminary study. J Evid-Based Psychother 17:1–20. https://doi.org/10.24193/jebp.2017.1.1

Papezová H, Yamamotová A, Uher R (2005) Elevated pain threshold in eating disorders: physiological and psychological factors. J Psychiatr Res 39:431–438. https://doi.org/10.1016/j.jpsychires.2004.10.006

Pérez-Pérez A, Vilariño-García T, Fernández-Riejos P et al (2017) Role of leptin as a link between metabolism and the immune system. Cytokine Growth Factor Rev 35:71–84. https://doi.org/10.1016/j.cytogfr.2017.03.001

Pintér E, Pozsgai G, Hajna Z et al (2014) Neuropeptide receptors as potential drug targets in the treatment of inflammatory conditions. Br J Clin Pharmacol 77:5–20. https://doi.org/10.1111/bcp.12097

Platisa MM, Nestorovic Z, Damjanovic S, Gal V (2006) Linear and non-linear heart rate variability measures in chronic and acute phase of anorexia nervosa. Clin Physiol Funct Imaging 26:54–60. https://doi.org/10.1111/j.1475-097X.2005.00653.x

Porges SW (1995) Orienting in a defensive world: mammalian modifications of our evolutionary heritage. A Polyvagal Theory Psychophysiology 32:301–318. https://doi.org/10.1111/j.1469-8986.1995.tb01213.x

Procaccini C, La Rocca C, Carbone F et al (2017) Leptin as immune mediator: interaction between neuroendocrine and immune system. Dev Comp Immunol 66:120–129. https://doi.org/10.1016/j.dci.2016.06.006

Raymond NC, Dysken M, Bettin K et al (2000) Cytokine production in patients with anorexia nervosa, bulimia nervosa, and obesity. Int J Eat Disord 28:293–302. https://doi.org/10.1002/1098-108x(200011)28:3<293::aid-eat6>3.0.co;2-f

Reul JM, de Kloet ER (1985) Two receptor systems for corticosterone in rat brain: microdistribution and differential occupation. Endocrinology 117:2505–2511. https://doi.org/10.1210/endo-117-6-2505

Reul JM, Gesing A, Droste S et al (2000) The brain mineralocorticoid receptor: greedy for ligand, mysterious in function. Eur J Pharmacol 405:235–249. https://doi.org/10.1016/s0014-2999(00)00677-4

Rodríguez L, Graniel J, Ortiz R (2007) Effect of leptin on activation and cytokine synthesis in peripheral blood lymphocytes of malnourished infected children. Clin Exp Immunol 148:478–485. https://doi.org/10.1111/j.1365-2249.2007.03361.x

Rytter MJH, Kolte L, Briend A et al (2014) The immune system in children with malnutrition--a systematic review. PLoS One 9:e105017. https://doi.org/10.1371/journal.pone.0105017

Sachs KV, Harnke B, Mehler PS, Krantz MJ (2016) Cardiovascular complications of anorexia nervosa: a systematic review. Int J Eat Disord 49:238–248
Salem AM (2021) Variation of leptin during menstrual cycle and its relation to the hypothalamic-pituitary-gonadal (HPG) Axis: a systematic review. Int J Women's Health 13:445–458. https://doi.org/10.2147/IJWH.S309299
Sánchez-Muniz FJ, Marcos A, Varela P (1991) Serum lipids and apolipoprotein B values, blood pressure and pulse rate in anorexia nervosa. Eur J Clin Nutr 45:33–36
Schaumberg K, Welch E, Breithaupt L et al (2017) The science behind the academy for eating disorders' nine truths about eating disorders. Eur Eat Disord Rev J Eat Disord Assoc 25:432–450. https://doi.org/10.1002/erv.2553
Schwartz MW, Peskind E, Raskind M et al (1996) Cerebrospinal fluid leptin levels: relationship to plasma levels and to adiposity in humans. Nat Med 2:589–593. https://doi.org/10.1038/nm0596-589
Sekaninova N, Bona Olexova L, Visnovcova Z, Ondrejka I, Tonhajzerova I. (2020) Role of neuroendocrine, immune, and autonomic nervous system in anorexia nervosa-linked cardiovascular diseases. Int J Mol Sci, 21(19):7302. https://doi.org/10.3390/ijms21197302. PMID: 33023273; PMCID: PMC7582625
Selye H (1956) The stress of life. McGraw-Hill, New York, NY
Solmi M, Veronese N, Favaro A et al (2015) Inflammatory cytokines and anorexia nervosa: a meta-analysis of cross-sectional and longitudinal studies. Psychoneuroendocrinology 51:237–252. https://doi.org/10.1016/j.psyneuen.2014.09.031
Sorrells SF, Sapolsky RM (2007) An inflammatory review of glucocorticoid actions in the CNS. Brain Behav Immun 21(3):259–272
Spaulding-Barclay MA, Stern J, Mehler PS (2016) Cardiac changes in anorexia nervosa. Cardiol Young 26:623–628. https://doi.org/10.1017/S104795111500267X
Steinman L (2007) A brief history of T(H)17, the first major revision in the T(H)1/T(H)2 hypothesis of T cell-mediated tissue damage. Nat Med 13:139–145. https://doi.org/10.1038/nm1551
Steinman J, Shibli-Rahhal A (2019) Anorexia nervosa and osteoporosis: pathophysiology and treatment. J Bone Metab 26:133–143. https://doi.org/10.11005/jbm.2019.26.3.133
Sterling P, Eyer J (1988) Allostasis: a new paradigm to explain arousal pathology. In: Handbook of life stress, cognition and health. John Wiley & Sons, Oxford, pp 629–649
Stoltenborg I, Peris-Sampedro F, Schéle E et al (2022) TRAPing ghrelin-activated circuits: a novel tool to identify, target and control hormone-responsive populations in TRAP2 mice. Int J Mol Sci 23:559. https://doi.org/10.3390/ijms23010559
Takimoto Y, Yoshiuchi K, Ishizawa T et al (2014) Autonomic dysfunction responses to head-up tilt in anorexia nervosa. Clin Auton Res Off J Clin Auton Res Soc 24:175–181. https://doi.org/10.1007/s10286-014-0250-1
Tan BHL, Fearon KCH (2008) Cachexia: prevalence and impact in medicine. Curr Opin Clin Nutr Metab Care 11:400–407. https://doi.org/10.1097/MCO.0b013e328300ecc1
Taub DD (2008) Novel connections between the neuroendocrine and immune systems: the ghrelin immunoregulatory network. Vitam Horm 77:325–346. https://doi.org/10.1016/S0083-6729(06)77014-5
Taylor EB (2021) The complex role of adipokines in obesity, inflammation, and autoimmunity. Clin Sci Lond Engl 1979 135:731–752. https://doi.org/10.1042/CS20200895
Thayer JF, Ahs F, Fredrikson M et al (2012) A meta-analysis of heart rate variability and neuroimaging studies: implications for heart rate variability as a marker of stress and health. Neurosci Biobehav Rev 36:747–756. https://doi.org/10.1016/j.neubiorev.2011.11.009
Tonhajzerova I, Mestanik M (2017) New perspectives in the model of stress response. Physiol Res 66:S173–S185. https://doi.org/10.33549/physiolres.933674
Tonhajzerova I, Mestanik M, Mestanikova A, Jurko A (2016) Respiratory sinus arrhythmia as a non-invasive index of "brain-heart" interaction in stress. Indian J Med Res 144:815–822. https://doi.org/10.4103/ijmr.IJMR_1447_14

Tonhajzerova I, Mestanikova A, Jurko A et al (2020) Arterial stiffness and haemodynamic regulation in adolescent anorexia nervosa versus obesity. Appl Physiol Nutr Metab Physiol Appl Nutr Metab 45:81–90. https://doi.org/10.1139/apnm-2018-0867

Treasure J, Zipfel S, Micali N et al (2015) Anorexia nervosa Nat Rev Dis Primer 1:15074. https://doi.org/10.1038/nrdp.2015.74

Tzartos JS, Friese MA, Craner MJ et al (2008) Interleukin-17 production in central nervous system-infiltrating T cells and glial cells is associated with active disease in multiple sclerosis. Am J Pathol 172:146–155. https://doi.org/10.2353/ajpath.2008.070690

Vignaud M, Constantin J-M, Ruivard M et al (2010) Refeeding syndrome influences outcome of anorexia nervosa patients in intensive care unit: an observational study. Crit Care Lond Engl 14: R172. https://doi.org/10.1186/cc9274

Viltart O, Vanbesien CC (2013) Neural basis of fear and anxiety. In: Ely EW, Stevens RD, Sharshar T (eds) Brain disorders in critical illness: mechanisms, diagnosis, and treatment. Cambridge University Press, Cambridge, pp 84–96

Viltart O, Duriez P, Tolle V (2018) Metabolic and neuroendocrine adaptations to undernutrition in anorexia nervosa: from a clinical to a basic research point of view. Horm Mol Biol Clin Investig 36(1). https://doi.org/10.1515/hmbci-2018-0010

Wade S, Bleiberg F, Mossé A et al (1985) Thymulin (Zn-facteur thymique serique) activity in anorexia nervosa patients. Am J Clin Nutr 42:275–280. https://doi.org/10.1093/ajcn/42.2.275

Waisman A, Hauptmann J, Regen T (2015) The role of IL-17 in CNS diseases. Acta Neuropathol (Berl) 129:625–637. https://doi.org/10.1007/s00401-015-1402-7

Wang Q, Yang C, Gelernter J, Zhao H (2015) Pervasive pleiotropy between psychiatric disorders and immune disorders revealed by integrative analysis of multiple GWAS. Hum Genet 134: 1195–1209. https://doi.org/10.1007/s00439-015-1596-8

Watson HJ, Bulik CM (2013) Update on the treatment of anorexia nervosa: review of clinical trials, practice guidelines and emerging interventions. Psychol Med 43:2477–2500. https://doi.org/10.1017/S0033291712002620

Watson HJ, Yilmaz Z, Thornton LM et al (2019) Genome-wide association study identifies eight risk loci and implicates metabo-psychiatric origins for anorexia nervosa. Nat Genet 51:1207–1214. https://doi.org/10.1038/s41588-019-0439-2

Westcott EB, Segal SS (2013) Perivascular innervation: a multiplicity of roles in vasomotor control and myoendothelial signaling. Microcirculation 20:217–238

Wong CK, Cheung PF-Y, Lam CWK (2007) Leptin-mediated cytokine release and migration of eosinophils: implications for immunopathophysiology of allergic inflammation. Eur J Immunol 37:2337–2348. https://doi.org/10.1002/eji.200636866

Wu R, Zhou M, Das P, Dong W, Ji Y, Yang D, Miksa M, Zhang F, Ravikumar TS, Wang P (2007a) Ghrelin inhibits sympathetic nervous activity in sepsis. Am J Physiol Endocrinol Metab 293(6): E1697–E1702. https://doi.org/10.1152/ajpendo.00098.2007. Epub 2007 Oct 2

Wu R, Dong W, Cui X, Zhou M, Simms HH, Ravikumar TS, Wang P. (2007b) Ghrelin down-regulates proinflammatory cytokines in sepsis through activation of the vagus nerve. Ann Surg 245(3):480–486. https://doi.org/10.1097/01.sla.0000251614.42290.ed. PMID: 17435556; PMCID: PMC1877017

Further Recommended Reading

Anorexia Nervosa

Bulik CM et al (2006) Prevalence, heritability, and prospective risk factors for anorexia nervosa. Arch Gen Psychiatry 63:305–312. https://doi.org/10.1001/archpsyc.63.3.305

Fichter MM et al (2017) Long-term outcome of anorexia nervosa: results from a large clinical longitudinal study. Int J Eat Disord 50(9):1018–1030

Gorwood P, Blanchet-Collet C, Chartrel N, Duclos J, Dechelotte P, Hanachi M, Fetissov S, Godart N, Melchior JC, Ramoz N, Rovere-Jovene C, Tolle V, Viltart O, Epelbaum J (2016

Jun 29) New insights in anorexia nervosa. Front Neurosci 10:256. https://doi.org/10.3389/fnins.2016.00256

Hormones Involved in Regulation of Food Intake

Francisco V, Pino J, Campos-Cabaleiro V et al (2018) Obesity, fat mass and immune system: role for leptin. Front Physiol 9:640. Published 2018 Jun 1. https://doi.org/10.3389/fphys.2018.00640
Kojima M, Hosoda H, Date Y, Nakazato M, Kangawa K (1999) Ghrelin is a growth-hormone-releasing acylated peptide from stomach. Nature 402:656–660. https://doi.org/10.1038/45230
Wren AM, Small CJ, Ward HL, Murphy KG, Dakin CL, Taheri S, Kennedy AR, Roberts GH, Morgan DG, Ghatei MA, Bloom SR (2000 Nov) The novel hypothalamic peptide ghrelin stimulates food intake and growth hormone secretion. Endocrinology 141(11):4325–4328. https://doi.org/10.1210/endo.141.11.7873

Endocrine Alterations in Anorexia Nervosa

Bou Khalil R, Souaiby L, Fares N (2017) The importance of the hypothalamo-pituitary-adrenal axis as a therapeutic target in anorexia nervosa. Physiol Behav 171:13–20
Germain N, Galusca B, Grouselle D, Frere D, Tolle V, Zizzari P, Lang F, Epelbaum J, Estour B (2009 Apr) Ghrelin/obestatin ratio in two populations with low bodyweight: constitutional thinness and anorexia nervosa. Psychoneuroendocrinology 34(3):413–419
Schorr M, Miller KK (2017 Mar) The endocrine manifestations of anorexia nervosa: mechanisms and management. Nat Rev Endocrinol 13(3):174–186
Tolle V, Kadem M, Bluet-Pajot MT, Frere D, Foulon C, Bossu C, Dardennes R, Mounier C, Zizzari P, Lang F, Epelbaum J, Estour B (2003 Jan) Balance in ghrelin and leptin plasma levels in anorexia nervosa patients and constitutionally thin women. J Clin Endocrinol Metab 88(1):109–116

Animals Models of Anorexia Nervosa

Lucas S, Taront S, Magnan C, Fauconnier L, Delacre M, Macia L et al (2012) Interleukin-7 regulates adipose tissue mass and insulin sensitivity in high-fat diet-fed mice through lymphocyte-dependent and inde- pendent mechanisms. PLoS One 7(6):e40351
Macia L, Viltart O, Delacre M, Sachot C, Héliot L, Di Santo JP, Wolowczuk I (2010 Apr 1) Interleukin-7 plasma levels in human differentiate anorexia nervosa, constitutional thinness and healthy obesity. PLoS One 5(4):e9953. https://doi.org/10.1371/journal.pone.0009953
Méquinion M, Chauveau C, Viltart O (2015a May 19) The use of animal models to decipher physiological and neurobiological alterations of anorexia nervosa patients. Front Endocrinol (Lausanne) 6:68. https://doi.org/10.3389/fendo.2015.00068
Méquinion M, Caron E, Zgheib S, Stievenard A, Zizzari P, Tolle V, Cortet B, Lucas S, Prévot V, Chauveau C, Viltart O (2015b Feb 1) Physical activity: benefit or weakness in metabolic adaptations in a mouse model of chronic food restriction? Am J Physiol Endocrinol Metab 308(3):E241–E255. https://doi.org/10.1152/ajpendo.00340.2014. Epub 2014 Dec 2. PMID: 25465889

Immune System in Anorexia Nervosa

Gautron L, Layé S (2010 Jan) Neurobiology of inflammation-associated anorexia. Front Neurosci 8(3):59

Germain N, Viltart O, Loyens A, Bruchet C, Nadin K, Wolowczuk I, Estour B, Galusca B (2016 Sep 9) Interleukin-7 plasma levels in human differentiate anorexia nervosa, constitutional thinness and healthy obesity. PLoS One 11(9):e0161890

Autonomic Nervous System

Sequeira H, Hot P, Silvert L, Delplanque S (2009) Electrical autonomic correlates of emotion. Int J Psychophysiol 71:50–56. https://doi.org/10.1016/j.ijpsycho.2008.07.009

Autonomic Nervous System in Anorexia Nervosa

Clarke J, Ramoz N, Fladung A-K, Gorwood P (2016) Higher reward value of starvation imagery in anorexia nervosa and association with the Val66Met BDNF polymorphism. Transl Psychiatry 6: e829–e829

Couton C, Gorwood P, Pham-Scottez A et al (2022) Pupil psychosensory reflex in response to own and standardised silhouettes in patients with anorexia nervosa. Eur Eat Disord Rev 30:135–145. https://doi.org/10.1002/erv.2881

Fairburn CG, Cooper Z, Shafran R (2003) Cognitive behaviour therapy for eating disorders: a "transdiagnostic" theory and treatment. Behav Res Ther 41:509–528. https://doi.org/10.1016/s0005-7967(02)00088-8

Hardy SGP, Naftel JP (1997) Viscerosensory pathways. Fundamental neuroscience. Churchill Livingstone, New York

Lechin F, van der Dijs B, Pardey-Maldonado B, Rivera JE, Baez S, Lechin ME (2010) Anorexia nervosa depends on adrenal sympathetic hyperactivity: opposite neuroautonomic profile of hyperinsulinism syndrome. Diabetes Metab Syndr Obes 3:311–317. https://doi.org/10.2147/DMSOTT.S10744

Mosmann TR, Coffman RL (1987) Two types of mouse helper T-cell clone: implications for immune regulation. Immunol Today 8(7):223–227. https://doi.org/10.1016/0167-5699(87)90171-X

Peyser D, Scolnick B, Hildebrandt T, Taylor JA (2021) Heart rate variability as a biomarker for anorexia nervosa: a review. Eur Eat Disord Rev 29(1):20–31. https://doi.org/10.1002/erv.2791

Glossary

Activational Effects of Hormones Influence of hormones that is acute and sustained only in the presence of the hormone.

Adaptive Immune System Slower-acting arm of the immune system characterized by highly specific actions of B cells and T cells.

Adhesion Molecules Cell-surface proteins that mediate the interaction between two cells or between cells and the extracellular matrix. Different families of adhesion molecules exist, which include immunoglobulin-like molecules, integrins, cadherins, and selectins.

Anorexia Nervosa Eating disorder mainly characterized by distorted body image, excessive dieting leading to severe weight loss with a pathological fear of becoming fat, and cognitive rigidity.

Antagonistic Pleiotropy Principle formulated by George C. Williams in 1957, according to which the same genes can "have opposite effects on fitness" at different ages of organisms.

Antigen-Presenting Cells A heterogeneous group of immune cells mediating an immune response by presenting antigens for recognition by lymphocytes and more precisely T cells. Concerning the latter, antigen-presenting cells include dendritic cells, macrophages, Langerhans cells, and another type of lymphocytic cells, B cells.

Autoimmune Antibodies (Autoantibodies) Autoimmune antibodies or autoantibodies are the immunoglobulin molecules that have the capacity to bind to self-structures. Each immunoglobulin molecule is composed of two heavy and two light polypeptide chains. Five different types of heavy chains in humans define five classes of immunoglobulins: A, D, G, E, and M.

Autonomic Nervous System The autonomic nervous system is a two-neuron chain connecting preganglionic neurons through ganglia to visceral target tissues (cardiac muscle, smooth muscles, secretory glands, metabolic cells, cells of the immune system). The system controls involuntary physiological functions, which ensure the body regulation.

Behavioural Fever It is the behavioural response of ectothermic animals to infection and is characterized by a change in the preferred thermal zone and movement of infected animals to an environment with a higher temperature.

J. P. Konsman, T. M. Reyes (eds.), *Neuroendocrine-Immune System Interactions*, Masterclass in Neuroendocrinology 13,
https://doi.org/10.1007/978-3-031-21358-8

Behavioural Immune System (BIS) It is a coordinated set of emotional and cognitive mechanisms that allow animals to detect a potential source of pathogens and to modify behaviour so as to distance them from this source. The idea of BIS was developed by Schaller in 2006.

Biological System Notion referring to a combination of body tissues or organs having the same characteristics in physiology.

Brain Homeostasis It refers to a self-regulating process that preserves constancy in the internal cerebral environment in view of maintaining optimal conditions for cell survival and functions. It includes maintenance of a stable inorganic and organic composition of brain fluids, of osmolarity, and of the concentration of carbonic gas, the regulation of the acid–base balance, and the control of blood flow.

Brain-Gut Microbiota Axis Complete array of interactions between the central nervous system and the gut microbiota that involves neural activation, endocrine responses, immune activity, and intestinal bacteria and their metabolites.

Cell Theory Theory stipulating that cells are the basic structural, functional, and organizational units of living organisms, that cells divide to pass on hereditary information and that energy flows within and between cells of an organism or different organism(s) to maintain life.

Circadian Rhythms. Naturally occurring internal rhythms of an organism's physiology or behavior with a 24-h period, synchronized to the environment (e.g., Earth's cycling of available light, or solar day).

Cytokine Storm Exaggerated inflammatory response generated by the innate immune system that can result in multi-organ damage.

Cytokines A class of intercellular messenger molecules that can be distinguished from hormones and neurotransmitters (although no absolute separations exist between these categories) based on their mostly local mode of action, biological activity at very low concentrations and pleiotropic (one molecule having many different effects) and redundant (the same effect brought about by different molecules) actions.

Diurnal. Peak activity, physiology or behavior naturally occurring during the Earth's day (i.e., period of light between sunrise and sunset). Compare nocturnal.

DSM-V This manual is the standard classification of mental disorders used by mental health professionals in the United States. It is classically used in all clinical settings by clinicians of different theoretical orientations. It can be used by mental health and other health professionals, including psychiatrists and other physicians, psychologists, social workers, nurses, occupational and rehabilitation therapists, and counselors. *DSM*-5 can also be used for research in clinical and community populations. It is also a necessary tool for collecting and communicating accurate public-health statistics (http://psychiatry.org/psychiatrists/practice/dsm).

Dural Lymphatics This specific lymphatic network is localized in the dura mater of the meninges. Dural lymphatics may transport interstitial fluid/CSF or CSF components from the subarachnoid space into deep cervical lymph nodes via foramina at the base of the skull.

Dysbiosis Any change in microbial community composition that may be causally linked to an effect on host physiology.

Eating Disorders These concern behavioral conditions characterized by severe and persistent disturbance in eating behaviors and associated distressing thoughts and emotions. They can be very serious conditions affecting physical, psychological, and social function. Types of eating disorders include anorexia nervosa, bulimia nervosa, binge eating disorder, avoidant restrictive food intake disorder, and other specified feeding and eating disorders, such as pica and rumination disorder (American Psychological Association or APA, https://www.psychiatry.org/patients-families/eating-disorders/what-are-eating-disorders).

Eco-Evo-Devo Ecological evolutionary developmental biology (eco-evo-devo) is an area of biology that focuses on the ecological context of development and evolution. Eco-evo-devo acknowledges that organism–environment interactions, including symbiosis, niche construction, plasticity, etc., are important sources of variation, inheritance, development, and natural selection. Eco-evo-devo is a branch of evo-devo (evolutionary developmental biology), which treats development, not genes, as the fundamental unit of evolution. Evolution from an evo-devo perspective is characterized as changes in the heritable properties of development. This is contrary to the standard definition of evolution as changes in the genetic frequencies of a population.

Electrodermal Activity (EDA) This activity corresponds to electrical skin variations related to the functioning of eccrine sweat glands, mainly located in the hypodermis of palmar and plantar regions. These glands are under the control of sympathetic innervation, and eccrine sweating has been used as a reliable physiological indicator of neural activations evoked by events of a novel, significant, or intense nature. Sympathetic discharges induce increases in sweating, which are recorded as variations of skin conductance (SC) or resistance. Following the guidelines on EDA research by the Society for Psychophysiological Research, the SC method became the international standard technique to record and analyze the EDA. The EDA is constituted of two main components: skin conductance level (SCL) and skin conductance responses (SCRs). Both components can index variations of central activations related to main cognitive, emotional, and executive functions.

Energy Balance State in which energy intake and energy expenditure are compared and that can describe modifications of body weight and composition after changes in energy intake and expenditure.

Evolutionary Medicine A discipline that uses the principles of evolutionary biology. It applies this knowledge to questions of acute and chronic diseases in humans.

FKBP5 A 51 kDa protein that acts as a co-chaperone for heat-shock protein 90 (Hsp90) and binds to the glucocorticoid receptor, thereby reducing glucocorticoid receptor binding affinity for cortisol, inhibiting normal GR translocation to the cell nucleus and thus preventing transcription.

Four-Core Genotype Genetically modified mouse model that facilitates isolation of sex differences due to sex steroids or sex chromosome complement.

Function Usually considered to be closely related to their corresponding structures, functions often describe what certain structures (or systems) do. Sometimes, researchers have in mind *why* they do something, i.e., they want to explain why it has been evolutionarily selected for. Philosophers have amply debated these various conceptions of "function" and the relationships between them. Biologists frequently use "knock-out" experiments or investigate pathological cases to study what happens when a certain part is removed/dysfunctional and infer its normal function from how the system changes.

Gender According to the World Health Organization, gender refers to "the characteristics of women, men, girls and boys that are socially constructed. This includes norms, behaviours and roles associated with being a woman, man, girl or boy, as well as relationships with each other. As a social construct, gender varies from society to society and can change over time." It is not appropriate to refer to gender in Non-human animals. While many organisms have a sex, only humans can have an associated gender. Gender is not a biological construct. Reference: https://www.who.int/health-topics/gender#tab=tab_1

Genome-Wide Association Study (GWAS) Genome-wide association studies are a way to identify genes involved in human diseases. This method searches the genome for small variations, called single nucleotide polymorphisms or SNPs that occur more frequently in people with a particular disease than in people without the disease. Each study can look at hundreds or thousands of SNPs at the same time. Researchers use data from this type of study to pinpoint genes that may contribute to a person's risk of developing a certain disease.

Glucocorticoid Resistance A phenomenon by which the glucocorticoid receptor has sufficient ligand available but no longer engages in the expected anti-inflammatory actions.

Heart Rate Variability Heart rate variability (HRV) is defined as the variation in duration between successive heartbeats, based on measurement of the intervals between two subsequent R waves of an electrocardiogram. HRV can be analyzed through three methods: temporal, frequency, and nonlinear. The time-domain methods are applied straight to the series of successive RR interval values. In the frequency domain methods, a power spectrum density (PSD) estimate is calculated for the RR interval series. Finally, the nonlinear properties of HRV reflecting the dynamical structure of the signal can also be analyzed using measures such as Poincaré plot, approximate and sample entropy that quantifies the predictability of fluctuations in the time series, detrended fluctuation analysis, correlation dimension and recurrence plots.

History of Scientific Ideas A history of science that highlights the role of concepts, hypotheses, and theories to account for changes over time in a field of science.

History of Scientific Techniques A history of science that focuses on and emphasizes the role of observational and measuring instruments, experimental approaches, and technologies employed in a field of science to describe its evolution.

Holism Holism is a theoretical position, often opposed to reductionism, according to which the properties of a system cannot be explained exclusively through its individual components, since the functional sum of the parts is always greater than (or in any case different from) the same parts taken individually. For instance, in biology a holistic approach conceives an organism as such, as not reducible to a simple assembly of its constituent parts. Some types of holism are more tangible than others.

Hormones Signaling molecules released by glands in the bloodstream and transported to distant organs within the body. They allow communication between different tissues and cell types.

Hypothalamic-Pituitary-Adrenal (HPA) Axis The neuroendocrine response to psychological and physical stress. During stress corticotropin-releasing hormone (CRH) and arginine vasopressin (AVP) neurons in the medial parvocellular region of the paraventricular nucleus of the hypothalamus become activated and release CRH and AVP into the median eminence, where they activate corticotrophs to release adrenocorticotropic hormone (ACTH) into the bloodstream. ACTH circulates in the blood before acting on the adrenal cortex to stimulate the release of cortisol. Cortisol co-ordinates a global non-specific stress response and negatively feeds back onto the pituitary gland and hypothalamus to curtail further HPA axis activation in the short term.

Immune Privilege A concept referring to a specific site that protects foreign-tissue grafts from rejection contrary to other body sites. The brain, the testis, and the eye are some of these immune privileged sites. These sites are characterized by a special microenvironment where the immune responses are reduced and highly controlled.

Inflammation Inflammation is classically considered as the response to tissue injury and infection. It is triggered by sentinel cells that monitor tissue stress and malfunction (disruption of homeostasis) and molecules that participate in the inflammatory process to restore normal homeostasis (Box 14.2). Recently, molecules and cells associated with inflammation have also been shown to be activated or expressed at high concentrations during psychological and metabolic stress, and in the absence of tissue injury or infection. The concept of inflammation must be enlarged to include classic innate immune responses to potentially harmful stimuli such as pathogens or injury, and to metabolic and psychiatric stress.

Innate Immune System The generalized, immediate, non-specific immune system response to invading pathogens. It acts in parallel with the acquired immune system, which mounts a pathogen-specific response dependent upon acquired recognition of the pathogen.

Intercellular Communication Signals exchanged between cells that allow them to regulate and coordinate one another's behavior.

Linkage Disequilibrium This is the correlation between nearby variants such that the alleles at neighboring polymorphisms (observed on the same chromosome) are associated within a population more often than if they were unlinked.

Lipopolysaccharide A common non-infectious immune stimulus used in laboratory assessment of immune function.

Melatonin. A hormone produced primarily by the pineal gland within the brain during the night. Extrapineal sites of production include the gut, skin, Harderian gland, and leukocytes. Numerous biological functions including regulation of sleep/wake cycles and circadian rhythms, antioxidant properties, immunomodulator, etc.

Microbiome The combined genetic material of all microbes (i.e., the microbiota) in a given niche.

Microbiota All of the microbes (including bacteria, viruses, fungi/yeast, and archaea) in a given niche.

Microglia Immunocompetent cells of the central nervous system, microglia are glial cells derived from mesoderm with a similar function to peripheral nervous system macrophages.

Microglial Activation It is a state of activation of microglia upon infection and/or neuroinflammation. In the milieu of pro-inflammatory mediators, they change their morphology and phenotype. Activated microglia have three prominent features: great numbers, an enlarged cell body and fewer branches. In mammals the most used protein marker of microglia activation is an elevated level of calcium-binding protein IBA-1.

Microglial Priming Sensitization of microglia such that they display more spherical cell bodies, de-ramification, fragmented cytoplasm, and an exaggerated inflammatory response to an immune stimulus (without a basal inflammatory profile).

Molecular Circadian Clock. Internal, self-sustaining biological oscillator within an organism's cells that generates 24-h rhythms in gene transcription and protein translation that afford temporal organization of cellular function.

NEIMS *N*euro-*e*ndocrine-*i*mmune-*m*icrobiota *s*ystems, a term used to indicate a recomposition of elements of historically distinct biological systems (which were largely based on anatomical criteria or the mapping of one function to one structure) based on functional criteria at a systemic level.

Neuroendocrine–Immune Interactions The relationships between and integration of the nervous, endocrine and immune system.

Neuroendocrine Communications Neuroendocrine communications involve chemical substances produced by the nervous system and acting as endocrine signals, i.e., acting on distant targets after release into systemic circulation. Among these neuroendocrine signals are classical transmitters (amino acids, monoamines), neuropeptides, etc.

Neuroimmune System The systems and processes involved in the interaction between the immune system and the nervous system.

Neuropeptides Neuropeptides are small proteinaceous substances that are expressed and released by neurons in a regulated fashion and that mediate or modulate neuronal communication as signaling molecules. About 90 neuropeptide gene precursors are known in humans. (From Encyclopedia of Neuroscience, Acad. Press, 2009)

Neurotransmitters Chemical compounds produced and released by neuronal cells that transmit a signal across the synapse to an adjacent target cell, which can be a neuron, a muscle cell, or a gland cell. They normally act directly after release, but they can also be transported in the bloodstream, although they are in general unstable and quickly degraded.

Neurovascular Unit A relatively recent concept defined as an anatomical and functional structure composed of closely linked components around brain microvessels (endothelial cells, pericytes and smooth muscle cells, basal lamina and extracellular matrix, astrocytes, neurons, and interneurons).

Nocturnal. Peak activity, physiology, or behavior naturally occurring during the Earth's night (i.e., period of dark between sunset and sunrise). Compare diurnal.

Nuclear Factor Kappa B A major transcription factor that mediates inflammatory responses and engages with glucocorticoid and sex steroid receptors.

Organizational effects of hormones Effects of sex steroids that occur during early developmental periods and are not reversible.

PLWH People Living With HIV and adherent to antiretroviral therapy who experience chronic inflammation. Study of their condition provides valuable insight to neuroendocrine-immune interactions.

Progenitor Cells Cells that are descendants of stem cells and differentiate to form specialized cell types. Progenitor cells can differentiate into different cells of a same tissue or organ. Contrary to stem cells, the cell potency of progenitor cells is more restricted since their self-replication is not infinite and they are not pluripotent.

Reductionism In philosophy, the term "reductionism," with respect to any science, holds that the entities, methodologies, or concepts of a research field can be traced back to a more fundamental level, sufficiently to explain the facts of the theory in question. For example, according to reductionism, psychology can be reduced and explained in biological terms. In turn, biology can be reduced and explained in chemical-physical terms. Reductionism, therefore, argues that it is possible to formulate the concepts and language of a scientific theory in terms of another theory considered more fundamental.

SABV Sex As a Biological Variable is a critical aspect of rigor and reproducibility and the 2015 NIH mandate stipulates that SABV must be considered in experimental design and interpretation.

Selfish Organs or Systems Bodily organs or systems, such as the brain or the immune system, that under certain circumstances have the capacity to drain important energy flows, at the potential expense of other body parts.

Sex Differences The dimensions in which non-reproductive differences between the sexes can occur are most frequently in the category of differences as opposed to dimorphism such that the sexes occupy divergent but overlapping aspects of a continuum.

Sickness Behaviour It is a synchronized set of behavioural changes induced by infections and mediated by pro-inflammatory cytokines. It includes depression of locomotor, social and sexual activity, anxiety, malaise, loss of appetite, sleepiness and failure to concentrate. It is often accompanied by fever and it aids survival. Sickness behaviour is a motivational state and enhances recovery by conserving energy to combat pathogens.

Stress Although no consensus definition exists because this term is employed in different fields of research, it is safe to say that in psychology stress refers more to a state of an organism, whereas in physiology it is more considered as a response of an organism to a threat to homeostasis.

Structure Roughly, anything that can be ordered with respect to some criteria can be considered "structured." On the molecular level, primary, secondary, and tertiary structure denote sequences or three-dimensional folding patterns, respectively. On a "higher" level, cellular structures or objects at higher levels of organization are often divided into individual structures (e.g., organelles, organs, specific systems).

Suprachiasmatic Nucleus (SCN). A bilateral collection of cell bodies located above the optic chiasm in the anterior hypothalamic region of the brain. In mammals, the SCN functions as the master or central clock, which is necessary for the generation of circadian rhythms.

System Although the term is widely used in science, it is hardly ever defined, and thus ends up referring to many different things. Here it is considered as a collection of things that are either spatially connected, i.e., for structural reasons, or working together, i.e., for functional reasons, is frequently subsumed under the umbrella of being a system. Whether these are divisions that correspond to something in nature or rather arbitrary divisions through the researchers' perspective is often an issue of debate. Sometimes systems will be characterized in a circular manner, e.g. the immune system is anything of interest to the immunologist.

Trier Social Stress Test A common laboratory stressor to induce an acute stress response in humans which consists of a recorded impromptu speech in front of an impassive audience and counting backward from 100 in increments of 13.

GPSR Compliance
The European Union's (EU) General Product Safety Regulation (GPSR) is a set of rules that requires consumer products to be safe and our obligations to ensure this.

If you have any concerns about our products, you can contact us on

ProductSafety@springernature.com

In case Publisher is established outside the EU, the EU authorized representative is:

Springer Nature Customer Service Center GmbH
Europaplatz 3
69115 Heidelberg, Germany

www.ingramcontent.com/pod-product-compliance
Lightning Source LLC
Chambersburg PA
CBHW060801290525
27408CB00004BA/81
9783031213571